This book is a unique overview of cardiovascular development from the cellular to the organ level across a broad range of species. Leading experts in the fields of anatomy, embryology, experimental and molecular biology, comparative physiology, pediatric cardiology, and fetal heart surgery have contributed to this integrated text. The book is divided into three parts. The first focuses on the molecular, cellular, and integrative mechanisms that determine cardiovascular development. It discusses the molecular biology, intracellular and extracellular environment, functional maturation, vasculogenesis, and regulation of developing cardiovascular systems. The second part summarizes cardiovascular development in invertebrate and vertebrate systems. The third provides an overview of environmental factors and the effects of disease on cardiovascular development with discussions of the effects of environmental and morphogenetic influences on nonmammalian and mammalian development. It offers strategies for the management of congenital cardiovascular malformations in utero and postnatally.

The book will interest those who work in the fields of developmental biology, physiology, and molecular and pediatric cardiology.

DEVELOPMENT OF CARDIOVASCULAR SYSTEMS

DEVELOPMENT OF CARDIOVASCULAR SYSTEMS

MOLECULES TO ORGANISMS

Edited by

WARREN W. BURGGREN
University of Nevada, Las Vegas

BRADLEY B. KELLER
University of Rochester

PUBLISHED BY THE PRESS SYNDICATE OF THE UNIVERSITY OF CAMBRIDGE
The Pitt Building, Trumpington Street, Cambridge CB2 1RP, United Kingdom

CAMBRIDGE UNIVERSITY PRESS
The Edinburgh Building, Cambridge CB2 2RU, United Kingdom
40 West 20th Street, New York, NY 10011-4211, USA
10 Stamford Road, Oakleigh, Melbourne 3166, Australia

First published 1997

Printed in the United States of America

Typeset in Times

A catalog record for this book is available from the British Library.

Library of Congress Cataloging-in-Publication Data

Development of cardiovascular systems : molecules to organisms /
edited by Warren W. Burggren, Bradley B. Keller.
p. cm.
Includes bibliographical references.
ISBN 0-521-56072-1 (hardback)
1. Cardiovascular system – Growth. I. Burggren, Warren W.
II. Keller, Bradley B.
[DNLM: 1. Cardiovascular System – embryology. 2. Cardiovascular
System – growth & development. WG 201 D4887 1997]
QP101.D385 1997
573. 1′2138 – dc21
DNLM/DLC
for Library of Congress 96-44920

ISBN 0-521-56072-1 hardback

Contents

Part II Species diversity in cardiovascular development

Contributors

Thomas K. Borg
Department of Developmental Biology and Anatomy
University of South Carolina School of Medicine
Columbia, SC 29208
USA

George B. Bourne
Department of Biology
University of Calgary
Calgary, Alberta T2V 1N4
Canada

Warren W. Burggren
Department of Biological Sciences
University of Nevada, Las Vegas
4505 Maryland Parkway
Las Vegas, NV 89154-4004
USA

Wayne Carver
Department of Developmental Biology and Anatomy
University of South Carolina School of Medicine
Columbia, SC 29208
USA

Jau-Nian Chen
Cardiovascular Research Center
Massachusetts General Hospital–East
149 13th Street
Charlestown, MA 02129
USA

Ka Hou Chu
Department of Biology
Chinese University of Hong Kong
Hong Kong

Edward B. Clark
Department of Pediatrics
Medical Director's Office
Primary Children's Medical Center
100 North Medical Drive
Salt Lake City, UT 84113
USA

Anthony P. Farrell
Department of Biological Sciences
Simon Fraser University
Burnaby, BC V5A 1S6
Canada

Mark C. Fishman
Developmental Biology Laboratory and Cardiovascular Research Center
Massachusetts General Hospital–East
149 13th Street
Charlestown, MA 02129
USA

Regina Fritsche
University of Göteborg
Department of Zoophysiology
Medicinaregatan 18A
41390 Göteborg
Sweden

George D. Giraud
Departments of Physiology, Medicine, and Pediatrics and the Congenital Heart Research Center
Oregon Health Sciences University
3181 SW Sam Jackson Park Road
Portland, OR 97201
USA

Adriana C. Gittenberger–de Groot
University of Leiden
P.O. Box 9602
2300 RC Leiden
The Netherlands

Frank L. Hanley
Department of Surgery
University of California, San Francisco
505 Parnassus Avenue, Box 0118

San Francisco, CA 94143-0118
USA

Ping-Chun Lucy Hou
Department of Biology
National Cheng Kung University
Tainan, Taiwan
ROC 700

José M. Icardo
Department of Anatomy and Cell Biology
Faculty of Medicine
University of Cantabrai
Poligono de Cazona, s/n
39011 Santander
Spain

Fusae Kajio
Division of Pediatric Cardiology
The Heart Institute of Japan
Tokyo Women's Medical College
8-1 Kawada-cho, Shinjuku-ku
Tokyo 162
Japan

Bradley B. Keller
NIH Specialized Center of Research in Pediatric Cardiovascular Diseases
Department of Pediatrics
University of Rochester School of Medicine
601 Elmwood Avenue, Box 631
Rochester, NY 14642
USA

Lynn Mahony
Department of Pediatrics
University of Texas Southwestern Medical Center at Dallas
5323 Harry Hines Boulevard
Dallas, TX 75235-9063
USA

Brian R. McMahon
Department of Biology
University of Calgary
Calgary, Alberta T2V 1N4
Canada

Mark J. Morton
Departments of Physiology, Medicine, and Pediatrics and the Congenital Heart Research Center
Oregon Health Sciences University
3181 SW Sam Jackson Park Road
Portland, OR 97201
USA

Anne M. Murphy
Ross Laboratories 1144
The Johns Hopkins University School of Medicine
720 Rutland Ave
Baltimore MD 21205
USA

Makoto Nakazawa
Division of Pediatric Cardiology
The Heart Institute of Japan
Tokyo Women's Medical College
8-1 Kawada-cho, Shinjuku-ku
Tokyo 162
Japan

Bernd Pelster
Institut für Zoologie
Leopold-Franzens-Universität
A-6020 Innsbruck
Austria

Alan W. Pinder
Department of Biology
Dalhousie University
Halifax, NS B3H 4H7
Canada

Robert E. Poelmann
University of Leiden
P.O. Box 9602
2300 RC Leiden
The Netherlands

Anna Ratajska
Department of Anatomy and The Cardiovascular Center
Bowen Science Building
University of Iowa
Iowa City, IA 52242-1109
USA

V. Mohan Reddy
Division of Cardiothoracic Surgery
University of California, San Francisco
505 Parnassus Avenue, M593
San Francisco, CA 94143-0118
USA

Mark D. Reller
Departments of Physiology, Medicine, and Pediatrics and the Congenital Heart Research Center
Oregon Health Sciences University
3181 SW Sam Jackson Park Road
Portland, OR 97201
USA

Peter J. Rombough
Department of Zoology
Brandon University
Brandon, MB R7A 6A9
Canada

Hiroshi Tazawa
Department of Electronic Engineering

Muroran Institute of Technology
27-1 Mizumoto
Muroran
Japan

Louis Terracio
Department of Developmental Biology and Anatomy
University of South Carolina School of Medicine
Columbia, SC 29208
USA

Kent L. Thornburg
Department of Physiology
School of Medicine
Oregon Health Sciences University
Portland, OR 97201
USA

Robert J. Tomanek
Department of Anatomy
University of Iowa
1402 Bowen Science Building
Iowa City, IA 52242-1109
USA

Stephen J. Warburton
Department of Biology
New Mexico State University
Las Cruces, NM 88003
USA

Constance Weinstein
National Institutes of Health
National Heart, Lung, and Blood Institute
Two Rockledge Center, Suite 9044
6701 Rockledge Drive, MSC 7940
Bethesda, MD 20892-7940
USA

Jackson Wong
Pediatrix Medical Group, Inc.
Broward General Medical Center
1455 Northpark Drive
Fort Lauderdale, FL 33326
USA

Foreword

CONSTANCE WEINSTEIN

Several features distinguish this book from others on development of the cardiovascular system and therefore establish it as an important contribution to literature on the subject. Among these features is the careful description of the anatomy and physiology of various animal models of development commonly used in genetic and molecular biological studies of the heart and vascular system. Understanding the differences among species is crucial to proper interpretation of findings, particularly when those findings form the basis for inferences on their significance for human malformations and diseases.

A laudable emphasis on the physiology of the developing cardiovascular system appears throughout the book. This draws attention to the fact that, although form and function are closely related, function may be perturbed without visible perturbations of form, and because the physiology of the embryo and fetus differs from that of the postnatal organism, malformations may have deleterious physiological effects only after birth. The text sets out clearly that among species there are important functional differences that are reflected in the expression of proteins characteristic of the individual species and that change with maturation of the cardiovascular system. All these factors must be taken into account when considering the significance of experimental results.

To my knowledge, this is the first book to cover comparative embryology of the cardiovascular system in invertebrates and vertebrates in a text that is intended to attract readers engaged in biomedical research, closing as it does with chapters on human disease and treatment strategies. Comparative embryology is an important branch of science, and its inclusion in this book may inspire investigators to explore less commonly used models for answers to difficult questions about more complex physiological systems. For example, scientists have long been puzzled by the fact that mature cardiomyocytes appear to be terminally differentiated; that is, they are unable to reenter the cell cycle and replace dead cells after injury to the heart. If the controls that govern the cardiomyocyte cell cycle were understood, it might be possible to treat myocardial infarction by stimulating cellular proliferation to replace dead muscle tissue. Despite the eagerness of investigators to overcome the obstacle of terminal differentiation, few have paid attention to the observation made by the Oberprillers more than 20 years ago (Oberpriller & Oberpriller, 1974) that cells of adult newt and frog hearts are able to reenter the cell cycle and undergo complete mitosis with cytokinesis in response to injury. Over the years they have identified a number of molecules that enhance or inhibit replication of cardiac myocytes in amphibians (Oberpriller et al., 1995). Such information provides a strong background for cell-cycle experts to initiate new studies of mammalian cardiac cells.

This book provides interesting reading but, more important, it will be a source of ideas for new approaches to the problems of perturbed cardiovascular development and treatment of acquired disease.

Introduction: Why study cardiovascular development?

WARREN W. BURGGREN AND BRADLEY B. KELLER

Socrates, Aristotle, Vesalius, and Galileo are just a few of the long list of notables who have written of their fascination with the rhythmic pulsing of the red spot evident in an opened bird egg. Our curiosity has not abated during the long history of human interest in cardiovascular development. Today we are still fascinated observers of that red spot as we attempt to identify the scientific underpinnings of cardiovascular development and correlate these findings with disease states in humans. Fortunately, our tools have expanded far beyond those of the ancient observers to include phase-contrast microscopy, pulsed Doppler flow, gene sequencing, genetic manipulation, high-performance liquid chromatography, and gel electrophoresis, to name but a few of the techniques the reader will encounter in this book. We are also beginning to appreciate the power of a comparative approach that employs a variety of animal species with different, yet similar, cardiovascular characteristics. Where have these observations of ever increasing sophistication led us?

The primacy of the developing cardiovascular system

We know that the cardiovascular system is the first system to begin functioning in a developing animal. So many of the contributors to this book began their first drafts with these words that, as editors, we felt that we must instead highlight this point once, prominently, at the beginning of the book. The developing heart and circulation do deserve particularly close scrutiny because of their fundamental role in the developmental process. As the system for delivery of raw metabolic substrates and oxygen to the rapidly growing embryo, any constraint on, or limitation of, early cardiovascular performance is likely to constrain all other subsequently developing systems because of their utter dependence on the timely receipt of these materials. Put another way, the embryonic cardiovascular system is an early and potentially severe "choke point" in development.

But study of the developing cardiovascular system is important for more reasons than "just" its role as a vital component of embryonic development. The cardiovascular system is unlike almost all other body systems in that *the cardiovascular system must perform even as it develops and grows*. This situation is emphasized when one looks at the growing mammalian embryo and fetus. Many of the roles of the lungs, liver, urogenital system, and a host of other organ systems are carried out by the corresponding maternal organs. These embryonic organs can undergo prolonged periods of reconfirmation and growth without the need to perform their eventually intended function. Similar examples are found in embryonic and larval fishes, amphibians, reptiles, and birds, where certain organ systems do not come "on-line" until shortly before, or at, hatching or birth. The cardiovascular system, however, enjoys no such luxury as a period of growth without need for performance. We still have a great deal to understand about how the heart can coil, twist, and differentiate into chambers as it simultaneously generates ever increasing unidirectional flow of blood into a proliferating vascular bed. The relation between developing structure and developing function remain enigmatically entangled as we try to understand, for example, whether heart shape determines flow patterns or flow patterns determine heart shape.

The rationale for this book

Books in the genre of heart development are generally produced either as a set of published symposium papers when the field is emerging (and will benefit from some direction and synthesis, however incomplete) or when the field is maturing (enough is known for a comprehensive forms). As editors, we speak for each contributor to this book when we highlight the former circumstances as our justification – that our knowledge of cardiovascular development is fragmentary in many important areas, and rudimentary at best in most others. By indicating what *is* known in our respective fields, we all hope that we will clearly highlight what *is not.* To this end, each chapter concludes with a section that outlines some of the important unanswered questions and fruitful future directions to be followed in the next several years. In short, this book is intended to educate: not just with the facts as we know them today but also by highlighting the vast areas of our ignorance so as to challenge future investigators.

The scope and content of this book

This book represents a determined effort to explore the development of the cardiovascular system with a comprehensive approach spanning

Organizational hierarchy (molecules to organisms)
Scientific paradigms (basic science to clinical applications)
Systematic perspectives (simple invertebrates to humans)

The scope is purposefully broad in order to bring together groups of investigators, representing differences in approach, that rarely interact, let alone contribute articles to the same books. The editors encouraged communication between authors during the preparation of the chapters. The result has been not only a consistency of style and cross-referencing between chapters but also an appreciation for the power of this broad approach, which comes across clearly in each chapter, regardless of its own focus.

Development of Cardiovascular Systems: Molecules to Organisms is divided into three parts. In Part I, "Molecular, Cellular, and Integrative Mechanisms Determining Cardiovascular Development," authors explore the mechanisms that form the basis of cardiovascular development. With topics ranging from gene-regulated expression of contractile proteins (Chapter 3) to endothelial cell proliferation (Chapter 6) to a sequential approach to determining cardiovascular function (Chapter 7), this first part lays the groundwork for an understanding of cardiovascular development. The editors are particularly pleased that one of our initial secret fears – that the molecularly oriented chapters would prove to be highly "self-focused" – never materialized. Even the chapters on the most molecular of topics and approaches indicate how the data within complement studies carried out at tissue, organ, and organismal level and contribute in concert to a greater overall understanding of the heart and vascular system.

After the groundwork is laid in the first section of the book, Part II, "Species Diversity in Cardiovascular Development," explores the great variety of "pumps and pipes" associated with cardiovascular development found within the animal kingdom, and the approaches by which this diversity can be investigated and categorized. Chapters 9 and 10 set the tone with considerations of evolution, development, and their interaction. They are followed by a six-chapter survey of animal cardiovascular systems and their development. Ironically, even as diversity of patterns is emphasized and celebrated in these chapters, what emerges as well is prominent commonalities in cardiovascular development. It would appear that the earlier in development one conducts experiments, the more broadly applicable are the findings to all species. This means that experimenters focusing on early development can take a cosmopolitan approach, employing any combination of species that facilitates the acquisition of new and useful data. For example, the explosion of zebra fish onto the developmental scene in the last few years is not because they are interesting animals in their own right (although see Chapter 12 on fishes' cardiovascular development) but because a few key features of zebra fish biology, particularly the ease with which genetic mutagenesis can be induced, lend themselves to answering developmental questions, including those concerning cardiovascular development (see Chapter 1), not easily garnered from other sources. These data, so easily acquired, are generally transferable to vertebrates as a whole, including humans. Of course, much of what we know of hemodynamic development comes from the study of the chick embryo (see Chapters 7 and 15). The long-standing assumption that findings on the chick are broadly applicable to all vertebrates has been verified by studies of fish (Chapter 12), amphibians (Chapter 13), and rodents (Chapter 7), which show that in many respects there is a "vertebrate" pattern of early physiological development that will almost certainly hold for larger mammals, including humans. (This topic is discussed more fully in Chapter 13.)

In Part III, "Environment and Disease in Cardiovascular Development," attention is on external factors influencing cardiovascular development and how the resulting malformations can be managed clinically. Following two chapters that deal with environmental influences on cardiorespiratory development (Chapter 17) and its modeling (Chapter 18), perspectives are provided on more clinically oriented topics that range from principles of abnormal cardiac development (Chapter 19), through therapeutic treatments (Chapter 20), to management of human patients (Chapter 21).

Development of Cardiovascular Systems: Molecules to Organisms concludes with a synthesis of, and perspective on, a dynamic field as it currently stands, and provides projections for the future.

The book also offers a compendium of nearly 1500 references. Reflecting the explosive growth in this field, more than half the references have been published just in the 1990s. Having a single reference list was an active pedagogical decision to allow for paper cross-referencing. Thus, each reference concludes with an indication of each chapter in which it is cited. This should prove a useful tool that complements the Index as the reader tracks down information on particular topics. Additionally, our hope is that a biomedically oriented student searching for a reference on development of, for example, starling relationships will be intrigued by seeing this paper cited in the chapters on fish and amphibian development and will then read these additional chapters to gain a comparative perspective. Similarly, our hope is that the curiosity of the comparative zoologist will be sufficiently aroused by seeing a familiar paper cited in a clinically oriented chapter that he or she will read that chapter as well.

As is evident from even a quick scan of this book, the advanced tools and approaches being brought to bear on all aspects of cardiovascular development are yielding new information at an enormous rate. The field of cardiovascular development, however, is in an early era of growth and development, when new questions are unearthed more quickly than we can answer them. In many ways, that rhythmically pulsing "red spot" remains as enigmatic a beacon to us now as it was to the philosopher-scientists of history. Our intention is that this book will help to remedy that situation through the organization and display of current information and through the stimulation of continued interest.

Part I

Molecular, cellular, and integrative mechanisms determining cardiovascular development

1

Genetic dissection of heart development

JAU-NIAN CHEN AND MARK C. FISHMAN

Introduction

Along with the evolution of the closed chordate circulation, which contains blood at relatively high pressures, came the evolution of a highly muscular heart with valves separating low- and high-pressure chambers and of a vasculature lined throughout by endothelium (see Chapter 9). How is this seamless tubular system fashioned?

The myocyte has been the focus of much of the molecular work to date concerned with heart development. Yet the heart is constituted of cells with many fates, varying in specialized function from conduction to secretion to contraction. Furthermore, it has not been feasible to address at a molecular level essential issues of cardiovascular morphogenesis, especially with reference to larger-scale organotypic decisions (Fishman & Stainier, 1994). For example, are there genes that determine heart size? Are there genes that demarcate chamber borders? Are there single genes crucial to the fashioning of organotypic structures, such as endocardium or valves? Are different vascular beds assembled differently? Although many approaches might be envisioned, genetics has already proven to be powerful in revealing binary decisions during development in *Drosophila* and *Caenorhabditis elegans.* We explore here the possibilities offered by three genetic systems–*Drosophila,* mouse, and zebra fish–to discover the earliest molecular decisions that fashion the cardiovascular system.

Drosophila: The power of genetic screens in invertebrates

Drosophila *heart development*

The fly heart is a simple tubelike organ that is located at the dorsal midline beneath the epidermis and that extends nearly the length of the body. The circulation is open. Although there are regions with "vessels" containing hemolymph in the wings and in other body regions, it is not clear whether the lining cells can be likened to endothelium or rather represent spaces between tissue planes. Hemolymph is sucked into the heart through posterior ostia and ejected anteriorly. Although a valvelike function has been ascribed to the region between the posterior and the narrower, anterior part of the tube, no distinctive structure has been identified. There are three cell types associated with the heart of *Drosophila:* cardial cells, pericardial cells, and alary muscles (Figure 1.1) (Bate, 1993; Rugendorff, Younossi-Hartenstein, & Hartenstein, 1994). Alary muscles connect the heart to the epidermis. Cardial cells express myofibrillar proteins such as actin, myosin, and tropomyosin, and are the

Figure 1.1 Diagram of a transverse section of *Drosophila* embryonic heart. The embryonic fly heart consists of three cell types: cardial cells, pericardial cells, and alary muscle. The cardial cells (*gray areas with stripes*), surrounded by pericardial cells (*gray areas with dots*), form a tube on the dorsal side of the embryo. The tube is connected to the epidermis (*gray*) by the alary muscles (*stars*).

contractile cells. Pericardial cells are believed to be secretory. Unlike the vertebrate heart, there are no endocardial cells or specific chamber boundaries in the heart of *Drosophila.* The circulation is not responsible for oxygen delivery, which is the purview of the elaborate tracheal system.

As in higher vertebrates, heart cells in *Drosophila* are mesodermal in origin. During gastrulation, the mesoderm is generated from ventral cells of the blastoderm, which invaginate into the interior of the embryo along the ventral midline, flatten, and form a single cell layer that spreads dorsally along the inner surface of the ectoderm (Figure 1.2). The dorsal-most cells of the advancing mesoderm are the cardiac progenitor cells. The mesoderm splits into two just ventral to these cells. The inner, or visceral, mesoderm will provide gut musculature, and the outer, or somatic, mesoderm will provide muscle to the body wall. After this separation, molecular markers that characterize the different muscle types begin to be expressed. The two dorsal rows of cardiac mesodermal cells then move toward the dorsal midline and fuse into a tube underneath the dorsal epidermis (Bodmer, 1995). Hence, at a functional and embryological level, *Drosophila* is similar to vertebrates in that its heart contains autonomously contractile cells arranged in a tube arising from bilateral precursors in the mesoderm. It differs in that the heart of *Drosophila* is a dorsal structure, whereas vertebrate hearts are ventral; the heart of *Drosophila* lacks endocardium and closed, endothelium-lined vasculature; the heart of *Drosophila* is not separated into chambers and lacks distinctive valves and coronary circulation. Some, but not all, insect hearts are neurogenic, in that neural stimuli initiate each contraction. To our knowledge this has not been determined for *Drosophila.*

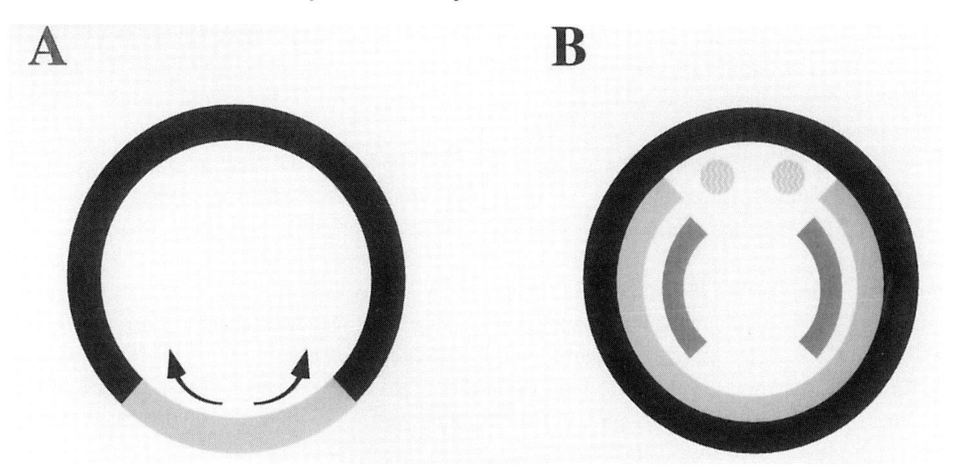

Figure 1.2 Gastrulation and heart formation in *Drosophila*. Diagrams are oriented with the dorsal side to the top. (*A*) Cross section of blastula-stage embryo. The mesodermal cells invaginate into the embryo along the ventral midline and spread dorsally during gastrulation. The black area represents the ectoderm, and the light gray area represents the mesoderm. (*B*) Cross section of the 7-hour *Drosophila* embryo. By this time, the mesodermal primordia of somatic muscle (*light gray*), visceral muscle (*dark gray*), and heart (*black*) have separated.

Drosophila genetics

Most of our knowledge about the genetic cascade responsible for the development of the *Drosophila* heart relates to isolation of a homeobox gene – *tinman* – that lies genetically downstream of the genes that initiate mesoderm determination and upstream of the cascade that distinguishes heart from other mesodermal derivatives. Absent *tinman,* the *Drosophila* embryo does not develop a heart.

The mesoderm arises from ventral cells of the blastoderm, which are distinguished from the dorsal cells by localized activity of the zygotic protein, Dorsal, in their nuclei. Dorsal is actually expressed uniformly around the blastoderm, but ventrally it is transported into the nuclei, where it acts as a transcription factor. Dorsal activates the ventral expression of two genes, *twist* and *snail,* which are expressed in all mesodermal precursors. The product of *twist* is a helix-loop-helix protein, and the product of *snail* is a zinc-finger protein. Mutation in either of these genes prevents mesodermal differentiation (St. Johnston & Nüsslein-Volhard, 1992). The gene *tinman* appears to be necessary for subsequent distinction of mesodermal sublineages and is first expressed soon after *twist* in all mesodermal primordia at the blastoderm stage. The gene *tinman* is dependent upon *twist* for its expression, and *twist* binding sites are found in the promoter region of *tinman.* It is therefore reasonable to assume that *twist* controls *tinman* (Bodmer, 1993, 1995).

The expression of *tinman* persists during gastrulation as the mesoderm invaginates ventrally. Just before the subdivision of somatic and visceral mesoderm lineages, *tinman* expression becomes restricted to the visceral and cardiac mesodermal precursors. By the germ-band-extended stage, *tinman* expression is restricted to the cardiac mesoderm (Bodmer, Jan, & Jan, 1990). Mutations in *tinman* prevent differentiation of visceral and cardiac mesoderm. Genes normally expressed in the early visceral mesoderm, such as *bagpipe,* and those expressed in the cardiac mesoderm, such as *zfh,* are missing from *tinman* mutants, suggesting that these are genetically downstream of *tinman* (Azpiazu & Frasch, 1993; Bodmer, 1993). Thus, it is currently believed that *tinman* is the crucial link between the general determination of mesoderm by *twist* and the specific differentiation of the visceral and the cardiac mesoderm.

Vertebrate homologues of *tinman* have been isolated from the frog *Xenopus* (*XNkx2.5*) and from the mouse (*Csx/mNkx2.5*) (Komuro & Izumo, 1993; Lints et al., 1993; Tonissen et al., 1994). As expected, there are regions of significant homology between *Drosophila tinman, XNkx2.5,* and *Csx/mNkx2.5.* The amino acid sequences of *XNkx2.5* and *Csx/mNkx2.5* homeodomain show 67% and 65% identity with *Drosophila tinman,* respectively. A 10 amino acid region at the N-terminus (TN domain) is identical in *Drosophila tinman, XNkx2.5* and *Csx/mNkx2.5.* However, an 18 amino acid sequence conserved in the NK2 family is present in *XNkx2.5* and *Csx/mNkx2.5* but absent from *Drosophila tinman.* The structure of the coding region and the homologies between *Drosophila tinman* and its vertebrate homologues are illustrated in Figure 1.3.

The expression patterns of *XNkx2.5* and *Csx/mNkx2.5* include regions of cardiac precursors, although the expression patterns described in mouse and frog begin relatively later in development than does that of *tinman* in *Drosophila.* Transcripts of *XNkx2.5* are first detected in frog embryos at the neurula stage in the bilateral mesodermal regions corresponding to the cardiogenic mesoderm. Expression persists in the heart throughout the embryonic stages to the adult (Tonissen et al., 1994). In the mouse, *Csx/mNkx2.5* transcripts are first detected in the early headfold stage in a crescent of anterior and lateral plate mesoderm, a region corresponding to that of the heart progenitors. Expression persists in the heart tube, embryonic heart, and fetal heart. Pharyngeal endoderm, which is adjacent to the cardiogenic crescent, also expresses *tinman* (Komuro & Izumo, 1993; Lints et al., 1993). Except for the tongue, other muscles do not express *tinman.* Mice mutated in the *Csx/mNkx2.5* gene have recently been obtained by homologous recombination (Lyons et al., 1995). Unlike the *tinman*-deficient fly, in which the heart is absent, the *Csx/mNkx2.5*-deficient mouse does develop a heart, but it fails to loop. The relative normalcy of early heart development without *Nkx2.5* may reflect compensation, perhaps from other members of the *Nkx* family, or may suggest that the singular role of *Drosophila tinman* in heart development has no analogy in higher vertebrates. In terms of functional relatedness, it will be useful to know whether a vertebrate *Nkx2.5* can rescue the *tinman*-mutant *Drosophila.*

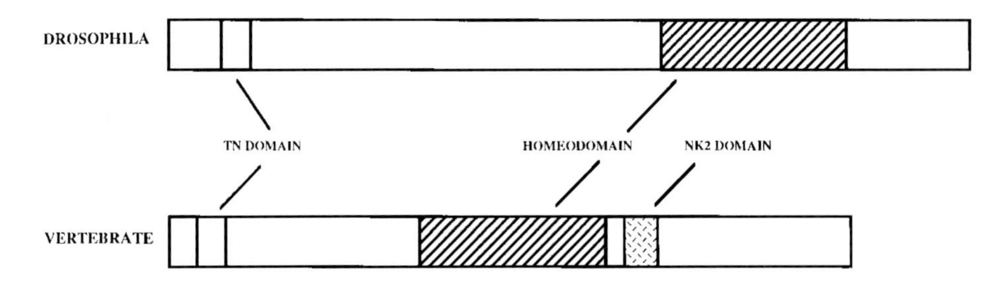

Figure 1.3 Diagrammatic comparison of the cDNA structure of *Drosophila tinman* and its vertebrate homologues. A 10 amino acid sequence at the N-terminus, the *TN* domain, is identical in *Drosophila tinman, XNkx2.5* and *Csx/mNkx2.5.* The homeodomain is highly conserved in *Drosophila, Xenopus,* and mouse. An 18 amino acid sequence downstream of the homeodomain, the *NK2* domain, is found in *XNkx2.5* and *Csx/mNkx2.5* but not in *Drosophila tinman.*

It is not clear how *tinman* expression is progressively restricted from ventral mesoderm, first to visceral and cardiac mesoderm and finally to cardiac mesoderm alone. One possibility is that *tinman* expression is patterned by ectodermal signals, acting across germ layers. A likely signaling candidate is *decapentaplegic* (*dpp*), a member of the bone morphogenic class of TGF-β-like signaling molecules. Early in development, *dpp* expression is restricted to the dorsal ectoderm, and a putative *dpp* receptor, *thickvein,* is expressed in the mesoderm. By the time of the subdivision of the somatic and visceral mesodermal lineages, the expression of *tinman* is restricted to the dorsal portion of the mesoderm, directly below the dorsal ectodermal cells, which express *dpp* (St. Johnston & Gelbart, 1987; Immergluck, Lawrence, & Bienz, 1990; Panganiban et al., 1990). In mutants lacking *dpp, tinman* expression is present in the early mesoderm, as in the wild type, but disappears at the time when *tinman* expression normally is restricted to the dorsal mesoderm, and no heart is formed. Restoration of *dpp* in ectopic ectodermal domains in the *dpp* mutants induces *tinman* in adjacent mesoderm and some heart progenitors. Interestingly, ectopic expression of *dpp* in ventral domains of wild-type flies, although expanding the *tinman* domain ventrally, does not induce new heart tissue in the new location. This suggests that *dpp* and *tinman* alone are not sufficient for development of cardiac mesoderm and that additional regulators are involved (Staehling-Hampton et al., 1994; Frasch, 1995).

What genes are genetically downstream of *tinman* in heart generation? One candidate is *Dmef2.* The *mef2* genes encode transcription factors. In mammals there are four *mef2* genes, which are expressed in a variety of tissues, including the heart, skeletal muscle, and brain. MEF2 proteins bind A-T-rich DNA sequences of skeletal, smooth muscle, and cardiac genes (Edmondson et al., 1994). They have been of interest as potential key regulatory proteins with regard to heart muscle cell development. *Drosophila* contains only one *mef2* gene, expressed first in the ventral mesodermal precursors, then throughout the mesoderm, and subsequently in all cells of the cardiac, visceral, and somatic muscle lineages (Lilly et al., 1994). Mutation of the *Dmef2* gene blocks myogenic differentiation in all three lineages, although the myoblasts assemble in the correct position (Lilly et al., 1995).

The mouse: The power of gene knockout

The cardiovasculature of the fruit fly differs in many fundamental aspects from that of vertebrates. Therefore, the relevance of *Drosophila* is predicted to reside chiefly in revealing decisions that drive cell fate rather than those that fashion higher-order structure (Kern, Argao, & Potter, 1995). To understand the determinants of cardiovascular structure relevant to human diseases, a vertebrate system is essential.

Targeted gene mutation by homologous recombination ("gene knockout") in mice is a powerful genetic tool for the study of vertebrate development. From such studies, several genes appear to be crucial for normal cardiovascular development. These are listed in Table 1.1. Equally interesting are those genes expressed in the early heart that when mutated do not alter cardiac phenotype. For example, *Msx-1* (*Hox-7*) is expressed in the endocardial cushions (Chan-Thomas et al., 1993) and *Hoxa-2* is expressed in the embryonic mouse vasculature and neonatal heart and smooth muscle. Targeted disruption of both results in no visible cardiovascular abnormalities (Kern, Argao, & Potter 1995). Clearly, expression pattern does not correlate in a simple manner with the effects of gene ablation. Reasons may be that in some cases the protein may not be expressed, and in others there may be compensatory or redundant pathways brought into play.

Table 1.1. *Examples of cardiovascular effects of murine targeted gene knockouts*

Gene	Role	Effect on cardiovascular system	First heart effect noted	Lethal	References
connexin 43	Gap junctions	Enlarged conus filled with intraventricular septae that obstruct right ventricular outflow tract	e 16.5	Birth	Reaume et al., 1995
neurofibromatosis type-1	Regulation of *ras*	Thin myocardium at e 13.5; ventricular chamber hypoplasia, ventricular septal defects, and persistent truncus arteriosus; delayed cushions	e 12.5	e 11.5–13.5	Brannan et al., 1994
retinoic acid receptor α (*RXR*α)	Ligand-dependent transcription factor	Thin compacta and poor trabeculation; ventricular septal defects	e 12.5	e 13.5–16.5	Sucov et al., 1994
retinoic acid receptor $\alpha\gamma$ (*RAR*$\alpha\gamma$)	Ligand-dependent transcription factor	Thin compacta; persistent truncus arteriosus; ventricular septal defects; aortic arch abnormalities	Before e 18.5	During gastrulation or immediately after birth	Mendelsohn et al., 1994b
fibronectin	Cell adhesion, migration	Variable: bilateral cardiac primordia not fused; cardiac jelly deficient; endocardium absent; dorsal aortae absent	e 8.0	e 10.5	George et al., 1993
α_5 *integrin*	Fibronectin receptor	Vascular defects, including dorsal aortae not closed (heart not grossly abnormal)	e 9.5	e 10.5–11.5	Yang, Rayburn, & Hynes, 1993
α_4 *integrin*	Fibronectin, VCAM-1 receptor	Epicardium not formed; no coronary vessels; rest of heart normal	e 12.5	e 9.5–15.5	Yang, Rayburn, & Hynes, 1995
VCAM-1	Cell surface protein, binds α_4 integrin	Epicardium not formed; no coronary vessels; rest of heart normal	e 11.5	e 10.5–12.5	Kwee-Isaac et al., 1995

N-myc	Proto-oncogene	Thin compacta; ventricular septal defects	Before e 11.5	e 11.5–12.5	Charron et al., 1992; Stanton et al., 1992; Moens et al., 1993; Sawai et al., 1993
c-myc	Proto-oncogene	Enlarged heart; pericardial effusion	Before e 9.5	e 10.5	Davis et al., 1993
Nkx2.5	Homeobox gene	Heart tube fails to loop, poor trabeculation	e 8.5	e 9–10	Lyons et al., 1995
Hoxa-3	Homeobox gene	Defects of major arch vessels and aortic and pulmonary semilunar valves	Before e 13.5	Birth	Chisaka & Capecchi, 1991
flk-1	Receptor tyrosine kinase	No blood vessels, severely reduced blood cells	e 7.5	e 8.5–9.5	Shalaby et al., 1995
flt-1	Receptor tyrosine kinase	Disorganized vasculature, thicker and disorganized endocardial lining of the heart ventricle	e 8.0	e 9–11.5	Fong et al., 1995
tie-1	Receptor tyrosine kinase	Defect in the integrity of vessels, hemorrhage, edema	e 13.5	Birth	Sato et al., 1995
tek (*tie-2*)	Receptor tyrosine kinase	Diminished number of endothelial cells and defects in sprouting and branching of the vessels, thin myocardium	e 8.5	e 9.5–10.5	Dumont et al., 1994; Sato et al., 1995

Note: e = embryonic day.

Several phenotypes recur among the gene knockouts that do perturb the cardiovasculature. One is the failure to finish growth of the ventricular septum and of the ventricular compact layer. Beginning at embryonic day 12.5, the growth of the compact zone is responsible for most of the wall thickness of the heart. As a consequence of these knockouts, the ventricular wall remains thin and the septum incomplete. This is the phenotype with mutation of *tek, neurofibromatosis type-1, N-myc, Nkx2.5, retinoic acid receptor* α *(RXR*α*),* and *retinoic acid receptor* α γ and (RAR$\alpha\gamma$) double knockout. Loss of either *VCAM-1* or α_4 integrin interferes with formation of epicardium, a tissue that arises from extracardiac sources and brings the angioblasts that generate the coronary arteries. These mice therefore lack coronary vessels. Perturbation of "cardiac" neural crest (Kirby & Waldo, 1990) may be responsible for failure to develop the major arch vessels (*Hoxa-3*) and for persistent truncus arteriosus (e.g., *neurofibromatosis type-1* and *RAR*$\alpha\gamma$). The only reported lesion affecting relatively early cardiac development is the knockout of fibronectin, which leads to failure of fusion of the bilateral cardiac primordia. Therefore, it remains uncertain how often specific gene perturbation will disrupt specific structural processes or functional maturation during early cardiac morphogenesis. It is clear that late cardiac development may be accessible to investigation using homologous recombination in the mouse.

The zebra fish: The power of genetic screens of a vertebrate

Zebra fish heart development

Although fish and mammals diverged about 400 million years ago, the essential components of early heart and vascular development are identical. Later in cardiogenesis the mammals, as well as amphibians, reptiles, and birds (see Chapters 13–15), differ from fish by the acquisition of a separate pulmonary circulation for air-breathing. In fish, as in mammals, cardiac progenitors in the lateral plate mesoderm generate bilateral tubes that merge at the midline, generating the primitive heart tube, which has an inner, endocardial, and an outer, myocardial, layer. As the thickness of the myocardium increases, the tube loops to the right, and cushions appear, separating chambers referred to as sinus venosus, atrium, ventricle, and bulbus arteriosus. Blood circulates through the ventral aorta to the gills and thence to the dorsal aorta and the body (see Chapter 12; Rombough, 1994).

The zebra fish, *Danio rerio,* is a small bony fish (Osteichthyes) of the Cyprinidae family, originally described from rivers of the Far East. It is of paramount interest today because it is the only vertebrate widely studied that is amenable both to embryonic manipulation and to large-scale genetic screens. It also is well suited to studies of internal organ development because, as with many fish and amphibians, the embryo is transparent.

The clarity of the zebra fish embryo makes feasible the determination of the origin and migratory paths of cardiac progenitor cells by injection of fluorescent dextran into individual blastomeres. A discrete zone of the mid- to late blastula stage provides cardiac progenitors (R. Lee et al., 1994), as shown in Figure 1.4. Therefore, decisions directly relevant to generation of the heart begin as soon as transcription is initiated. Patterning information is already in place, because progeny of individual blastomeres evidence chamber restriction (Stainier, Lee, & Fishman, 1993; R. Lee et al., 1994).

Development in the zebra fish is rapid (Figure 1.5). The primitive heart tube is formed by 22 hours postfertilization (hpf). The heart is beating and the circulation becomes evident at 24 hpf. By 30 hpf, the heart tube starts to loop to the right side of the embryo. By 36 hpf,

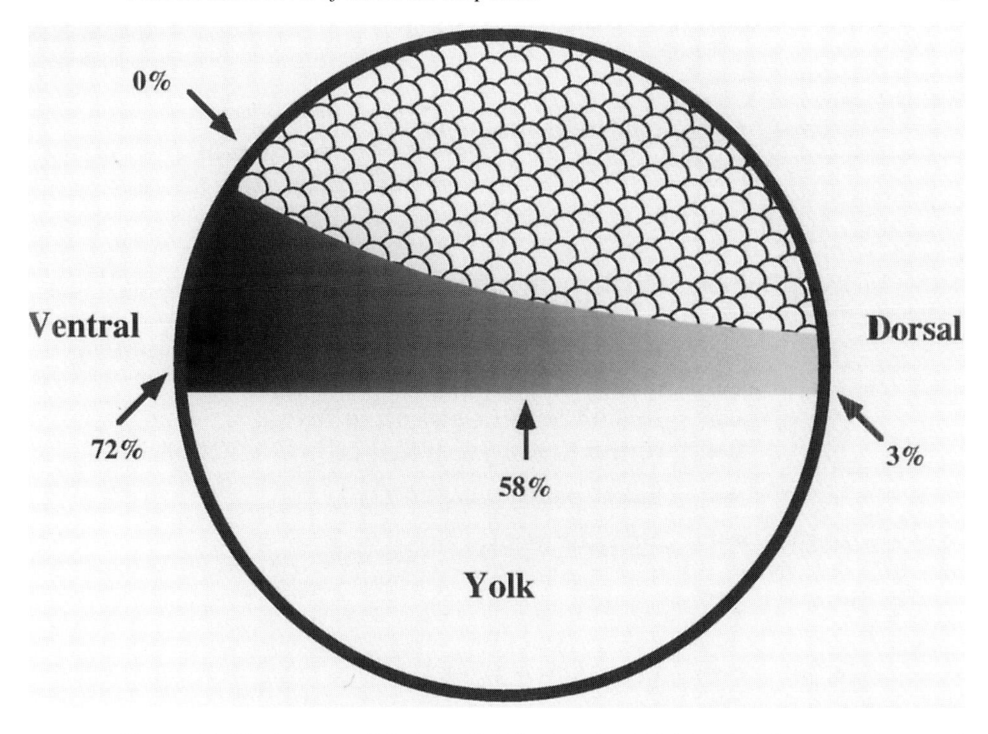

Figure 1.4 The location of the heart field in the zebra fish blastula. Lineage studies reveal that most zebra fish heart precursor cells are located at the ventral side of the embryo at the blastula stage. The intensity of gray in the diagram represents the propensity to form heart. The percentages indicate the probability of having heart cells included among the progeny.

chamber boundaries are evident, although molecular markers can distinguish them in the primitive heart tube. (Stainier & Fishman, 1992). Looping places the atrium toward the left and the ventricle toward the right side of the embryo. At about this time, endocardial cells located at the chamber boundaries undergo epithelial-to-mesenchymal transition to form cushions. By 60 hpf, the valves are present. By the fifth day, the heart has assumed its adult configuration, with the atrium sitting dorsally with respect to the ventricle (Figure 1.5) (Stainier & Fishman, 1994). Myocardial cells and endocardial cells can be clearly distinguished at single-cell resolution in the living embryo (Figure 1.6). This contrasts with other vertebrates, such as the chick and the mouse, where little cellular resolution is possible. As with other fish (see Chapter 12), it appears that the embryo's small size and aqueous environment mean that gas exchange is not limiting, so that the zebra fish embryos can survive without a functioning cardiovascular system for up to 5 days after fertilization (Chen & Fishman, unpublished data). Therefore, despite mutations that reduce or eliminate cardiac function, it is feasible to study the progressive effects of a failing cardiovascular system. In contrast, mouse embryos with severe cardiac defects degenerate rapidly.

Zebra fish genetics

Saturation mutagenesis aims to identify all genetic loci that can be mutated and that reveal an informative phenotype. A dose of mutagen, such as ethylnitrosourea, is chosen by effects on known loci, such as for pigmentation, and for generating one to two recessive lethal

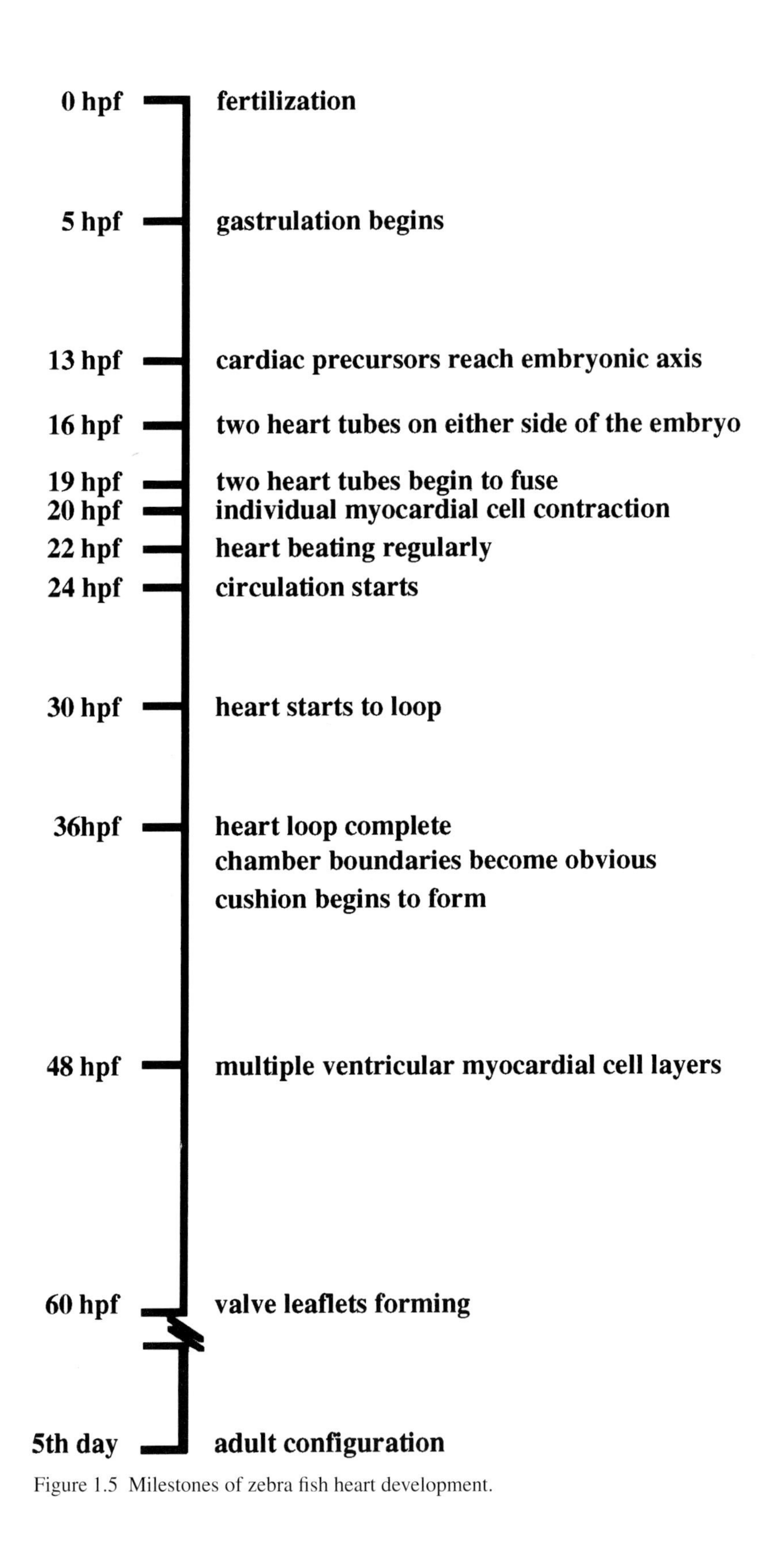

Figure 1.5 Milestones of zebra fish heart development.

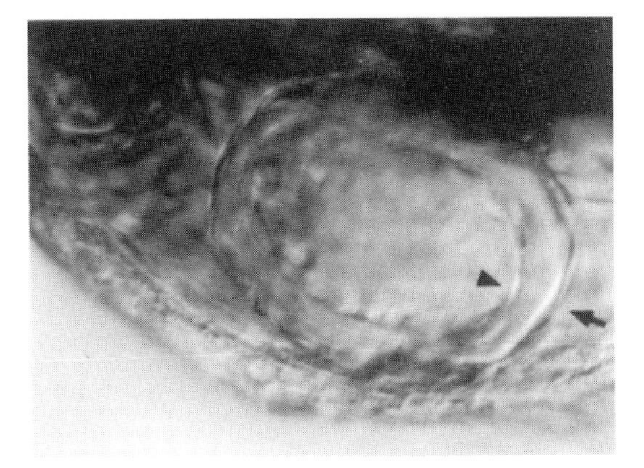

Figure 1.6 Zebra fish embryonic heart. The heart of zebra fish is located at the ventral side of the embryo on top of the yolk. The myocardium (*arrow*) and the endocardium (*arrowhead*) are distinguishable under Nomarski optics. The picture shown here is taken from a second-day silent-heart mutant, in which the heart does not beat but the morphology of the heart is normal. The embryo is oriented with the head to the left.

mutations per haploid genome (Mullins et al., 1994; Solnica-Krezel, Schier, & Driever, 1994). After breeding for two generations, the F_3 progeny are scored for phenotype, which should be evident in 25% of the offspring of two heterozygotes. This large-scale screen is achievable for the zebra fish because they are small and hardy, and because an adult female lays hundreds of eggs each week, which are externally fertilized. Two large-scale zebra fish screening projects are currently under way, one at the Cardiovascular Research Center of the Massachusetts General Hospital and one at the Max-Planck-Institut für Entwicklungsbiologie, in Tübingen. Preliminary observations suggest that many mutations will be identified that perturb many of the structural and functional "milestones" of heart development.

Although chemical mutagenesis results in a wide range of phenotypically interesting mutations, the cloning of such mutations is arduous. Genomic maps (H. Jacob, personal communication; Postelthwait et al., 1994) and large insert genomic libraries are under construction. Insertional mutagenesis techniques are being developed (Lin et al., 1994). What is already clear, however, is that crucial bimodal molecular decisions of early vertebrate development can be clearly and informatively analyzed, just as they have been for *Drosophila* and *C. elegans.*

Summary

Genetics offers more than a first step toward gene isolation. Genetic hierarchies of gene action can be determined by appropriate breeding. Whether a mutation perturbs a cell autonomously or by causing deficits in signaling from other cells can be revealed by transplantations between mutants and wild-type embryos. All of these attributes of developmental decision making can be discovered prior to cloning the gene. Once cloned, homology screens and targeted gene mutation in the mice will provide access to the gene's role in mouse and human.

2

Cardiac membrane structure and function

LYNN MAHONY

Cardiac function is critically dependent on the movement of ions across membranes. Contraction of the heart begins when an action potential depolarizes the plasma membrane of the myocyte, the sarcolemma. Depolarization is caused by sequential inward currents of sodium ions and calcium ions. The slow inward calcium current enters through voltage-gated channels in the sarcolemma and serves several purposes. First, this current supplies a small amount of the activator calcium for binding to troponin C. Second, the sarcoplasmic reticulum (SR) takes up a considerable portion of this calcium, where it forms part of the internal store for release in subsequent contractions. Finally, this calcium triggers the release of a large amount of calcium from the SR. The released SR calcium initiates contraction by binding to troponin C, resulting in formation of cross-bridges between the myosin head and actin. Relaxation begins with the removal of calcium from the myofilaments, which is mediated primarily by the SR calcium pump but also by the Na^+–Ca^{2+} exchanger. Removal of calcium from the myofilaments results in the breaking of actin–myosin cross-bridges and return to the resting state.

Important changes in the structure and function of cardiac membranes occur during maturation of the mammalian heart (and probably in all vertebrate hearts). These changes result in significant age-related differences in myocardial function. This chapter discusses recent work that has increased our understanding of developmental changes in cardiac membranes and of the impact of these changes on myocardial performance.

The developing sarcolemma

The sarcolemma contains ion pumps, channels, and exchangers that maintain chemical and charge differences between the intracellular and extracellular spaces of the myocyte (Figure 2.1). Ion flow across the sarcolemma controls depolarization and repolarization. The sarcolemma also contains receptors for various hormones and enzymes.

The sodium pump (sodium–potassium ATPase) uses energy derived from adenosine triphosphate (ATP) hydrolysis to maintain the sodium and potassium gradients across the sarcolemmal membrane. Sodium is pumped out of the cell in exchange for potassium. The enzyme consists of two subunits. The catalytic subunit (α) contains the binding site for cardiac glycosides. The function of the smaller glycoprotein subunit (β) is unknown. Different isoforms of the α-subunits exist in rats (Sweadner, 1979; Charlemagne et al., 1987) and humans (Shull & Lingrel, 1987). Developmental changes in isoform expression have been described in rats (Orlowski & Lingrel, 1988; Lucchesi & Sweadner, 1991). Sodium pump

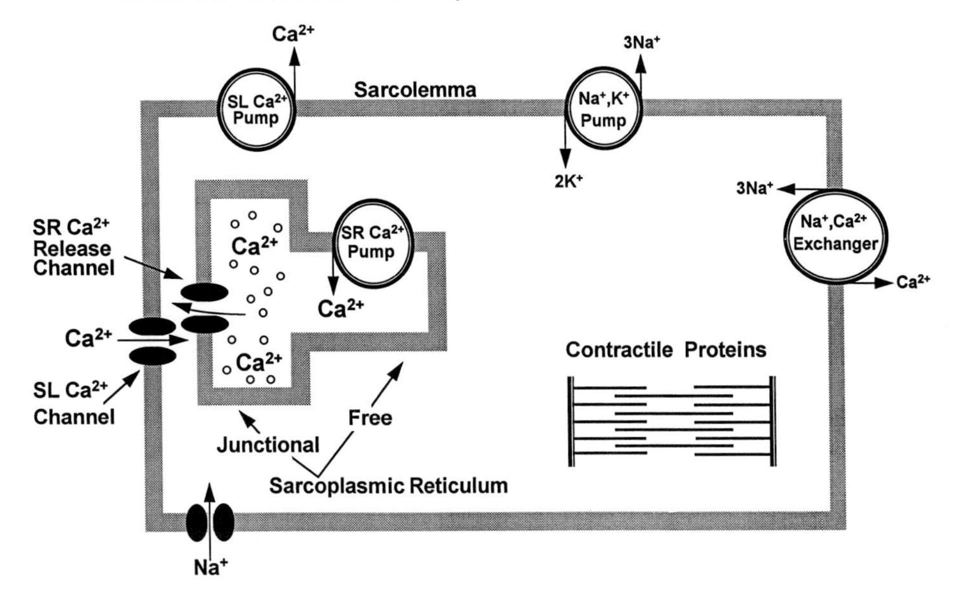

Figure 2.1 Schematic representation of myocyte membrane systems (SR = sarcoplasmic reticulum, SL = sarcolemma).

activity is less in sarcolemmal membranes isolated from immature hearts than in membranes from mature hearts (Mahony & Jones, 1986; Artman, 1992).

During the plateau phase of the action potential, Ca^{2+} enters the myocyte through voltage-dependent Ca^{2+} channels in the sarcolemma. Ca^{2+} channel density increases during embryonic and postnatal development of chick and rat hearts (Kazazoglou et al., 1983; Renaud et al., 1984; Marsh & Allen, 1989; Renaud et al., 1989). It is important to note that, at least in the embryonic chick heart, the number of ligand binding sites is not related to the function of Ca^{2+} channels (Aiba & Creazzo, 1993). Peak Ca^{2+} current density decreases during embryonic life in the chick (Tohse, Meszaros, & Sperelakis, 1992). In contrast, peak current density increases twofold to threefold during maturation of the rabbit heart. However, voltage-dependent activation properties do not differ (Osaka & Joyner, 1991; Huynh et al., 1992; Wetzel, Chen, & Klitzner, 1993).

The sarcolemmal Na^+–Ca^{2+} exchanger is another important component of the sarcolemma. Three sodium ions are exchanged for one Ca^{2+} ion, and thus charge moves across the membrane. The exchanger carries Ca^{2+} out of the myocyte after each contraction to maintain an appropriate intracellular Ca^{2+} concentration (Bridge, Smolley, & Spitzer, 1990; Cannel, 1991). The driving force for this Ca^{2+} efflux is the sodium gradient between the intracellular and extracellular spaces, which is maintained by the sodium pump. Ca^{2+} also enters the myocyte via the Na^+–Ca^{2+} exchanger. The large sodium entry during the fast sodium current raises subsarcolemmal sodium concentration, and the inside of the myocyte becomes positively charged during the plateau phase of the action potential. This favors "reversal" of the exchanger, with resulting Ca^{2+} influx and sodium efflux. Inasmuch as the exchanger can move Ca^{2+} into or out of the myocyte, the exchanger may play a role in mediating both myocardial contraction and relaxation.

Some studies suggest that the exchanger is distributed throughout the sarcolemma (Kieval et al., 1992), but others report preferential localization to the transverse tubules (Frank et al., 1992). In the developing rabbit heart, indirect immunofluorescence studies show localization of the exchanger to the peripheral sarcolemma in immature cells without transverse tubules (Chen et al., 1995). The exchanger is also seen in transverse tubules as these structures develop in maturing hearts (Chen et al., 1995).

Important species differences in the activity of the Na^{+}–Ca^{2+} exchanger have been reported (Sham, Hatem, & Morad, 1995). Inward Na^{+}–Ca^{2+} exchange currents elicited by caffeine-induced SR Ca^{2+} release are smallest in rats, intermediate in guinea pigs and humans, and largest in hamsters. These differences may reflect differences in exchanger density, regulation, and/or spatial relationship to SR Ca^{2+} release channels (Sham, Hatem, & Morad, 1995).

Striking developmental changes in exchanger density and function occur during maturation of the rabbit heart. Exchanger activity measured in sarcolemmal vesicles isolated from hearts of variously aged rabbits is more than threefold higher in fetal and newborn than in adult vesicle preparations (Artman, 1992). Similarly, the relative amount of exchanger protein detected immunologically is about 2.5 times as high in fetal and newborn vesicles as in adult preparations. The fact that steady-state exchanger mRNA levels are 2.5 times as high in hearts from newborn rabbits as in adult rabbit hearts indicates that these differences are not related to differential purity of the membrane preparations (Boerth, Zimmer, & Artman, 1994). These data suggest that transcriptional – not translational – processes regulate exchanger expression in the developing heart.

An ATP-dependent Ca^{2+} pump in the sarcolemma contributes to Ca^{2+} efflux from the myocyte. Binding of Ca^{2+} and the Ca^{2+}-binding protein, calmodulin, stimulate the pump by increasing both Ca^{2+} sensitivity and maximal velocity (Caroni et al., 1983). Although Ca^{2+} is removed from the myocyte by the sarcolemmal calcium pump, it is likely that the rapid, high-capacity exchanger is primarily responsible for maintaining steady-state intracellular Ca^{2+} concentration (Cannel, 1991; Langer, 1992). Little information is available regarding changes in the structure or regulation of the sarcolemmal Ca^{2+} pump during maturation of the heart.

The sarcolemmal sodium–hydrogen exchanger uses the energy of the sodium gradient to transport protons out of the myocardial cell. The exact role of the exchanger in maintaining intracellular pH is not known. Sodium–hydrogen exchange in the presence of a pH gradient is higher in sarcolemmal vesicles isolated from newborn rabbit hearts than that measured in vesicles from adult rabbit hearts (Meno, Jarmakani, & Phillipson, 1989). This may contribute to the greater recovery of mechanical function in the newborn heart than in the adult heart after intracellular acidosis.

The transverse tubule system

The transverse tubule system is a continuation of the sarcolemma that extends transversely into the central regions of the cell and runs longitudinally between adjacent sarcomeres. As myocytes enlarge and cell volume increases during development, acquisition of transverse tubules facilitates transmission of the action potential to the interior of the myocardial cell, which results in rapid activation of the entire cell (Sheldon, Friedman, & Sybers, 1976; Page, 1978; Gotoh, 1983). At birth, there is considerable variation in the degree of development of the transverse tubule system. Myocytes from animals relatively mature at birth (guinea pigs and lambs) have well-developed transverse tubules, but myocytes from less mature newborns (rabbits and rats) do not. Variations in the development of the transverse tubule system likely contribute to species-related differences in the mechanisms regulating excitation–contraction coupling in immature animals.

Mitochondria

The size and relative volume of mitochondria increase during development (Smith & Page, 1977; Legato, 1979; Sheridan, Cullen, & Tynan, 1979; Hoerter, Mazet, & Vassort, 1981; Smolich et al., 1989; Barth et al., 1992), and the mitochondrial cristae lengthen and become more densely packed. In several mammalian species, striking maturation of the mitochondria occurs immediately after birth (Hoerter, Mazet, & Vassort, 1981; Rolph, Jones, & Parry, 1982; Smolich et al., 1989).

Age-related increases in mitochondrial-dependent aerobic metabolism are explained in part by the increases in mitochondrial volume and in the density of the inner cristal system. Long-chain free fatty acids are the primary myocardial energy substrate in adult hearts. ATP is produced when activated free fatty acids are transported into the mitochondria and then metabolized by β-oxidation. Carnitine palmitoyl CoA transferase is responsible for the transport of activated free fatty acids from the cytosol into the mitochondria. In immature hearts, the activity of this enzyme is decreased. As a result of these and other factors, the primary energy substrates in the immature heart are lactate and/or carbohydrates (Fisher, Heymann, & Rudolph, 1980, 1981).

The sarcoplasmic reticulum

The SR is a tubular membrane network that surrounds the myofibrils and has two components: the junctional SR and the free, longitudinal SR (Figure 2.1) (Lytton & MacLennan, 1992). The junctional SR (terminal cisternae in skeletal muscle) is a flat tubule that makes contact with the sarcolemma and transverse tubules. Corbular SR, a form of junctional SR that has no contact with the plasmalemma, is located at the Z lines of the sarcomere. The cytosolic membrane surfaces of the SR contain the "foot" structures that span the gap between the SR and sarcolemma at the dyad coupling. These foot structures are the SR Ca^{2+} release channels and are the binding site for ryanodine, a plant alkaloid that selectively interferes with Ca^{2+} release from the SR (Bers, Bridge, & MacLeod, 1987). Ca^{2+} flows through these channels to the myofilaments to initiate contraction. The various parts of the junctional SR are connected by anastomosing strands of free SR. This portion of the SR is rich in ATP-dependent Ca^{2+} pumps and the regulatory protein, phospholamban. Active transport of Ca^{2+} into the SR by Ca^{2+} pumps facilitates myocardial relaxation.

Dramatic changes in the structure and function of the SR occur during maturation of the heart. The content of the SR is decreased and the SR is less organized in the immature mammalian heart (Maylie, 1982; Nassar, Reedy, & Anderson, 1987; Nakanishi et al., 1987). In particular, the separation of the junctional and free regions is less well defined (Nassar, Reedy, & Anderson, 1987). Expression of sarcolemmal voltage-gated Ca^{2+} channels and SR Ca^{2+} release channels is coordinately regulated during maturation of the rabbit heart (Brillantes, Bezprozvannaya, & Marks, 1994). The mRNA for both channels doubles at the time of birth and then gradually increases to adult levels. The mechanism for the marked increase in expression at the time of birth remains to be defined.

Developmental changes in SR Ca^{2+} transport have been studied in isolated SR vesicles. Ca^{2+} uptake, activity of the Ca^{2+} pump, the number of Ca^{2+} pumps, and the efficiency of the Ca^{2+} pump are decreased in vesicles from immature hearts compared with that measured in vesicles from mature hearts in rabbits (Nakanishi & Jarmakani, 1984; Fisher, Tate, & Phillips, 1992), guinea pigs (Kaufman et al., 1990), and sheep (Mahony & Jones, 1986; Pegg & Michalak, 1987; Mahony 1988). The amount of Ca^{2+} pump mRNA parallels the increase in pump protein in adult rabbits (Fisher, Tate, & Phillips, 1992). Thus, transcriptional and/or posttranscriptional regulation of the Ca^{2+} pump gene contributes to age-related changes in SR Ca^{2+} transport.

Resting and action potentials

The resting potential of the myocardial cell is determined in large part by the potassium ion gradient across the cell membrane. During maturation of the heart, the resting potential increases (becomes more negative) in both birds and mammals (Bernard, 1975; Fozzard & Sheu, 1980; Danilo et al., 1984; Rosen & Danilo, 1992). Inasmuch as the potassium gradient is maintained primarily by the sodium pump, developmental changes in the activity of this pump likely contribute to the observed changes in resting potential (Mahony & Jones, 1986; Orlowski & Lingrel, 1988; Lucchesi & Sweadner, 1991; Artman 1992; Rosen & Danilo, 1992).

The action potential results from a series of ion fluxes down concentration and electrical gradients. Sodium ions entering the cell by way of voltage- and time-dependent sodium channels are responsible for the rapid upstroke of the action potential. The maximal rate of rise and the amplitude of the action potential increase during maturation of the heart (Danilo et al., 1984; Rosen & Danilo, 1992). This results in part from age-related changes in the number and properties of the sodium channels, although experimental results are somewhat disparate. For example, some investigators have examined the effects of sodium channel blocking agents to assess whether channel inactivation properties change during maturation of the heart. The rate of recovery from lidocaine- or phenytoin-induced channel block is slower in neonatal rat ventricular myocytes than in adult cells (Xu, Pickoff, & Clarkson, 1991, 1992). Similar results are found in canine ventricular myocytes (Jeck & Rosen, 1990). However, an opposite pattern is seen in canine Purkinje cells (Morikawa & Rosen, 1995), and no developmental differences are found in guinea pig myocytes (Jeck & Rosen, 1990). Thus, age-related changes in sodium channel number and properties may be species and/or tissue specific. Furthermore, methodologic differences may account for inconsistencies in experimental results (Xu, Pickoff, & Clarkson, 1992).

A slow inward current carried by Ca^{2+} ions is largely responsible for the plateau phase of the action potential. Ca^{2+} enters the cell by way of the sarcolemmal L-type Ca^{2+} channels, which are gated by both membrane potential and intracellular Ca^{2+} concentration. Flow of Ca^{2+} into the cell is passive because the concentration of Ca^{2+} in the extracellular space (1 mM) is greater than in the cytosol of the resting myocyte (2×10^{-4} mM). (The ontogeny of Ca^{2+} currents in the developing heart was discussed in a previous section.)

Currents carrying potassium ions out of the cell through several potassium-selective channels result in repolarization. Developmental changes occur in these currents, which include the transient outward, inward (anomalous) rectifier, and outward (delayed) rectifier currents. The transient outward current increases during maturation in rats (Kilborn & Fedida, 1990; Wahler et al., 1994) and dogs (Jeck & Boyden, 1992). The peak density of the inward rectifier current also increases during maturation in rabbits (Huynh et al., 1992; Sanchez-Chapula et al., 1994) and rats (Wahler, 1992; Masuda & Sperelakis, 1993). In studies of human atrial myocytes, no differences in the inward rectifier are seen between myocytes from young (<2.5 yr) and adult patients (Crumb, Pigott, & Clarkson, 1995). However, the current density of the transient outward current is higher and the recovery kinetics are faster in cells from older patients (Escande et al., 1985; Crumb, Pigott, & Clarkson, 1995). These data are consistent with the concept that age-related changes in potassium currents contribute to the observed developmental differences in the shape and duration of the action potential.

Excitation–contraction coupling

The process by which depolarization of the myocyte results in myocardial contraction is the subject of intense study and remains incompletely defined. Ca^{2+} enters the myocyte across voltage-gated channels during the plateau phase of the action potential. However, the absolute Ca^{2+} flux is much less than that required for maximal contraction in mature myocytes (Fabiato, 1983; Wier, 1992; Barry & Bridge, 1993). A large increase in intracellular Ca^{2+} concentration is critical for binding of sufficient Ca^{2+} to troponin C. Although the contribution of the SR varies among species, the source of this activator Ca^{2+} is the SR, and thus an intact SR is required for normal excitation–contraction coupling in mature mammalian myocytes (except under nonphysiologic experimental conditions) (Nabäuer et al., 1989; Sham, Cleeman, & Morad, 1992). Ca^{2+} entering the myocyte after depolarization likely interacts with the SR Ca^{2+} release channel, leading to local accumulation of Ca^{2+} in the region of the SR release channel. Locally accumulated Ca^{2+} then triggers release of Ca^{2+} to the myofilaments (Ca^{2+}-induced Ca^{2+} release) (Fabiato & Fabiato, 1978; Beukelmann & Wier, 1988; Fabiato, 1989; Nabäuer et al., 1989; Niggli & Lederer, 1990; Wier, 1992; Stern, 1992).

The relative contribution of Ca^{2+} influx by way of the sarcolemmal Ca^{2+} channel versus the Na^+–Ca^{2+} exchanger in activating SR Ca^{2+} release is controversial. The fact that the processes controlling excitation–contraction vary among mammalian species complicates interpretation of available data. Traditionally, Ca^{2+} influx through the sarcolemmal Ca^{2+} channel has been thought to provide the majority of the Ca^{2+} necessary to trigger SR Ca^{2+} release under physiologic conditions in the mature heart (Cleeman & Morad, 1991; Sham, Cleeman, & Morad, 1992; Wier, 1992). Sarcolemmal Ca^{2+} channels and SR Ca^{2+} release channels in rat cardiac myocytes are coupled functionally in that Ca^{2+} flux through either type of channel alters the activity of the other type (Sham, Cleeman, & Morad, 1995). The concept of a close relationship between these two channels is also supported by the finding that Ca^{2+} entering through sarcolemmal Ca^{2+} channels is 20–160 times as effective as Ca^{2+} entering through "reverse" Na^+–Ca^{2+} exchange in gating SR Ca^{2+} release (Sham, Cleeman, & Morad, 1995). In contrast, relatively fast Ca^{2+} release mediated by the exchanger is documented in ventricular myocytes from guinea pigs (Leblanc & Hume, 1990; Levesque, Leblanc, & Hume, 1994; Kohmoto, Levi, & Bridge, 1994; Levi et al., 1994). Species differences in the activity of the exchanger may explain some of these results (Sham, Hatem, & Morad, 1995).

The relative contribution of Ca^{2+} influx across the sarcolemma to activation of myocyte contraction may be larger in immature hearts than in adult hearts because the SR is structurally and functionally underdeveloped in immature hearts. Immature hearts are more sensitive than adult hearts to the negative inotropic effects of Ca^{2+} channel antagonists (Boucek et al., 1984; Artman, Graham, & Boucek, 1985; Seguchi et al., 1986). Furthermore, ryanodine produces a decreased negative inotropic response in immature rabbit hearts compared with that in mature hearts (Seguchi, Harding, & Jarmakani, 1986). This suggests that a large amount of the Ca^{2+} bound to troponin C comes from somewhere other than the SR in immature hearts. In addition, the early peak of tension in response to voltage-clamp controlled depolarization, which is present in mature myocardium and is thought to result from release of Ca^{2+} from the SR, is absent in immature myocardium (Klitzner & Friedman, 1988, 1989). Finally, myocytes isolated from neonatal rabbits are more dependent on a rapidly exchangeable compartment of Ca^{2+} (consistent with transsarcolemmal Ca^{2+} influx) than are myocytes from adult hearts (Chin, Friedman, & Klitzner, 1990).

Thus, Ca^{2+} influx through Ca^{2+} channels may play a relatively important role in transsarcolemmal Ca^{2+} influx in the immature heart in some species. However, peak Ca^{2+} current density is actually decreased in immature rabbit myocytes compared with that measured in mature cells (Osaka & Joyner, 1991; Huynh et al., 1992; Wetzel, Chen, & Klitzner, 1993). Similar results are described for human atrial myocytes (Hatem et al., 1995). Furthermore, immature myocytes do not increase Ca^{2+} influx through a delayed time course of inactivation or an increase in sustained current (Wetzel, Chen, & Klitzner, 1993). Finally, in ventricular myocytes from adult rabbits, contraction amplitude is very dependent on Ca^{2+} entry via voltage-gated Ca^{2+} channels in that the tension–voltage relationship is bell-shaped and contractions are nearly completely suppressed by nifedipine (Wetzel, Chen, & Klitzner, 1995). In contrast, the contraction amplitude in neonatal cells correlates poorly with voltage-gated Ca^{2+} currents and is much less sensitive to the dihydropyridine Ca^{2+} channel blocking agent nifedipine (Wetzel, Chen, & Klitzner, 1995). Thus, in immature hearts from rabbits, Ca^{2+} influx through Ca^{2+} channels provides only a minor portion of the activator Ca^{2+} to the myofilaments.

The immature rabbit heart may rely on Ca^{2+} entry by way of "reverse" Na^{+}–Ca^{2+} exchange. The contractile response to increases or decreases in extracellular sodium is larger in newborn than in adult hearts (Hoerter, Mazet, & Vassort, 1981; Nakanishi & Jarmakani, 1981). Indirect immunofluorescent studies show that the Na^{+}–Ca^{2+} exchanger is prominent in the sarcolemma of neonatal rabbit hearts (Chen et al., 1995). As discussed earlier, the exchanger activity and amounts of immunoreactive protein and mRNA are relatively increased in immature rabbit hearts (Artman, 1992; Boerth, Zimmer, & Artman, 1994). Further evidence supporting the importance of Na^{+}–Ca^{2+} exchange in immature hearts comes from measurement of exchange current density (Artman et al., 1995). Current density is about 4 times as high in newborn as in adult rabbit myocytes. Interestingly, no difference is detected in current density in newborn and adult guinea pig hearts. Compared to newborn rabbits, newborn guinea pigs are relatively mature; the SR and transverse tubular system are fairly well developed (Hirakow & Gotoh, 1980; Goldstein & Traeger, 1985). Thus, although Ca^{2+} influx by way of the Na^{+}–Ca^{2+} exchanger is less important for excitation–contraction coupling in mature myocardium, Na^{+}–Ca^{2+} exchange may play an important role in myocytes from relatively immature rabbit hearts.

In contrast to immature rabbit hearts, Ca^{2+} channel-gated Ca^{2+} release is present in atrial myocytes from a 3-day-old human infant (Hatem et al., 1995). The rate of Ca^{2+} release from intracellular pools normalized for maximal Ca^{2+} release increases with age, suggesting that the functional coupling of the sarcolemmal and SR Ca^{2+} release channels increases during maturation of the human atrium. These results may reflect increased maturity of human neonates compared with newborn rabbits. The mechanism of contraction activation in the immature human ventricle remains to be defined.

Relaxation

Relaxation, an active process by which the myocardium returns to steady state after contraction, depends on rapid removal of Ca^{2+} from troponin C. This is mediated primarily by active transport of Ca^{2+} back into the SR. The SR Ca^{2+} pump ATPase (SERCA2a) couples hydrolysis of ATP to active Ca^{2+} transport. The rate of SR Ca^{2+} uptake correlates well with the observed rate of myocardial relaxation.

Regulation of SR Ca^{2+} pump activity is mediated by the intrinsic SR protein, phospholamban. Phospholamban transcripts occur in the developing mouse heart at the time when contractions are first observed (Ganim et al., 1992). However, the expression of phospholamban is regulated independently of expression of the Ca^{2+} pump (Arai et al., 1992). The function of the Ca^{2+} pump is inhibited by direct interaction with phospholamban (James et al., 1989). Phosphorylation of phospholamban reverses this inhibition by increasing both Ca^{2+} sensitivity and the turnover rate (V_{max}) of the Ca^{2+} pump (Tada et al., 1983; Sasaki et al., 1992), thus facilitating removal of Ca^{2+} from troponin C and accelerating myocardial relaxation. Phospholamban is phosphorylated at three distinct sites by three different protein kinases: cyclic-AMP-dependent protein kinase, Ca^{2+}-calmodulin-dependent protein kinase, and protein kinase C (Simmerman et al., 1986). The first two kinases are involved in phosphorylation of phospholamban in response to β-adrenergic stimulation (Wegener et al., 1989). Targeted ablation of the phospholamban gene in mice results in augmentation of basal myocardial contractility and loss of responsiveness to β-adrenergic stimulation (Luo et al., 1994). A direct correlation between isoproterenol-induced stimulation of the SR Ca^{2+} pump and enhancement of isovolumic relaxation has been shown in isolated guinea pig hearts (Lindemann et al., 1983).

Ca^{2+} is also removed from the myofilaments by extrusion across the cell membrane. In the steady state, the amount of Ca^{2+} removed from the myocyte equals the amount entering through the Ca^{2+} channels (Bridge, Smolley, & Spitzer, 1990). Most of the Ca^{2+} is extruded by the Na^+–Ca^{2+} exchanger, because the sarcolemmal Ca^{2+} pump plays only a minor role in removing Ca^{2+} from the myocyte in the mature heart (Bers, Lederer, & Berlin, 1990; Cannel, 1991; Langer, 1992). The exchanger is sensitive to both membrane potential and the intracellular Ca^{2+} concentration.

The relative participation of the SR pump and the Na^+–Ca^{2+} exchanger in removing Ca^{2+} from the cytosol during relaxation differs among mammalian species. Bassani, Bassani, and Bers (1994) studied the effects of inhibiting the SR Ca^{2+} pump with thapsigarin and of inhibiting Na^+–Ca^{2+} exchange by removing extracellular sodium on the decline of the Ca^{2+} transient after an electrically stimulated twitch in isolated rabbit and rat ventricular myocytes. Sequestration of Ca^{2+} by the SR is dominant in both species; however, a greater fraction of Ca^{2+} is extruded by the exchanger in rabbits (28%) than in rats (7%). These results are consistent with other studies showing species-related differences in exchanger activity (Sham, Hatem, & Morad, 1995).

Age-related changes in myocardial relaxation and in the relaxation response to various stimuli occur in the dog (Park, Michael, & Driscoll, 1982), rabbit (Okuda et al., 1987), guinea pig (Kaufman et al., 1990), and chicken embryo (Cheanvechai, Hughes, & Benson, 1992). For example, the percentage decrease in the time constant of isovolumic relaxation (τ) in response to isoproterenol is twice as high in adult as in newborn guinea pig hearts (Kaufman et al., 1990). Furthermore, in the same hearts, Ca^{2+} uptake, Ca^{2+} pump activity, and Ca^{2+} pump density are all decreased in SR vesicles isolated from newborn hearts compared with values measured in SR vesicles from adult hearts. Thus, the decreased ability of the SR to sequester Ca^{2+} described in this and other studies (Mahony & Jones, 1986; Pegg & Michalak, 1987; Mahony, 1988; Fisher, Tate, & Phillips, 1992) may contribute to limitations in the ability of the immature myocardium to augment relaxation.

Infusion of ryanodine impairs relaxation and confirms the dependency of relaxation on SR Ca^{2+} sequestration. Interestingly, ryanodine decreases isovolumic relaxation to a greater extent in adult than in newborn hearts (Kaufman et al., 1990). The observations that Na^+–Ca^{2+} exchanger activity, protein content, and mRNA levels are all higher in immature than in mature rabbit hearts support the hypothesis that removal of Ca^{2+} from the myofilaments across the sarcolemma by way of the Na^+–Ca^{2+} exchanger contributes greatly to relaxation in the immature heart (Artman, 1992; Boerth, Zimmer, & Artman, 1994; Artman et al., 1995). Further support for a compensatory role for the Na^+–Ca^{2+} exchanger in the presence of diminished SR function comes from studies of human hearts (Studer et al., 1994). SR Ca^{2+}–ATPase protein and mRNA levels are decreased, but Na^+–Ca^{2+} exchanger protein and mRNA levels are increased, in failing, as compared with nonfailing, adult hearts.

Conclusion

Maturation of the mammalian myocardium is marked by profound changes in the structure and function of the various cardiac membranes. These changes result in important age-related differences in myocardial function. Cellular and molecular biology provide powerful tools with which we can further characterize the control systems that integrate developmental changes in membrane structure and function into global myocardial performance. This knowledge should allow us to optimize the pharmacologic management of neonates and young children with congenital and acquired heart disease.

3

Development of the myocardial contractile system

ANNE M. MURPHY

During the course of maturation the developing heart is subjected to variations in hemodynamic load and in the availability of oxygen and other substrates, so it is not surprising that the constituents responsible for the force generated during myocardial contraction will vary as well. The investigation of contractile protein variation during cardiac development has focused on qualitative and quantitative changes in protein isoforms. Information is also emerging about the regulatory signals that control these processes. However, much remains unknown about the functional consequences of the shifts in cardiac contractile proteins that occur with development. This chapter will outline the developmental changes in the contractile apparatus that mediate the cellular mechanics of contraction.

Mechanisms for diversity in contractile protein isoforms

The protein content of the myofibril in general reflects the transcript levels for individual proteins. Variation in myofibril composition during development is mediated by two mechanisms: alternative splicing of transcripts or differential expression of members of a multigene family (Andreadis, Gallego, & Nadal-Ginard, 1987; Nadal-Ginard et al., 1991; Chien et al., 1993; Sartorelli, Kurabayashi, & Kedes, 1993). For example, alternative splicing of transcripts regulates the troponin T (TnT) isoforms in the developing heart and determines the tissue-specific isoforms of the α-tropomyosin (α-TM) gene (Nadal-Ginard et al., 1991). The transcriptional regulation of contractile protein genes in the heart is the subject of active investigation. A few general themes have emerged from the study of the transcriptional regulation of these genes. First, although some regulatory elements may be shared between skeletal muscle and heart, the regulatory regions controlling gene expression in heart are often distinct from those regulating expression in skeletal muscle. For example, both troponin C (TnC) and myosin light chain 2 ($MLC2_V$) are expressed in both slow-twitch skeletal muscle and heart, but their expression in these distinct fiber types is regulated by different regions of the gene (Henderson et al., 1989; Lee et al., 1992; Parmacek et al., 1992; Parmacek et al., 1994). A second theme that has emerged in gene regulatory studies is that multiple elements and factors can interact and collaborate in the control of gene expression. This is illustrated by recent detailed studies of the $MLC2_V$ gene. The ventricular myocyte specificity of this gene is determined by at least three distinct positive and negative regulatory elements, and the positive element (HF-1a/HF-1b) interacts with both a ubiquitous factor and a muscle factor (Navarkasattusas et al., 1992; K. Lee et al., 1994). A third theme is that although there are many similarities in regulation in skeletal and cardiac muscle, there are distinct cardiac regulatory factors. One example of a cardiac-specific transcription factor is GATA-4, a zinc-finger protein that is a member of a multi-gene family of regulatory factors important for hematopoietic gene expression (Arceci et al., 1993). GATA-4 is not expressed in cells of skeletal muscle lineage, although it is expressed in other cells of endodermal origin (Arceci et al., 1993). GATA-4 appears to have a role in the cardiac-specific expression of several genes in the heart, including B-natriuretic peptide (BNP), TnC, α-myosin heavy chain (MHC), and cardiac troponin I (TnI) (Grepin et al., 1994; Molkentin, Kalvakolanu, & Markham, 1994; Ip et al., 1994; Murphy & Thompson, 1995).

The contractile process

The contractile apparatus is organized into the functional unit of the sarcomere, which consists of interposed lattices of thick and thin filaments, the components of which are described later in this chapter. During the process of contraction of a myocardial cell, activator calcium is released from the sarcoplasmic reticulum (see Chapter 2) and thus becomes available to the myofilaments. Calcium binds to a regulatory calcium site on TnC, changing the affinity of TnC for TnI and resulting in changing interactions between other thin-filament proteins such that cross-bridging of myosin heads to actin becomes more favorable (reviewed in El-Saleh, Warber, & Potter, 1986; Zot & Potter, 1987; Lehrer, 1994). When a cross-bridge is bound, the myosin head may complete an ATP-dependent cycle that results in sliding of the thick and thin filaments in opposite directions and muscle shortening.

The relationship between free-calcium concentration and muscle tension, shortening velocity, or ATPase activity is not fixed and may vary in the short term (i.e., beat to beat) as well as in the long term, such as with developmental variation in protein isoforms. Interactions between thin-filament proteins and between the thin filament and the thick filament may modulate contractility by altering the amount of calcium required for a given level of activation (Solaro, 1991; Moss, 1992). In addition, the slope of the relationship between tension, velocity of shortening, or ATPase activity versus free calcium is variable. Mechanical factors – such as sarcomere length, changes in the chemical milieu (e.g., pH), and phosphorylation of contractile proteins – can alter these relationships (Cole & Perry, 1975; Hartzell, 1984; Solaro, 1991; Moss, 1992; Sweeney, Bowman, & Stull, 1993). In summary, all of these regulatory processes contribute to the dynamic nature of the contractile process. Developmental variation in individual contractile proteins is one level in which this modulation can occur.

Thick-filament proteins

Myosin heavy chain. Hoh et al. (1978) described three MHC isoforms in rat heart based on their electrophoretic mobility in nondenaturing gels. Subsequently, the molecular basis of this variation was found. The fast-mobility isoform (V1) is composed of a homodimer of the α-MHC gene transcript, and the slower-mobility isoform (V3) is a homodimer of the β-MHC gene transcript and is also the isoform found in slow-twitch skeletal fibers (Mahdavi, Chambers, & Nadal-Ginard, 1984). The V2 isoform is an αβ heterodimer.

In the mouse embryo, the cardiac developmental sequence for MHC begins with coexpression of α- and β-MHC transcripts in the ventricle between postcoital days 7.5 and 8

(Lyons et al., 1990; Lyons, 1994). In the fetal rodent heart, the β-MHC gene transcript predominates in the fetal ventricle, whereas the α-MHC transcript increases after birth and predominates in the adult cardiac ventricle (Hoh et al., 1978; Lompre, Nadal-Ginard, & Mahdavi, 1984). Thyroid hormone augments this developmental switch, and the postnatal surge in thyroid hormone may result in the completion of the developmental switch (Hoh, McGrath, & Hale, 1978; Chizzonite & Zak, 1984; Izumo, Nadal-Ginard, & Mahdavi, 1986; Mahdavi, Izumo, & Nadal-Ginard, 1987). In the atrium, the α-MHC is the predominant transcript in both rodent and human hearts, though β-MHC is coexpressed in the region of the sinoatrial node and in the portion of the right atrium derived from the atrioventricular canal (Kurabayashi et al., 1988; Wessels et al., 1991; Lyons, 1994).

In the mature ventricles of mammals with slower heart rates, including the human, β-MHC is the dominant isoform throughout development (Schier & Adelstein, 1982; Mercadier et al., 1983; Wessels et al., 1991). In the human embryo, coexpression of α-MHC is seen in the ventricle at Carnegie stage 14 (31–35 days of gestation) and in the developing outflow tract up to stage 19 (51 days of gestation) (Wessels et al., 1991). Some fibers in the fetal and adult human ventricle contain α-MHC, particularly those associated with the conduction system (Bouvagnet et al., 1987; Wessels et al., 1991). Mercadier et al. (1981) examined ventricular myosin content by quantitative immunoassay and found that α-MHC content was generally low (<5%) in normal human adult and fetal ventricles.

The myosin isoforms differ in their mechanical properties. The α-MHC isoform is associated with high Ca^{2+}-ATPase activity of the myosin as well as with increased shortening velocity of the muscle, whereas β-MHC has lower Ca^{2+}-ATPase activities and shortening velocities (Barany, 1967; Schwartz et al., 1981). The lower velocity of β-MHC allows it to be more energy efficient, perhaps offering some physiologic advantage in its expression in the rodent fetus.

Myosin light chain 1. The heart expresses two MLC1 genes in a regionally and developmentally specific manner. The presence of a fetal MLC1 isoform in the rat ventricle was described by Whalen and Sell (1980). This isoform is identical to the isoform in embryonic skeletal muscle and was later identified as the adult atrial myosin light chain ($MLC1_A$) (Barton et al., 1988). In contrast, the mature ventricle expresses a different isoform, $MLC1_V$, that is also present in slow-twitch skeletal fibers (Barton et al., 1985). The developmental isoform shift from expression of $MLC1_A$ to $MLC1_V$ is present in both the rat and the human ventricle (Price, Littler, & Cummins, 1980; Rovner, McNally, & Leinwand, 1990).

A role for MLC1 in the modulation of muscle function has been suggested by experiments showing that skeletal myosin lacking MLC1 has a reduced ability to develop force, and by in vitro motility assays in which removal of MLC1 decreases the velocity of actin movement by myosin (Lowey, Waller, & Trybus, 1993; VanBuren et al., 1994; Trybus, 1994). Although the functional differences between $MLC1_V$ and $MLC1_A$ in cardiac tissue remain unclear, a correlation between velocity of shortening and MLC1 isoform type was found in skeletal muscle (Sweeney et al., 1988), suggesting that the isoform shift in heart may be functionally significant.

Myosin light chain 2. Two MLC2 genes are expressed in a regionally specific manner in the heart. In both rodent and human myocardium, $MLC2_V$, the predominant isoform in the adult ventricle, is identical to the slow-twitch skeletal isoform (Kumar et al., 1986). $MLC2_V$ is expressed in the embryonic mouse at day 8 in the region of the heart tube destined to be the ventricle (O'Brien, Lee, & Chien, 1993) and remains predominant throughout development. The atria express a separate isoform, $MLC2_A$ (Hailstones et al., 1992), though a small amount of $MLC2_A$ transcript is also found in human ventricles. The regional specificity of MLC2 appears to be higher during rabbit heart development (Hailstones et al., 1992). Unlike $MLC1_A$, the $MLC2_A$ isoform does not appear to be expressed at any stage of skeletal myogenesis (Hailstones et al., 1992).

Although the functional implications of the regional specificity of MLC2 isoforms during development are unclear, phosphorylation of $MLC2_V$ results in increased isometric twitch tension and increased calcium sensitivity in striated muscle, indicating a regulatory role for this protein (Sweeney, Bowman, & Stull, 1993). Partial extraction of MLC2 reduces the shortening velocity of muscle in the high-velocity phase (Hofmann et al., 1990). In addition, proteolysis of MLC2 noted in human cardiomyopathic hearts is correlated with reduced maximal myofibrillar ATPase activity (Margossian et al., 1992), indicating the importance of this protein to myofibrillar function.

C-protein. The C-protein isoform expressed in mature ventricle appears to be cardiac specific based on electrophoretic mobility (Reinach et al., 1982; Yamamoto & Moos, 1983) as well as cDNA sequence (Kasahara et al., 1994). Only slightly reduced amounts of cardiac C-protein indexed to total myofibrillar protein were noted in embryonic human cardiac ventricle (Morano et al., 1994). In addition, the extent of phosphorylation of C-protein by protein kinase A in neonatal and adult rat hearts was nearly identical (A. M. Murphy & R. J. Solaro, unpublished data). These findings suggest that there may not be developmental variation in C-protein in the heart.

Although C-protein was previously thought to have primarily a structural role in the myofilament, more recent findings support a modulatory role for C-protein in contractile function. Hartzell (1985) demonstrated that cardiac muscle relaxation rate is increased by C-protein phosphorylation. In addition, extraction studies have demonstrated that partial removal of C-protein increased the calcium sensitivity of skinned cardiac fibers (Hofmann, Hartzell, & Moss, 1991) and increased the low-velocity phase of shortening (Hofmann, Greaser, & Moss, 1991).

Thick-filament summary. The patterns of developmental isoform transitions in rodent and human heart are summarized in Tables 3.1 and 3.2. In the thick filament, variation of MHC during development appears to be functionally significant in the rodent heart (Hoh, McGrath, & Hale, 1978; Lompre, Nadal-Ginard, & Mahdavi, 1984; Chizzonite & Zak, 1984; Mahdavi, Chambers, & Nadal-Ginard, 1984; Izumo, Nadal-Ginard, & Mahdavi, 1986; Mahdavi, Izumo, & Nadal-Ginard, 1987; Lyons et al., 1990; Lyons, 1994). However, in human heart MHC variation does not contribute substantially to physiologic variation (Mercadier et al., 1983). Isoform transitions in the fetal ventricle from the expression of $MLC1_A$ to $MLC1_V$ could influence force development, shortening velocity, or calcium sensitivity, but we lack sufficient data to judge the functional impact of this developmental shift. At this time, it does not appear that variation during development of MLC2 or C-protein contributes to differences in ventricular function during development.

Thin-filament proteins

α-actin. Three actin genes – cardiac, skeletal, and smooth muscle α-actin – contribute to the sarcomere during various stages of heart development. In chick heart, smooth muscle α-

Table 3.1. *Expression of isoforms in rodent heart*

Protein	Adult ventricles	Adult atria	Fetal ventricles
MHC	α-MHC	α-MHC	β-MHC α-MHC
MLC1	$MLC1_V$	$MLC1_A$	$MLC1_A$
MLC2	$MLC2_V$	$MLC2_A$	$MLC2_V$
C-protein	cC-P	cC-P	cC-P
Actin	α-cardiac	α-cardiac	α-skeletal α-cardiac
Tropomyosin	α-TM	α-TM	α-TM β-TM
Troponin I	cTnI	cTnI	ssTnI
Troponin C	cTnC	cTnC	cTnC
Troponin T	cTnT (– exon 4)	cTnT	cTnT (+ exon 4)

Note: MHC = myosin heavy chain; MLC = myosin light chain. References are cited in the text for individual contractile proteins.

Table 3.2. *Expression of isoforms in human heart*

Protein	Adult ventricles	Adult atria	Fetal ventricles
MHC	β-MHC α-MHC (<5%)	α-MHC	β-MHC α-MHC (<5%)
MLC1	$MLC1_V$	$MLC1_A$	$MLC1_A$
MLC2	$MLC2_V$	$MLC2_A$	$MLC2_V$
C-protein	cC-P	cC-P	cC-P
Actin	α-skeletal α-cardiac	—	α-cardiac α-skeletal (<20%)
Tropomyosin	α-TM β-TM (9%)	α-TM β-TM (9%)	α-TM β-TM (4–5%)
Troponin I	cTnI	cTnI	ssTnI cTnl
Troponin C	cTnC	cTnC	cTnC
Troponin T	$cTnT_3$	$cTnT_?$	$cTnT_{1,2,3,4}$

Note: MHC = myosin heavy chain; MLC = myosin light chain. References are cited in the text for individual contractile proteins.

actin mRNA and protein are detected between Hamburger–Hamilton stages 8 and 9 in the cardiac tube, followed by appearance of cardiac α-actin at stage 9 and skeletal α-actin at stage 12 (Ruzicka & Schwartz, 1988; Sugi & Lough, 1992). In the mature rodent and chicken atrium and ventricle, the α-cardiac isoform predominates (Mayer et al., 1983; Ruzicka & Schwartz, 1988; Sugi & Lough, 1992; Carrier et al., 1992). The transcript for the skeletal α-actin gene is present in large amounts in the fetal rodent heart and is present in decreasing amounts after birth (Mayer et al., 1983; Carrier et al., 1992; Boheler et al., 1992). The pattern of actin expression is quite different in human heart. Boheler et al. (1991) found that the two sarcomeric actin transcripts are coexpressed in the cardiac ventricles at different stages of development and in heart failure. In fetal and neonatal human hearts, however, the α-cardiac transcript predominated, whereas the predominant transcript in the ventricle of adults is the α-skeletal isoform, which accounted for approximately 60% of the transcript (Boheler et al., 1991).

The sarcomeric actin isoforms differ by only 1% of the primary sequence, and smooth muscle actin functions indistinguishably from skeletal α-actin when studied by in vitro motility assays (Harris & Warshaw, 1993). However, despite the similarities in primary sequence, a strain of mice that expresses increased levels of skeletal α-actin in the heart was found to have higher levels of contractility as measured by $+dP/dt$ (maximal rate of contraction), suggesting a functional role for this isoform shift in rodents (Hewett et al., 1994).

Tropomyosin. Two sarcomeric tropomyosin genes – tropomyosin (α-TM) and β-tropomyosin (β-TM) – are expressed in cardiac muscle. Both of these genes have alternatively spliced exons, allowing for the tissue-specific expression of a number of isoforms. In the hearts of rodents, the α-TM gene predominates (Cummins & Perry, 1973). A recent study, however, demonstrates that there is β-TM transcript in embryonic and adult mouse heart, although the relative α/β transcript ratio increases from 4.5:1 to 60.5:1 from embryonic day 11 to adult (Muthuchamy et al., 1993). In adult human heart, as much as 9% of atrial and ventricular tropomyosin is the β-TM isoform; fetal and neonatal human ventricles contain a smaller amount of this isoform (Humphreys & Cummins, 1984). Thus there is a difference in the relative expression of the β-TM isoform in human and rodent hearts.

Tropomyosin is involved in the translation of the calcium signal in the thin filament and may have a role in length-dependent activation of muscle (El-Saleh, Warber, & Potter, 1986; Zot & Potter, 1987; Earley, 1991). The physiologic consequences of tropomyosin variation are largely unknown. To date, despite species differences in tropomyosin isoforms, developmental regulation of alternatively spliced tropomyosin transcripts in heart has not been documented. As with other contractile proteins, even subtle variation in spliced tropomyosin transcripts might result in physiologically different isoforms.

Troponin I. Two troponin I genes are developmentally regulated in heart. Protein gel electrophoresis and immunoblotting suggested the presence of a fetal isoform in dog, rat, and chick heart that comigrated with the slow-twitch skeletal isoform of troponin I (ssTnI) (Solaro et al., 1986; Sabry & Dhoot, 1989; Saggin et al., 1989). Molecular cloning, RNA measurements, and in situ experiments have subsequently demonstrated that the ssTnI transcript is first expressed in cardiac tube, followed by coexpression of the ssTnI isoform with the cardiac isoform of troponin I (cTnI) in fetal heart. Late-gestation fetal heart is characterized by decreasing ssTnI and increasing cTnI protein and mRNA content (Murphy et al., 1991; Gorza et al., 1993). The developmental switch appears to complete postnatally. Studies of human heart demonstrated a similar pattern of isoform switching (Hunkeler, Kullman, & Murphy, 1991; Bhavsar et al., 1991).

The cardiac and slow skeletal troponin I isoforms are functionally different. The cardiac isoform is subject to phosphorylation by protein kinase (PK) A and C (Cole & Perry, 1975; Noland & Kuo, 1991). Phosphorylation of TnI by PKA is associated with desensitization of the myofibrils to calcium, and phosphorylation by PKC is associated with a reduction in maximal actomyosin ATPase activity (Holroyde, Howe, & Solaro, 1979; Venema & Kuo, 1993). The ssTnI isoform does not contain the amino terminal region containing the PKA phosphorylation sites, suggesting that this modulatory effect of phosphorylation should be diminished in the fetal myocardium (Murphy et al., 1991). This prediction has been confirmed by in vitro phosphorylation of myofibrils reconstituted with a recombinantly produced cTnI lacking the PKA sites (Wattanapermpool, Guo, & Solaro, 1995). However, there are conflicting data on the precise sites for PKC phosphorylation of TnI, so the developmental differences in PKC effects are unknown (Venema & Kuo, 1993). In addition to differences in phosphorylation effects, the presence of the cTnI isoform in the myofibrils is correlated with desensitization of the myofibrils to calcium upon exposure to acidosis, whereas the ssTnI isoform appears to be protective against this effect (Solaro et al., 1986; Ball, Johnson, & Solaro, 1994). Experiments conducted with a recombinantly produced cTnI that lacks the amino terminal region suggest that the developmentally variable sensitivity to acidosis is not mediated by the cardiac-specific amino terminal region of TnI (Guo et al., 1994).

Troponin C. The cardiac TnC(cTnC) gene is the same gene expressed in slow-twitch skeletal fibers. This gene is expressed in both the atria and the ventricles throughout fetal, neonatal, and adult life (Toyota & Shimada, 1981; Toyota, Shimada, & Bader, 1989) and is relatively invariant during development. Recent in situ hybridization studies have demonstrated that fast-twitch skeletal TnC (fsTnC) mRNA is detectable (transiently) only in the truncus arteriosus of chicken cardiac tubes at Hamburger–Hamilton stage 20 (Toyota, 1993).

TnC is a member of the EF-hand superfamily of calcium-binding proteins. The molecule has a dumbbell shape with Ca^{2+}-binding sites I/II and III/IV separated on opposite ends of a central helix (Herzberg, 1986). The cTnC isoform differs from the fsTnC isoform in that calcium regulates contraction by binding to both sites I and II in fsTnI whereas only a single regulatory site (II) is active in cTnC. (Putkey, Sweeney, & Campbell, 1989). Although TnC is invariant through much of cardiac development, the calcium affinity of the troponin complex is influenced by other thin-filament proteins and by the thick-filament proteins as well (reviewed in Moss, 1992; Solaro, 1991).

Troponin T. The developmental variation of TnT in rat and chicken heart is due to alternative splicing (Cooper & Ordahl, 1985; Jin et al., 1992). The predominant adult TnT transcript in mature heart excludes exon 4 in rat or exon 5 in chicken, which encodes a 10 amino acid fragment in the amino terminal region of the molecule. Anderson and colleagues have detected additional cardiac TnT isoforms in rabbit heart and human heart (Anderson & Oakeley, 1989; Nassar et al., 1991; Anderson et al., 1991; Greig et al., 1994; Anderson et al., 1995). In the case of rabbit heart, the predominant isoform in the immature rabbit heart is generated by a transcript (TnT_2) that includes an alternatively spliced 10 amino acid insert analogous to that in chicken and rat heart (Greig et al., 1994). The predominant adult transcript (TnT_4) does not contain this exon. Although additional alternatively spliced transcripts are also present, the majority of TnT isoforms contain the 10 amino acid insert and are associated with a higher calcium sensitivity and increased calcium binding of the myofibrils (McAuliffe, Gao, & Solaro, 1990; Nassar et al., 1991). In human heart a slightly different developmental isoform profile is seen. The human fetal heart is characterized by high contents of TnT_1, which contains both a 10 amino acid and a 5 amino acid insert, and TnT_4, which contains neither of these inserts (Anderson et al., 1995). In normal human adult heart the TnT_3 isoform, which contains only the 5 amino acid insert, predominates.

Thin-filament summary. The patterns of developmental isoform transitions are summarized in Tables 3.1 and 3.2. Variation in thin-filament proteins contributes to physiologically significant transitions in the heart. The recent work of Hewett et al. (1994) has suggested that an increased content of α-skeletal actin in the heart may result in an increased *+dP/dt*. In this context it is of interest to note the difference between rodent and human heart in the trend of mRNA contents for the α-actin transcripts. In rodent fetal ventricle, α-skeletal actin is the dominant transcript, whereas in human fetal ventricle, α-cardiac actin is the dominant transcript (Mayer et al., 1983; Boheler et al., 1991; Boheler et al., 1992; Carrier et al., 1992). The developmental transitions in TnI and TnT expression in avian, rodent, and human hearts are qualitatively similar (Cooper & Ordahl, 1985; Sabry & Dhoot, 1989; Saggin et al., 1989; Anderson & Oakeley, 1989; Hunkeler, Kullman, & Murphy, 1991; Nassar et al., 1991; Anderson et al., 1991; Jin et al., 1992; Greig et al., 1994; Anderson et al., 1995). Both of these transitions also appear to be functionally significant. The predominance of the ssTnI isoform in fetal heart is predicted to produce a myocardium that is resistant to the deleterious effects of acidosis on calcium sensitivity of the myofibrils and to blunt the effects of PKA-mediated phosphorylation, since the ssTnI isoform lacks these sites (Solaro et al., 1986; Ball, Johnson, & Solaro, 1994; Guo et al., 1994; Wattanapermpool, Guo, & Solaro, 1995). The isoform profile of TnT in fetal ventricle is predicted to produce a contractile apparatus that has a higher calcium sensitivity (Nassar et al., 1991). The functional consequences of tropomyosin variation in heart are not clear. As discussed, TnC is relatively invariant during development.

Conclusions and future directions

Major phenotypic changes occur in the myofilaments during the maturation of the heart. The isoform profile and function of sarcomeric units change throughout development temporally and spatially and exhibit a significant amount of variation between species. Clearly there are many gaps in our knowledge about the nature of functionally relevant contractile protein phenotypic changes during ontogeny. The quantitative aspects of variation of individual proteins remain undefined in many species. Several recent studies demonstrating the functional effects of isoform shifts during development suggest that newer techniques will contribute substantially to our understanding of contractile protein variation (Harris & Warshaw, 1993; Hewett et al., 1994; Guo et al., 1994; Wattanapermpool, Guo, & Solaro, 1995). Elegant structural and mechanical studies conducted with native variants, as well as with recombinant and transgenic variants, are likely to further elucidate the molecular basis of contractile protein functional diversity.

Contractile protein variation has broader implications than those related to development. Hypertrophic cardiomyopathy has been defined as a disease of the myofibril, as indicated by the identification of mutations in myosin heavy-chain, troponin T, and tropomyosin genes (Thierfelder et al., 1994). It is quite possible that additional diseases or variations of the myofilaments will be elucidated.

Contractile protein variations may exist with more subtle phenotypes that influence the clinical course of common disorders such as ischemic heart disease or hypertension. Insight gained from the investigation of protein isoform changes and their functional consequences during heart development may ultimately lead to new therapeutic approaches for the management of cardiac contractile failure through alterations of the contractile protein phenotype in the adult.

4

Vasculogenesis and angiogenesis in the developing heart

ROBERT J. TOMANEK AND ANNA RATAJSKA

The coronary vasculature begins to develop during primary cardiac morphogenesis, as the ventricular chamber walls thicken. Endothelial precursor cells reach the heart via the epicardium, where they form vascular tubes, a process termed "vasculogenesis." These tubes coalesce to form a microvascular network that expands by sprouting (i.e., vessels are formed from preexisting vessels), a process termed "angiogenesis." Subsequently, larger vessels are formed via unification of microvessels, and several of these vessels form a ring at the root of the aorta, coalesce, and then grow into the aorta, forming the two main coronary arteries. These events occur during a relatively short portion of the gestation period. Despite the importance of this aspect of heart development, our understanding of coronary vasculogenesis and angiogenesis has lagged behind our knowledge of other aspects of heart development. This chapter reviews the current status of this topic by considering precoronary, as well as coronary, events, and it explores the putative regulators of coronary vascular growth.

Precoronary events

Formation of the heart tubes occurs as endothelial cells migrate from the cardiogenic plate to the basal surface of the endoderm (Virágh et al., 1990). These endocardial angioblasts have been shown to arise from mesodermal tissue that lies just anterior to Hensen's node (Coffin & Poole, 1991). DeRuiter and colleagues (1993) postulated that endocardial endothelial cells originate from precursors that are independent of the cardiogenic plate. Recent data from Markwald (1995) on early expression of genes in cardiovascular development support this hypothesis. Subsequently, myocytes proliferate to form a myocardium, the cardiac jelly diminishes, and the endocardial surface becomes progressively more trabeculated. This establishes what has been termed a lacunary or sinusoidal system consisting of a highly trabeculated myocardium with endocardial-lined sinusoids, described nearly a century ago (Lewis, 1904), which minimizes diffusion distances. Thus, during early development the ventricle receives most of its nourishment directly from its lumen (some O_2 may also be available from the epicardial surface). Flow through the ventricles occurs far in advance of coronary vascularization. For example, flow can be measured as early as Hamburger–Hamilton (HH) stage 12 in the chicken (Hu & Clark, 1989). The coronary vasculature begins to form when vasculogenesis is triggered by the thickening of the myocardium. During this period of myocardial growth and vascularization, part of the ventricle is compact and part is spongy.

In most fishes the spongy ventricular myocardium persists throughout life (Agnisola & Tota, 1994). However, in some fish species the ventricular wall is composed of both spongy and compact tissue. The fishes are unique in cardiac phylogeny since they are the lowest vertebrate group to develop a coronary vasculature. Tota and colleagues (1983) documented four different levels of cardiac organization in fishes (Figure 4.1). Type I consists entirely of a spongy myocardium, lacks capillaries, and may or may not have epicardial vessels. Types II–IV show a progressive increase in the proportion of the compact layer of myocardium with a proportional decrease in the spongy layer. In type IV, characteristic of elasmobranchs, a transmural arterial system is evident, and connections between the coronary capillaries and the ventricular lumen have been recognized, thereby providing evidence for arterioluminal connections (Tota, 1989). In amphibians the coronary circulation and sinusoidal system each serve part of the ventricular wall (Ošťádal, Rychter, & Poupa, 1970).

Precursor cells and early vascularization

The origin of precursor cells

In vertebrates the epicardium develops from proepicardial tissue that lies between the sinus horns and the liver primordium (Virágh et al., 1993). As demonstrated in quail, proliferating proepicardium consists of glandular-like structures containing angioblasts, fibroblasts, and blood cells; it involutes once the epicardium completely covers the heart (Virágh et al., 1993). This tissue layer is key to the formation of coronary vascular channels since vascularization of the heart begins as soon as the epicardium is formed. Evidence regarding the origin of coronary endothelial precursor cells has been provided by studies on chicken hearts. Formation of the epicardium occurs as mesothelial protrusions elongate from the ventral surface of the sinus venosus and spread over the surface of the ventricles (Hiruma & Hirakow, 1989). Mikawa and Fischman (1992) used retroviral labeling to study the formation of coronary vessels. When the retrovirus was injected at HH stage 15 or earlier, all stained cell clusters were myocardial myocytes. In contrast, when the injections were performed at stage 17 or later, connective tissue and endothelial and smooth muscle cells were labeled as well, especially when the retrovirus was injected into the region of the mesocardium. Thus, the authors suggested that endothelial precursor cells migrate to the heart with the mesothelial cells that form the epicardium. The origin of these cells was determined by Poelmann et al. (1993), who used anti-MB1 antibody specific for quail endothelial cells and their precursors. They showed that the first endothelial precursors in the epicardium were detected at HH stage 19 near the sinus venosus and thereafter spread through the epicardium and into the myocardium. These cells formed a dense network of microvessels in the myocardium but only incidentally contacted the endocardium. Further experiments were then performed using quail–chicken chimeras. When pure epicardial primordium transplants were performed, endothelial cell formation did not occur. However, a liver graft, with or without epicardial primordium, resulted in vascularization. These data indicate that coronary endothelial cells originate in the liver region and migrate to the epicardium during heart development.

Formation of the microvascular network

Endothelial precursor cells, first appear in the epicardium, in conjunction with erythroblasts and erythrocytes to form blood islands. This constitutes the onset of vasculogenesis and has been observed in many species, including chickens (Hiruma & Hirakow, 1989), rats

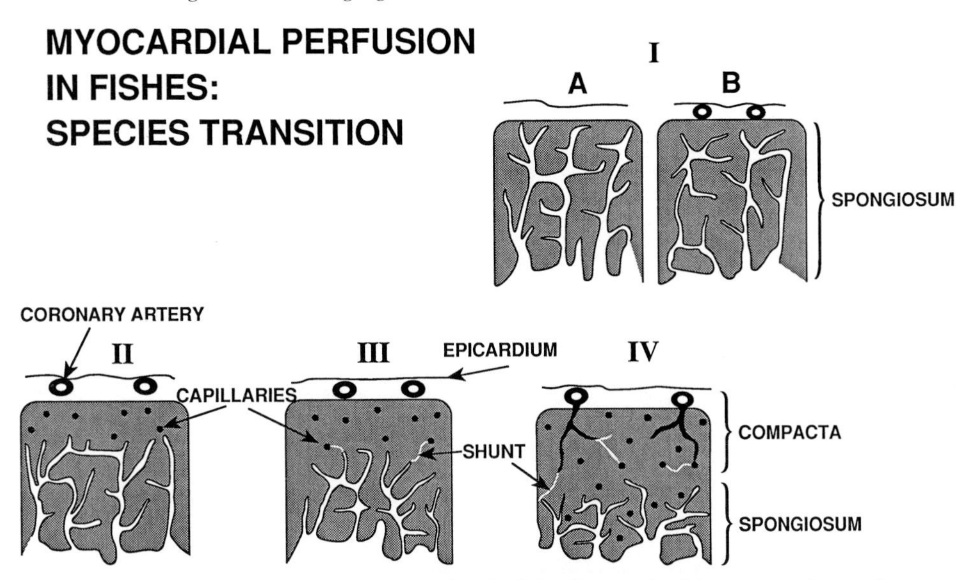

Figure 4.1 The ventricular myocardium in fishes is organized into a spongiosum (i.e., trabeculated structure of the entire wall) (type I) or into a combination of spongiosum and compacta (nontrabeculated) (types II–IV). The trabeculated arrangement allows blood from the ventricular lumen to come close to myocardial cells via the endocardium-lined spaces (sinusoids). Coronary vessels may or may not be present in the epicardium. The presence of capillaries and coronary arteries characterizes types II–IV. A progressive increase in the thickness of the compact region and a decrease in the thickness of the spongiosum is evident in types II–IV. In types III and IV the shunts from the capillaries to the sinusoids are found, and in type IV transmural coronary arteries are present (based on the work of Tota et al., 1983).

(Rongish et al., 1994), and humans (Hirakow, 1983; Hutchins, Kessler-Hanna, & Moore, 1988). These isolated cell clusters first appear in the epicardium and the underlying compact myocardium of the ventricles (Figure 4.2). In humans they are first seen in the interventricular sulcus and then in the atrioventricular sulcus (Hutchins, Kessler-Hanna, & Moore, 1988). A capillary plexus then spreads over the ventricles and most likely constitutes the onset of angiogenesis (sprouting). This early vascularization occurs at HH stage 23 in chickens (Hiruma & Hirakow, 1989), embryonic day 13 in the rat (Heintzberger, 1978; Rongish et al., 1994), and Carnegie stages 14–16 (32–37 days of gestation) in humans (Hirakow, 1983; Hutchins, Kessler-Hanna, & Moore, 1988). These structures give rise to a subepicardial capillary plexus that spreads over the surface of the heart (Bennet, 1936). As vascularization proceeds, capillaries are present in various stages of maturation, the endothelial cell cytoplasm thins, plasmalemma vesicles increase in number, and specialized junctions develop (Porter & Bankston, 1987a, 1987b; Rongish et al., 1994). We (Rongish et al., 1996) have noted that (1) a rich complement of fibronectin is present even prior to vessel formation, (2) laminin is deposited as the capillary tube forms, and (3) the incorporation of collagen IV closely follows the deposition of laminin. Pericytes appear as the capillary endothelium matures, consistent with cessation of proliferation (Porter & Bankston, 1987a, 1987b). Coronary capillary interendothelial cell junctions are, surprisingly, impermeable to carbon and ferritin during development, as shown in experiments on

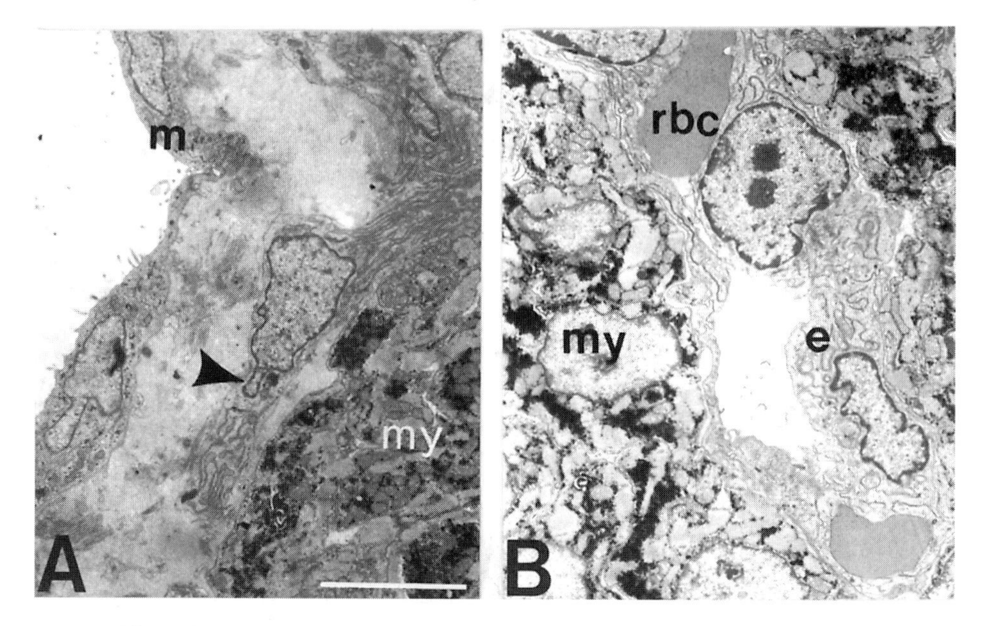

Figure 4.2 (*A*) Epicardium with precursor cell (*arrowhead*) and underlying myocardial cell (my) in a 13-day-gestation rat heart. A mesothelial cell (m) is seen. (*B*) A blood-island-like structure consisting of immature endothelial (e) cells and red blood cells (rbc). Scale bar = 5 μm.

rats (Porter & Bankston, 1987a, 1987b). Ferritin and myoglobin are transported via plasmalemmal vesicles, coated vesicles, and transendothelial cell channels. Thus, the permeability characteristics of newly formed capillaries resemble those found in the adult.

Historically, the origin of venous structures was commonly believed to be the sinus venosus (Grant, 1926; Goldsmith & Butler, 1937; Voboril & Schiebler, 1969; Rychter & Ošťádal, 1971). There was general agreement that the capillary plexus in the atriovenous region came in contact with the sinus venosus. Heintzberger (1983), who studied the rat, described the first signs of a venous system as follows: "the subepicardial vascular plexus contacts some endothelial offshoots from the wall of the left sinus horn" (p. 272). However, it was recently shown that the capillary plexus grows into the sinus venosus, rather than vice versa. Using quail–chicken chimeras, Poelmann and colleagues (1993) demonstrated that the grafted quail endothelial cells appeared as precursors in the myocardium prior to their incorporation into the sinus venosus. The contact of the capillary plexus with the coronary sinus in humans occurs during Carnegie stages 15–17 (i.e., 33–41 days of gestation) and closely follows the onset of vasculogenesis (Hirakow, 1983; Hutchins, Kessler-Hanna, & Moore, 1988).

The coronary arterial system

Although growth of the coronary arteries by sprouting from the aorta was never documented, this misconception persisted in the literature until relatively recently. Dbaly, Ošťádal, and Rychter (1968), who, like others (Heintzberger, 1983), noted vascular plexuses at the root of the aorta, reported that they could not find blind aortic evaginations that would constitute evidence that coronary arteries originated in the aorta. Several studies published during the last decade have provided key findings documenting the origin of the major coronary arteries. Bogers and colleagues (1989) followed vessel development by using the monoclonal antiendothelium antibody in quail and found that coronary artery formation involved ingrowths rather than outgrowths. Subsequently, Waldo, Willner, and Kirby (1990) studied chick embryos that had been made translucent by a clearing agent after being injected with ink and found that a capillary ring around the aorta penetrated its root at specific points in the aortic sinuses to form the left and right coronary arteries. They also noted that this capillary ring was connected to subepicardial endothelial channels. More recently we have verified these findings in the rat using serial sections (Tomanek et al., 1996). Mikawa and Fischman (1992) injected a retrovirus into chick embryos and clearly demonstrated that discontinuous colonies of microvessels coalesce to form coronary arteries. Finally, Poelmann et al. (1993) showed in quail–chicken chimeras that quail endothelial cells have several ingrowth sites into the aorta, of which two generally form a lumen. Surprisingly, the pulmonary orifice is not penetrated. The time required for a microvascular network to form and attach to the aorta is short. For example, the time between the onset of vasculogenesis and the formation of the roots of the left and right coronary arteries is only about 3–4 days in chickens, mice, and rats (Dbaly, Ošťádal, & Rychter, 1968; Voboril & Schiebler, 1969; Rychter & Ošťádal, 1971; Virágh & Challice, 1981; Heintzberger, 1983; Hiruma & Hirakow, 1989; Waldo, Willner, & Kirby, 1990; Rongish et al., 1994) and is about 2 weeks in humans (Hirakow, 1983). The left coronary ostium consistently appears prior to the right (Hirakow, 1983; Bogers et al., 1989; Mandarim-de-Lacerda, 1990).

The development of the tunica media occurs after the ingrowth of the vascular plexus and proceeds from the ventricle's base to its apex as a continuous arborization (Hood & Rosenquist, 1992). Observations of serial sections of rat fetal hearts indicate that the anastomotic nature of the coronary arteries persists even as the media is developing (Tomanek et al., 1996). Thus, these vessels must undergo remodeling, even after birth, to establish the branching system characteristic of the adult coronary arterial tree. Some of the endothelial channels apparently disappear (Poelmann et al., 1993). Parasympathetic ganglia filled with neural crest cells are associated with coronary arteries and have been suggested to be essential for the development of these vessels (Hood & Rosenquist, 1992; Waldo, Willner, & Kirby, 1994). These authors also noted neural crest cell association with neurofilament-positive fibers and capillaries in the region of the coronary ostia and suggested that they may serve an endocrine or sensory function. However, the importance of the neural crest–derived parasympathetic ganglia has recently been questioned because coronary arteries are able to grow after partial neural crest ablation (Gittenberger–de Groot et al., 1995).

Postnatal vascular growth

The rapid growth of the myocardium and the increase in its metabolic demand during the early postnatal period require a substantial capillary proliferation. At the time of birth in humans, the myocyte/capillary ratio (number of myocytes per capillary) is 6:1; the ratio falls to 1:1 by maturity (Roberts & Wearn, 1941). Similarly, in the rat there are four myocytes for every capillary at birth, but capillary angiogenesis results in one myocyte per capillary by 30–40 days (Rakusan & Poupa, 1963). Capillary volume density increases by 187% in the first 5 days of extrauterine life in the rat (Smolich et al., 1989); during the first month of life, both capillary and myocyte numbers increase, but during the next 3 weeks, only the capillaries continue to proliferate (Rakusan et al., 1964). Rakusan and Turek

(1985) stress the importance of the neonatal period for capillary growth and indicate that nearly half of all adult capillaries in the rat are formed during the first 3–4 weeks following birth. Although growth after this period is less dramatic and capillary density falls, the capillary bed nevertheless continues to grow substantially until the time of maturity, as evidenced by a more than twofold increase in total capillary length between 5 and 13 weeks (Mattfeldt & Mall, 1987). This growth involves both an increase in length and the splitting of capillaries to increase the number of channels (van Gronigen, Wenink, & Testers, 1991). Arteriolar postnatal changes in humans include increases in medial thickness, number, and segment length (Kurosawa, Kurosawa, & Becker, 1986).

Regulation of coronary vascularization

Although there is considerable evidence implicating a number of growth factors, extracellular matrix molecules, and mechanical events in angiogenesis (Risau & Lemmon, 1988; Hudlicka, 1988; Klagsbrun & D'Amore, 1991; Carey, 1991), little is known about mechanisms that control various events in coronary vascularization. One of the fascinating observations concerning coronary vascularization is the spatial and temporal preciseness of the process. Once angioblasts have migrated to the epicardium, a transmural vascularization is triggered and venous connections are formed in a very precise way. Subsequently, a capillary plexus penetrates the aorta at very predictable locations, and the newly formed coronary arteries develop a media. Such events must involve tightly controlled local mechanisms.

Recent interest has focused on two growth factors, vascular endothelial growth factor (VEGF) and basic fibroblast growth factor (bFGF), because they have been shown to enhance collateral blood flow in the ischemic or hypoxic canine heart. Intracoronary injection of bFGF in hearts with infarctions increased the number of capillaries and arterioles (Yanagisawa-Miwa et al., 1992). In dogs with ischemia, daily intracoronary injections increased myocardial collateral flow (Unger et al., 1994). Similarly, VEGF is effective in enhancing the development of small coronary arteries supplying the ischemic myocardium and thereby enhancing collateral flow (Banai et al., 1994).

A number of studies have shown that bFGF is immunolocalized in myocytes, fibroblasts, smooth muscle cells, and endothelial cells of the embryonic and fetal heart (Spirito et al., 1989; Joseph-Silverstein et al., 1989; Parlow et al., 1991; Consigli & Joseph-Silverstein, 1991; Tomanek et al., 1996). This growth factor is able to stimulate key events in vascularization, for example, endothelial cell proliferation and migration, release of proteases, and tube formation (Schott & Morrow, 1993). Both bFGF and acidic fibroblast growth factor (aFGF) are widespread in the developing rat heart (Spirito et al., 1989). It has been hypothesized that aFGF, which is also expressed during the prenatal period in the rat, functions in capillary angiogenesis in the neonate (Engelmann, Dionne, & Jaye, 1993).

Recent in situ experiments have shown that bFGF mRNA transcripts are at peak levels during the 14th–15th day of gestation and during the first week of postnatal life of the rat (Tomanek et al., 1996). These periods correspond to the stages of early vascularization and marked capillary proliferation, respectively, thus suggesting a possible role for bFGF in these vascularization events. We have also examined proliferation and tube formation in cultured embryonic hearts (Ratajska et al., 1995). Although bFGF is effective in increasing cell migration, significant tube formation occurs only when VEGF is added to the medium. Accordingly, these data suggest roles for both of these growth factors in prenatal vascularization.

VEGF mRNA is up-regulated by ischemia and hypoxia, as shown by experiments using coronary ligation (Hashimoto et al., 1994) and isolated cells (Schweiki et al., 1992), respectively. Hypoxia increases VEGF levels by enhancing its mRNA transcription rate (Risau & Lemmon, 1988) and by increasing the stability of its mRNA (Luscinskas & Lawler, 1994). Although hypoxia has not yet been demonstrated in embryonic hearts, up-regulation of VEGF by lower P_{O_2} is a possibility. This hypothesis is based on the fact that the myocardium adjacent to the epicardium becomes progressively more remote from the ventricular lumen as the compact zone of the myocardium expands via myocyte hyperplasia (Figure 4.3). It is in this region that we have first observed VEGF immunoreactivity, which subsequently spreads toward the endocardium. Thus, it is plausible that the VEGF in this model is triggered by hypoxia.

Although other growth factors should not be excluded from consideration, evidence for their possible role in vascularization during heart development has not yet been provided. Platelet-derived endothelial cell growth factor (PD-ECGF) and platelet-derived growth factor (PDGF) are angiogenic and chemoattractant (Schott & Morrow, 1993); however, there are no data at this time implicating them in coronary vascularization during development. Transforming growth factor-β (TGF-β) is a widely distributed peptide that may be impor-

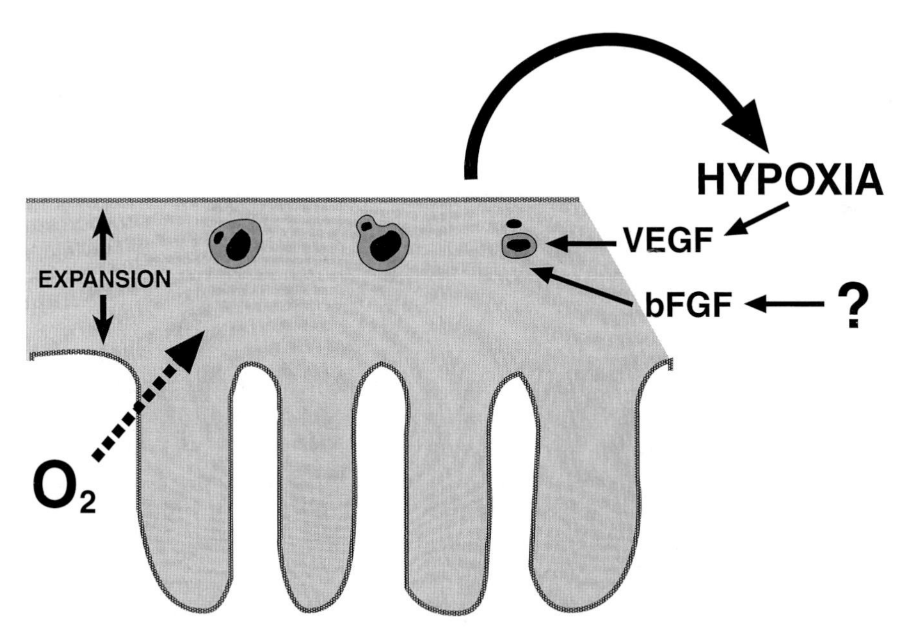

Figure 4.3 Growth factors and vascularization. As cardiocyte growth expands the compact zone in the embryonic heart, the diffusion distance for O_2 from the ventricle is increased as well. It is hypothesized that this myocardial growth compromises the O_2 available to the cardiocytes adjacent to the epicardium. Under this condition of relative hypoxia, VEGF is stimulated and facilitates vasculogenesis in this region of the ventricle. Although bFGF may also be important in embryonic vascular growth, the mechanism for its expression is not known.

tant from the standpoint of regulating extracellular matrix formation and proteolysis (Schott & Morrow, 1993).

The roles of mechanical factors and the extracellular matrix have not been discussed in this review because these factors have yet to be specifically explored in the developing heart. Nevertheless, they have been shown to be important in vascular growth in other systems and models. The role of physical forces in angiogenesis has recently been reviewed (Hudlicka & Brown, 1993). For example, stretch is a factor that is considered to be a stimulus for endothelial growth. One would expect that the various cells in the rapidly growing myocardium would be subjected to this mechanical force, which in turn could trigger the release of growth factors. The extracellular matrix also plays a key role in vascular growth. For example, fibronectin provides a migratory pathway for endothelial cells, and laminin is associated with tube formation and maturation (Risau & Lemmon, 1988). Integrins mediate endothelial cell attachment and migration (Luscinskas & Lawler, 1994). Many other components of the extracellular matrix also modulate vascular growth. For example, during heart development in the rat, the early appearance of fibronectin provides a primary scaffolding for vasculogenesis, and tube formation coincides with laminin deposition (Rongish et al., 1996). Thus, these events are consistent with data from other systems. Given the scarcity of data on the regulation of coronary vascularization, it should be evident that future studies need to focus on various aspects of this critical issue.

Unanswered questions

Our knowledge about the development of the coronary vasculature is inadequate for a number of reasons, including the relative inaccessibility of the mammalian heart and the fact that few studies have utilized experimental perturbations to answer questions about mechanisms of development. The use of avian models should be useful in circumventing the problem of inaccessibility and in allowing perturbations to be performed. In vitro studies are an important adjunct to, rather than a substitute for, in vivo experiments.

The major challenge for research in coronary vessel development is to define the mechanisms that regulate specific events directing coronary vasculogenesis and angiogenesis. The first question concerns the stimuli for the initiation of vasculogenesis; that is, what signals direct angioblasts to form tubes? Although growth factors are undoubtedly involved, their secretion or availability must be a consequence of some upstream signaling mechanism. A second question regards spatial direction. For example, the formation of veins from vascular plexuses in the region of the sinus venosus requires a regional signal. Similarly, the fusion of plexuses and their ingrowth into the aorta to form the two major coronary arteries are selectively regulated. Finally, vascular specialization constitutes another dilemma. What directs some microvascular channels to form venous structures and others to form arterial channels, and what regulates the addition of smooth muscle to endothelial channels? The answers to these questions are certain to be complex and can be found only by exploring a wide range of signaling mechanisms, including mechanical, metabolic, chemotactic, as well as growth factors.

5

Extracellular matrix maturation and heart formation

WAYNE CARVER, LOUIS TERRACIO, AND
THOMAS K. BORG

Introduction

The extracellular matrix (ECM) provides a structural framework for the formation of the complex 3-dimensional organization of tissues of the body. The ECM ranges from little more than a fluid medium to a rather rigid scaffolding. The composition and organization of the ECM are very dynamic in most tissues, particularly during embryogenesis, wound repair, and some disease states (Borg & Terracio, 1990; Borg & Burgess, 1993). Cellular interactions with the ECM influence a diverse array of processes not only through the ECM's role as a physical scaffold but also through specific signal transduction mechanisms (see Hynes, 1992, and Juliano & Haskill, 1993, for reviews).

The ECM of the heart is intimately associated with development and maintenance of heart structure and function. The heart ECM is composed of collagens, particularly types I, III, IV, and VI, proteoglycans, and noncollagenous glycoproteins such as laminin and fibronectin (Table 5.1; see also Kitten, Markwald, & Bolender, 1987; Borg & Terracio, 1990; Bashey, Martinez-Hernandez, & Jiminez, 1992; Borg & Burgess, 1993; Carver, Terracio, & Borg, 1993). These ECM components are expressed in precise temporal and spatial patterns in the developing and mature heart (see, e.g., Little et al., 1989; Iruela-Arispe & Sage, 1991; Sinning, Krug, & Markwald, 1992; Zagris, Stavridis, & Chung, 1993). Laminin, for example, appears localized to basement membranes of specific cells of the developing embryo and is particularly abundant in the heart (Figure 5.1). Laminin continues to be localized to the basement membranes associated with cardiomyocytes and vessels in the ventricular wall in the mature heart (Figure 5.1C). Expression of laminin is altered during cardiovascular development and with disease (Lipke et al., 1993).

Examination of the expression, accumulation, and organization of ECM components has resulted in a wealth of information regarding the ECM and its potential roles in cardiogenesis and heart disease. However, identification of the roles of the ECM in the heart has, until recently, remained largely speculative. Recently, a number of studies using either in vitro models or perturbable in vivo systems have begun to test for the ECM's structural role and its role as a transducer of chemical and mechanical information in the heart. Although much remains to be determined regarding the roles of the ECM in normal cardiogenesis and cardiovascular defects, significant strides are being made in this area.

Several aspects of the ECM in heart development will be discussed in this chapter. The focus will be on areas of development where functional data concerning the ECM have been obtained. Significant data have been accumulated on the ECM during cardiac cushion formation, myofibrillogenesis, and myocyte differentiation. The emphasis here is in no way

Table 5.1. *Matrix glycoproteins identified in the heart*

Matrix component	Location and comments	References[a]
Collagens		
Collagen I	Fibrillar collagen; predominant interstitial matrix component in the heart	Borg & Burgess, 1993; Eghbali-Webb, 1995
Collagen II	Fibrillar collagen abundant in cartilage; transient expression in the embryonic heart	Cheah, Lau, & Au, 1991; Swiderski et al., 1994
Collagen III	Fibrillar collagen often collocated with type I collagen; more distensible than type I collagen	Borg & Terracio, 1990; Eghbali-Webb, 1995
Collagen IV	Sheet-forming collagen localized to basement membranes of myocytes and vessels of heart	Yurchenco & Schittny, 1990
Collagen V	Fibrillar collagen localized perivascularly; putatively involved in matrix assembly	Honda et al., 1993; Andrikopoulos et al., 1995
Collagen VI	Found in most connective tissues, particularly associated with blood vessels; significantly increased in diabetic cardiomyopathy	Bashey, Martinez-Hernandez, & Jiminez, 1992; Spiro & Crowley, 1993
Collagen VIII	Transient expression in developing heart	Iruela-Arispe & Sage, 1991
Collagen IX	Facet-type collagen; transient expression in embryonic heart	Bork, 1992; Eghbali-Webb, 1995
Noncollagenous glycoproteins		
Elastin	Highly insoluble component of central amorphous region of elastic fibers; implicated in supraventricular aortic stenosis associated with Williams syndrome	Rosenbloom, Abrams, & Mecham, 1993; Ewart et al., 1994
Entactin	A 150-kilodalton molecular weight glycoprotein found in basement membranes complexed with laminin; involved in matrix organization and directional migration	Chung et al., 1993; Levavasseur et al., 1994
Fibronectin	Multidomain glycoprotein found in most connective tissues abundant in cardiac valve progenitors and other regions of the heart abundant in ECM components; important in migration of many cell types	Linask & Lash, 1988; Potts & Campbell, 1994
Fibrillin	A 350-kilodalton glycoprotein found in the outer portion of elastic fibers; strongly implicated in Marfan syndrome	Hurle et al., 1994
Fibulin	At least two forms identified thus far; found in most connective tissues in fibrils with elastin and fibrillin; particularly abundant in cardiac cushions, heart valves, and blood vessels	Spence et al., 1992; Roark et al., 1995; Zhang et al., 1995
Laminin	Multichain basement membrane component; abundant surrounding cardiac myocytes and blood vessels	Little et al., 1989; Price et al., 1992
Tenascin	At least three variants identified; localized particularly at sites of tissue remodeling in developing and pathological conditions; may play roles in migration and matrix organization	Erickson, 1994; Chiquet-Ehrismann, Hagios, & Matsumoto, 1994; Mackie, 1994

[a]References are, in most cases, recent reviews of the ECM.

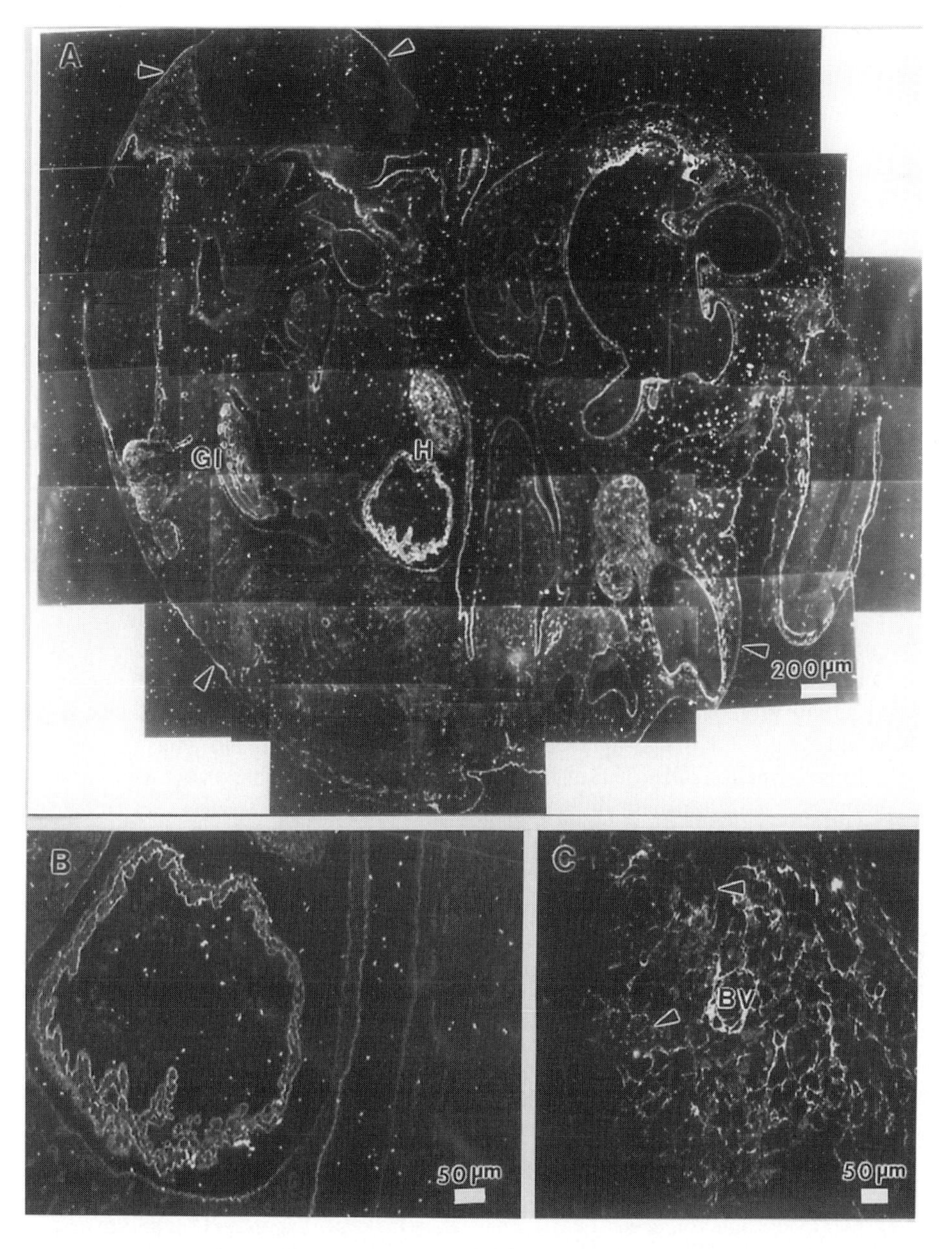

Figure 5.1 (*A*) Confocal micrograph of a laminin-stained section of a 13.5-day rat embryo. Intense staining for laminin is seen in basement membranes throughout the embryo, including the skin (*arrowheads*), heart (*H*), and gastrointestinal tract (GI). (*B*) Higher magnification micrograph of the laminin-stained embryo section seen in *A* illustrating intense staining in the developing heart. (*C*) Confocal micrograph of a laminin-stained heart ventricle from a 3-day-old rat. Intense staining is seen associated with vessels in the heart wall (BV) and along the surfaces of cardiomyocytes (*arrowheads*).

intended to deny the ECM's role in other aspects of heart development, as the ECM undoubtedly influences most of these.

Cardiac cushion formation

Cardiac cushion formation and their maturation into valvular and septal components of the heart are areas of intense investigation, largely focusing on the ECM. The primitive heart tube is composed of myocardial and endocardial layers separated by an acellular ECM termed the "cardiac jelly," which is secreted predominantly by the myocardium (Markwald, Fitzharris, & Manasek, 1977; Markwald et al., 1985). In the atrioventricular (AV) canal and outflow tract regions of the heart, this ECM expands to form localized swellings termed "endocardial cushions" or "cardiac cushions." Population of the cardiac cushions with mesenchymal cells begins with an epithelial–mesenchymal transformation of specific endothelial cells of the heart tube (Markwald et al., 1985). The mesenchymal cells invade the ECM of the future cushion regions in response to inductive signal(s) generated by the myocardium (Runyan & Markwald, 1983; Markwald et al., 1985; Krug, Mjaavedt, & Markwald, 1987). Recent work indicates that transforming growth factor-β (TGF-β) may be one such signaling molecule involved in this process (Potts & Runyan, 1989; see also Chapter 1).

The epithelial–mesenchymal transformation responsible for cell migration into the cushion regions can be replicated in vitro by the culture of AV endothelium on collagen gels and either (1) coculture with AV myocardium, (2) culture in cardiocyte conditioned medium, or (3) culture with proteins from EDTA extracts of embryonic myocardial basement membrane (Markwald et al., 1985; Krug, Mjaavedt, & Markwald, 1987; Markwald, Mjaavedt, & Krug, 1990). Thus, formation of cardiac cushion mesenchyme is dependent on the regionally specific production of myocardially derived ECM proteins. Further studies have determined that the inductive potential of the myocardial ECM is localized to "adheron" particles containing a number of EDTA-soluble proteins (Krug, Mjaavedt, & Markwald, 1987; Mjaavedt, Lepera, & Markwald, 1987; Mjaavedt & Markwald, 1989; Markwald, Mjaavedt, & Krug, 1990; Sinning, Krug, & Markwald, 1992). These adheron particles are sufficient for induction of epithelial–mesenchymal transformation in vitro (Mjaavedt & Markwald, 1989). Moreover, antibodies to adheron particles (ES antibodies) are capable of inhibiting this process regardless of the inductive source (Mjaavedt, Krug, & Markwald, 1991). A number of data indicate an inductive role for the myocardially derived ECM in the initial epithelial–mesenchymal transformation, which results in the population of the cardiac cushions. To date, the identity of these specific inductive factor(s) in the ECM is unclear.

A number of ECM components involved in the transformation of endothelial cells and/or their subsequent migration are expressed in specific spatial and temporal patterns. Some of these, such as fibronectin, several collagens (types I, II, III, IV, and VIII), and hyaluronic acid, appear to have relatively uniform distributions in the cardiac jelly of the early avian heart (Swiderski et al., 1994, and references therein). Other ECM components have more restricted localization patterns during cardiac cushion development. Cytotactin and fibulin, for example, appear to surround presumptive mesenchymal cells that have begun to migrate into the cardiac cushion (Crossin & Hoffman, 1991; Spence et al., 1992). Fibrillin appears to be expressed by a subset of endothelial cells and by all mesenchymal cells following transformation.

Determining the functional roles of each of these ECM components has proven somewhat more difficult than examining their expression patterns during heart development. In vitro systems such as the collagen gel model and whole-embryo culture are being used as functional assays of ECM components. Injection of antibodies to fibronectin abolishes the migration of cushion mesenchymal cells within the ECM following transformation in chick embryos (Icardo et al., 1992). Elegant studies are now beginning to determine the individual ECM components that are present during specific stages of cardiac cushion formation and to elucidate their importance in this morphogenetic process.

Immunostaining data such as those for fibrillin and other experimental data have led to the hypothesis that the endocardium is composed of at least two subpopulations of endothelial cells, only one of which is able to transform into mesenchymal precursors and migrate into the cardiac cushion (Markwald, Mjaavedt, & Krug, 1990; Wunsch, Little, & Markwald, 1994). The initial role of the ECM in cardiac cushion formation may be inductive. Following transformation, ECM components, including fibrillin, fibronectin, and vitronectin, also appear to play roles in the directed migration of mesenchymal cells in the cardiac cushion and subsequent maturation of the cushion into valvular and septal components. Although the mechanisms of cardiac cushion development are far from understood, intense investigations have made great strides in elucidating the underlying mechanisms of this critical cardiogenic process.

Matrix receptors

Interactions of cells with surrounding ECM components are critical during morphogenesis of the cardiac cushions and other regions of the heart. Several groups of molecules have been identified that mediate interactions of cells with the ECM (Akiyama, Nagata & Yamada, 1990; Loeber & Runyan, 1990). The integrins are the predominant family of ECM receptors (Albelda & Buck, 1990; Humphries, 1990; Hynes, 1992). Their potential roles in heart development are supported by the recent production of "knockout" mice lacking specific integrin chains or their ligands (Yang, Rayburn, & Hynes, 1993; Kwee-Isaac et al., 1995). In each of these recombinant knockouts, the absence of specific integrins resulted in severely abnormal cardiovascular development and embryonic lethality.

The integrins are heterodimers of α and β chains that have large extracellular regions, single transmembrane domains, and, generally, short cytoplasmic tails (Hynes, 1992). The integrins are loosely grouped into subfamilies based on the constituent β chain, the $\beta 1$ subfamily being the most abundant in the developing heart described thus far (Borg, Rasso, & Terracio, 1990; Terracio et al., 1991; Carver & Terracio, 1993; Baldwin & Buck, 1994; Carver et al., 1994). The combination of α and β chains determines the ligand specificity of each individual integrin complex. There appears to be significant multiplicity among the integrins in that several different complexes may bind the same matrix ligand. For instance, the $\alpha_1\beta_1$, $\alpha_2\beta_1$ and several other complexes all potentially bind laminin (Hall et al., 1990). Furthermore, several of the integrin complexes bind multiple matrix ligands. This repetitiveness has made it very difficult to determine the functional roles of specific matrix–ligand interactions. However, it is clear from the integrin knockout studies and from in vitro studies described below, that integrin-mediated cell–matrix interactions are critical for normal heart development.

Cardiac cushion maturation

Following formation of the cardiac cushions, these tissues normally fuse and differentiate into valvuloseptal components of the heart. These events transform the primitive heart tube into the multichambered mature organ. Although the process of initial formation of the cardiac cushions has been intensely investigated, the development of these tissues into mature structures has received much less attention. Several studies of rat and chick embryos have described the accumulation of matrix components, including elastin, fibrillin, and collagens, in the maturing valve leaflets (Carver, Terracio, & Borg, 1993; Hurle et al., 1994). However, the roles of these components remain to be determined.

An in vitro collagen gel model has been used to investigate the mechanisms whereby fibroblasts modify the organization of the ECM. This culture model provides a matrix remodeling system that is reminiscent of wound healing and cardiac cushion maturation (Guidry & Grinnell, 1985). In this "collagen gel contraction" assay, fibroblasts interact with collagen, reorganize the collagen fibers, and contract the gel severalfold over time. Studies with this model have indicated that integrin receptors of the β_1 subfamily are essential for collagen gel contraction (Gullberg et al., 1990; Burgess et al., 1994). In particular, the α_1 and α_2 chains of this subfamily play critical roles in this process (Schiro, Chan, & Roswit, 1991; Carver et al., 1995). Incubation of cells with antibodies to the α_1 integrin chain alters the ability of these cells to migrate within the gel and to subsequently cause contraction. It is clear then that cellular interactions with the ECM components are essential for matrix reorganization.

The mechanisms regulating matrix modifications in cardiogenesis are currently undefined. The changes in the ECM that result in structural remodeling of the developing heart appear to involve extracellular proteolytic enzymes that modify existing ECM components (Alexander & Werb, 1989). The mechanisms of matrix remodeling have not been fully elucidated, but interactions of cells with matrix ligands stimulate the expression of specific genes, including matrix proteases (Werb et al., 1989; Damsky & Werb, 1992). These matrix proteases may be critical in morphogenetic processes of matrix remodeling associated with cardiac cushion maturation.

Immunohistochemical studies on rat embryos indicate the presence of collagenase in the developing heart valves and in the myocardial trabeculae (Nakagawa et al., 1992). These studies clearly show collagenase on some endocardial cells prior to epithelial–mesenchymal transformation. Also, migrating cells in the cardiac jelly are associated with collagenases, particularly on the surfaces of the cells nearest the myocardium. Collagenases are also expressed by fibroblasts cultured in the collagen gel contraction system (Nakagawa et al., 1990), indicating that expression of collagenase may be important in cardiac cushion formation and maturation in vivo as well as in ECM remodeling in vitro.

Further studies with the collagen gel model indicate potential roles for chemical factors in modulating matrix reorganization. Platelet-derived growth factor (Gullberg et al., 1990) and angiotensin II (Burgess et al., 1994) stimulate gel contraction by heart fibroblasts. These chemical factors may play a role in modulating matrix remodeling through increased expression of matrix proteases, regulation of matrix receptor cell surface levels, or some other undetermined mechanism.

Intriguing results have recently been obtained comparing the abilities of trisomy 16 and normal mouse heart fibroblasts to contract collagen gels. Trisomy 16 mouse embryos have consistent cardiovascular defects, including AV cushion and outflow tract abnormalities thought to arise from abnormal cardiac cushion formation or maturation (Miyabara, 1990). Fibroblasts isolated either from whole hearts or from the cardiac cushions of trisomy 16 embryos contract collagen gels very poorly compared with age-matched normal mouse fibroblasts (Figure 5.2). This suggests that cardiovascular defects seen in the trisomy 16 mouse may involve abnormal cell–matrix interactions such as those required to contract

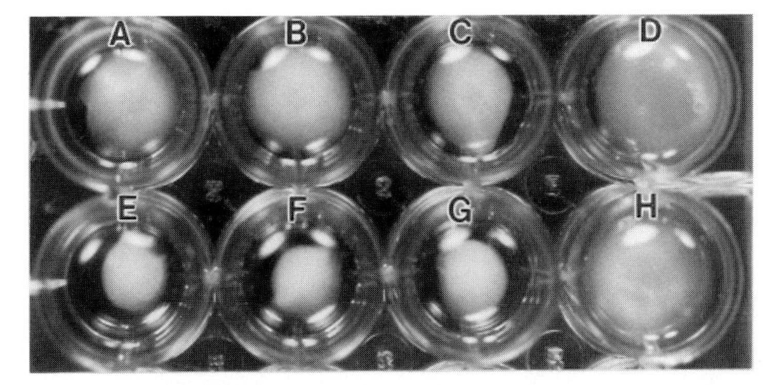

Figure 5.2 Collagen gels after 24 hours of culture containing trisomy 16 (*A–D*) and normal (*E–H*) mouse embryo fibroblasts. Equal numbers of fibroblasts from age-matched embryos (13.5 days of gestation) were cultured in either the presence (*A–C, E–G*) or the absence (*D, H*) of serum. Trisomy 16 cells contract collagen gels much less efficiently than normal cells.

collagen gels in vitro. Intense investigations are under way to determine the mechanisms of these abnormalities.

Three-dimensional organization of heart cells

The heart is composed of several distinct cell types, including myocytes, fibroblasts, endothelial cells, and smooth muscle cells. The component cells of the heart are precisely arranged during cardiogenesis, establishing distinct delineations between connective tissue and cardiomyocytes in several areas. Although it is clear that the 3-dimensional organization of heart cells is critical to cardiovascular function, the mechanisms underlying the formation and maintenance of this architecture are unclear (see also Chapter 1).

Immunostaining experiments have indicated accumulations of ECM components in connective tissue regions of the heart. The valve leaflets and outflow tract, for instance, contain abundant interstitial collagen (Carver, Terracio, & Borg, 1993) and fibronectin (Figure 5.3). Studies using an in vitro aggregate model have indicated that differential accumulation of matrix components, as well as cell interactions with the ECM, are critical in establishing the 3-dimensional organization of heart cells (Armstrong & Armstrong, 1984, 1990). In particular, interactions of embryonic chick fibroblasts with fibronectin are, at least in part, responsible for sorting of these cells from cardiomyocytes in vitro (Armstrong & Armstrong, 1990). These interactions with fibronectin are mediated through the β_1 integrin family of cell surface receptors. Recent work with rat heart fibroblasts and myocytes indicated that the $\alpha_5\beta_1$ integrin complex, the "classic" fibronectin receptor, was critical for the in vitro sorting of heart cells (T. Bacro, L. Terracio, T. K. Borg, & W. Carver, unpublished observations). Although the $\alpha_5\beta_1$ integrin may be essential for in vitro sorting, the significance of these findings for in vivo cardiogenesis is not yet clear. In support of the role of specific ECM and receptor components in the developing heart are recent data that mice lacking either fibronectin or the α_5 integrin have severe cardiovascular defects (George et al., 1993; Yang, Rayburn, & Hynes, 1993). Work with the whole-embryo culture system and transgenic animals should begin to address the role of cell–matrix interactions in establishing the 3-dimensional organization of cells during cardiovascular development.

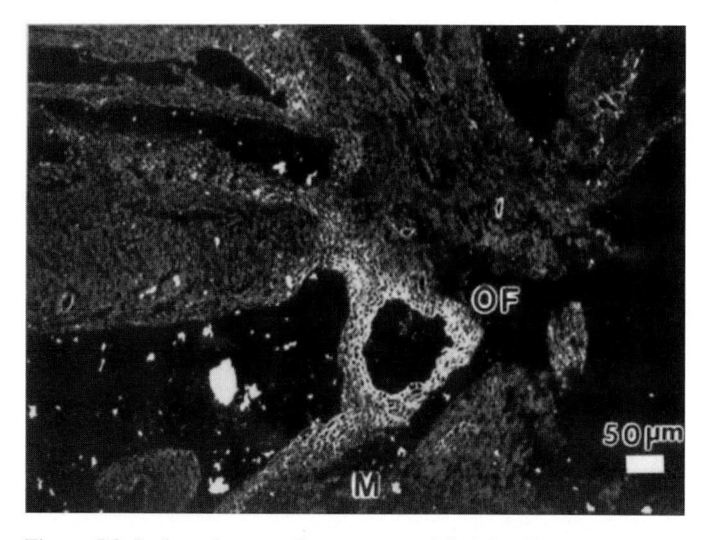

Figure 5.3 Indirect immunofluorescent staining for fibronectin in the 3-day neonatal rat heart. Staining is particularly prominent in the outflow region of this section (OF). Weak staining is also seen in the ventricular myocardium (M).

Myofibrillogenesis

Interactions of various types of cells with the ECM are known to influence cellular differentiation (Tournier et al., 1992; Adams & Watt, 1993). A number of approaches have been used to investigate the role of cell–matrix interactions in the organization and maintenance of the cardiocyte contractile phenotype. ECM components, including interstitial collagen (Caulfield & Borg, 1979; Borg & Caulfield, 1979; Borg, Johnson, & Lill, 1984) and basement membrane laminin (Price et al., 1992), accumulate at specific sites along the myocyte surface. Also localized at these sites are matrix receptors of the β_1 integrin family (Hilenski et al., 1989; Hilenski, Terracio, & Borg, 1991) and myofibrillar components, including vinculin (Terracio et al., 1990).

Upon isolation, neonatal heart myocytes lose their elongated shape and myofibrillar organization. During culture, these cells "redifferentiate": myofibrils are reorganized and intrinsic contractile activity returns. These cells increase their expression of matrix components and matrix receptors in precise patterns during redifferentiation (Figure 5.4). The composition and organization of the ECM have a significant effect on myofibrillar organization in these cells (Hilenski, Terracio, & Borg, 1991). Incubation of isolated myocytes with antibodies that block β_1 integrin–mediated cell–matrix interactions indicate that these interactions are necessary for spreading of the cells on the substratum and myofibril formation (Borg et al., 1990; Hilenski et al., 1992). These results clearly indicate a critical role for integrin-mediated cell–matrix interactions in myocyte differentiation in vitro.

Cardiomyocyte differentiation in vivo results in the formation of an elongated, rod-shaped phenotype. Although most in vitro systems result in stellate, flat cardiomyocytes, a novel system in which aligned collagen is used as the culture substratum produces elongated cells. Isolated rat myocytes plated on this aligned collagen form elongated, rod-shaped cells that parallel the collagen fibers (Simpson et al., 1994). During redifferentiation, myocytes in this system express matrix components and matrix receptors in precise

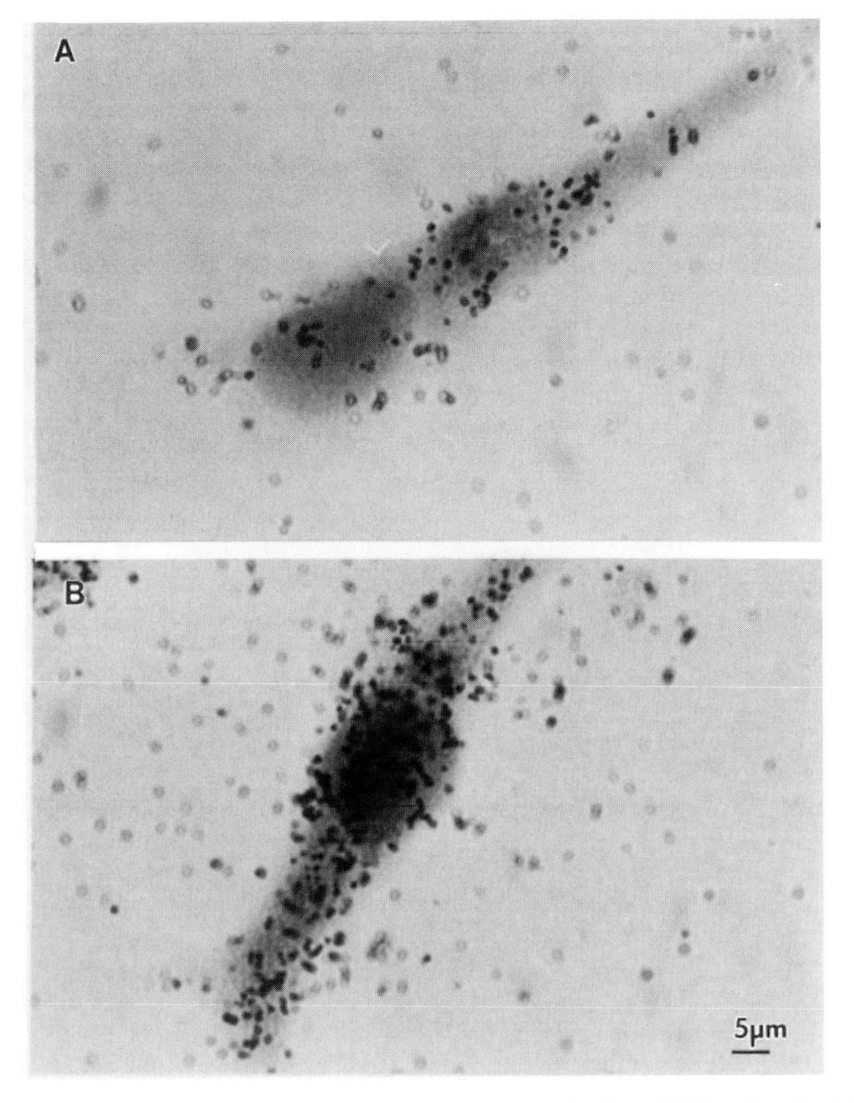

Figure 5.4 Micrographs of isolated heart myocytes hybridized to a cRNA probe to laminin B1. Myocytes were cultured for 3 days (*A*) or 6 days (*B*) on collagen-coated slides. Hybridization to laminin mRNA is significantly increased over the culture period.

temporal patterns (B. Burgess, W. Carver, & L. Terracio, unpublished observations) that reflect in vivo differentiation. Incubation of the cells with anti-β_1 or anti-α_1 integrin IgGs inhibits the production of the in vivo–like phenotype. These results indicate that both matrix composition and matrix organization play roles in the determination of heart myocyte morphology and function. The use of this novel in vitro aligned culture system will be invaluable in determining the functional roles of specific components of the ECM, cell surface, and cytoskeleton in myocyte differentiation.

Neonatal heart development

A number of changes occur in the mammalian heart in the period immediately following birth (Zak, 1974). These structural and biochemical alterations are thought to occur largely in response to altered loading conditions on the heart during this period. The composition and organization of the ECM change dramatically during the neonatal period (Borg & Caulfield, 1979; Borg, 1982; Borg, Gay, & Johnson, 1982). Interstitial collagen, in particular, increases rapidly following birth (Borg & Caulfield, 1979; Carver, Terracio, & Borg, 1993). An elaborate weave of collagen forms that interconnects adjacent myocytes and connects myocytes and capillaries in the ventricular myocardium during this period.

The collagen weave is thought to be important as a structural framework in the myocardium and in the dissipation of forces throughout the heart wall. Inhibition of collagen cross-linking in the developing rat heart results in aneurysms in the heart wall (Borg & Caulfield, 1979), clearly indicating the structural significance of this network.

The regulation of matrix expression and accumulation is quite complex, with transcriptional, translational, and posttranslational levels of modulation. Although it is currently not clear what controls ECM accumulation during the neonatal period, a number of chemical factors affect the expression of matrix components in vitro (Table 5.2). These include several factors identified in the developing heart, such as transforming growth factor-β (Ignotz, Heino, & Massague, 1989) and angiotensin II (Rogers & Lokuta, 1994). Increased expression of growth factors, including basic fibroblast growth factor (bFGF), have been seen concurrent with the increased production of collagen in the neonatal heart (Engelmann, 1993). A regulatory role for these factors has not yet been demonstrated in the neonatal heart.

Neonatal heart fibroblasts also increase collagen expression in response to mechanical forces in vitro (Carver et al., 1991). It is currently not clear if collagen gene expression is increased directly in response to mechanical forces or if this occurs in response to production of unidentified chemical factors stimulated by stretch. Recent studies using cardiomyocytes indicate that angiotensin II mediates the response of these cells to mechanical forces (Komuro & Yazaki, 1994). Further work with smooth muscle cells has shown that platelet-derived growth factor may modulate the response of these cells to physical forces (Wilson et al., 1993) and that this chemical factor is directly stimulated by these forces through a stretch response element in the genomic sequence. Similar studies are required to determine the molecular mechanisms whereby interstitial collagen is increased in the neonatal heart.

Proteoglycans

Proteoglycans are macromolecules that consist of a core protein and covalently bound carbohydrates termed glycosaminoglycans (for reviews of proteoglycans see Gallagher, Lyon, & Steward, 1986; Kjellen & Lindahl, 1991; Jackson, Busch, & Cardin, 1991). Proteoglycans occur intracellularly, at the cell surface, and in the ECM. These multidomain macromolecules have been loosely grouped into "families" based on their core protein structure. The biological roles of proteoglycans are diverse and range from structural support functions to cellular processes, including adhesion, migration, and differentiation. Evidence implicating proteoglycans in specific functional roles during heart development has been largely correlative and arises from the identification of specific patterns of expression of particular proteoglycan species. A number of these expression patterns have been identified in the heart, such as those for perlecan-1 and n-syndecan (Carey et al., 1992; Murdoch et

Table 5.2. *Chemical factors known to modulate expression and accumulation of matrix components*

Factor	Effect	Cell issue	Reference
TGF-β	Increased collagen synthesis	Cardiac fibroblasts	Eghbali et al., 1991; Chua et al., 1991
		Dermal wounds	Pierce et al., 1992
	Increased collagen I, III, and FN	Fat-storing cells	Casini et al., 1993
	Increased laminin and FN	Astrocytes	Baghdassarian et al., 1993
	Increased sulfated proteoglycans	Kidney epithelial cells	Kovacs et al., 1994
	Increased tenascin production		Pearson et al., 1988
	Increased elastin production	Skin of transgenic mice	Katchman et al., 1994
bFGF	Decreased collagen synthesis	Smooth muscle cells	Majors & Ehrhart, 1993
	Decreased collagen IV synthesis	Corneal endothelial cells	Kay et al., 1993
	Increased collagen deposition	Chondrocytes	Hill et al., 1992
	Decreased elastin production	Lung fibroblasts	Brettell & McGowan, 1994
	Decreased tenascin production		Tucker et al., 1993
PDGF	Decreased collagen IV synthesis	Arterial smooth muscle	Okada et al., 1993
	Increased collagen V synthesis	Arterial smooth muscle	Okada et al., 1993
	Increased laminin and collagen IV	Mesangial cells	Floege et al., 1993
	Increased GAG deposition	Dermal wounds	Pierce et al., 1992
	Increased FN	Dermal wounds	Pierce et al., 1992
	Increased collagen synthesis	Dermal fibroblasts	Gessin et al., 1991
	Increased collagen synthesis	Vascular smooth muscle	Amento et al., 1991
Angio II	Increased collagen I	Cardiac fibroblasts	Brilla, Maisch, & Weber, 1993
	Increased FN	Heart (fibrosis)	Crawford, Chobanian, & Brecher, 1994
	Increased collagen expression	Smooth muscle cells	Kato et al., 1991
	Increased laminin expression	Smooth muscle cells	Regenass et al., 1994
	Increased tenascin	Smooth muscle cells	Sharifi et al., 1992
	Decreased elastin synthesis	Smooth muscle cells	Tokimitsu et al., 1994
IL-1	Increased collagen	Lung fibroblasts	Elias et al., 1990
	Increased collagen IV	Mammary epithelial cells	Matsushima et al., 1985
	Decreased collagen synthesis	Embryo skin fibroblasts	Bodo et al., 1994
	Decreased GAG synthesis	Embryo skin fibroblasts	Bodo et al., 1994
	Increased collagen expression	Gingival fibroblasts	Irwin, Schor, & Ferguson, 1994
		Lung fibroblasts	Diaz et al., 1993
	Decreased hyaluronon insulin	Gingival fibroblasts	Irwin et al., 1994
	Increased laminin synthesis	Rat mesangial cells	Abrass, Spicer, & Raugi, 1994
	Decreased FN	Rat mesangial cells	Abrass, Spicer, & Raugi, 1994
	Increased laminin and collagen	Rat glomeruli	Fukui et al., 1992
	Increased collagen synthesis	Chondrocytes	Hill et al., 1992
RA	Increased laminin production	Rat mesangial cells	Ross, Ahrens, & De Luca, 1994
	Increased elastin synthesis	Lung fibroblasts	Liu, Harvey, & McGowan, 1993
IGF-I	Increased collagen synthesis	Chondrocytes	Hill et al., 1992
	Increased elastin production	Smooth muscle cells	Wolfe et al., 1993
	Increased collagen synthesis	Skin fibroblasts	Goldstein et al., 1989
TNF-α	Decreased collagen II expression	Cartilage	Inagaki et al., 1995
IFN-α	Increased laminin expression	Lung epithelial cells	Maheshwari et al., 1991
NO	Increased laminin synthesis	Mesangial cells	Trachtman, Futterweit, & Singhal, 1995

Note: Angio II = angiotensin II; bFGF = basic fibroblast growth factor; FN = fibronectin; GAG = glycosaminoglycan; IFN-α = interferon-α; IGF-I = insulin-like growth factor I; IL-1 = interleukin-1; NO = nitric oxide; PDGF = platelet-derived growth factor; RA = retinoic acid; TGF-β = transforming growth factor-β; TNF-α = tumor necrosis factor-α.

al., 1994); however, the biological roles of these molecules are largely unexplored. Proteoglycans, in general, appear to modulate developmental events through their interactions with matrix macromolecules, cell adhesion molecules, or growth factors (Kjellen & Lindahl, 1991).

A number of proteoglycans have been demonstrated to interact with matrix components, including collagen, fibronectin, and laminin (Scholzen et al., 1994; Salmivirta et al., 1994). These interactions may modulate the cell–substratum binding critical in cell adhesion, cell migration, and tissue morphogenesis. ECM-bound proteoglycans may also influence the ability of cells to interact with ECM components. For instance, the binding of heparin sulfate proteoglycan to fibronectin may modulate cell attachment by exposing the cell-binding domain of this complex matrix macromolecule (Couchman et al., 1988). Syndecan-1 has also been implicated in modulating the invasion of cells into a collagen matrix (Liebersbach & Sanderson, 1994). B lymphoid cells expressing syndecan-1 are inhibited from invading collagen, whereas those cells not expressing this proteoglycan readily invade the matrix. Glycosaminoglycans, such as hyaluronate and chondroitin sulfate, may facilitate the migration of fibroblasts in collagen gels by altering the arrangement of collagen fibrils within the gel (Docherty et al., 1989). These aspects of proteoglycan function are largely unexplored during heart development.

Proteoglycans also bind cell adhesion molecules and may regulate cell behavior through these interactions. For instance, phosphacan binds neural cell adhesion molecules and may compete for ligands of transmembrane phosphatases (Milev et al., 1994). Phosphacan has thus been suggested to modulate neuronal adhesion, neurite outgrowth, and signal transduction in the developing central nervous system, at least in part through the interaction with cell adhesion molecules.

Growth factors bind to the ECM and cell surfaces largely through interactions with proteoglycans. Many proteoglycans in the ECM function as modulators of growth factors; the role of heparin sulfate proteoglycans as binders of FGF has been known for some time (Burgess & Maciag, 1989). Binding to proteoglycans may be important in protecting the FGFs from degradation and also in creating a pool of matrix-bound growth factors (Kjellen & Lindahl, 1991). Recent evidence has indicated that binding of bFGF to its cell surface receptor requires prior binding to heparin sulfate. This proteoglycan–growth factor interaction is mediated through the glycosaminoglycan component of the proteoglycan. TGF-β, on the other hand, binds to proteoglycans through the core protein. The role of proteoglycan–growth factor interactions in the developing heart also requires further investigation.

The inhibition of proteoglycan synthesis by β-xyloside during late gastrulation/early neurulation in embryos of *Xenopus laevis* results in a range of cardiovascular defects (Yost, 1990). The most severe defect seen in these experiments is the absence of early heart tube looping. These results suggest that proteoglycan synthesis and accumulation play a fundamental role in left–right axial formation and directed cardiac cell migration in the early embryo.

As mentioned previously, several glycosaminoglycans have been shown to enhance the migratory ability of fibroblasts in collagen gels. A number of proteoglycans have been identified in the developing cardiac cushions and have been implicated in cell–matrix interactions necessary for cushion formation (Markwald et al., 1984). Cytotactin-binding proteoglycan and cytotactin have recently been localized to the cardiac jelly, temporally correlating with epithelial–mesenchymal transformation and mesenchymal cell migration (Crossin & Hoffman, 1991). The presence of these two molecules has also been correlated with migration of neural crest cells and may be associated with similar events in the cardiac cushion. A number of studies indicate that proteoglycans, particularly through glycosaminoglycan side chains, play roles in morphogenesis either via their modifying action on the ECM or by binding growth factors in the matrix. However, the specific functional roles of most of these interactions during heart development remain to be determined.

Conclusions and future directions

This chapter, while by no means exhaustive, illustrates that a wealth of information is currently being generated on the ECM in embryonic development. However, a number of critical questions remain regarding the ECM in heart development and maturation. Future work in this area must address a number of issues, including (1) the mechanisms through which matrix, matrix receptors, and matrix-modifying enzymes are regulated in the developing heart; (2) the potential modifications of matrix components and interactions between matrix components in the heart; and (3) the functional roles of the ECM in heart development. With ongoing generation of specific reagents, new molecular methodologies, and novel culture systems, the information base will undoubtedly continue to rapidly increase in these areas and others relevant to the ECM in heart development.

The ECM contains a complex, highly organized spectrum of macromolecules. The ECM plays roles in morphogenesis, tissue maintenance, and disease through its structural and signal transduction properties. It is apparent that the expression, accumulation, and organization of the ECM are important in vertebrate embryogenesis. It is also becoming increasingly evident that these macromolecules consist of multiple domains that, when exposed by proteolytic enzymes at the protein level or by splicing at the mRNA level, may provide different signals to cells. Future studies must address not only the localization of matrix components in the developing heart but also what form of the matrix component the surrounding cells are actually "seeing." It is already clear that laminin isolated from the heart provides different cues to heart cells in vitro than does laminin isolated from the mouse Engelberth–Holm–Swarm (EHS) tumor (T. A. Reaves & T. K. Borg, unpublished observations). It is thought that these different signals are related to the relative amount of carbohydrate associated with the isolated laminin. It is essential then to determine where and when ECM components are expressed in the heart, the splice variants that may be present, and the modifications that exist on the matrix molecules. Subtle differences in any of these may be important in regulating cellular processes involved in heart formation.

Localization studies have illustrated that ECM components are expressed in specific temporal and spatial patterns that correlate with developmental events. ECM accumulation and organization are regulated not only by the expression of ECM components but also by the expression of ECM receptors and expression and activation of matrix proteases and their inhibitors. In light of this fact, critical examination of the spatial and temporal expression of not only ECM but matrix receptors, matrix proteases, and protease inhibitors during heart development is critical. Precise definition of the relative levels of all these components in the normal heart is necessary in attempting to understand the roles that they may play in normal heart development and how their alterations may contribute to congenital heart defects.

The regulation of the expression of matrix components in the heart by chemical and physical factors is only now beginning to receive attention. It is clear from other systems that many chemical factors modulate the expression of matrix components. Complicating this area is the fact that different cell types may have varied responses to these chemical factors (see Table 5.2). A critical analysis of these specific growth factors and cytokines is necessary using both isolated heart cell and whole-animal models.

In vitro assays have elegantly illustrated potential roles of ECM components in cellular processes related to cardiac development. During cardiogenesis, the ECM influences structural and functional development by modulating cell migration, proliferation, differentiation, and gene expression. Most of these in vitro studies have been carried out by plating isolated cells onto different matrix components and analyzing relative cell behavior such as adhesion, migration, and proliferation. Further studies have utilized blocking antibodies to begin to determine the roles of specific cell–matrix interactions in these cellular processes in vitro. These studies have been very informative, but it is clear that the in vitro environment, including the matrix composition, does not exactly mimic that seen in vivo. As further studies define the in vivo extracellular milieu, the challenge will be to more closely recreate this environment in vitro by combining multiple matrix components and/or physically organizing the matrix. As described previously, this is being done with an aligned collagen culture system that produces cardiac myocytes with an in vivo–like morphology. Even with novel in vitro systems that more closely mimic the in vivo environment, many questions remain that can be addressed only in vivo.

The actual in vivo functions of ECM components are only now beginning to be determined. Novel molecular methodologies combined with whole-embryo culture, the production of transgenic animals, and other animal models are essential for further understanding the roles of the ECM in normal development. Whole-animal model systems that can be manipulated during specific time periods of development hold a great deal of promise in addressing the functional roles of matrix components in specific cardiogenic events. Continued work in this area will undoubtedly indicate the roles of altered ECM expression and organization in congenital defects of the cardiovascular system.

6

Endothelial cell development and its role in pathogenesis

JACKSON WONG

Introduction

The endothelium is a single layer of cells that lines the intimal surface of the circulatory system of vertebrates. Initially, the endothelial cell (EC) was thought to function as only a passive inert barrier between the circulating blood and surrounding tissue. However, it is now evident that the endothelium is an important physiological and metabolic organ with broad paracrine and endocrine activity (Pearson, 1991). The EC population is not homogeneous. It displays remarkable phenotypic and functional differences not only between species but also within the same organism (Fajardo, 1989). This chapter will review some current knowledge on the morphogenesis, development, and maturation of the EC and its physiologic and metabolic roles. Also, it will discuss the potential role of developmental aberrations in endothelial function in pathogenesis.

Endothelial cell development

Vasculogenesis and angiogenesis

The embryonic development of the endothelium has been closely studied in conjunction with the development and maturation of blood vessels. Embryonic ECs and their precursors are very different from mature adult cells. The metamorphosis of embryonic ECs may occur in less than 24 hours, whereas adult cells may not change for more than 20 years (Pardanaud & Dieterlen-Lievre, 1993). Endothelial precursors arise from the differentiation of mesodermal aggregates called blood islands found in the yolk sac (Risau et al., 1988). The outer layers of the island develop into endothelial precursors, and the inner cells form the hematopoietic stem cells.

During the embryonic period, extensive proliferation and remodeling of the ECs are required in the development of the regional circulation (Risau & Lemmon, 1988). Despite limited techniques, by the beginning of this century the vascular endothelium was noted to originate from both extra- and intraembryonic sources (Evans, 1909). Currently these processes are respectively termed "vasculogenesis" and "angiogenesis" (Poole & Coffin, 1989; see also Chapter 4, this volume).

During organogenesis, both vasculogenesis and angiogenesis are active, and a vessel can be formed by either or both processes (Pardanaud, Yassine, & Dieterlen-Lievre, 1989). Vasculogenesis is the formation of vessels by the development of endothelial precursors (angioblasts) into mature ECs in situ (Noden, 1989). Vasculogenesis can be subdivided into types I and II. In type I, endothelial precursors assemble into the vessels close to the original mesoderm from which they differentiate. Type II precursors migrate to a distant site before assembling into a vessel. Angiogenesis is the formation of new vessels from preexisting vessels. New vessels form from a branch of a preexisting vessel called a sprout. This sprout is a highly specialized organ designed for the invasion into the extracellular matrix (ECM) to form a new vessel. Angiogenesis continues to be active throughout the normal life span of an organism, for instance, as in the female reproductive system (Klagsbrun & D'Amore, 1991).

The study of the development and regulatory control of endothelial precursors has been greatly enhanced by the discovery of specific surface antigens and molecular markers (Labastie et al., 1986; Korhonen et al., 1994). Transplantation and heterotypic mesoderm graft studies have revealed that angioblasts, like neural cell precursors, have great migratory capacity. Their migration patterns are dependent on signals derived from the microenvironment (Poole & Coffin, 1989; Noden, 1989; Poelmann et al., 1990; Coffin & Poole, 1991). The transplanted angioblast is pluripotent and can differentiate itself into the appropriate regional vessels (Poelmann et al., 1990; Coffin & Poole, 1991). Transplantation studies suggest that the maturation of the EC may in part be controlled by angiogenesis factors produced by the microenvironment.

Angiogenesis factors

The precise orchestration of EC maturation is dependent on multiple angiogenesis factors. Angiogenesis factors regulate the induction and suppression of gene expression and the protein and metabolic products produced by the EC throughout its life span (see also Chapter 4). Different angiogenesis factors are required during the different stages of the life span of the cell. Interactions as early as those among endodermal, mesodermal, and ectodermal layers may direct the development of the endothelial precursors' lineage (Risau, 1991a; Pardanaud & Dieterlen-Lievre, 1993). The influences of early angiogenesis factors can be suppressed in the mature cell but reactivated during wound healing or pathological conditions (Torry & Rongish, 1992; Renter et al., 1993). These angiogenesis factors are as diverse as physical cell contact, ECM, receptors, growth factors, and mechanical forces.

Cell contact

Many different experimental models, including the formation of the blood–brain barrier, have been used to study the effects of the local environment on the maturation of the EC (Coffin & Poole, 1991; Bertossi et al., 1993). ECs exclusively form the blood–brain barrier, a specialized organ with unique physiological properties and surface proteins (Janzer & Raff, 1987; Bertossi et al., 1993; Schlosshauer, 1993). As endothelial precursors develop to form the blood–brain barrier, a special luminal cell surface protein named neurothelin is produced (Schlosshauer, 1991, 1993). It is postulated that neurothelin may be important in the regulation of the permeability of the blood–brain barrier. The development of these specialized ECs is in part dependent on signals derived from surrounding astrocytes. Transplanting astrocytes to other sites in an embryo can induce surrounding ECs to express neurothelin and form a regional functional blood–brain barrier (Schlosshauer & Herzog, 1990). Therefore, intracellular physical communications and soluble factors between developing

ECs and their surrounding cells may in part determine the physiological properties of mature ECs.

Extracellular matrix

The ECM is a complex structure composed of many different substances (see also Chapter 5). The composition of each regional ECM is unique for each species and individual organ system (Har-el & Anzer, 1993). The ECM is essential to cellular differentiation and gene regulation (Lin & Bissell, 1993). The ECM interacts with the basal surface of ECs. It can influence the migration patterns and physiological development of ECs. Like neural crest cell migration, the migration pattern of endothelial precursors depends on the components of the surrounding ECM (Risau et al., 1988; Venstrom & Reichardt, 1993). For example, in vasculogenesis of the dorsal aorta, endothelial precursors in the blood islands and early intraembryonic capillaries are surrounded by a matrix rich in fibronectin and poor in laminin. During vessel maturation, the amount of laminin in the ECM increases. Similarly, in the angiogenesis of blood vessels in the brain, the sprouts initially migrate into a fibronectin-rich matrix that is devoid of laminin. During the assembly of a new vessel from a sprout, the amount of laminin in the surrounding matrix also increases. These results suggest that the appearance of laminin in the ECM may be a sign of vessel maturation and a negative-feedback signal to the ECs to prevent further vessel formation (Drake et al., 1990; Sumida, Nakamura, & Satow, 1991). However, the mechanisms providing negative feedback to the ECs are unknown.

The ECM influences ECs throughout their life span. There are specific spatial and temporal relationships between the effects of individual matrix components and the maturing and adult ECs. For example, ECs become less responsive to the effects of collagen as they mature (Aciniegas, Pulido, & Pereyra, 1988). However, the phenotypic expression of mature ECs is not static and can be altered. Nonembryonic ECs can be induced to form new vessels by exposure to different types of ECM in vitro (Madri & Williams, 1983; Nicosia, Bonanno, & Smith, 1993). It is clear that mature ECs retain the embryonic capacity to rapidly proliferate and differentiate, and their dynamic interaction with the ECM continues to be an important modulator of these changes.

Cell adhesion molecules

The ECM and surrounding cells communicate with ECs through specific cell adhesion molecules (CAMs). There are at least three specialized subgroups of CAMs, and within each subgroup there are regional differences (Albelda & Buck, 1990; Albelda, 1991). Each subgroup of CAMs can transmit specific extracellular signals across the endothelial membrane to initiate a cascade of signal transduction to alter gene expression (Damsky & Werb, 1992; Elisabetta, 1993).

Cell–substratum adhesion molecules, a subgroup of CAMs, allow the basal surface of the EC to interact with the components of the basement membrane and ECM. These substratum molecules and the other subgroups of CAMs undergo developmental changes during the maturation of the ECs. For example, during vascular morphogenesis, immature ECM and mature ECM require different substratum molecule groups of integrins (Albelda & Buck, 1990; Albelda, 1991). Endothelial precursors have specific integrins for fibronectin, and mature ECs produce other integrins specific for collagen and laminin. Likewise, the initial attachment of endothelial precursors to the immature ECM is dependent on specialized focal contacts composed of unique integrin molecules. After attachment, other integrin molecules maintain the adhesion between the mature ECs and the ECM (Elisabetta, 1993; Davis, 1993).

There are also developmental changes of cell–substratum adhesion molecules similar to developmental integrin subgroup changes found in endothelial precursors. For example, there is decreased production of endothelium-derived adhesion molecules during the formation of endocardial cushion tissue (Crossin & Hoffman, 1991). These observed changes are important in the development and the physiological role of the immature and mature ECs. Aberrations of CAMs have been shown to alter vasculogenesis and prevent the formation of vessels (Drake, Davis, & Little, 1992; McCormick & Zetter, 1992).

Cell–cell adhesion molecules allow adjacent ECs to communicate with each other to maintain the integrity of the vascular lining (Albelda, 1991). Aberrations of these molecules will alter the cytoskeleton of the ECs (Takeichi, 1990). This will lead to distortion of the shape of the EC and disrupt the vascular permeability of the lining. As with cell–substratum molecules, different cell–cell adhesion molecules appear during vascular morphogenesis (Albelda, 1991; Newman et al., 1990). A platelet–endothelial cell adhesion molecule-1 may be involved in regulating cell–cell contact in immature, but not mature, ECs (Albelda, 1991; Newman et al., 1990).

Luminal–cell adhesion molecules on the apical surface of the mature EC regulate the interactions of the circulating cells, hormones, and growth factors with the endothelium (Smith, 1993). During inflammation, specific molecules signal circulating cells to recognize and bind to the injured ECs; then other molecules regulate the cell transendothelial migration into the ECM. Spatial and temporal expression of luminal molecules by mature ECs may differ from those of endothelial precursors (Bevilacqua et al., 1989). One can postulate that there are also developmental subgroups of luminal–cell adhesion molecules for the different hormones and growth factors that are essential for the normal maturation of the EC.

Growth factors

Besides interacting with the ECM and circulating cells, CAMs also interact with growth factors to modulate the development and physiological function of the ECs (Schwartz & Liaw, 1993). A variety of growth factors and inhibitors have been isolated (Klagsbrun & D'Amore, 1991; Schwartz & Liaw, 1993). Some of these growth factors are produced by the ECs. This suggests that the developing EC may promote its own growth during vessel formation, regeneration, and healing and may influence the maturation of other cells and ECM. Certain growth factors, such as Int 2 and HST, appear to be expressed only during embryonic vascular morphogenesis, whereas other factors, such as angiogenin, are more essential for the mature EC and reparative processes (Folkman & Klagsbrun, 1987; Schwartz & Liaw, 1993).

There is also a highly precise spatial and temporal relationship between growth factors and growth factor receptors during EC differentiation and maturation (Schwartz & Liaw, 1993; Peters, De Vires, & Williams, 1993; Eichmann et al., 1993). An example is the relationship of vascular endothelial growth factor (VEGF) with its receptors (Weiner, Weiner, & Swain, 1987; Breier et al., 1992). VEGF may be a specific angiogenic mitogen for vascular ECs (Peters, De Vires, & Williams, 1993; Wilting et al., 1993; L. B. Jackman et al., 1993). VEGF transcription levels are highest during embryonic and postnatal brain development. During these periods, two VEGF receptors (Quek 1 and Quek 2) have distinct patterns of expression during vascular morphogenesis and maturation that regulate the effects of VEGF (Eichmann et al., 1993).

Besides growth factors, there may also be different growth inhibitors that influence the proliferation of immature and mature ECs (Klagsbrun & D'Amore, 1991). These growth inhibitors may in part provide an important mechanism preventing the development of an abnormal circulation during vascular morphogenesis and reparative processes.

Mechanical forces

The development and maturation of ECs are regulated by soluble factors and influenced by pressure, shear stress, and cyclic stretch (Bevan & Laher, 1991; Nerem et al., 1993; Langille, 1993). The apical surface of the EC senses the changing mechanical forces of the circulating blood, and the basal surface senses the force produced by the ECM. Each EC experiences a different intensity of these forces depending on its location in the regional vasculature (Nerem et al., 1993). A normal hemodynamic blood flow pattern is required for embryonic vessel formation (Clark, 1918; Rychter, 1962). Aberrations in shear stress can alter the morphology of the EC, cytoskeleton of the EC, gene expression of the EC, and metabolic function of the EC (Levesque & Nerem, 1985; Lamontagne, Pohl, & Busse, 1992; Girard & Nerem, 1993). The orientation of embryonic ECs can be changed by altering the pattern of blood flow (Icardo, 1989; Davies et al., 1992). Alterations in venous blood return are also associated with abnormal heart development (A. C. Gittenberger–de Groot, personal communication). The sensors and mediators of mechanical forces may involve multiple mechanisms, including ionic channels and vasoactive products that may in turn alter gene expression (Rychter, 1962; Rubanyi et al., 1990; Bevan & Laher, 1991; Davies et al., 1992). Changes in gene expression may lead to alterations in the development, maturation, and function of the EC.

Mechanical forces continue to influence and remodel mature vessels throughout the life span of the organism. Mechanical forces may play an important role in pathogenesis (Levesque et al., 1986; Dzau & Gibbons, 1993). Normal shear stress helps to regulate the basal production of vasoactive substances by ECs. Acute changes in blood flow elicit different endothelium-derived substances that alter the diameter and tone of the vessel (Frangos et al., 1985; Rubanyi, Romero, & Vanhoutte, 1986). In addition, chronic changes in shear stress will alter the ECM and the function of the vascular smooth muscle (Moncada, Palmer, & Higgs, 1991; Wang & Prewitt, 1993). Therefore, the integrity of the endothelium and vessels is dependent on normal mechanical forces. Derangements of normal forces may result in a pathological state.

The role of the extracellular matrix in pathogenesis

The complexity of angiogenesis factors that are produced by the microenvironment and that direct the development of ECs helps explain the heterogeneity of mature ECs. ECs differ not only in morphology but also in physiological function. Endothelium-derived biological substances, such as endothelium-derived nitric oxide (EDNO, formerly called endothelium-derived relaxing factor; Moncada, Palmer, & Higgs, 1991), can produce multiple and overlapping effects. Therefore, the alteration of one endothelium-derived substance can produce a cascade of abnormal effects in a pathological condition. The remaining text will highlight the developmental aspects of the diverse physiological roles of the endothelium and its potential role in diseases.

Coagulation and hemostasis

The endothelium plays an important role in both the coagulation and the hemostasis system. The endothelium has pro- and anticoagulant and pro– and anti–platelet aggregation properties to maintain the patency of the blood vessel and repair injuries (Crutchley, 1987; Fajardo, 1989; Crossman & Tuddenham, 1990). Although we have extensively studied these pathways in the mature organism, little is known of their developmental aspects. Even in the mature organism the production of coagulation factors varies in different regions of the vasculature. For example, mature small vessels produce more tissue-type plasminogen activator (t-PA) than large vessels (Kluft, 1990). Therefore, in mature organisms the regional differences in production of t-PA and other endothelium-derived substances may predispose certain vessels to thrombotic or bleeding disorders. This predisposition may have occurred during the normal maturation and development of the organism. For example, during hemostasis, platelet adhesion is dependent on von Willebrand factor (vWf), which is produced by ECs (Jahroudi & Lynch, 1994). The expression of vWf, like t-Pa, varies between species and even within the same organism (Turner, Beckstead, & Warnke, 1987). The mechanism for the differences in regional production of vWf is not completely understood, but during development vWf expression is limited to certain embryonic vessels. This suggests that heterogeneity of vWf expression and perhaps of other biological endothelium-derived substances may be dependent on the developmental lineage of EC precursors.

Immunity

Similar to the coagulation and hemostasis cascades, the endothelium plays an important role in the immune system. The endothelium can express histocompatibility leukocyte class I and class II antigens (HLA). HLA expression exhibits both species and intraspecies differences and may be both developmentally and nondevelopmentally regulated (Page et al., 1992). The endothelium appears to synthesize class I antigens constitutively, whereas the expression of class II antigens requires stimulation (Suzuki & Kashiwagi, 1989; Prober & Cotran, 1991). The endothelium can present processed antigens to circulating T cells to initiate an immune response. Therefore, the endothelium is in a unique position to recognize and process antigens and thereby contribute to the pathogenesis of autoimmune disease and rejection of organ transplants and grafts of vessels (Cherry et al., 1992; Taylor et al., 1992; Frampton et al., 1993). The investigation of developmental induction of class I and II antigens by the endothelium may lead to a better understanding of self-tolerance and graft-versus-host diseases.

ECs are also important in the development and migration of the pre–T cells to memory cells (Ager, 1990; Prober & Cotran, 1991). During the maturation of pre–T cells, ECs composing the high endothelial venules (HEVs) in the lymphatic system direct the transmigration and transformation of the pre–T cells. Both the morphology and the phenotype of T cells are altered depending on whether they are influenced by the apical or the basal surface of the HEV. Therefore, there are even differences between physiological roles of the apical and basal surfaces of the ECs.

Vascular tone

The endothelium produces many vasoactive substances, both vasodilators and vasoconstrictors, to regulate the basal tone of a vessel. The production and effects of these vasoactive substances change during the development of regional circulations. The endothelium can synthesize and release many different endothelium-derived vasodilating factors, EDNO (Moncada, Palmer, & Higgs, 1991).

Production of EDNO by endothelium is important in the regulation of pulmonary and systemic vascular tone, and its effects are developmentally dependent. In experimental models, endogenous EDNO production is lower in the pulmonary circulation in the late gestational fetus and rapidly increases during the neonatal period (Abman et al., 1991; Shaul, Farrar, & Magness, 1993). Differences of endogenous production of EDNO may regulate the physiological role of EDNO in the pulmonary circulation. Aberrations of EDNO endogenous production by pulmonary ECs during the transitional circulation from fetus to adult may result in persistent pulmonary hypertension of the newborn (PPHN) (Abman et al., 1990; Fineman et al., 1994). In fetal lamb models the inhibition of endogenous EDNO production leads to marked physiological changes that are similar to those seen in PPHN. In experimental models this condition can be partially reversed by stimulating endogenous EDNO production. Perhaps in the future we can detect which infants are at risk for deficiency in endogenous production of EDNO and in utero or postnatally alter the natural course of PPHN.

Whereas decreased production of EDNO in the transitional pulmonary circulation can lead to marked physiological aberrations, excess production of EDNO can contribute to pathogenesis. For example, in septic shock, there is an abnormal vasodilation of the systemic circulation. Endotoxin can stimulate ECs to produce more EDNO and augment systemic hypotension (Weitzberg, 1993). This hypotension may be partially reversed by specific EDNO synthesis inhibitors. We do not know how excess endogenous EDNO affects the developing organism in utero. For example, in the chick embryo acute exogenous nitric oxide results in dramatic venodilation and reduced cardiac output (Bowers, Tinney, & Keller, 1996). However, we do not know how a chronic increase in EDNO production will affect the morphological or physiological development of regional vasculature.

The endothelium can also synthesize endothelium-derived vasoconstricting factors, such as endothelin-1 (ET-1) (Filep, 1992). It can also degrade ET-1 to regulate its local concentration and its effects (H. L. Jackman et al., 1993). The hemodynamic effects of ET-1 are very complex and are regulated through a cascade of different receptor subtypes found on the surface of the EC, intracellular pathways, and vasoactive mediators (Wong et al., 1993a, 1993b; Wong, Fineman, & Heymann, 1994). ET-1 can produce both vasodilation and vasoconstriction. Its effects are dependent on the development and maturation of the regional circulation and vascular tone. In the pulmonary vasculature, ET-1 produces vasodilation in the fetal circulation vasoconstriction in the adult circulation. However, ET-1 can produce vasodilation when the adult pulmonary vascular tone is increased. The complex physiological role of ET-1 may in part be dependent on the maturation and availability of the specific ET-1 receptor subtypes (Wong et al., 1993c; Wong, Fineman, & Heymann, 1994).

When the endothelium is injured, regulation of the hemodynamic effects of vasoactive substances such as ET-1 is altered, and a pathological response occurs. For example, when the endothelium is denuded or injured, the pulmonary vasodilation effects of ET-1 are lost and are replaced by vasoconstriction (Wong et al., 1994). Another example is the development of pulmonary hypertension associated with congenital heart disease (Rabinovitch et al., 1986). Here, chronic increase of blood flow induces EC injury and dysfunction. In an in utero fetal lamb model, simulating increased pulmonary blood flow resulted in marked EC histological and physiological dysfunction (Fineman, Reddy, & Wong, 1994). In these lambs, the effects of pulmonary endothelium-dependent vasodilators, such as EDNO, were attenuated (Fineman, Reddy, & Wong, 1994). Thus, the effects of pulmonary endothelium-dependent vasoconstrictors, such as ET-1, were augmented in these animals (Wong et al., in press). Therefore, the physiological aberrations in ECs secondary to increased mechanical forces and to responses to vasoactive products produced by the ECs may in part mediate the pulmonary hypertension found in children with congenital heart disease. Future studies are required to detect the cellular aberrations that occur in these injured ECs. It is also not known whether these injured ECs resume their normal physiological functions after the injury is repaired.

Neovascularization

The proliferation and growth of mature ECs at rest are usually slow, but the ECs retain the capacity to proliferate and alter their phenotypic expression during wound healing. When the vessel is repaired, antiangiogenesis factors direct the ECs back to their resting state. However, this tight control of vascular formation is disrupted in microangiopathic disorders and in neovascularization of tumors. With malignant transformation, a tumor requires a vascular network to provide nutrients to sustain its rapid growth. Multiple angiogenesis factors and CAMs have been found in tumor ECs that are not found in normal ECs. Some of these molecules found on tumor cells can be found in EC precursors (Ruiter et al., 1993; Berger et al., 1993, Leenstra et al., 1993; Tang et al., 1993). Most new blood vessels produced by tumors are initiated in venules and do not form a competent or organized network (Denekamp, 1993). Future studies of tumor ECs that have many characteristics of EC precursors may lead to new avenues for cancer therapy by prevention of tumor growth and selective cytotoxic chemotherapy. These studies may also provide insights into angiogenesis and vasculogenesis during normal development.

The unanswered questions

The maturation of the EC is a complex process, involving many different mechanisms molding each individual EC into a specialized cell with specific physiological functions. The endothelium is not a homogeneous organ but has many different regional activities. The role and effects of each type of EC must be evaluated in the EC's own microenvironment. The future challenges are to determine how different lineages of ECs develop and mature. Questions that need to be answered include whether and how the timing of developmental changes of EC lineage can influence the future health of the organism. In diseases that are secondary to EC dysfunction, will in utero alteration of the developing pluripotent endothelium provide treatment for disease? In addition, can the metabolic capacity of EC precursors be reactivated in mature endothelium in vivo to prevent or help treat diseases? Lastly, can the injured EC regain normal physiological function after healing?

7

Embryonic cardiovascular function, coupling, and maturation: A species view

BRADLEY B. KELLER

Introduction: Scope and definitions

Many investigators have stated that structure and function are related during embryogenesis (Spranger et al., 1982; Clark, 1990). This common insight reflects centuries of investigation of developmental mechanisms using descriptive methods and through the testing of hypotheses related to biological interactions. Over this long period of time, much has been described regarding the basic physiologic parameters of circulating blood (or hemolymph) volume, heart rate (HR), and blood pressure and flow. Few, however, of the underlying mechanisms that determine these relationships have been defined. One explanation for the lack of fundamental insights into cardiovascular functional maturation is that most experimental protocols are designed and analyzed with respect to a single functional cardiac parameter (e.g., HR) of a single cardiac segment (e.g., atrium, ventricle, or vascular bed) (Keller, 1995).

Current research on the mature circulation supports the definition of the cardiovascular system as a closed system, where the input, function, and output of each of the elements are interrelated to accomplish cardiovascular regulation (Sunagawa, Sagawa, & Maughan, 1984). We still need to define embryonic cardiovascular function within an integrated, interactive framework to expose the fundamental mechanisms that regulate embryonic cardiovascular function and influence cardiovascular form. The goal of this chapter is to provide the reader with a multifaceted framework for the investigation, interpretation, and correlation of cardiovascular function during primary cardiovascular morphogenesis.

The embryonic cardiovascular system is a biomechanical system, and, as such, its function is defined by a set of intrinsic variables (e.g., blood pressure, circulating blood volume), a set of relationships that define interactions between these variables (e.g., Laplace's relationship between vessel radius and vascular resistance), a set of morphogenetic determinants (e.g., embryonic stage of development, species), and extrinsic environmental variables (e.g., temperature, oxygen partial pressure), as indicated in Table 7.1. It is important to note that physiologic data in the embryo must be interpreted within the context of species, stage of development, and both intrinsic and extrinsic physiologic variables. As important, the method by which the physiologic data are obtained is crucial in the correlation of results between studies. In this fashion, much has been discovered regarding embryonic cardiovascular functional maturation, providing a fascinating window into one of the central developmental events in embryogenesis.

Historically, the primary experimental model for the investigation of cardiac morphogenesis and function has been the chick embryo (Patten & Kramer, 1933; Barry, 1948), and

Table 7.1. *Embryonic cardiovascular functional determinants*

Intrinsic factors	Extrinsic factors	Interactions	Morphogenetic determinants
Blood volume	Temperature	Biomechanical	Species
Red cell mass	Energy resources	Hydraulic	Developmental stage
Heart rate	Oxygen supply	Biochemical	Genetic heterogeneity
Blood pressure	Teratogens	Veno–atrial	Stochastic events
Blood flow	—	Atrial–ventricular	—
Vascular impedance	—	Ventricular–vascular	—

despite the small size of the embryonic chick heart, many investigators have developed experimental methods to accurately measure blood pressure, blood flow, and chamber size and to alter cardiac function or form (Tazawa & Mochizuki, 1976; Clark & Van Mierop, 1989; Keller et al., 1991a). During these early stages of cardiac development, the vertebrate heart has no functional innervation, has primitive atrioventricular (AV) and conotruncal cushions, and has no functioning conduction system or coronary arteries. Therefore, the embryonic vertebrate cardiovascular system responds to increasing hemodynamic demand using load-sensitive, integrated tissue and cellular mechanisms (Keller, Hu, & Tinney, 1994). In contrast to the early focus on the avian embryonic circulation, cardiovascular functional maturation is now under investigation in a broad range of species, including the human embryo. As might be expected of critical processes during embryogenesis, avian and mammalian embryonic cardiovascular physiology are similar at comparable stages of development (Nakazawa et al., 1988). In the following discussion, chick embryos are referred to at specific stages of development selected to represent doublings of embryo weight (Hamburger & Hamilton, 1951; Clark & Hu, 1982). At stage 12, wet embryo weight is 2.22 mg and the chick embryo heart is just beginning to loop; by stage 27, the wet embryo weight has increased 67-fold to 149 mg. The geometric increases in embryo body mass and heart mass that occur with development can be described using power functions relating embryo and heart mass to incubation time (Romanoff, 1967; see also Chapter 16, this volume). Mouse and rat embryos are referred to at specific days of development selected to parallel chick development (Sissman, 1970; Thieler, 1989).

Embryonic cardiovascular morphometry

A rigorous description of cardiovascular structure is crucial for the interpretation of functional data in the developing embryo. The morphometry of the developing vertebrate heart has been extensively studied in several species by scanning electron microscopy (Pexieder & Janecek 1984; Pexieder, 198( ). The microstructure of the embryonic heart has been extensively detailed using confocal laser microscopy (Borg, Gay, & Johnson, 1982; Terracio & Borg, 1988). In the chick embryo, the embryonic right and left atria increase in size and have a large intra-atrial communication until stage 29 (Hay & Low, 1970). In the normal right-side-up position in ovo, the posterior, superior embryonic atrium is obscured by the anterior, inferior, and rightward ventricle (Figure 7.1). However, in order to image the beating atrium, the chick embryo can be repositioned left side up without altering embryonic HR. Coincident with ventricular growth, atrial maximum and minimum cross-sectional areas increase

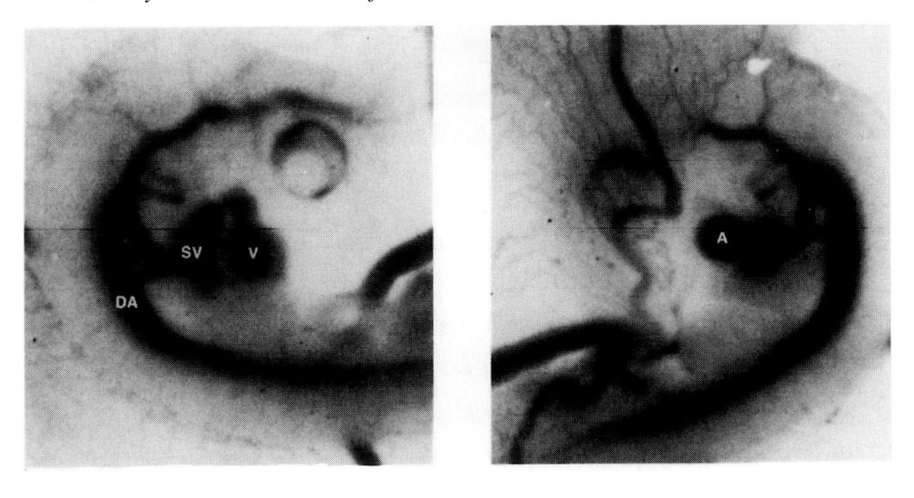

RIGHT SIDE UP LEFT SIDE UP

Figure 7.1 Stage 18 chick embryo, right side up and left side up (DA = dorsal aorta; SV = sinus venosus, V = ventricle, A = atrium).

linearly ($y = 0.10x - 1.41$, $r = .89$, $P < .05$; and $y = 0.05x - 0.67$, $r = .82$, $P < .05$; respectively) in the chick embryo from stage 16 to 24 (Campbell et al., 1992).

Video microscopy and planimetry of the beating embryonic chick left atrium reveal a smooth-walled endocardium, a spherical shape, and a distinct atrial contraction phase (Van Mierop, 1979; Campbell et al., 1992). The embryonic ventricle has a smooth-walled endocardium until stage 17, when the heart wall begins to trabeculate. Dynamic ventricular dimensions are accurately measured by cinephotography and videography (Faber, Green, & Thornburg, 1974; Keller, Hu, & Clark, 1990), and video recordings can be combined with simultaneous measures of ventricular pressure to quantify diastolic and systolic chamber mechanics (Keller, Hu, & Clark, 1991; Keller, Hu, & Tinney, 1994). The embryonic vascular bed can also be imagined and measured, revealing a progressive increase in vessel segment length and diameter during vascular bed expansion (Shafer et al., 1992). Imaging of the embryonic chick vascular bed is then useful for quantifying vasodilation in response to paracrine stimulation by compounds such as atrial natriuretic peptide or isoproterenol (see Chapter 8).

Depending on the species, the embryonic heart rapidly transforms from a simple, pulsatile tube into the mature, three-, four-, or five-chambered structure before any other organ begins to function. From scanning electron microscope studies, we know that the architecture of the developing ventricular wall in numerous species progresses through several phases (Pexieder et al., 1984; Pexieder & Janecek, 1984; Vuillemin & Pexieder, 1989a, 1989b). During cardiac looping, the endocardium-lined tube is smooth, with cushions at the AV orifice and at the junction of the primitive ventricle and the conotruncus (see Chapter 10). As development progresses, the common ventricular chamber expands, the smooth endocardial luminal surface is transformed into a meshwork of trabeculae, but the outer, compact layer of myocardial cells remains relatively thin. It is important to note that there are species differences in the internal and the external morphology of the embryonic heart; for example, an interventricular sulcus is present in the murine embryo but absent from the avian embryo at comparable developmental stages (Figure 7.2). At the completion of ventricular septation, the trabeculae have condensed and the free walls have thickened to achieve the characteristics of the mature ventricle. To date, the tissue and cellular mechanisms that trigger the morphologic transformation from a smooth-walled to a trabecular heart are unknown.

Embryonic cardiovascular functional maturation: An overview

Although the embryonic heart is morphologically different from the mature heart, intriguing physiologic similarities and differences exist. In addition, at comparable early points in cardiovascular development, diverse species exhibit functional similarities. As in the mature heart, the embryonic cardiac cycle includes temporally overlapping atrial filling and ejection phases, ventricular filling and ejection phases, and, in some species, pulsatile vascular flow. The following discussion approaches the embryonic cardiovascular system segmentally, following the circulation of blood from the venous to the arterial beds. The reader must recognize, however, that blood flow to and from the developing heart is directly related to the functional demands of the embryo and also to the structural and functional transitions that occur at metamorphosis, hatching, or birth.

Embryonic cardiovascular function is influenced by simultaneous maturation at numerous organizational levels: gene, protein, cellular organelle, cell membrane, extracellular matrix, and tissue. Thus, several chapters in this book directly relate to the functional maturation of the cardiovascular system. The embryonic myocardium is electrochemically active, and the interactions between the developing myocardial sarcolemma, sarcoplasmic reticulum, and contractile architecture are reviewed in Chapter 2. The myocardial ionic environment affects ventricular diastolic and systolic function, HR, and rhythm. It is extremely important to recognize the electrochemical heterogeneity between species, and further investigation of the mechanisms and rationale for this heterogeneity is needed. For example, Na^+–Ca^{2+} exchanger activity in rat myocytes differs from that in guinea pig myocytes (see Chapter 2). The cascades of gene switching and protein isoform transitions that transform the immature myocyte into the mature contractile unit are described in Chapter 3. Again, there are temporal and species-specific transformations in the developing contractile apparatus that determine the functional characteristics of the myocyte and the heart. Once the embryonic myocyte contracts, it is equally important to understand how the generated force is transmitted to the surrounding environment (Chapter 5), how the system of blood vessels coupled to the heart arises (Chapters 5 and 6), and how circulating vasoactive substances regulate cardiovascular performance (Chapter 8). Finally, a fundamental understanding of the functional impact of the cardiovascular system on oxygen uptake and delivery is crucial in evaluating normal and experimentally altered cardiovascular function (Chapter 10). With these broad areas related to the maturing cardiovascular system detailed elsewhere in our book, the following text provides an overview of the functional maturation of embryonic cardiovascular systems. Limited examples are provided for several species to highlight the similarities and differences that exist in cardiovascular systems developing in unique environments.

Atrial function

Atrial conduit, reservoir, and pump functions directly affect ventricular function (Yellin, Peskin, & Frater, 1972). Because atrial pressure remains low during avian cardiogenesis

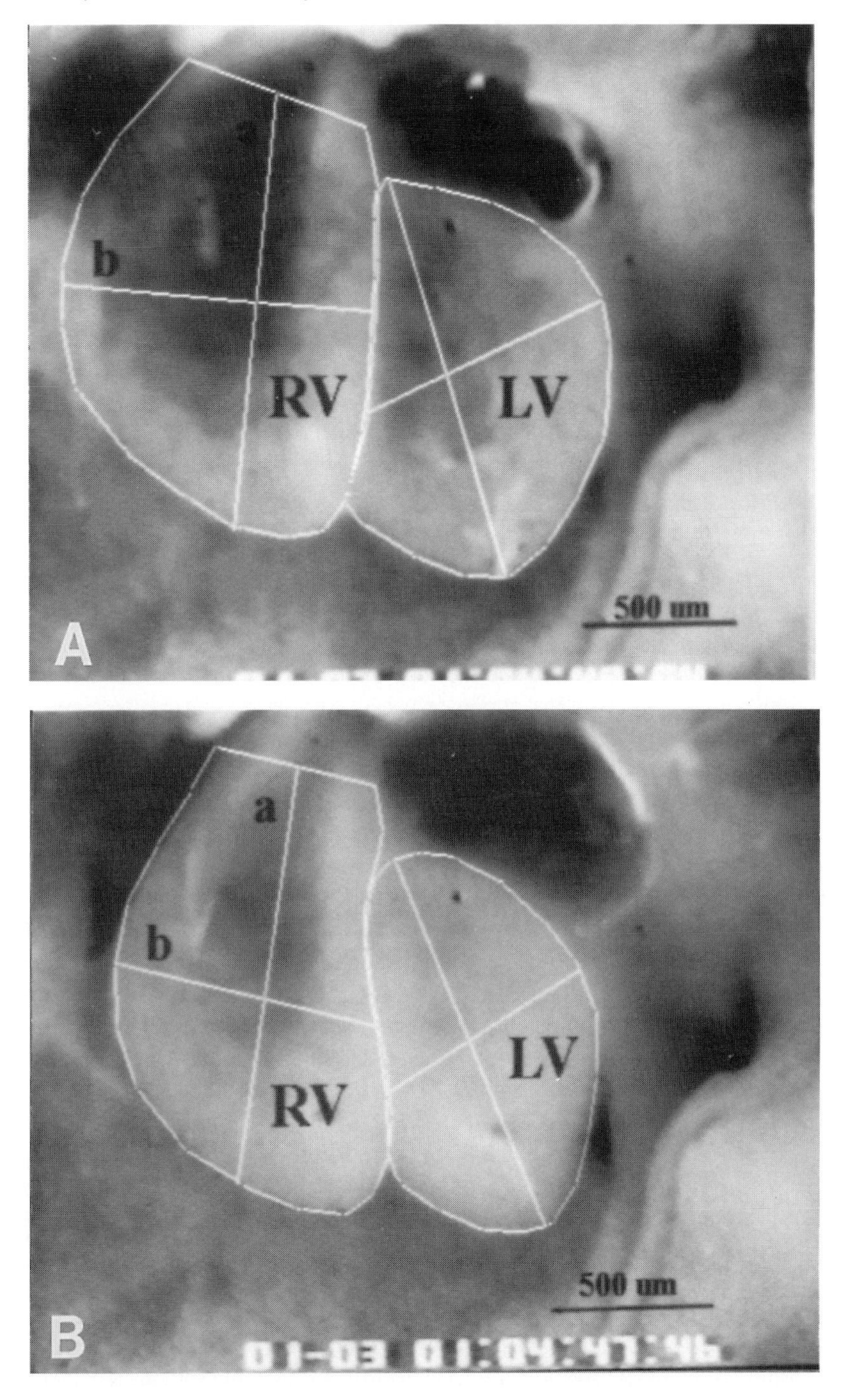

Figure 7.2 Representative photomicrographs of ventral view of the embryonic day 12.5 embryonic mouse heart at end-diastole (a) and end systole (b). The ventricular epicardial borders and both major (*A*) and minor (*B*) axes have been manually planimetered (RV = right ventricle, LV = left ventricle). Scale bar = 500 μm.

and the embryonic atrium is obscured by the anterior ventricle in the chick embryo, we have little quantitative information on the influence of atrial function on ventricular diastolic and systolic function in the embryonic circulation (Hu, Tinney, & Keller, 1994). Clearly, functional parameters such as HR, atrial–ventricular conduction velocity, atrial compliance, and atrial contractility influence atrial–ventricular coupling and the diastolic filling of the embryonic heart.

Atrial morphogenesis

Blood flows into the embryonic chick atrium through multiple venous entrances and influences atrial septum formation (Sol, 1981). Venoatrial anatomic relationships change during cardiac morphogenesis with the progressive absorption of the sinus venosus into the right atrium. Prior to atrial septation, the sinoatrial chamber orifice (the future right atrium) functions as a conduit and the left atrium as a reservoir.

Atrial pacemaker function

HR is the single parameter of cardiovascular function determined across the broadest range of species (see also Chapter 14). This is due to the crucial role of HR and activation sequence in ventricular function and the ease and reproducibility of measures of HR. The embryonic sinus venosus is the dominant pacemaker during cardiac development in the chick embryo (Kamino, Hirota, & Fujii, 1981). At the onset of contraction, the HR is approximately 60 bpm (beats per minute) (Kamino, Hirota, & Fujii, 1981), and from stage 12 to 29, HR increases from 103 to 208 bpm (Hu et al., 1991). Of interest, both systolic and diastolic time intervals change proportionately with cycle length (Hu, Keller, & Clark, 191). Alterations in HR significantly affect cardiac performance (Nakazawa et al., 1986; Benson et al., 1989; Cuneo, Hughes, & Benson, 1991). It is also important to note that embryonic pacemaker function may be affected by the method of HR determination: noninvasive measures of HR yield higher rates than measures performed following opening of the eggshell (Haque et al., 1994). HR and AV activation sequence can be acutely altered in the chick embryo by environmental hyperthermia and hypothermia, sinus venosus pacing, thermal probe application to the sinus venosus, and digoxin-induced heart block.

Environmental hypothermia decreases HR in the chick embryo (Nakazawa et al., 1985) and mouse embryo (B. Keller, unpublished data), and hyperthermia increases HR (Nakazawa et al., 1986) in the chick embryo, consistent with findings in other vertebrate embryos and in larvae (Chapter 17). Environmentally induced alterations in HR decrease net cardiac output. Sinus venosus pacing above the intrinsic rate decreases cardiac output by reducing passive ventricular filling, and pacing from the ventricle reduces output drastically (Dunnigan et al., 1987). In response to environmental hypothermia in chick embryos, cycle length increases, though there is a fixed relationship between systolic and diastolic time intervals and cycle length (Hu, Keller, & Clark 1991). With the increase in diastolic filling time associated with bradycardia, active filling time remains constant, increasing the percent passive filling time versus active filling time. As with many studies in the developing embryo, these cardiac time intervals have not been investigated at ventricular preloads and afterloads, which also change in response to altered cycle length.

Thermal probe application to the sinus venosus acutely increases or decreases HR without altering activation sequence in the chick embryo (Cuneo, Hughes, & Benson, 1991).

Predictably, end-diastolic pressure is related to cycle length (Zimmerman et al., 1991). The maximum rate of ventricular pressure fall is also related to cycle length, and as cycle length decreases, the rate of pressure fall reaches an asymptote, consistent with reaching the maximum rate of calcium reuptake or the time constant of viscoelastic relaxation (Cheanvechai, Hughes, & Benson, 1992). The thermal probe technique can also be used to determine ventricular pressure–volume (PV) characteristics during alterations in cycle length. In response to acutely altered HR, end-diastolic volume and stroke volume are linearly related to cycle length (Casillas et al., 1993). In contrast to the mature heart, developed pressure is inversely related to end-diastolic volume at large end-diastolic volumes.

The serial application of digoxin to the embryonic heart produces progressive heart block, beginning at the conotruncal cushions and proceeding proximally (Paff, Boucek, & Klopfenstein, 1948). Regional variations in myocardial conduction velocities facilitate forward blood flow in the embryonic heart, though the role of altered conduction velocity on embryonic cardiovascular morphogenesis and function has not been determined. Recognizing that cycle length shortens with development in most species, the finding in numerous studies that maximum cardiac output occurs at the intrinsic HR of each embryonic stage suggests a dynamic optimization of the timing of atrial–ventricular and ventricular–vascular coupling. In contrast to the mature circulation, the preinnervated embryonic heart does not rely on HR regulation to acutely adjust cardiac output (Benson et al., 1989). The lack of compensatory cycle length change during acute preload alteration suggests that embryonic pacemaker function may be developmentally regulated and may not be acutely load sensitive. However, a sudden increase in atrial preload is associated with abrupt bradycardia in the chick embryo (B. Keller, unpublished data). The specific mechanisms responsible for matching embryonic HR to the hemodynamic demand and mechanical characteristics of the rapidly developing cardiovascular system are largely unknown. HR variability has been recognized for almost a century to reflect autonomic regulation of pacemaker function. The chick embryo lacks HR variability during primary cardiac morphogenesis, then acquires HR variability following coupling of the parasympathetic nervous system to the heart (Kempski et al., 1993). HR variability increases significantly toward the completion of development (Tazawa et al., 1991; see also Chapter 16).

Atrioventricular blood flow

The AV cushions function as a valve to prevent retrograde flow during ventricular systole (Barry, 1948). Atrial peak systolic pressure increases with stage and exceeds ventricular pressure throughout ventricular filling (Hu & Keller, 1995). Multiple separate flow streams course from the venous circulation through the atria and AV canal into the developing ventricle (Nishibatake et al., 1990). Unequal maturation and division of the AV cushions drastically alter ventricular growth and morphogenesis (Sweeney, 1981). As in the mature heart, ventricular filling can be partially characterized by measuring the velocity profile of flow across the AV cushions. Ventricular filling has distinct passive and active phases, and the ratio of passive to active filling increases sharply in the chick embryo from stage 12 to 18 and then increases gradually to stage 27 (Hu et al., 1991). Total diastolic filling volume increases by 200%, and passive ventricular filling decreases from 92% to 24% of total volume (Hu et al., 1991). In rat embryos at comparable developmental stages, pulsed Doppler measures of blood flow reveal similar results (Nakazawa et al., 1988). Interestingly, in contrast to chick and rat embryos, the mouse embryo does not have a prominent passive filling profile, which is perhaps related to differences in murine passive myocardial stiffness (B. Keller, unpublished data). The human embryo at comparable stages reveals distinct passive and active ventricular filling phases and a geometric increase in blood flow with development (Wladimiroff, Huisman, & Stewart, 1991; Wladimiroff et al., 1992).

Simultaneous atrial and ventricular pressure gradients

Ventricular diastolic filling is directly influenced by atrial systole and by a dynamic resistance to atrial ejection produced by changes in the orifice of the AV canal. Simultaneous atrial and ventricular blood pressures in stage 16–27 chick embryos reveal a pressure gradient between the atrium and ventricle throughout ventricular diastole (Figure 7.3; Hu & Keller, 1995). Peak atrial pressure increases geometrically from 0.38 to 1.21 mmHg, and ventricular end-diastolic pressure increases linearly from 0.18 to 0.55 mmHg. Ventricular diastolic passive and active phase mean pressure gradients increase from 0.23 and 0.20 mmHg at stage 16 to 0.52 and 0.62 mm Hg at stage 27, respectively. Thus, the pressure gradient across the AV orifice is a resistance point that may regulate passive atrial conduit function and may act as an atrial "afterload" to influence atrial maturation. Changes in AV resistance to forward flow likely influence ventricular development (Lau, Sagawa, & Suga, 1979).

Atrial pump function

Changes in the atrial systole profile are detailed in the atrial PV loop and reflect the dynamic coupling of the atrium to the ventricle (Lau & Sagawa, 1979). The loop consists of a clockwise passive filling curve corresponding to atrial filling followed by atrial contraction. Embryonic atrial pump function can be similarly defined by PV loops (Hu, Tinney, & Keller, 1994). When left atrial pressure of the stage 24 chick embryo is plotted against simultaneous left atrial volume, our preliminary data show that the result is a dual loop sim-

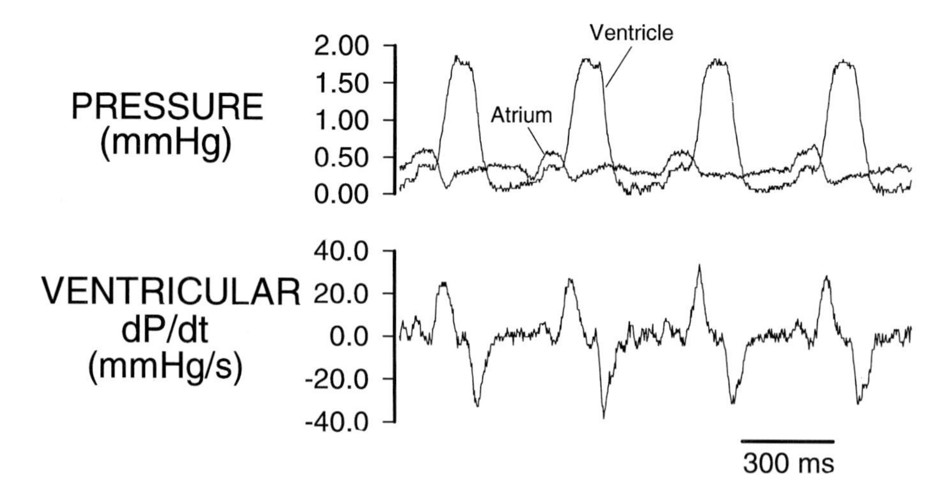

Figure 7.3 Simultaneous atrial and ventricular pressures in a stage 24 chick embryo. Note that atrial pressure exceeds ventricular pressure through ventricular diastole.

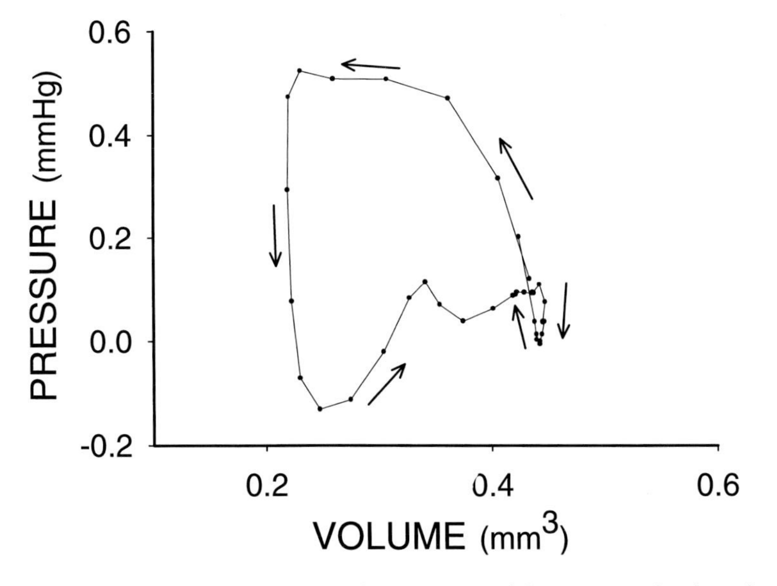

Figure 7.4 Relationship between simultaneous atrial pressure and volume in a stage 24 chick embryo. Arrows indicate the direction of change of atrial pressure and volume during the cardiac cycle. Cycle length = 405 ms.

ilar to the mature atrium (Figure 7.4). This represents the first application of PV mechanics to the embryonic atrium. Future studies are needed to apply this method of simultaneously determining atrial pressure and volume to the definition of atrial compliance, conduit, and reservoir function during developmental and experimentally altered atrial loading.

Vasoactive atrial natriuretic protein

In the mature circulation, atrial natriuretic polypeptide regulates renal and cardiovascular function (Pegram et al., 1986). The developing cardiovascular system is sensitive to exogenous atrial natriuretic peptide exposure (Nakazawa et al., 1990a; Hu et al., 1994; see also Chapter 8, this volume) and to the oversynthesis of atrial natriuretic protein (Barbee et al., 1994). Despite the wide distribution of atrial natriuretic protein in the embryonic atrium and ventricle, the endocrine effect of this protein is predominantly peripheral venodilation (Nakazawa et al., 1990a). In response to atrial natriuretic peptide, passive ventricular filling decreases 49%, active filling remains relatively constant, and dorsal aortic stroke volume decreases by 33% (Hu et al., 1994). Thus, venodilation increases venous capacitance such that blood from the periphery is partially sequestered, reduces atrial reservoir and conduit function, and decreases ventricular filling, with the net result of decreased cardiac output. Much smaller releases of atrial natriuretic protein in vivo likely modulate venous tone and venous–atrial coupling.

Ventricular function

Ventricular function depends on atrial–ventricular interactions, intrinsic ventricular properties, and ventricular–vascular interactions. Ventricular function has been studied in the chick embryo for decades, though the analysis of ventricular function coupled to measures of preload and/or afterload has occurred only recently (Keller, Hu, & Tinney, 1994). Embryonic ventricular diastolic and systolic properties have been studied in an ever expanding range of species, including the *Xenopus,* zebra fish, frog, chick, rat, and mouse, under various acute and chronic experimental protocols, as detailed in the following paragraphs.

Ventricular morphogenesis

One obvious prerequisite to understanding embryonic cardiovascular functional maturation is a basic understanding of cardiovascular morphogenesis (Pexieder, 1986). Embryonic cardiovascular structure and function are interdependent as the heart tube loops and septates, forming the three-, four-, or five-chambered heart, depending upon the species of interest. Blood streams into the embryonic chick atrium through multiple venous entrances, influencing the process of atrial septum formation (Sol, 1981). Venoatrial anatomic relationships change during cardiac morphogenesis with the progressive absorption of the sinus venosus into the right atrium. Prior to atrial septation, the sinoatrial chamber orifice functions as a conduit, and the left atrium as a reservoir. Ventricular morphogenesis in most species involves the rapid growth of the ventricular chamber associated with a transformation of the endocardial surface from a smooth wall to a spongy trabecular surface. Morphogenesis of vertebrate AV communications and ventricular chambers is also related to the coupling of the developing heart to vascular structures perfusing developing lungs (cutaneous exchange organs) and to the systemic circulation (Burggren & Just, 1992; see also Chapter 13, this volume). Alterations in the geometry, material properties, or contractile function of the structures attached to the embryonic ventricles (e.g., the conotruncus in the chick embryo) likely influence both ventricular morphology and ventricular function. Central to these variations in cardiac anatomy are the transitions from embryo to larva to mature animal that occur in many species. Thus, data on embryonic cardiovascular function *must always be correlated with morphologic data and with an understanding of the intact, coupled cardiovascular system for the specific species under investigation.*

Ventricular diastolic maturation

Although the definition of diastole is controversial, ventricular diastole can be generally divided into periods of ventricular relaxation and ventricular filling. The study of ventricular relaxation has been limited in the embryonic heart, whereas numerous studies have analyzed ventricular filling characteristics.

Ventricular relaxation

Following each cardiac systole, the contracted ventricular chamber undergoes a process of relaxation followed by ventricular filling. Diastolic relaxation is determined by both calcium sequestration and viscoelastic properties (Mirsky, 1974; Morgan et al., 1988; Gilbert & Glantz, 1989). The embryonic myocyte has well-developed sarcolemmal calcium transport but little functional sarcoplasmic reticulum (Vetter & Will, 1986; see also Chapter 2, this volume). With maturation of the embryonic myocyte, developed force increases, the systolic ejection period decreases, and the maximum rate of ventricular pressure fall (t) increases (Kaufman et al., 1990; Hu et al., 1991; Cheanvechai, Hughes, & Benson, 1992).

When cycle length is experimentally decreased at a given stage, the rate of pressure fall reaches an asymptote, consistent with reaching a maximum rate of calcium uptake or the time constant of viscoelastic relaxation (Cheanvechai, Hughes, Benson, 1992). Following conotruncal banding, developed ventricular pressure increases, ventricular growth accelerates, yet the maximum rate of ventricular pressure fall is unchanged, suggesting that the maximal rate of myocardial Ca^{2+} handling may be developmentally regulated (Hu et al., 1992). Chronic treatment with verapamil, a sarcolemmal Ca^{2+} channel blocker, slows ventricular growth rate without affecting morphology or distribution of subcellular organelles during primary cardiac morphogenesis (Clark et al., 1991). Ventricular relaxation rate is also prolonged following chronic verapamil treatment, suggesting that sarcolemmal Ca^{2+} transport is involved in embryonic ventricular relaxation (Hu, Keller, & Clark, 1992).

Ventricular filling

Ventricular filling is determined largely by ventricular preload and the viscoelastic properties of the developing ventricular chamber. The early embryonic heart contains a thick layer of cardiac jelly, and embryonic myocytes lack the organized-fiber geometry of the mature heart (see Chapter 5). The dramatic change in ventricular morphology from a smooth to a trabecular wall likely allows myocardial mass to increase while stiffness declines. Biomechanical modeling should be able to test hypotheses that this geometric transformation facilitates increased ventricular filling and ejection rates, thus maximizing performance.

In addition to the maturation of the ventricular endocardial surface, differentiation also occurs within and around myocytes. Myofibrils increase in number and alignment, as do ventricular dimensions (Manasek, 1970; Clark et al., 1986). Interactions of myocytes with extracellular matrix and cell-to-cell adhesions certainly affect the mechanical characteristics of the ventricle. Prior to the trabeculation of the heart, the biomechanical force from the myocardial mantle is transmitted through cardiac jelly to the blood contained in the ventricular lumen (Nakamura & Manasek, 1981). As the embryo grows, myocytes are woven together by intricate cell–matrix and cell–cell adhesions, limiting wall motion and thus altering the local myocardial properties of the heart (Borg, Gay, & Johnson, 1982).

Ventricular filling has been studied using pulsed Doppler, servonull pressure, and video microscopic techniques during normal development in both chick and human embryos (Hu et al., 1991; Wladimiroff et al., 1992). To date, the viscoelastic properties of the embryonic heart have not been quantified (Taber, Keller, & Clark, 1992). Average diastolic ventricular wall stiffness, calculated from the ratio of ventricular pressure and volume during diastolic filling, decreases geometrically with development (Hu et al., 1990). However, the embryonic ventricle is approximately 1000-fold stiffer than the mature left ventricle (Mirsky, 1974). As preload varies, the passive ventricular PV filling relationship is determined by ventricular geometry and material properties, physical constraints, and AV coupling. Ventricular filling begins with a negative intraventricular pressure (diastolic suction), and the diastolic PV curve reflects both passive and active ventricular filling (Keller, Hu, & Tinney, 1994). The AV canal is patent during passive ventricular filling and atrial contraction. As development progresses, end-diastolic ventricular PV relations reflect increasing ventricular compliance, and in contrast to the mature heart, the pseudoelastic properties of the embryonic heart resemble a rubber shell with a curvilinear filling profile that reaches an asymptotic pressure with increasing distension, in contrast to the increasing pressure associated with distension of the mature mammalian left ventricle.

Ventricular systolic maturation

One of the fundamental characteristics of the embryonic cardiovascular system is the coordination of global improvements in pump performance with dramatic structural maturation at all levels. HR, developed pressure, and stroke volume increase during cardiac morphogenesis across a broad range of species. Cardiac output increases geometrically to match increasing hemodynamic and metabolic demand. Many hemodynamic studies have been performed with the chick embryo (Van Mierop & Bertuch, 1967; Girard, 1973a; Tazawa, 1981b; Clark & Hu, 1982; Keller et al., 1991a, 1991b), and comparable results have been obtained with rat (Nakazawa et al., 1988), amphibian (Hou & Burggren, 1995a, 1995b), mouse (B. Keller, unpublished data), and reptile embryos (Warburton, Hastings, & Wang, 1995), as well as with invertebrates (Spicer, 1994) and fishes (Hoar, Randall, & Farrell, 1992). Physiologic data on the human embryonic cardiovascular system acquired by noninvasive techniques are now becoming available (Wladimiroff et al., 1992). Measures of cardiovascular function were first developed and validated for the mature heart (Sagawa et al., 1988a). Ventricular pump function can be defined by preload-dependent measures such as cardiac output from "Starling" pump function curves; contractility-based indices (dP/dt_{max}, dr/dt_{max}); and time-varying systolic elastance (E_{max}) (Sagawa et al., 1988a). Equipment developed to measure blood pressure and flow in isolated vessels has been adapted to the embryonic heart and has enabled these functional measures to be applied to the embryonic circulation. The fact that the 1 mm diameter, stage 21 chick embryo ventricle generates a pressure waveform identical to that of the mature vertebrate left ventricle suggests a conservation of myocardial mechanics independent of scale. The embryonic cardiovascular system increases its performance geometrically during simultaneous rapid growth of the heart, though the mechanisms that directly regulate this interaction between growth and function have not yet been detailed.

Measures of systolic function

Isometric relations. The embryonic vertebrate heart contracts sequentially, from atrium to ventricle to conotruncus, facilitating forward flow. Yet atrial, ventricular, and arterial pressure and flow waveforms define the embryonic heart as pulsatile rather than peristaltic. Simultaneous PV measures show that the embryonic heart undergoes periods of isometric contraction prior to ejection, and isometric relaxation prior to filling, similar to the neonatal and mature heart. Just as anisotropy and nonuniform myofiber alignment cause the mature heart to twist during isovolumic contraction and relaxation, epicardial microsphere experiments show that surface deformations also occur during isometric contraction and relaxation in the stage 16 chick embryo (Taber, Keller, & Clark, 1992). This twisting of the embryonic heart during ventricular contraction and relaxation could be the result of nonuniform myocardial depolarization or nonuniform myocardial material properties due to regional differences in extracellular matrix content and myofiber distribution. Thus, the combined analysis of embryonic myocardial material properties and myofiber alignment will provide crucial data on the dynamic myocardial environment.

Preload-dependent measures. Contractile function is influenced by myocyte architecture, energy state, and load. Preload-dependent indices of embryonic ventricular function include peak systolic pressure and $+dP/dt_{max}$ (Clark et al., 1986), end-systolic pressure (Keller et al., 1991a), stroke volume (Faber, Green, & Thornburg, 1974; Clark et al., 1986; Keller et al., 1991b), the velocity of circumferential fiber shortening (Baldwin, Lloyd, & Solursh, 1994), and the slope of the stroke volume versus end-diastolic volume relationship (Keller, Hu, & Tinney, 1994). Of note, embryonic myocytes lack a functioning sarcoplasmic reticulum; thus, alterations in Ca^{2+} handling do not explain increased systolic performance (Vetter & Will 1986; see also Chapter 2, this volume). In addition, during early myocardial hyperplasia, embryonic myocytes have limited myofiber maturation and orientation (Clark et al., 1986; see also Chapter 3, this volume). To use the chick embryo as an example, during normal cardiovascular development, HR, developed pressure, $+dP/dt_{max}$, and stroke volume increase to match metabolic demand (Clark & Hu 1982, see also Chapter 16, this volume). Using both ventricular pressure and dimension measures allows a more accurate determination of end-systole and reveals that by stage 18, end-systole occurs significantly later than peak systolic pressure (Figure 7.5; Keller et al., 1991a; Keller, Hu, & Tinney, 1994). At baseline, peak velocities of circumferential shortening are similar in the chick embryo from stage 16 to 21 (B. Keller, unpublished data). Baldwin & Solursh (1994) measured ventricular diameter and reported velocities of fiber shortening in day 9.5 rat embryos after 24 hours of culture, noting significant increases in shortening velocity after hyaluronidase degradation of the cardiac jelly (Baldwin & Solursh, 1994).

End-systolic PV relations. Owing to rapid changes in HR and ventricular load during cardiovascular development, the application of a load-independent time-varying elastance model to describe embryonic ventricular function would be particularly useful (Maughan et al., 1985a, 1985b; Sagawa et al., 1988a). In contrast to the relatively linear end-systolic PV relationship in the mature mammalian left ventricle, the end-systolic PV relationship of

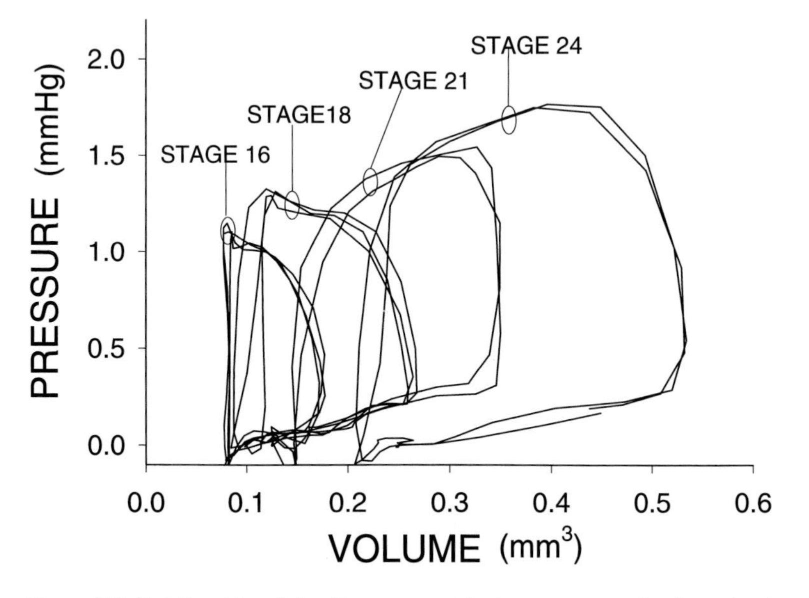

Figure 7.5 Relationship of simultaneous ventricular pressure and volume in stage 16–24 chick embryos.

the puppy left ventricle is markedly curvilinear (Suga et al., 1986). In fact, the mature end-systolic PV relationship is nonlinear when measured over large preload ranges, and the curvilinearity of the relationship is contractility dependent (Kass et al., 1989). Not surprisingly, the embryonic end-systolic PV relation is extremely curvilinear, much more so than for the puppy or the newborn lamb (Teitel et al., 1991). In the embryonic chick, large increases in end-diastolic volume result in decreased end-systolic pressure (Keller et al., 1991b). Increasing end-diastolic volume results in a greater increase in stroke volume than in developed pressure (Casillas et al., 1993; Keller, Hu, & Tinney, 1994). The extreme curvilinearity of the embryonic end-systolic PV relationship likely reflects rapid changes in afterload (Keller, Tinney, & Hu, 1992). When ventricular ejection is prevented by acutely constricting the conotruncus with a 10-0 nylon suture, end-systolic pressure increases dramatically versus the "sham operated" embryos at matched end-systolic volumes (Figure 7.6; B. Keller, unpublished data). Thus, the embryonic chick heart has significant contractile reserve produced by a combination of increased ventricular ejection and reduced ventricular afterload. In fact, embryonic ventricular chamber function and arterial tone change so rapidly that the end-systolic PV relationship cannot be used as an isolated measure of embryonic myocardial contractility and is of limited value in quantifying global cardiovascular performance in the embryo.

Energetics. The calculation of embryonic myocardial energetics has been limited by the small size of the developing heart. In the mature heart, the area within the PV loop and end-systolic pressure volume relationship correlates linearly to direct measures of myocardial oxygen consumption (Suga, Yamada, & Goto, 1984). Embryonic ventricular PV loop area

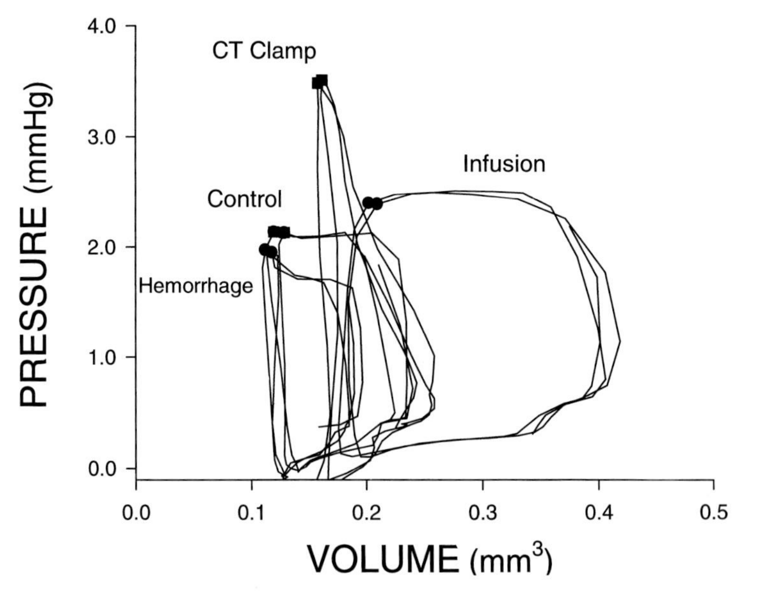

Figure 7.6 Ventricular PV loops during volume infusion, venous hemorrhage, and acute conotruncal constriction in a stage 21 chick embryo. Note that ventricular volume increases significantly with volume infusion, resulting in an increase in stroke volume without a dramatic increase in developed pressure. In contrast, acute conotruncal constriction results in a significant increase in developed pressure and a reduction in stroke volume.

increases from 0.07 to 0.31 mmHg·mm^{-3} from stage 16 to 24 (Keller, Hu, & Tinney, 1994). As the embryonic cardiovascular system performs "work," the mechanical efficiency of the system as a pump can be assessed in terms of energy consumed for "work" performed. PV loop area per minute indexed for embryo mass is similar for these stages, suggesting that myocardial energy consumption is indexed to embryo mass (Keller, Hu, & Tinney, 1994). In the mature mammalian heart, the energy supply to the myocardium is always limited (Sagawa et al., 1988b); however, it is important to recognize that myocardial energy supply occurs via simple diffusion in the thin-walled embryonic myocardium prior to trabecular compaction. Further studies regarding the coupling of embryonic cardiovascular function to myocardial energetics are needed to define the functional range and regulation of myocardial energy supply during primary cardiovascular morphogenesis.

Cycle length effects

As discussed earlier, embryonic HR increases with development and cardiac output is maximal at intrinsic HR for a given stage. Environmental temperature alters embryonic HR dramatically (Nakazawa et al., 1985, 1986), providing the ideal experimental paradigm to enchant elementary-school children with the wonders of science (Keller & Keller, 1995). Sinus venosus pacing 25% above the intrinsic rate decreases cardiac output by 50%, and pacing from the embryonic ventricle is rapidly lethal (Dunnigan et al., 1987). The major effect of altered cycle length is to alter diastolic filling time, subsequently influencing ventricular filling (Figure 7.7). End-diastolic volume is linearly related to diastolic filling time (Casillas et al., 1993).

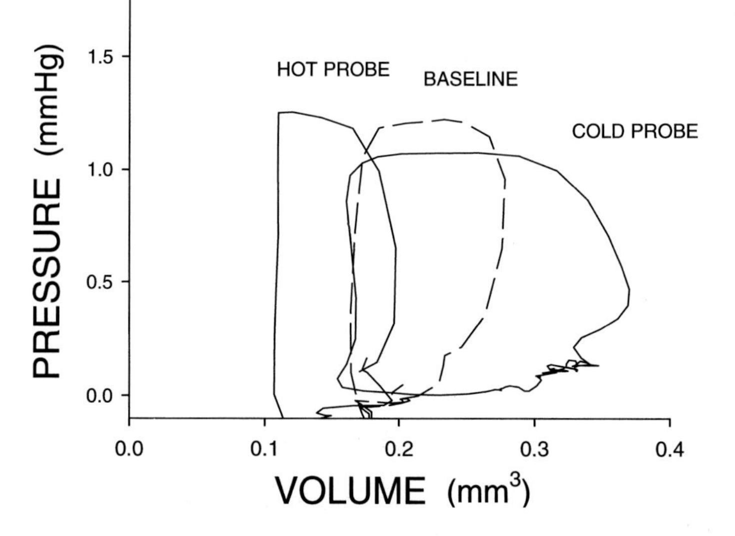

Figure 7.7 Stage 21 ventricular PV loops during thermal probe. Dashed line represents PV data at baseline cycle length, and solid lines represent data after cold-probe-induced bradycardia or hot-probe-induced tachycardia.

Preload effects

Preload can be acutely altered experimentally in the chick embryo by volume infusion, volume withdrawal, venous hemorrhage, and peripheral vasodilation. Infusion of a physiologic buffer into the vitelline vein increases ventricular developed pressure and stroke volume (Wagman, Hu, & Clark, 1990), and infusion into the sinus venosus allows the calculation of end-diastolic and end-systolic pressure–area and PV relations (Figure 7.8; Keller et al., 1991b; Keller, Hu, & Tinney, 1994). Preload infusion reverses the decreases in stroke volume produced by pacing-induced tachycardia (Benson et al., 1989). Alterations in preload define the length dependence of embryonic myocardium and provide data for the calculation of myocardial material properties.

Afterload effects

Ventricular systole includes two interrelated events: force generation and the ejection of blood into the arterial tree. Both the extent of force generation and the duration of systolic ejection are influenced by afterload. The embryo acutely decreases afterload to increase cardiac output in response to increased preload (Keller, Tinney, & Hu, 1992). This adaptation of afterload to maintain or increase stroke volume occurs almost instantaneously in the absence of a functional autonomic nervous system. Thus, tissue-level stretch-sensitive mechanisms must be involved in the regulation of beat-to-beat cardiovascular function. Because of the profound influence of alterations in afterload on embryonic cardiovascular

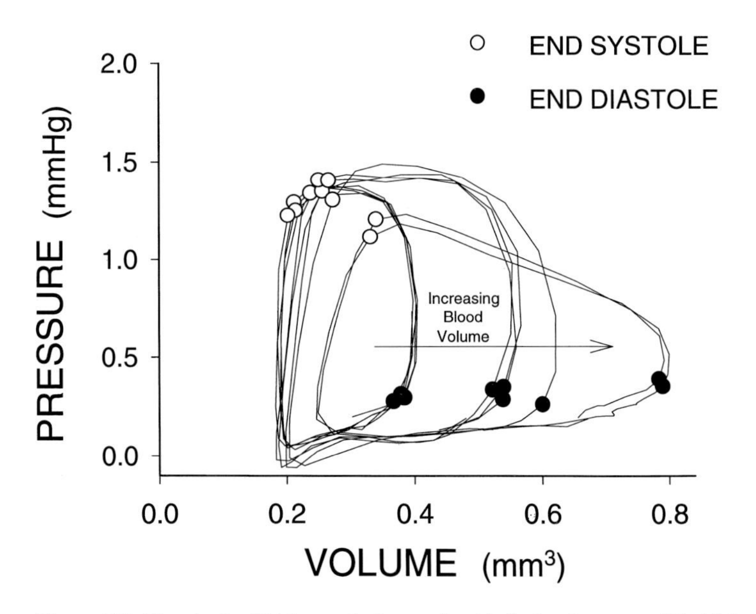

Figure 7.8 Ventricular PV loops during preload infusion in a stage 21 chick embryo. Individual end-diastolic points are identified by filled circles, and end-systolic points by unfilled circles. Note the decrease in end-systolic pressure that occurs at large increases in end-diastolic volume. The end-diastolic PV relationship is linear, and the end-systolic PV relationship is curvilinear downward.

performance, the following section discusses afterload within the broader context of ventricular–vascular coupling.

Ventricular–vascular coupling

In vivo, the heart is dynamically coupled to the arterial tree, and thus, the definition of global cardiovascular function requires the definition of ventricular–vascular coupling. The ventricle develops pressure, which is transmitted to the arterial tree during the ejection of a volume of blood (stroke volume). Thus, ventricular interactions with the loading system are defined with the same PV measures that define ventricular function (Sunagawa, Sagawa, & Maughan, 1987; Yin, 1987).

Circulating blood volume

Coincident with primary cardiovascular morphogenesis is a rapid expansion of circulating blood volume. Initially, red cells develop in blood islands in the chorioallantoic membrane of avian embryos and in the mesoderm component of the visceral yolk sac of murine embryos (Kaufman, 1992). The blood islands become incorporated within developing blood vessels prior to the onset of an effective circulation. Circulating blood volume and red cell mass then expand geometrically (Barnes & Jensen, 1959; Kind, 1975; see also Chapter 15, this volume). The dramatic increase in circulating blood volume is associated with increased oxygen delivery to the rapidly growing embryo. As stated by Tazawa and Hou in Chapter 15, embryo mass increases faster than circulating blood volume, though the ratio of blood volume to body mass is much greater in the embryo than in the mature animal. As described in the following sections, acute alterations in circulating blood volume can affect global cardiovascular function.

Arterial hemodynamics

As with ventricular function, aspects of vascular hemodynamics, including arterial pressure, stroke volume, mean resistance, impedance, and pulse wave velocity, have been calculated in a wide range of species. The embryonic arterial pressure waveform contains a dicrotic notch, consistent with wave reflections, and displays a progressive increase in pulse pressure toward the periphery (Van Mierop & Bertuch, 1967; Clark & Hu, 1982). From stage 16 to 24, the ejection curve of the PV loop becomes progressively curvilinear, and peak ventricular pressure occurs earlier in systole, consistent with increased vascular compliance (Figure 7.6). The time of end-ejection is also influenced by vascular impedance. The end of ejection is also directly influenced by the muscular conotruncus, which is not unique to the developing vertebrate heart. The conotruncus contracts following ventricular systole, resulting in a collapse of the conotruncal cavity that augments the ejection of blood into the aortic sac. Although conotruncal contraction ceases during maturation of the ventricular outflow tract in vertebrates, the conotruncus maintains an important contractile function in fish and amphibians in ejecting blood into the arterial circulation (Chapters 12–13). As with the AV cushions, a pressure gradient exists between the ventricle and the dorsal aorta of the embryonic chick (Hu & Clark, 1989) and rat (Nakazawa et al., 1995). Cardiovascular development is associated with a sequence of aortic arch appearance and selective loss, and the dynamic interaction of pulsatile flow, changing geometry, and alterations in branchial arch extracellular matrix probably influences arch selection.

Arterial impedance

Mean arterial resistance, calculated from simultaneous measures of dorsal aortic pressure and flow, decreases from 5.16 to 0.63 Woods units from stage 12 to 29 (Hu & Clark, 1989). Arterial impedance, calculated from similar measures, reveals a similar drop in the zero-order modulus (bulk resistance) with stage and also reveals a progressive decrease in the high-order moduli (Zahka et al., 1989). These changes likely reflect morphologic changes in the dorsal aorta and vitelline bed that alter vessel caliber and distensibility. Similar correlations exist for the mature circulation (O'Rourke, 1982). Of note, the proportion of hydraulic power expended in pulsatile flow increases from 28% to 65% from stage 18 to 29, versus the 10% normally expended in the mature circulation (O'Rourke & Taylor, 1967).

Arterial impedance can also be determined from the simultaneous measurement of arterial pressure and flow during experimental alterations in circulating blood volume. Alterations in systemic vascular tone can be produced by a range of experimental interventions, including β-adrenergic stimulation (Clark, Hu, & Dooley, 1985; see also Chapter 8, this volume), environmental hyperthermia (Nakazawa et al., 1986), and altered ventricular preload (Keller, Hu, & Tinney, 1994). Recent studies have shown that the embryonic peripheral vascular bed in the stage 24 chick embryo can alter vascular resistance significantly following alterations in as little as 1.3% of circulating blood volume (Figure 7.9; Yoshigi, Hu, & Keller, 1996). In addition, the change in vascular resistance is due to alterations in peripheral vascular resistance, not to proximal arterial compliance. The consequence of altered resistance in response to altered circulating blood volume is to maintain intravascular pressure at the expense of forward blood flow (Yoshigi, Hu, & Keller, 1996). Further studies are needed to determine the local regulation of embryonic vascular tone.

Wave reflections

Pulse wave velocity is calculated from the simultaneous measurement of pressure or flow from two distinct positions in the embryonic arterial bed. Wave velocity is calculated from the time between zero crossings of separate, simultaneous waveforms and a video microscopic measurement of the distance between the sites of measurement. The existence of wave reflections in the vascular network and the site of the reflection(s) can be determined from knowledge of aortic wave velocity and the frequency of the first zero crossing of the impedance phase (O'Rourke & Taylor, 1967). Pulse wave velocity calculated from simultaneous pulsed-Doppler measures of dorsal aortic and vitelline artery blood flow increases from stage 18 to 29 (Zahka, Hu, & Clark, 1987). The increase in pulse wave velocity in the developing embryo may be secondary to increased systolic ejection power, increased diameter and stiffness of the proximal arterial tree, and/or decreased opposing reflected waves from the periphery.

Blood flow distribution

Embryonic blood flow serves both nutrient and morphologic roles (Rychter, 1962; Sweeney, 1981; Yoshida, Manasek, & Arcilla, 1983; Nishibatake et al., 1990). Embryonic

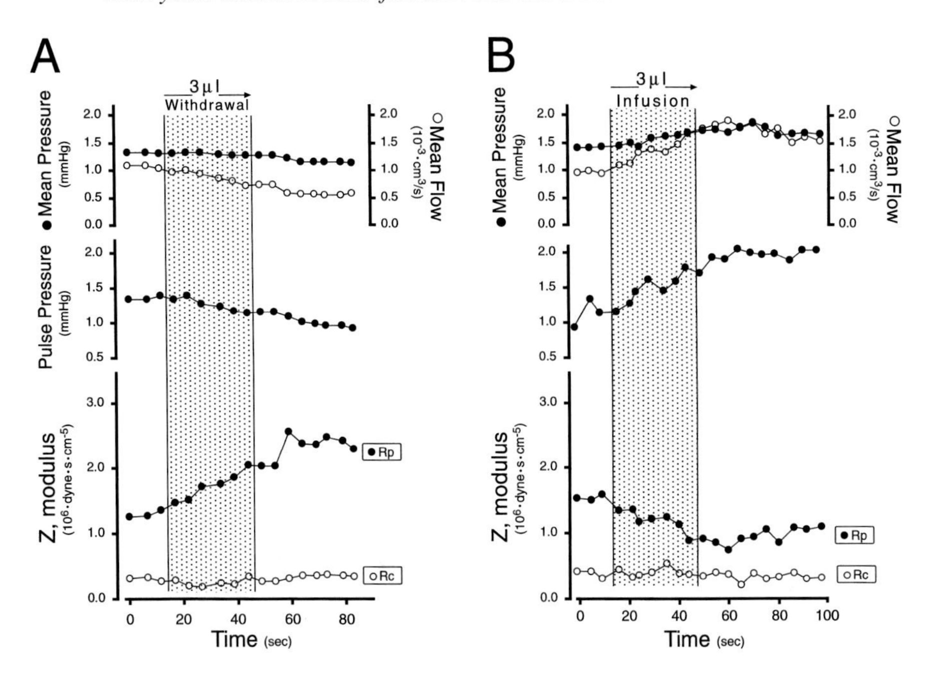

Figure 7.9 Dorsal aortic pressure and flow during a 3 μl alteration in circulating blood volume in the stage 24 chick embryo: (*A*) during a 3 μl withdrawal and (*B*) during a 3 μl infusion. Note that when volume is withdrawn, first blood flow decreases and then blood pressure, and that the magnitude of the decrease in flow exceeds that of pressure. When volume is infused, blood flow increases before blood pressure increases. Note that when pressure and flow data are analyzed for the components of vascular impedance (Z) represented by peripheral resistance (Rp) and proximal resistance (Rc), changes in circulating blood volume alter only peripheral resistance.

blood flow is distributed between the embryonic and extraembryonic circulations in all vertebrate embryos except those of mammals, where blood flow is distributed between the embryo and placenta. Experiments using dye-loaded microspheres and dorsal aortic blood flow measurement reveal that blood flow to the embryo increases from 18% to 34% of cardiac output from stage 18 to 24 (Hu, Ngo, & Clark, 1996). Expressed as blood flow normalized to tissue wet weights, blood flow per gram tissue is consistently higher in the embryo than in the vitelline bed (Hu, Ngo, & Clark, 1996). Following the completion of primary cardiac morphogenesis, allantoic blood flow can be measured in the vessels through the eggshell (Tazawa & Mochizuki, 1976; see also Chapter 15, this volume). Late in development, allantoic blood flow decreases as embryonic growth proceeds.

Arterial elastance

Arterial elastance, calculated from the slope between the end-diastolic volume intercept and the end-systolic PV point, reflects the energy required to transfer blood from the ventricle to the arterial tree (Sunagawa, Sagawa, & Maughan, 1987). In the chick embryo, arterial elastance decreases from –11.3 to –6.4 mm Hg·mm^{-3} from stage 16 to 24, consistent with the decrease in vascular impedance associated with vascular development (Keller, Tinney, & Hu, 1992). Arterial elastance also decreases with acute increases in end-diastolic volume, consistent with acute vasodilation to optimize ventricular–vascular coupling (Keller, Tinney, & Hu, 1992). Normalized arterial elastance versus normalized end-diastolic volume is linear in the embryonic heart, and interestingly, β-receptor stimulation with isoproterenol results in a greater shift in this coupling relationship than in HR. Mathematical modeling of mature ventricular–vascular relations predicts that decreases in peripheral resistance produce the largest increase in ventricular stroke volume (Sunagawa, Maughan, & Sagawa, 1985).

Vasoactive mediators

Many vasoactive drugs can be acutely or chronically delivered to the embryo, stimulating or blocking calcium channels, β-receptors, the nitric oxide pathway, and atrial natriuretic peptide receptors (Chapter 8). Importantly, the effects of these agents often differ from the mature cardiovascular response. The significance of β-receptors prior to functional autonomic innervation is unclear (Pappano, 1977). Topical β-receptor stimulation does not alter embryonic HR. However, acute β-receptor stimulation with isoproterenol increases $+dP/dt_{max}$ and peak ventricular pressure without changing HR (Clark, Hu, & Dooley, 1985). Topical application of isoproterenol causes a dramatic reduction in end-diastolic volume in the stage 21 chick embryo. However, following isoproterenol, end-systolic pressure is higher for matched end-diastolic volumes versus embryos where end-diastolic volume was reduced by venous hemorrhage (Keller, Heller, & Tinney, 1993). The nitric oxide pathway modulates vascular tone in the neonatal and mature cardiovascular systems. Acute stimulation with nitroprusside (2.5 mg, 0.6 mg·kg^{-1}) to the ventricular surface of stage 21 embryos reduced preload while maintaining afterload (Figure 7.10; Bowers, Tinney, & Keller, 1996). The PV ejection profile in response to nitroprusside differs markedly from

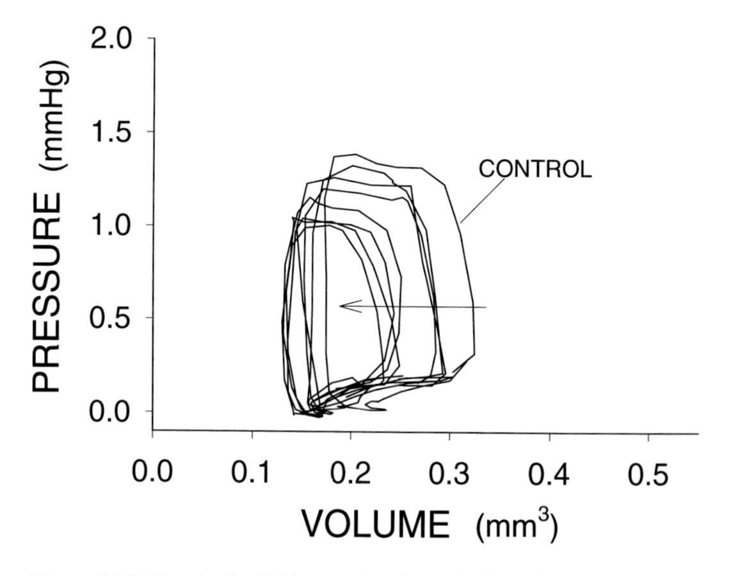

Figure 7.10 Ventricular PV loops after the topical application of 2.5 mg nitroprusside to a stage 21 chick embryo. Individual PV loops represent control baseline data and then data at 15–30 second intervals. Note the progressive decrease in end-diastolic volume that occurs over 180 seconds following nitroprusside dosing.

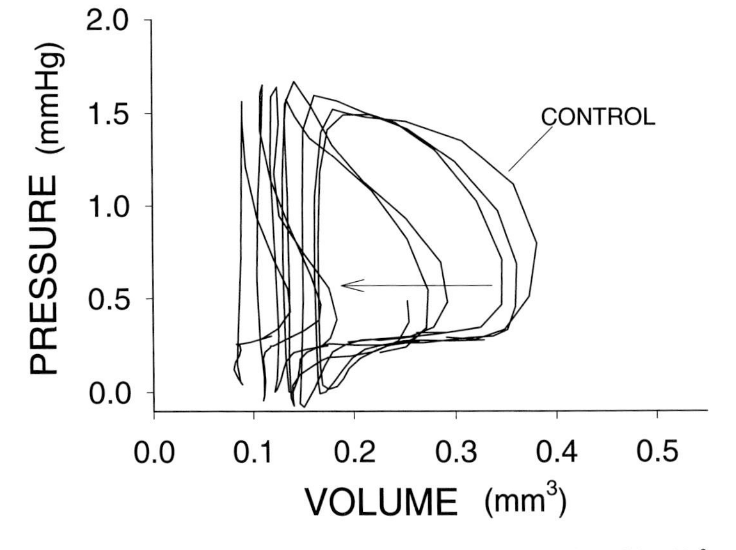

Figure 7.11 Ventricular PV loops after the topical application of 2×10^{-2} mg isoproterenol to a stage 21 chick embryo. Individual PV loops represent baseline data and then data at 15–30 second intervals. Note the progressive decrease in end-diastolic volume that occurs over 180 seconds following isoproterenol dosing. Note the progressive increase in end-systolic pressure following isoproterenol dosing in contrast to the decrease in end-systolic pressure that occurs following nitroprusside dosing.

the response to isoproterenol, suggesting that arterial impedance and venous capacitance may be independently modulated (Figure 7.11; Keller, Heller, & Tinney, 1993). Acute sarcolemmal calcium channel blockade with verapamil causes a dose-related decrease in peak ventricular systolic pressure and mean dorsal aortic blood flow in the stage 24 chick embryo (Thompson, Hu, & Clark, 1986). Thus, acute pharmacologic interventions are beginning to define the role of various ion channels and cell surface receptors for embryonic cardiovascular function and regulation (Chapters 8 and 13).

Functional consequences of ventricular growth acceleration

Many congenital cardiovascular defects include the partial or complete obstruction of ventricular outflow and result in altered chamber size and myocardial mass. In the embryonic heart, ventricular afterload can be surgically increased by conotruncal banding or by unilateral vitelline artery banding. The conotruncus can be acutely constricted by a surgical miniclip or by a loop of 10-0 nylon suture. In the mature heart, isovolumic pressure exceeds ejection pressure for any volume greater than the end-systolic volume (Maughan et al., 1985b). In response to conotruncal banding, the embryonic chick heart acutely increases developed pressure and accelerates ventricular growth velocity (Clark et al., 1989). Of note, the only morphologic event affected by growth acceleration is an increase in the number and thickness of ventricular trabeculae (T. Pexieder, unpublished data). Remarkably, cardiac output is unchanged (Clark et al., 1989; Keller, Hu, & Clark, 1991). The effect of conotruncal banding on atrial function and vascular hemodynamics is unknown. Unilateral vitelline artery banding may be a preferable experimental method to increase afterload because of the preservation of ventricular–conotruncal dynamic interactions. Ventricular hyperplasia results from chronic increased afterload produced by conotruncal banding. In contrast to the hypertrophic response of the mature heart, embryonic ventricular mass increases by myocyte hyperplasia during normal development and in response to conotruncal banding (Clark et al., 1989). As with ventricular hypoplasia, discussed later, quantitative changes in ventricular–vascular interactions and the impact of altered ventricular mass and material properties on cardiovascular regulatory mechanisms are unknown.

Functional consequences of ventricular growth deceleration

Underdevelopment of the right or left ventricular chamber is a common component of structural congenital heart disease (Chapter 19). During embryonic cardiac morphogenesis, ventricular hypoplasia is experimentally produced by chronically decreasing ventricular preload through surgical left atrium partial ligation or unilateral vitelline vein banding. Surgical miniclips (Rychter, 1962) or nylon suture (Sweeney, 1981) can be applied to the embryonic left atrium to reduce atrial compliance and to partially obstruct blood flow though the AV cushions. Atrial compliance directly correlates to ventricular filling in the mature heart (Suga, 1974; Lau & Sagawa, 1979), and thus, partial obstruction to left heart blood flow results in left heart hypoplasia. Because of distinct flow streams from the vitelline venous bed (Nishibatake et al., 1990), unilateral banding of a vitelline vein will also affect AV blood flow.

The result of chronic partial ventricular inflow occlusion is ventricular hypoplasia (Harh et al., 1973). Left ventricular hypoplasia in the chick embryo resembles the clinical spectrum of hypoplastic left heart defects (Sweeney, 1981). In the chick embryo, cardiac output is maintained via a redistribution of blood flow through the primitive right ventricle (N. Hu, unpublished data). However, quantitative changes in regional function and morphometry and the impact of altered blood flow patterns on ventricular–vascular interactions during this process are unknown.

The chronic delivery of vasoactive substances to the developing cardiovascular system has been best defined using the Ca^{2+} channel blocker verapamil (Clark et al., 1991). Chronic calcium channel blockade with verapamil at 1×10^{-3} mg·h^{-1} for 24 hours reduces developed ventricular pressure in stage 21 to 29 chick embryos and results in decreased ventricular growth through myocyte hypoplasia (Clark et al., 1991). However, following chronic verapamil infusion, basic cardiac morphogenesis is unaffected (Clark et al., 1991). Although ventricular function and morphometry have been described following chronic calcium channel blockade, little is known about the underlying mechanisms that maintain cardiac performance. Likewise, the effects of chronic β-receptor stimulation, nitric oxide pathway stimulation and blockade, and atrial natriuretic peptide stimulation on the rate of myocardial growth, on myocardial function, and on global cardiovascular performance are unknown.

Cardiovascular consequences of autonomic nervous maturation

Embryonic cardiovascular function becomes influenced, and likely regulated, by the maturation of sympathetic and parasympathetic nerve and receptor pathways. In the early chick embryo, β-receptors are present before completion of cardiac septation (Pappano, 1977). However, the embryonic cardiovascular response to adrenergic stimulation (e.g., isoproterenol in the stage 21 chick embryo) differs from the "mature" cardiovascular response found in the late embryo. Alterations in myocardial innervation have occurred following experimental neural crest ablation in the chick embryo and are associated with both structural and electrophysiologic cardiovascular anomalies (Nishibatake, Kirby, & van Mierop, 1987; Kirby, 1988a). In addition, numerous experimental models of abnormal cardiac morphogenesis are associated with abnormalities of neural crest migration (Chapter 19). Thus, the developmental time course of myocardial adrenergic and cholinergic receptor expression and coupling and the functional consequences of alterations in autonomic maturation require much additional investigation.

Embryonic cardiac mechanics

Experimental hemodynamic data provide crucial information for the refinement and validation of mathematical engineering models of the functioning cardiovascular system (Sagawa et al., 1988a). Many parameters of cardiovascular function can be directly measured, but because of the lack of experimental data on myocardial material properties, several crucial parameters currently require mathematical estimation (Taber, Keller, & Clark, 1992). These mathematical models allow the estimation of regional differences in wall stress, a parameter that may be critical in the regulation of myocardial growth and morphogenesis. These models also allow the testing of hypotheses related to the influence of alterations in material properties or geometry on myocardial growth or function for which experimental data are not currently available. Using acutely excised embryonic chick ventricles, the passive properties of the embryonic myocardium can be determined (Miller et al., in press). Acutely excised embryonic myocardium also responds to field stimulation, allowing the investigation of excitation–contraction coupling and mechanical restitution in the developing heart (P. Tsyvian, unpublished data). These experiments targeted at cardiac and tissue mechanics and their mathematical modeling will provide crucial data in determining structure–function relationships in the developing myocardium.

Significance and future directions of research

Independent of species, the cardiovascular system is the first functioning component of the developing embryo, and the geometric increase in embryonic hemodynamic demand is matched by increasing cardiovascular function. Despite over a century of investigation, insights into direct relationships between embryonic cardiovascular structure and function are only now becoming apparent. These tissue-level molecular and biomechanical mechanisms require further study in a broad range of experimental paradigms, from the genetically defined invertebrate to the targeted-gene mouse. The determination of cardiovascular function in these models is dependent upon a critical analysis of the interrelated components described in this chapter. This biomechanical foundation will serve as an interface between advances in our understanding of the molecular biology of heart development and interventions that alter the function of the beating, growing heart.

8

Hormonal systems regulating the developing cardiovascular system

MAKOTO NAKAZAWA AND FUSAE KAJIO

Introduction

The cardiovascular system is the first systemic organ to function in ontogenesis, and the stability of the system is essential for normal organogenesis of the embryo. In the mature individual, the autonomic nervous system and the endocrine system are vital for maintenance of the homeostasis of circulation, but both systems mature and begin to function at late stages of organogenesis or after birth. It has been thought that before these systems participate in cardiovascular control, the embryonic cardiovascular system is controlled exclusively by intrinsic characteristics of the heart, such as the Frank–Starling mechanism (Wagman, Hu, & Clark, 1990), and by changes of heart rate and vascular resistance to environmental temperature (Nakazawa et al., 1985; Nakazawa et al., 1986).

In addition to intrinsic mechanisms, atrial natriuretic peptide is present in the embryonic myocardium as early as day 11 of gestation in the rat (Toshimori et al., 1987; Scott & Jennes, 1988). The gene for atrial natriuretic peptide is first expressed at day 8 in mouse embryonic myocardial cells (Zeller et al., 1987), although the amount of the peptide is much less than that in late gestation or in the mature heart (Dolan & Dobrozsi, 1987). Atrial natriuretic factor is present in the human embryo heart at 8–9 gestational weeks, when ventricular septation is almost completed (Larsen, 1990). Atrial natriuretic factor is a potent vasodilator and reduces venous return in the early chick embryonic circulation (Nakazawa et al., 1990a; Hu et al., 1994); therefore, this peptide could have a hemodynamic role via an endocrine or paracrine pathway in the prenephric system.

Catecholamines may also serve a paracrine or endocrine role in the embryonic circulation. Tyrosine hydroxylase positive cells are first detected in the developing brain and in the cells migrating from the neural crest to the periphery on gestational day 11.5–12.5 in the rat embryo (Specht et al., 1981; Jonakait, Rosenthal, & Morrell, 1989; Hall & Landis, 1991). These tyrosine hydroxylase positive cells are potential sources of catecholamines in the embryo and act in a paracrine manner (Zhou, Quaife, & Palmiter, 1995). Dopamine-β-hydroxylase, a marker of noradrenergic and adrenergic phenotypes, is synthesized at gestational day 11.5 in the rat embryo (Black et al., 1981). Furthermore, catecholamines occur in the visceral yolk sac epithelium of the rat as early as gestational day 10 (Schlumpf & Lichtensteiger, 1979). Finally, maternal circulating catecholamines cross the placenta to the embryo, as evidenced from teratogenetic studies using epinephrine (Trend & Bruce, 1989). These facts indicate that catecholamines are present in the embryo soon after the establishment of the circulation and well before the completion of innervation or development of the adrenal gland.

Recently, the endothelin-1 (ET-1) gene has been targeted and knocked out in mice (Kurihara et al., 1994). The ET-1 –/– homozygous mice have ventricular septal defects, outflow anomalies, and anomalies of the aortic arches; these animals die of respiratory failure soon after birth. Heterozygous (ET-1 +/–) mice survive beyond birth and have elevated systolic and diastolic systemic blood pressures associated with lower serum endothelin levels than wild-type mice. The mechanism responsible for producing the congenital cardiovascular anomalies may relate to abnormal development of the cells of pharyngeal arches. It may also be possible, however, that a primary hemodynamic alteration caused the abnormal morphologies. The incidence of anomalies seen in ET-1 –/– homozygous mice increased by treating pregnant mothers with an ET-1–neutralizing monoclonal antibody or a selective ET-A receptor antagonist, BQ123 (Y. Kurihara, personal communication). Therefore, it is likely that maternal endothelin is passed transplacentally to the embryo, and so maternal ET-1 may also potentially contribute to the homeostasis of embryonic circulation. Further studies are needed to define the role of ET-1 in the regulation of embryonic hemodynamics.

Thus, because many hormones and peptides that have potent cardiovascular effects are found in the embryonic tissues, it is important to know if these hormones and peptides play a role in the maintenance of cardiovascular function during early developmental stages. In this chapter, we will review the effects of the above-mentioned catecholamines and peptides on embryonic cardiovascular function prior to functional innervation or development of the endocrine system.

Catecholamines and their effects

The chick embryo

The developmental change of the blood pressure response to epinephrine given topically in chick embryos has been studied from 2 days of incubation (Hamburger–Hamilton [HH] stages 12–13) (Hamburger & Hamilton, 1951) to hatching and in newborn chicks (Hoffman & van Mierop, 1971). The blood pressure response to epinephrine is biphasic: the blood pressure is lowered by the drug in embryos of 2–2.5 days of incubation (HH stages 12–17), it is not altered in 3-day embryos, and it is elevated in all older embryos. The pressor effect is largest in 4- to 6-day embryos, which may be due to the maximal adrenergic sensitivity at incubation days 6–7, which declines thereafter (Girard, 1973). Receptor function has been further characterized by defining the effects of norepinephrine, phenylephrine, and isoproterenol on the vitelline arterial and venous pressure in the day 3 chick embryo (HH stages 18–19; Petery & van Mierop, 1977). Both phenylephrine and norepinephrine elevate arterial pressure, but this effect is abolished in the presence of phenoxybenzamine. A low dose of norepinephrine and isoproterenol reduces the arterial and venous pressure; this effect is blocked by pretreatment with propranolol. The hypotensive effect of isoproterenol is enhanced by phenoxybenzamine. An intravenous injection of isoproterenol and lowers the vitelline vein pressure, decreases ventricular and atrial sizes, and reduces cardiac output (Clark, Hu, & Dooley, 1985; Nakazawa, Miyagawa, & Takao, 1986; Tomita et al., 1989) (Figures 8.1–8.3). These findings indicate that stimulation of the β-adrenergic receptors dilates the peripheral vessels, especially veins, causing blood pooling and reducing venous return to the heart (Keller, Heller, & Tinney, 1993). These hemodynamic effects are correlated with resultant cardiovascular anomalies (Bruyere, Folts, & Gilbert, 1984). These studies and studies by Pappano (1977) clearly demonstrate the presence of both α- and β-adrenergic receptors in the vascular system from very early stages of development and that

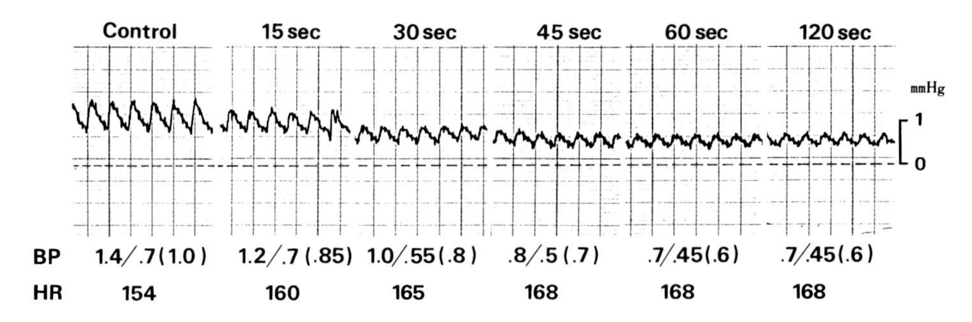

Figure 8.1 Isoproterenol, infused into the vitelline vein, lowers blood pressure in the vitelline artery of stage 21 chick embryos.

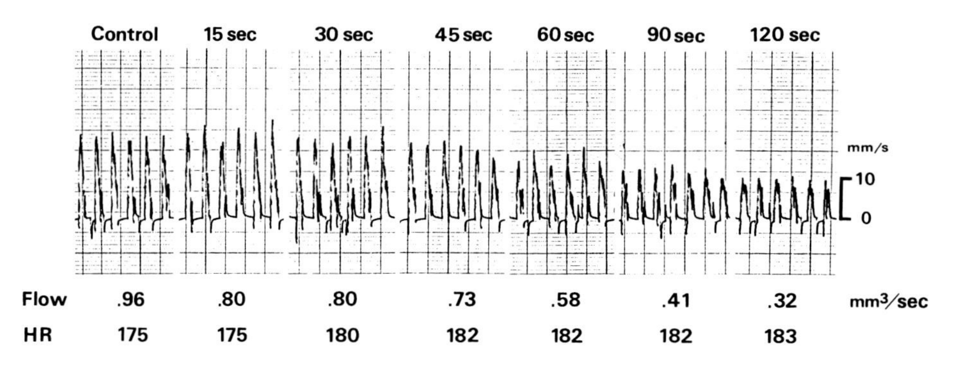

Figure 8.2 Isoproterenol infusion decreases dorsal aortic blood flow, a measure of cardiac output, in stage 21 chick embryos.

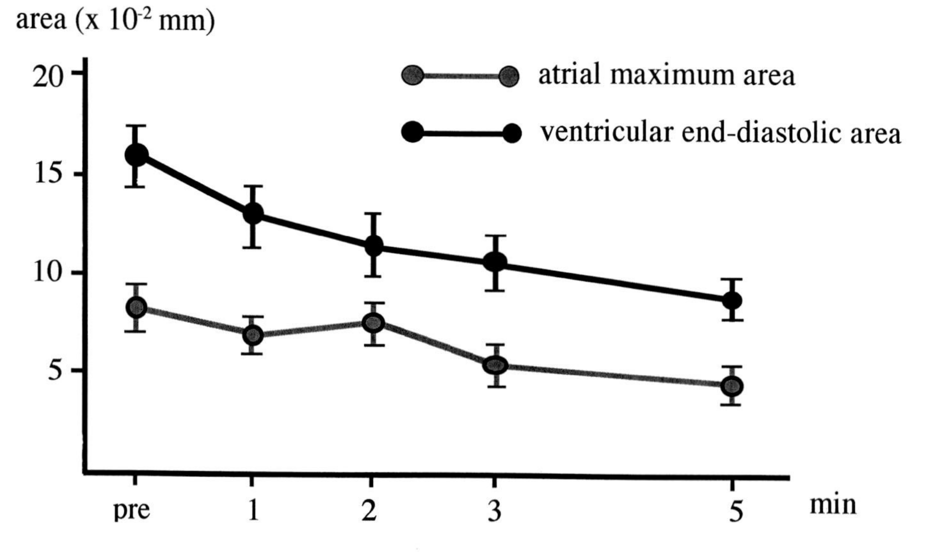

Figure 8.3 Isoproterenol infusion reduces maximum ventricular and atrial cross-sectional area in the stage 18 chick embryo. Cross-sectional area is determined by planimetry of the epicardial heart borders. The time scale shows 1, 2, 3, and 5 minutes after isoproterenol infusion.

β-receptors function earlier than α-receptors. Thus, the embryonic vascular system is capable of reacting to catecholamines, although it is not known how catecholamines actually contribute to homeostasis during these periods.

In contrast to the marked effect of catecholamines on blood pressure and cardiac output, the chronotropic effect of the catecholamines is not significant in the early embryo, though a subtle increase in heart rate following norepinephrine administration occurs in 5-day embryos (HH stage 26; Girard, 1973b). Chronotropic responsiveness is likely drug and dose specific, because a very large dose of isoproterenol results in only a small increase in heart rate in HH stage 21 embryos (Nakazawa, Miyagawa, & Takao, 1986). The lack of significant chronotropic response suggests that adrenergic receptors may not be completely coupled to intracellular second messengers during these stages. The systolic tension of isolated embryonic chick hearts immersed in oxygenated Tyrode's solution has been studied using a transducer to measure myocardial contractile tension. The addition of isoproterenol to the perfusate increases myocardial force in day 3 chick embryos, with increasing sensitivity by day 5 (Frieswick, Danielson, & Shideman, 1979), similar to previous studies where myocardial force was enhanced by β-adrenergic agonist in day 4 chick embryos (Higgins & Pappano, 1981). The positive inotropic effect of isoproterenol is blocked by propranolol. In isolated hearts of day 6 chick embryos, epinephrine increases heart rate, myocardial contractility, and cyclic AMP (Lipshultz, Shanfeld, & Chacko, 1981). Radioligand binding assay of β-adrenergic receptors shows that receptor density peaks at incubation day 9 and then decreases (Stewart, Kirby, & Aronstam, 1986). These studies indicate that β-adrenergic receptors are present in the embryonic ventricle as early as HH stages 18–19. Adrenergic receptor sensitivity and function gradually increase as development proceeds, peak a few days later, and then decrease thereafter. Quantitatively similar patterns occur in the developing amphibian heart (Burggren & Doyle, 1986b; Pelster et al., 1993; see also Chapter 13, this volume). In the mature circulation, the positive inotropic effect of β-receptor stimulation is usually expressed as an enhancement of cardiac performance, such as an increase in cardiac output, but this effect is not present in the embryonic system (Clark, Hu, & Dooley, 1985; Nakazawa, Miyagawa, & Takao, 1986), and the ejection fraction of the primitive ventricle does not increase after dosing with isoproterenol (Tomita et al., 1989). It seems that either the positive inotropism is offset by a decrease in preload or the dose used in the published studies is too small to produce an effect. Additional information on the responses of the chick heart to catecholamines is presented by Tazawa and Hou in Chapter 15.

The mammalian embryo

Isoproterenol enhances myocardial contractility in isolated rat embryonic hearts as early as gestational day 14, and the sensitivity peaks at day 18 and then declines thereafter (Frieswick, Danielson, & Shideman, 1979). Recently, β-adrenergic receptors were identified in gestational day 12 rat embryos, and the receptor concentration is comparable to that in mature target tissues, although the receptor expression is lower in the heart than in the liver (Slotkin, Lau, & Seidler, 1994). It is very intriguing that there is no or little coupling between stimulation of the receptor and activation of Gs-protein in the day 12 rat embryo, but by day 18 this coupling becomes significantly large and adenylate cyclase stimulation by isoproterenol is maximal (Slotkin, Lau, & Seidler, 1994). The role of the α-receptor in cardiac function is not large even in mature heart, and this seems to be the case also in the early embryonic heart since α_1-receptors are first found at gestational day 15 in the mouse embryo (which corresponds to day 16 in the rat embryo) (Yamada, Yamamura, & Roeske, 1980). Further studies on the ontogeny of the α-receptor and its function in the vascular system of the early mammalian embryo are needed.

Recently, the effect of adrenergic receptor stimulation on the mammalian embryonic cardiovascular system has been evaluated in situ (Nakazawa et al., 1988; Nakazawa et al., 1990b). The experimental setup is a small bath system that is perfused with Hanks solution. The perfusate is in a glass container and is oxygenated, warmed up with a heating pan, and circulated to the bath by a roller pump (see Nakazawa et al., 1988, for details). A part of the uterus containing one embryo is excised and placed in this bath, and the uterus wall and yolk sac are opened to expose the embryo and umbilical vessels. Hemodynamic parameters, including heart rate, cardiac output, blood pressure, and ventricular and atrial pressure, can be measured at gestational days 11–15. In contrast to the chick embryo, α-adrenergic stimulation with norepinephrine does not change blood pressure in the rat embryo at gestational day 12 (Figure 8.4). This stage of morphological development is comparable to the stage 21 chick embryo.

Since the circulation of the mammalian embryo is connected to the placenta, which has a large reservoir function, we speculate that modest changes in embryonic vascular tone are not manifested as a gross effect in the total circulation (Nakazawa et al., 1990b). Thus, stimulation of the α-adrenergic receptor may have effects in the intraembryonic vascular system of these embryos. However, following isoproterenol infusion, arterial blood pressure decreases and cardiac output increases; thus, vascular resistance must be lower (Figure 8.5) (Nakazawa et al., 1995). Despite the effect of β-receptor stimulation on the embryonic vasculature, there is no direct myocardial inotropic effect, because ventricular positive *dP/dt* was unchanged by isoproterenol infusion (Figure 8.6). Peak ventricular pressure increases from 2.3 ± 0.1 to 2.6 ± 0.1 mm Hg ($n = 7$, $P < .05$), and negative *dP/dt* increases from 80 ± 4 to 102 ± 6 mm Hg·s^{-1} ($P < .05$). The latter findings suggest that β-adrenergic receptors are present in the embryonic ventricle at this stage and that the intracellular signal transmission system has already started to function, although it may be very immature.

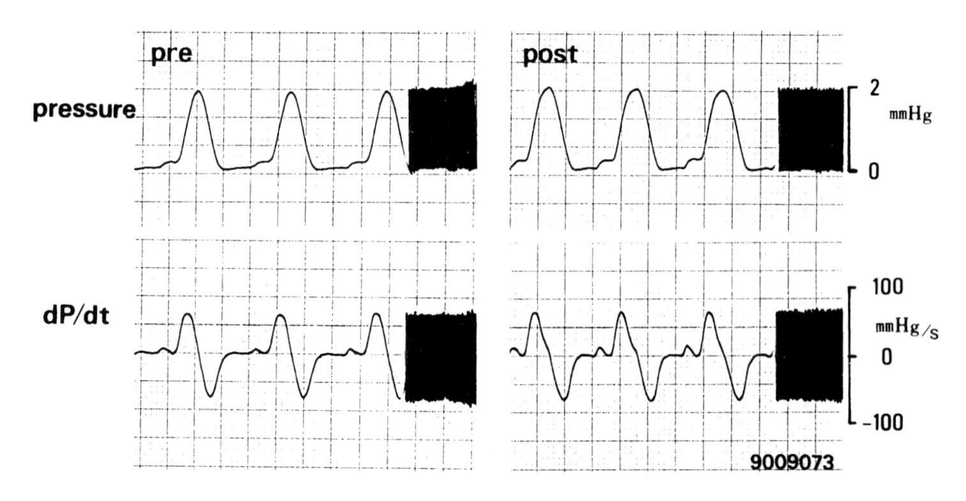

Figure 8.4 Norepinephrine infusion does not alter ventricular pressure or its first derivative (*dP/dt*) in the rat embryo at gestational day 12.

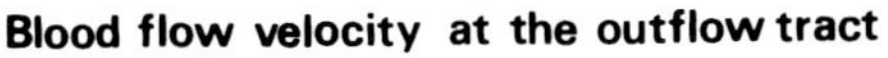

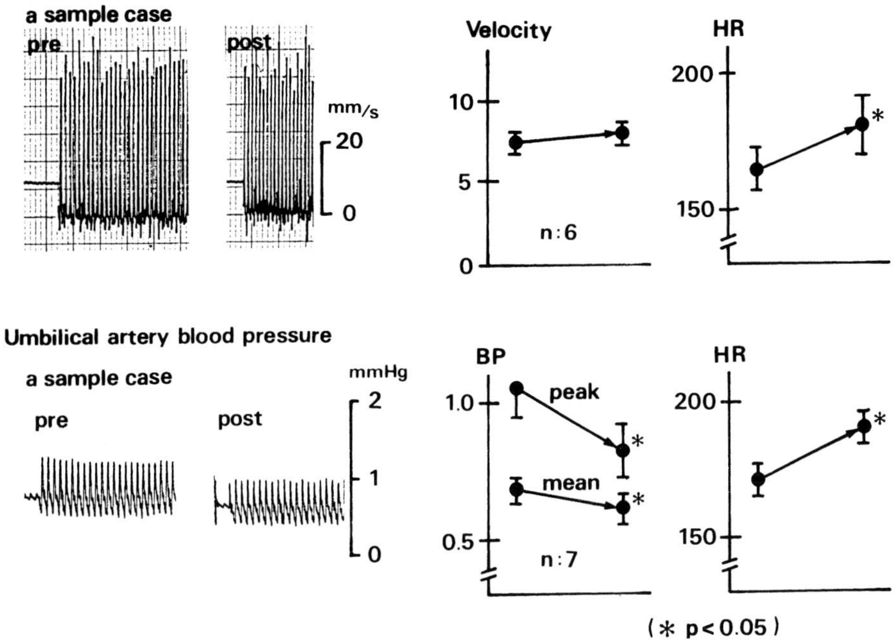

Figure 8.5 Isoproterenol infusion increases heart rate and reduces umbilical arterial peak and mean pressure without altering blood flow velocity in the gestational day 12 rat embryo. The decrease in blood pressure following isoproterenol infusion is consistent with a decrease in arterial resistance (HR = heart rate, BP = vitelline arterial blood pressure). (From Nakazawa et al., 1995.)

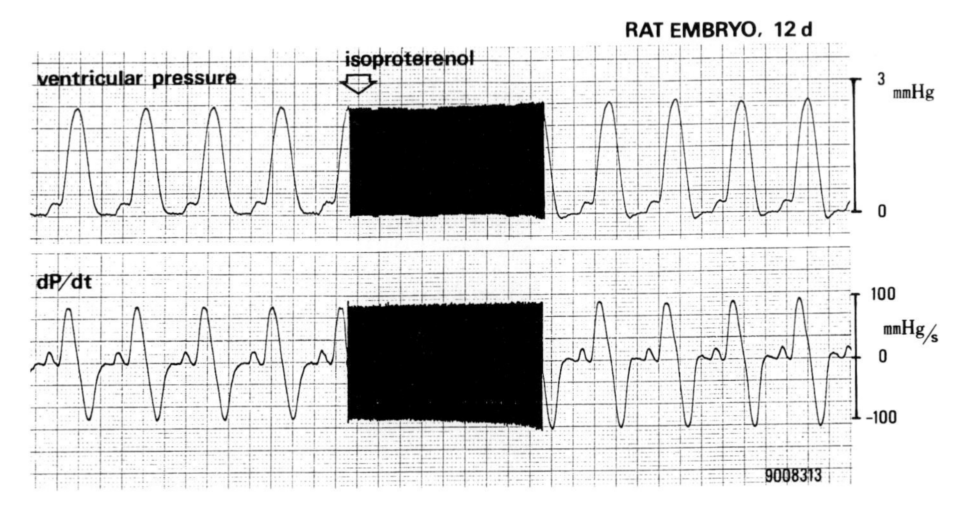

Figure 8.6 Ventricular pressure and *dP/dt* during isoproterenol infusion in the day 12 rat embryo. The intervention caused a subtle but statistically significant elevation of the ventricular pressure and increased negative *dP/dt*.

Atrial natriuretic peptide and its effects

Many studies have evaluated the cardiovascular function of atrial natriuretic peptide during mid- to late gestation, but only limited data are available for the period of primary cardiovascular morphogenesis (Nakazawa et al., 1990a; Nakazawa et al., 1992). As mentioned, atrial natriuretic peptide is present in the heart tissue at a relatively early stage of development in the chick embryo (stage 10), as suggested by the presence of multivesicular bodies (Manasek, 1969). The hemodynamic effects of atrial natriuretic peptide have been defined in chick embryos between HH stage 15 and 24 (Nakazawa et al., 1990a; Hu et al., 1994). When atrial natriuretic peptide is infused into the venous system, the vitelline vein dilates (Figure 8.7), which is associated with a lowering of venous pressure from 0.34 ± 0.05 to 0.10 ± 0.07 mm Hg ($n = 5$, $P < .01$). Atrial natriuretic peptide results in a dose-dependent decrease in arterial blood pressure (Figure 8.8) and in dorsal aortic blood flow, a parameter

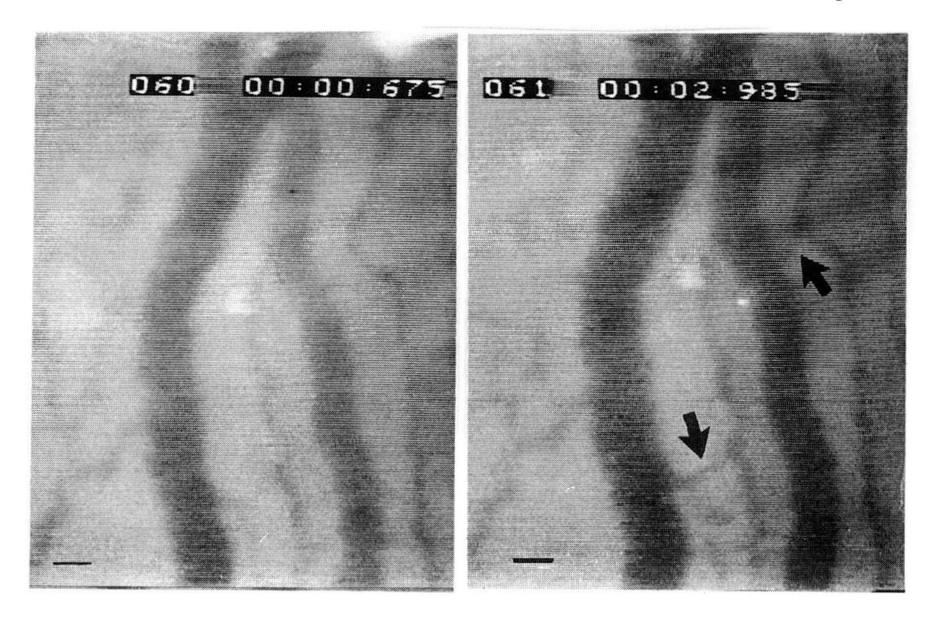

Figure 8.7 Atrial natriuretic peptide dilates the vitelline vein. The veins at control (*left*) become larger 3 minutes after peptide infusion (*right*). In addition, the small connecting veins are also dilated (*arrows*). Bars indicate 100 mm. (From Nakazawa et al., 1990a.)

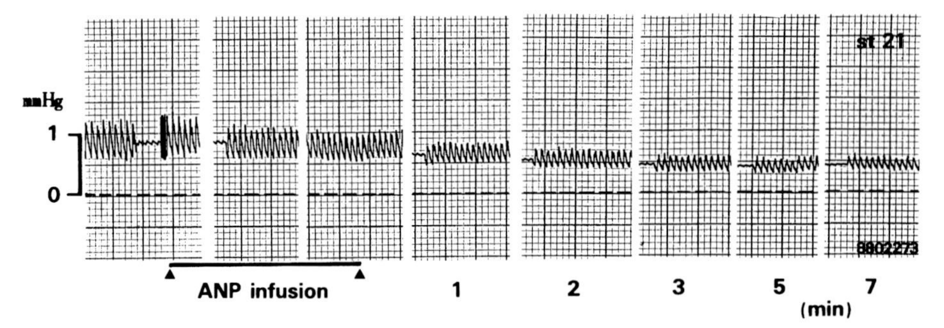

Figure 8.8 Atrial natriuretic peptide infusion decreases vitelline arterial blood pressure in a dose-dependent manner in the stage 21 chick embryo. (From Nakazawa et al., 1990a.)

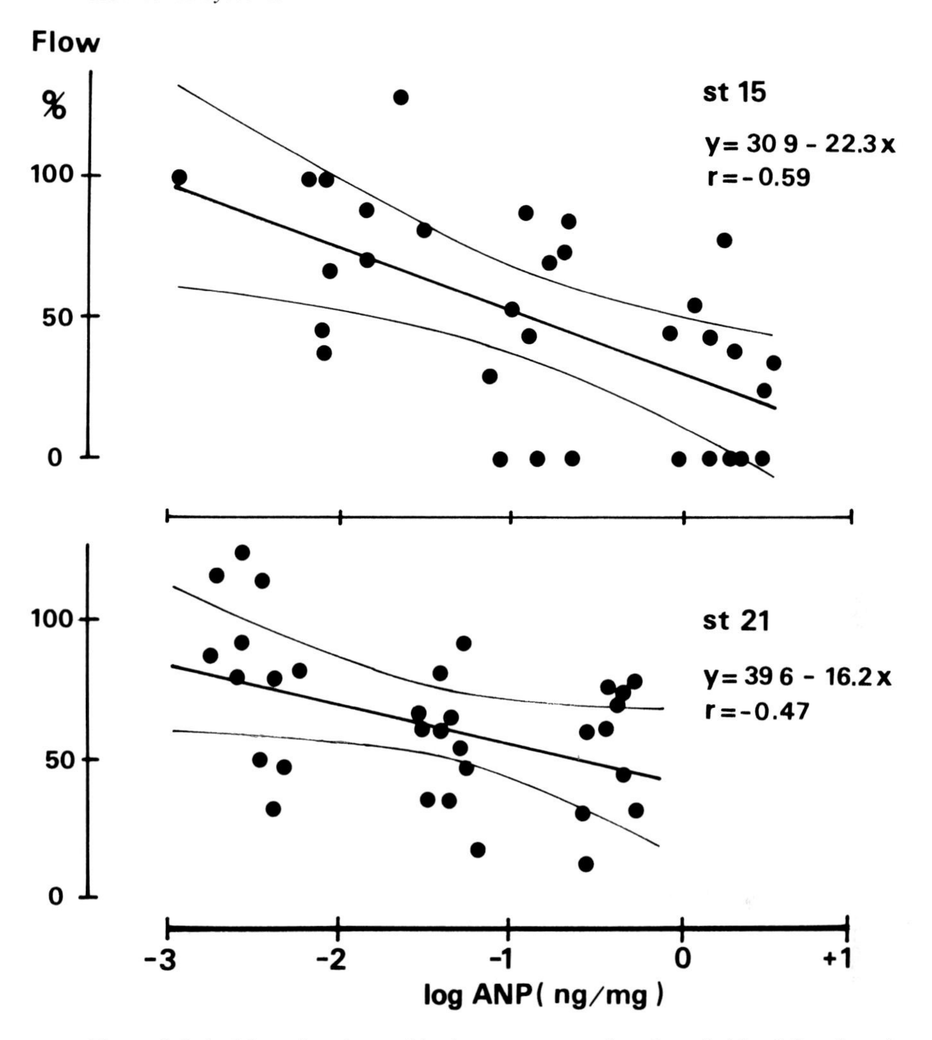

Figure 8.9 Atrial natriuretic peptide decreases mean dorsal aortic blood flow in a dose-dependent manner at stages 15 and 21. (From Nakazawa et al., 1990a.)

of cardiac output, likely because of increased blood volume fraction in the venous system (Figure 8.9). In contrast to the effect in the avian embryonic circulation, atrial natriuretic peptide causes no change in arterial blood pressure and blood flow velocity, a parameter of cardiac output ($P = .07$; Table 8.1) (Nakazawa et al., 1992), in the rat embryo. Although the change of cardiac output is not statistically significant, it may represent vasodilation of the resistance vessels of the rat embryo. There are no changes in ventricular function in these experiments, although atrial natriuretic peptide shows negative inotropic effect in cultured chick embryonic ventricular cells (Vaxelaire et al., 1989). These data indicate that atrial natriuretic peptide is capable of altering the vascular tone at very early stages of cardiovascular development and that the site and degree of the hemodynamic effect differ between the avian and the mammalian embryo at comparable developmental stages.

Table 8.1. *The effect of atrial natriuretic peptide (ANP) on blood pressure and outflow blood flow velocity in the gestational day 12 rat embryo*

Cardiovascular variable	*N*	Control	Post-ANP
Blood pressure (mm Hg)	6	0.60 ± 0.05	0.60 ± 0.03
Outflow blood velocity ($mm \cdot s^{-1}$)	8	5.4 ± 0.8	5.8 ± 0.8
Heart rate (bpm)	14	168 ± 10	160 ± 8

Endothelin-1

It is well known that ET-1 is a very potent vasoconstrictor, but the ontogeny and functional development of the ET-1 pathway during organogenesis have not been investigated. In stage 21 chick embryos, ET-1 at doses of 10^{-9}–10^{-7} M infused into the vitelline vein increases vitelline arterial blood pressure (Figure 8.10) and heart rate (M. Nakazawa, unpublished data). In order to characterize further the nature of the vasoconstrictive effect, changes in the diameter of vitelline arteries and veins of various sizes can be directly observed with a video microscopic system after ET-1 infusion (Figure 8.11). Veins between 100 and 200 μm in diameter constricted in response to ET-1 infusion, but veins smaller or larger than this size did not change. Arteries did not respond to the peptide (Figure 8.12), although arteries smaller than approximately 50 μm in diameter cannot be evaluated directly by this method (M. Nakazawa, unpublished data). The data suggest that ET-1 receptors function at very early stages of vascular development and that middle-sized venous vessels are the most sensitive. A similar differential effect is seen following ET-1 infusion in isolated pulmonary vessels from fetal lambs, and the threshold of vasoconstriction is almost 10 times lower in the vein (average diameter 273 μm) than in the artery (average diameter 169 μm) (Wang & Coceani, 1992). These data suggest that functional expression of the ET-1 pathway occurs first in veins. Interestingly, the vasodilatory effect of

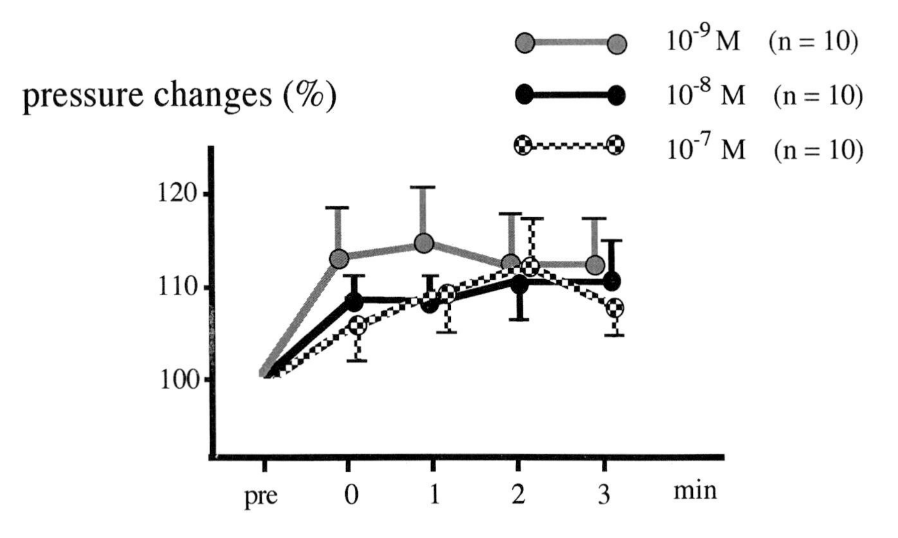

Figure 8.10 Intravenous endothelin-1 increases arterial blood pressure in stage 21 chick embryos. Endothelin-1 was infused in 1 minute. The time scale is pre (before infusion) and 1, 2, and 3 minutes after the start of infusion.

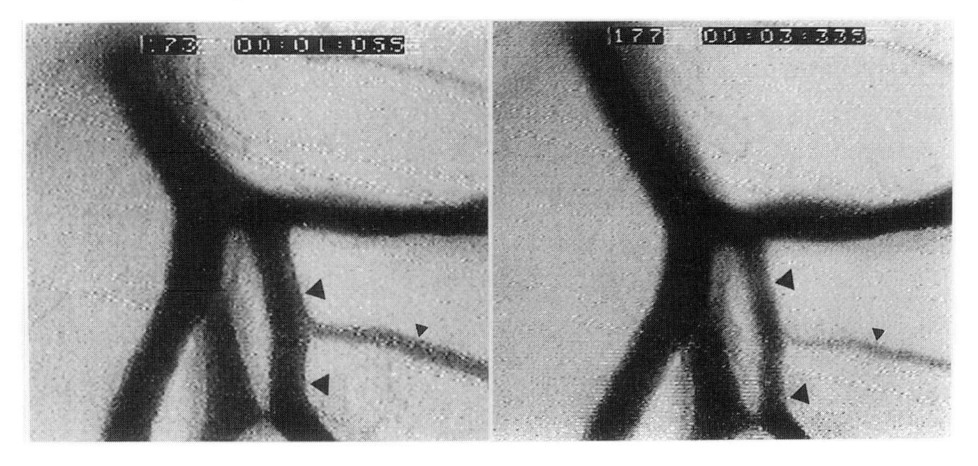

Figure 8.11 Endothelin-1 reduces the diameter of middle-sized vitelline veins in the stage 21 chick embryo. Compare vessels indicated by arrowheads at control (*left*) and at 2 minutes after endothelin-1 infusion (*right*).

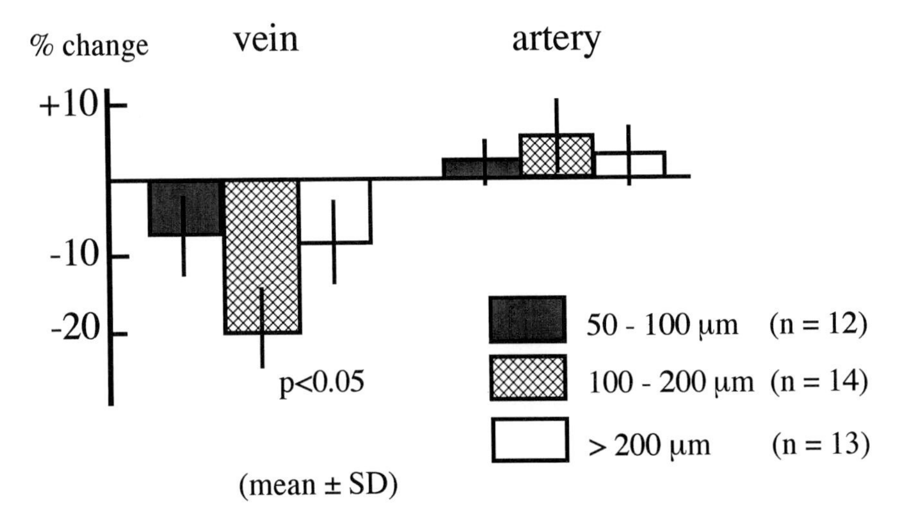

Figure 8.12 Summary of the effects of endothelin-1 on vitelline venous (*left*) and arterial (*right*) diameters in the stage 21 chick embryo. The endothelin-1 did not alter arterial diameter and constricted only the middle-sized vein.

endothelin on isolated pulmonary vessels of adult pigs is greater in veins than in arteries under various conditions (Zellers, McCormick, & Wu, 1994). Thus, there is likely a differential developmental sensitivity to endothelin between the artery and the vein. Endothelin has direct inotropic and chronotropic effects in cultured heart cells (Concas et al., 1989), and heart rate increases following ET-1 administration in the chick embryo (M. Nakazawa, unpublished data).

Conclusions and future directions

The embryonic circulation is controlled by the intrinsic mechanics of the heart and vessels, but cardiovascular function is also influenced by various paracrine and autocrine factors such as catecholamines, atrial natriuretic peptide, and endothelin. However, specific sensor--effector pathways have not yet been directly demonstrated in the early embryo, and further studies are necessary to define the significance of the hormonal system in homeostasis of the embryonic circulation. Since these hormonal systems also likely play an important role in morphogenesis, future research must determine the separate and/or cooperative influences of these paracrine and autocrine systems on embryonic functional and structural maturation.

Part II

Species diversity in cardiovascular development

9

Evolution of cardiovascular systems: Insights into ontogeny

ANTHONY P. FARRELL

Introduction

The chapters that follow in Part II deal with the diversity of cardiovascular development among the various phylogenetic groupings. The goal of this chapter, however, is to bring together some specific aspects of cardiovascular development beneath an evolutionary umbrella and to demonstrate that the study of cardiovascular change at an evolutionary level can provide insights into some of the basic concepts of cardiovascular development. Superficially, it may be difficult to see how this is possible. After all, evolutionary change is characterized as occurring either within or between populations over many generations and involving a change in gene distribution (Resnik, 1995). In contrast, developmental change occurs within an individual over one generation and need not involve a change in gene frequency. Hence, as recognized by Thompson (1917; see De Beer, 1958), ontogeny is concerned with the timing of the expression of certain genes, as well as with the relative rates of gene expression (expression of some genes may be accelerated, whereas for others it is retarded). Phylogeny is concerned more with whether or not specific genes are present and expressed. Yet it is clear that there are many similarities between certain aspects of cardiovascular change (sequences of events, changes in structures, and rearrangements of designs) occurring at an evolutionary level and those occurring during development. These similarities are explored in this chapter.

One false assumption that the reader may make about this chapter is that it resurrects the "theory of recapitulation." It does not. Haeckel's theory of recapitulation stated that "ontogeny is a short recapitulation of phylogeny" and "phylogeny is the mechanical cause of ontogeny" (1866; as quoted by De Beer, 1958). Thus, Haeckel proposed that the adult stages of the ancestors are repeated during the development of the descendants, but they are crowded back into the earlier stages of ontogeny. As an example, Haeckel suggested that the embryonic gill slits or pouches of embryonic birds were the gill slits of ancestral *adult* fish. On a broader scale, Haeckel also reasoned that the blastula and gastrula represented original ancestral *adult* stages because many animals pass through these two stages. We now know that these sorts of ideas are untenable.[1] In fact, because phylogeny is a scale pri-

[1] It is remarkable how the theory of recapitulation was introduced and accepted with relative ease in the mid–nineteenth century but was much harder to debunk. The writings of Sedgewick (1894) and Garstang (1922), as well as others in the 1920s and 1930s, did much to refute this theory (see De Beer, 1958), yet some 90 years later De Beer (1958) still felt the need to debunk the theory of recapitulation in his treatise entitled *Embryos and Ancestors*. Perhaps the irony of this saga is that von Baer (1828) had much earlier provided a more reasonable foundation for the study of development and phylogeny (see De Beer, 1958). His four conclusions are worth repeating: (1) In development from the egg, the general characters appear before the special characters. (2) From the more general characters, the less general and finally the special characters are developed. (3) During its devel-

marily derived from adult comparisons, phylogeny is more likely due to modified ontogeny (De Beer, 1958). Clearly, the magnificent and ongoing evolutionary experiment that we see reflected in phylogeny is not a "repetition of the ancestral adult conditions" as required by the theory of recapitulation.

The second false assumption the reader may gather from this chapter is that it engenders the idea of an evolutionary superiority for the class Mammalia. This is not the chapter's intent. In fact, each animal species has an effective cardiovascular solution for exploiting its own particular ecological niche. Thus, for example, the three-chambered turtle heart should not be considered as inferior to the four-chambered mammalian heart. Instead, the two types of hearts have evolved to perform somewhat different tasks effectively. Adult mammalian hearts generate much higher systemic blood pressures than turtle hearts. However, no stretch of the imagination could envisage an adult mammalian heart surviving anoxia and blood [Ca^{2+}] levels reaching 40 mM for 6 months. Yet this is exactly what certain adult turtles are capable of when they overwinter under ice (Herbert & Jackson, 1985). Furthermore, a vision of grandeur for the mammalian heart can be quite misleading. For example, staying with comparisons of pressure development by the heart, "lowly" jumping spiders (e.g., tarantulas) can generate hemolymph pressures in excess of 300 mm Hg to assist leg extension (Paul et al., 1994); even a giraffe heart would be hard-pressed to generate such high blood pressures. Obviously, we must encompass a more holistic view of cardiac function when comparisons are made (see Chapter 7).

Common solutions to common problems

Why is it that certain aspects of cardiovascular change occurring at an evolutionary scale are quite similar to those occurring during development? A change usually reflects a new solution to a problem, and so similarities often reflect common solutions to common problems. This being the case, one might expect that certain of the constraints and boundary conditions acting during cardiovascular development are equally important on an evolutionary time scale. Three types of constraints fall into this category and would tend to lead to similarities between developmental change and evolutionary change.

The genetic code is quite conservative

Because the genetic code is quite conservative, cardiovascular solutions will tend to be more similar in species that are phylogenetically closely related. In fact, many parallels can be found between developmental and evolutionary changes within the subphylum Vertebrata. The present chapter draws predominantly on these examples. In contrast, invertebrates show remarkable diversity in their form and function both within and between phyla. This diversity extends to cardiovascular design (McMahon, in press; see also Chapter 11, this volume). For example, within the phylum Mollusca can be found neurogenic and myogenic hearts and open and closed circulatory systems, representing the broadest possible range for circulatory mechanisms (see below). Consequently, similarities between cardiovascular change at the evolutionary and developmental levels are less prevalent among

---

opment an animal departs more and more from the form of other animals. (4) The young stages in the development of an animal are not like the adult stages of other animals lower down on the scale but are like the young stages of those animals.

invertebrates, and our present state of knowledge of invertebrate cardiovascular systems is in its infancy compared with vertebrates.

Even though substantial morphological differences exist in the cardiovascular design of vertebrates and invertebrates, there may be a genetic constraint of a broader nature. This relates to the clusters of homeotic (*Hox*) genes that sculpt the morphology of animal plans and body parts. For a more detailed discussion of *Hox* genes the reader is referred to the recent review by Carroll (1995). In this review, Carroll notes that arthropods and chordates share a family of *Hox* genes that regulate features such as segment morphology, appendage number and pattern in invertebrates and vertebral morphology, and limb and central nervous system pattern in vertebrates. Although homologies can be found between the *Hox* genes in arthropods and chordates, the chordates have more *Hox* genes, which are arranged in four *Hox* clusters. Furthermore, the single *Hox* complex found in *Amphioxus* (now *Branchiostoma*) appears to be the archetype of *Hox* clusters typical of modern vertebrates such as teleost fish and mammals (Carroll, 1995). Clearly, the large effects of *Hox* genes on axial morphology will influence to some degree the associated circulatory pattern.

Biomechanical processes operate within boundaries

Functionally the cardiovascular system generates flow and pressure. The biomechanical processes that underlie how muscle cells perform these tasks must ultimately constrain cardiovascular design equally well on developmental and evolutionary time scales. Although pressure and flow generation vary within a species during development and between species during evolution, flow rate is ultimately set by anatomical considerations and can be varied only through adjustments in pulsation rates and stroke volumes. Likewise, the quantity and physical arrangement of muscle cells will determine pressure development. It is perhaps surprising at first to learn that the ratio of heart to body mass (relative cardiac mass) of the skipjack tuna is quite similar to that of humans (i.e., around 0.4%). Yet, if we also recognize that the heart rates, blood pressures, and cardiac outputs are similar for the two species (Farrell, 1991), it becomes apparent that these two phylogenetically distant species share a common solution to a common problem. There is, however, a danger in oversimplifying such comparisons. For example, relative cardiac mass in hemoglobin-free Antarctic icefish is just as large as in the skipjack tuna and the human. However, because the cardiac muscle and vasculature are organized in a very different manner, high cardiac outputs are still possible in these icefish but only at relatively low blood pressures (Hemmingsen et al., 1972; Tota, Acierno, & Agnisola, 1991).

The cardiovascular system performs a limited number of primary functions

The cardiovascular system has four major functions: transport (gases, nutrients, waste products, hormones, heat, cells, defense molecules, etc.), storage (gases, nutrients, ions, buffers, heat, etc.), hydraulics (support and movement), and filtration (nutrients and wastes). The relative emphasis each organism places on these four cardiovascular functions is clearly species specific to enable it to adapt to its specific "ecological niche." An analogous situation exists during ontogeny, in which cardiovascular development reflects a functionally effective solution for a specific "developmental niche." Thus, a given stage of cardiovascular development is not merely aimed at becoming the "adult model"; developmental stages exhibit specific and changing support functions. For example, when the pupa of a winged insect undergoes metamorphosis, the cardiovascular system takes on a central function in the hydraulic unfolding of the wings. This hydraulic function, although a relatively brief event, has profound importance and survival value. Likewise, the cardiovascular system of amniotes switches from its initial embryonic function of nutrient transport to one of gas transport in adults.

Given these constraints (genetic, biomechanical, and primary function), it is hardly surprising to find parallels between the cardiovascular solutions that have evolved for a particular ecological niche at a phylogenetic level and those that are needed for a particular developmental niche at the ontogenic level. For example, the developmental switch from nutrient transport to gas transport already mentioned has an evolutionary parallel. The current dogma is that food distribution is the major selection pressure driving the evolution of the primitive vertebrate circulation. As shown in Figure 9.1, *Branchiostoma* lacks a chambered heart, but two of the three contractile vessels are notably associated with the circulation to the gut and liver. The subsequent evolutionary trend toward emphasis on gas transport occurred as vertebrates increased in size (Randall & Davie, 1980). The increase in size and complexity of organisms during both ontogeny and evolution seems to be an important common denominator in developmental changes and evolutionary transitions in cardiovascular design.

The evolution of cardiovascular systems

The focus of the chapter now shifts to the evolution of adult cardiovascular systems. The reader is also directed to more comprehensive descriptions of adult cardiovascular systems found in Romer, 1962; Robb, 1965; Bourne, 1980; Randall et al., 1981; Nilsson, 1983; Johansen and Burggren, 1985; Farrell, 1990a, 1990b; Burggren, Farrell, and Lillywhite, in press; and McMahon, in press. Some of the best illustrations are in Robb, 1965.

Circulatory fluids can be moved by one of six major mechanisms (Farrell, 1990a):

1. Intracellular movement
2. Ciliary, flagellar, or muscular activity moving external medium

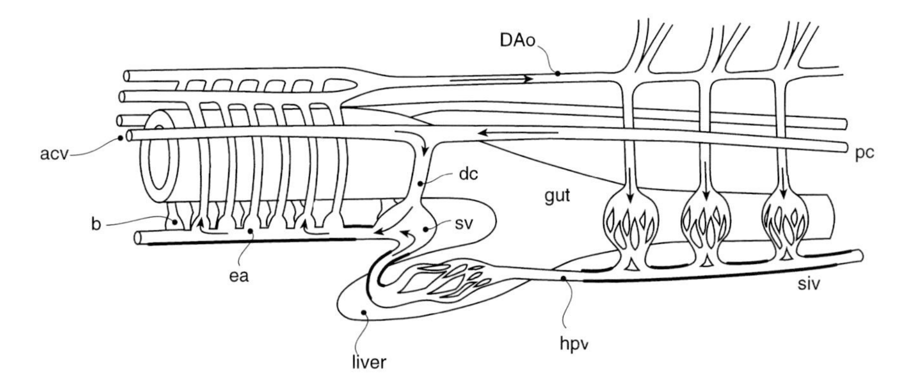

Figure 9.1 A schematic diagram of the circulatory system of the early vertebrate *Branchiostoma.* This animal lacks a centrally located, chambered heart and valves to separate the arterial and venous systems. However, there is unidirectional blood flow (the direction is indicated by the arrows), propelled in three regions of the circulation where the vessels are contractile (the contractile regions are indicated by a thicker vessel wall) (acv = anterior cardinal vein, b = bubulli, DAo = dorsal aorta, dc = ductus Cuveri, ea = endostylar artery, hpv = hepatic portal vein, pc = posterior cardinal vein, siv = subintestinal vein, sv = sinus venosus). (Redrawn and adapted from Romer, 1962; Randall & Davie, 1980.)

3. Somatic muscles moving body fluids
4. An open circulatory system moving hemolymph
5. A closed circulatory system (i.e., one with vessels at all levels lined with endothelium)[2] moving blood
6. Movement of lymph in vertebrates

The last four of these circulatory mechanisms all utilize morphological structure(s) loosely termed "hearts." The four types of heart are as follows.

1. *Chambered hearts.* These are the more familiar hearts found in all of the Vertebrata and in certain molluscs. The heartbeat is initiated in modified cardiac muscle cells termed the "myogenic pacemaker." Chambered hearts can be either single or double pumps, with one to four atria and one or two ventricles.
2. *Tubular hearts.* These are contractile tubes that are found in many invertebrates and the Protochordata. The heart usually, but not always, contains valves (ostia), and the heartbeat may be either a neurogenic one (initiated in nerve cells) or a myogenic one.
3. *Ampullar hearts.* These usually serve as accessory hearts to boost fluid through the peripheral circulation. Lymph hearts of amphibians and reptiles, caudal hearts of fishes, branchial hearts of cephalopods, and the accessory hearts of insects and some decapods fall into this category.
4. *Pulsating vessels.* Fluid can be propelled by peristaltic waves in many invertebrates, cephalochordates, the conus arteriosus of elasmobranchs, and veins in bat wings.

An important evolutionary trend has been toward increased aerobic capability and functional complexity. Associated with this trend has been the development of a central chambered heart. A central chambered heart along with regional vasoactivity apparently confers a greater degree of control over cardiac output and its distribution. There can be independent control of heart rate, stroke volume, and vasomotion. In addition, specialization of the chambered heart led to a differential control of the flow to and arterial pressure within the pulmonary and systemic circulations. In the case of birds and mammals, there was a shift toward higher blood pressures in the systemic circulation and lower blood pressures in the pulmonary circulation (Figure 9.2), although the flow through the two circulations was the same. The lower blood pressures in the pulmonary circulation allowed a thinner respiratory epithelium, which favored a more rapid rate of gas exchange. In the case of amphibians and reptiles, pulmonary blood flows and blood pressures can also be lower than those in the systemic circulation.

Five general characteristics distinguish vertebrate cardiovascular systems from those of invertebrates (Burggren, Farrell, & Lillywhite, in press). These five characteristics must be recognized as important evolutionary steps.

1. A single, ventrally located myogenic, chambered heart
2. Cephalically directed cardiac ejection of blood (invertebrate hearts can eject blood caudally, cephalically, and laterally)
3. Passive arterial and venous valves (invertebrates can have muscular arterial valves that are actively regulated, reflecting the need to regulate blood flow between the numerous outflow vessels from the heart)
4. Muscular vessels capable of vasomotor activity (most invertebrates lack smooth muscle in the peripheral vasculature)
5. A closed vascular system

[2] Although many invertebrates lack capillaries and endothelium and clearly have an open circulatory system, the historically clear-cut distinction between open and closed circulatory systems is now breaking down in light of the emerging information for invertebrates (see Burggren, Farrell, & Lillywhite, 1997; McMahon et al., 1997; and Chapter 13 in this volume).

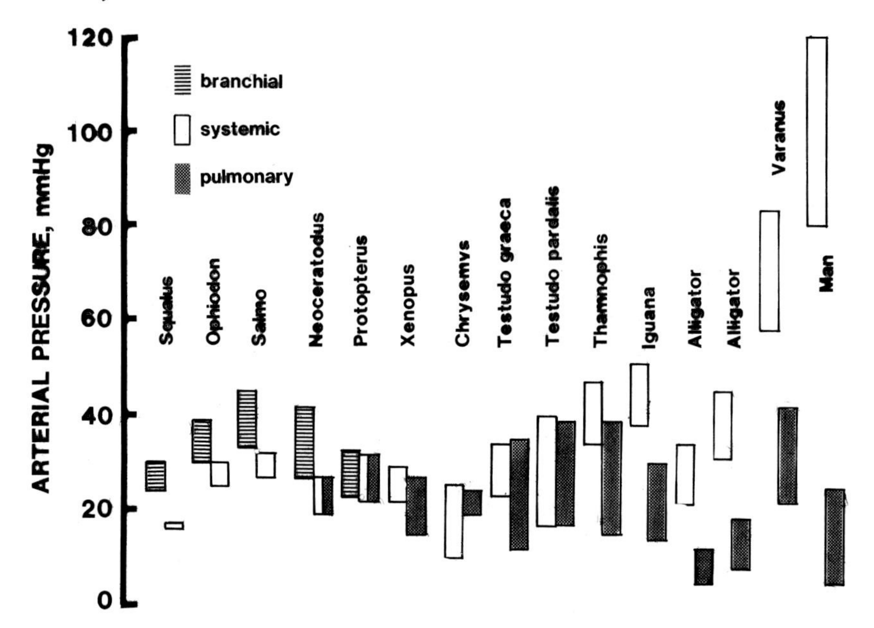

Figure 9.2 Selected blood pressures in vertebrates illustrating the evolutionary trends toward higher blood pressures in the systemic circulation and lower blood pressures in the respiratory (branchial and lung) circulation. (From Farrell, 1990a, with permission.)

There are, of course, exceptions to these generalizations. Perhaps the most significant is that the cardiovascular system of cephalopod molluscs shares many of the above vertebrates' characteristics. The octopus, for example, has a chambered, myogenic heart that generates blood pressures and flows comparable to those of many fish (Houlihan et al., 1987). The octopus heart also has a coronary circulation, although its arrangement is quite different from the typical vertebrate pattern. Again we see the theme of common solutions to common problems even in phylogenetically distant species. Relative to most invertebrates, the cephalopod mollusks are both large and active, especially the squids and octopuses.

Vertebrate cardiovascular evolution

The heart of the most primitive Protochordata (i.e., Urochordata) is a muscular tube that can pump blood in either direction in a reversible manner (Randall & Davie, 1980). As ancestral vertebrates increased in size and became more fusiform and segmental, the cardiac pump became more centrally located with a parallel segmental array of vessels. For example, *Branchiostoma* (Cephalopoda) pumps blood unidirectionally with three centrally located pulsating vessels (the endostylar artery, the hepatic vein, and the subintestinal vein; see Figure 9.1) that contain myogenically active myoepithelial cells. Subsequent evolution of one-way valves permitted (1) a more effective unidirectional flow and (2) separation of the heart from the arterial vessels during diastole and from the venous vessels during systole. This circulatory separation allowed the arterial system to serve as a pressure reservoir and the venous system to store blood (Burggren, Farrell, & Lillywhite, 1997). Cyclostomes (hagfishes and lampreys) have chambered hearts and valves in their circulatory systems. In addition, very large venous reservoirs and accessory hearts (to assist the low pressure-

A

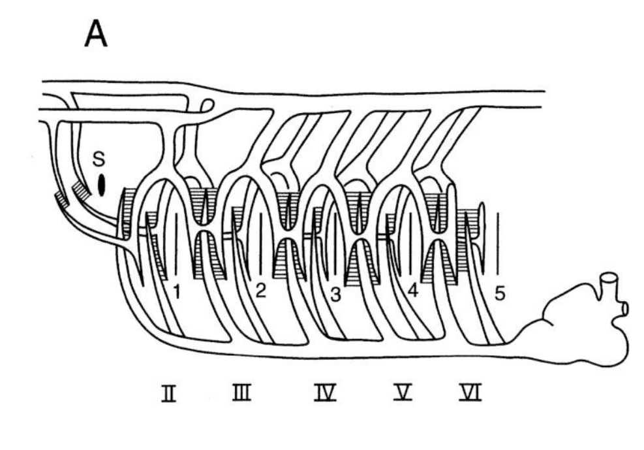

B

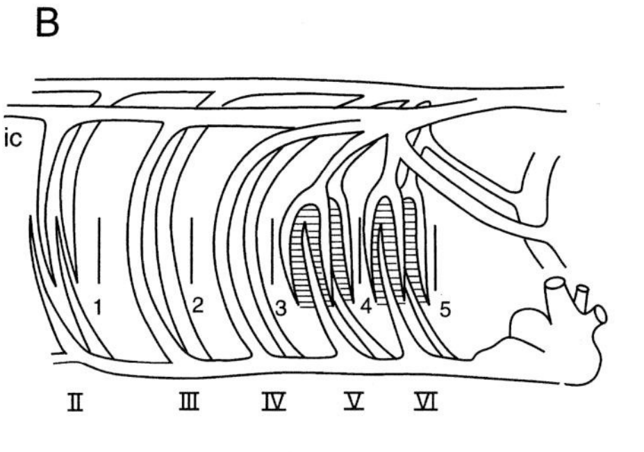

C

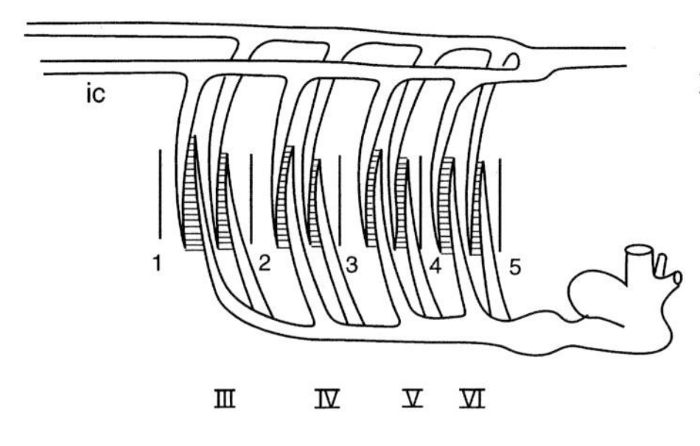

D

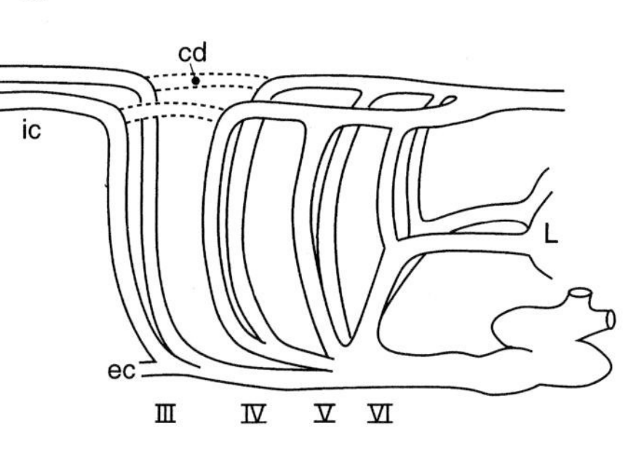

E F

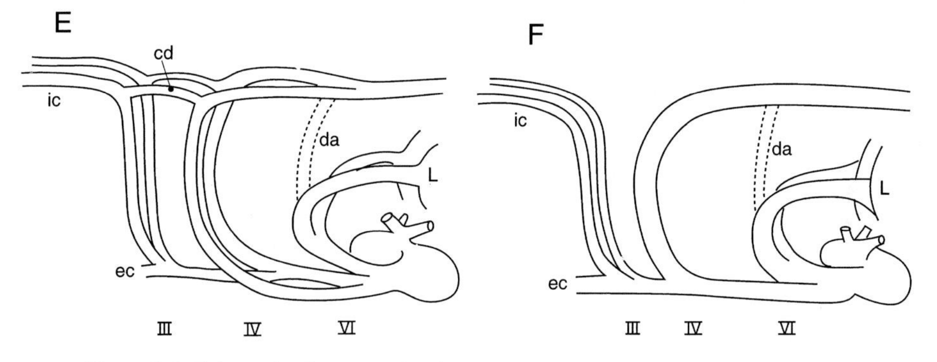

Figure 9.3 Schematic diagrams showing lateral views of the main arterial plan in various adult vertebrate types: (*A*) a shark; (*B*) a primitive air-breathing fish, *Protopterus;* (*C*) a teleost fish; (*D*) a terrestrial salamander; (*E*) a lizard; (*F*) a mammal. The diagrams illustrate that during evolution there has been a reduction in number and a shift away from the bilateral arrangement of the embryonic aortic arches (corresponding aortic arches have the same roman numeral; corresponding gill arches have the same arabic numeral). Note the similarity in arch design between *Amphioxus* (Figure 9.1) and the aquatic species (*A, B,* and *C*) in this figure. The avian arterial plan is similar to that of mammals except that the right IV aortic arch is retained in birds, whereas the left IV aortic arch is retained in mammals. Broken lines indicate the location of the embryonic ductus caroticus (cd) and ductus arteriosus (da) (ec = external carotid artery, ic = internal carotid artery, L = lung, S = spiracle). (Redrawn and adapted from Romer, 1962.)

generating ability of the main heart) feature prominently in hagfishes and elasmobranchs (Satchell, 1991). Low blood pressure and a large circulatory blood volume point to the fact that a fast circulation time was not a critical feature in their circulatory design. Furthermore, hagfishes have an additional myogenic heart associated with the hepatic portal circulation, again emphasizing the importance of the circulatory system in nutrient uptake rather than locomotion in these early vertebrates.

One-way valves and a chambered heart were clearly a critical preadaptation for the functional shift to placing greater demands on the circulatory system for gas transport. The appearance of these structures presumably set the stage for the single largest evolutionary radiation among vertebrates, that of the fishes. An important parallel event was the evolution of autonomic circulatory control, a topic covered in detail elsewhere (Nilsson, 1983).

The next major evolutionary change in cardiovascular design was related to the transition from water-breathing to air-breathing (Johansen & Burggren, 1980; Randall et al., 1981). Gills are better suited for water-breathing and, unless modified substantially, will collapse in air and increase branchial vascular resistance so as to restrict the passage of blood to the systemic circulation. The new vascularized sites dedicated for aerial gas exchange that evolved were diverse (e.g., the skin, the swim bladder, and even the stomach and intestine, as well as the formation of a true lung) and resulted in equally diverse circulatory arrangements and vascular controls (Randall et al., 1981; Farrell, 1990b). However, because these sites for aerial gas exchange were in parallel with the systemic circulation, there was presumably a strong selective advantage for central separation of the oxygenated and deoxygenated venous blood. As a result, we see a strong evolutionary progression toward (1) the separation of inflowing blood to the heart, (2) the separation of left and right atrial outputs within the ventricle, and (3) the separation of the systemic and pulmonary outflows from the ventricle.

Neoteny is an excellent example of developmental arrest on an evolutionary time scale. Kollmann (1882; see De Beer, 1958) introduced the term "neoteny" to describe animals that retain their larval form and habits either temporarily or permanently and become sexually mature in that condition. In other words, development of the reproductive organs proceeds at a faster rate than somatic development. Consequently, some of the structures that would normally appear briefly during development in related species are retained for either a large part of or the entire life history.

Among vertebrates perhaps the best examples of neoteny are provided by some, but not all, urodeles (salamanders). Permanent neoteny is shown by urodeles such as *Necturus,* the American mud puppy, which are characterized by having external gills throughout their life span. Nonneotenic amphibian larvae lose their gills at or before metamorphosis when lungs are established for the switch to air-breathing. The axolotl *Siredon mexicanum* is a special and interesting case. It may become mature while retaining its external gills and gill slits. Alternatively, it may undergo metamorphosis and lose the external gills under certain natural or experimental conditions. Likewise, the tiger salamander, *Ambystoma tigrinum,* can exist in a neotenic state or an adult state.

The loss of external gills in salamanders requires substantial vascular reorganization (both growth and regression) of the aortic arches and other vessels. Metamorphosis in *Ambystoma,* for example, is accompanied by a regression but not complete loss of the ductus arteriosus and the third gill arch, an enlargement of the pulmonary artery, and a coalescence of the main gill arteries to form the aortic loops to the dorsal aorta and carotid arteries (Figure 9.4). In the neotenic state, this arterial remodeling is arrested. Facultative neotenic amphibians may prove to be a particularly useful model if the arterial remodeling

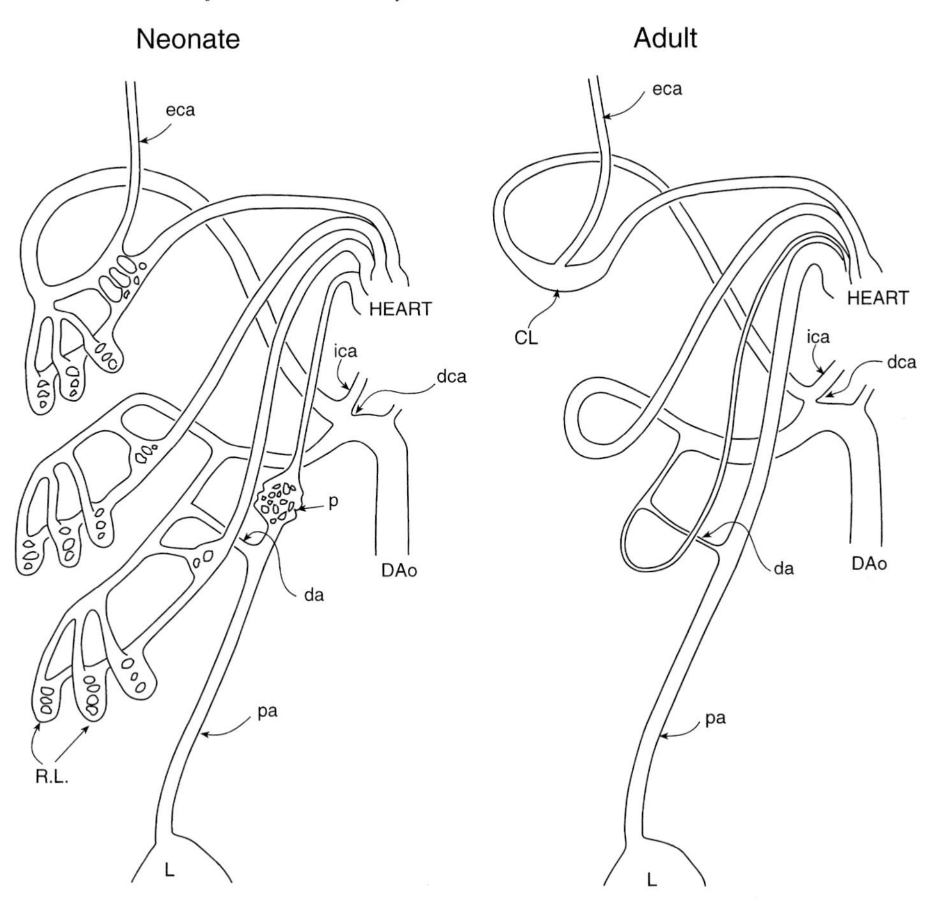

Figure 9.4 Schematic diagrams illustrating the major arteries in the neotenic tiger salamander, *Ambystoma tigrinum.* In the neonate, the respiratory lamellae (R.L.) of the larval gills are retained even though the lung (L) and lung circulation are functional. In the adult stage, the general arterial plan is retained but the respiratory lamellae have coalesced (CL = carotid labyrinth, DAo = dorsal aorta, da = ductus arteriosus, dca = ductus caroticus, eca = external carotid artery, ica = internal carotid artery, p = plexus, pa = pulmonary artery). (Redrawn and adapted from Malvin, 1985.)

can be experimentally switched on and off. Thus, comparative studies with urodeles may greatly ease the task of identifying the molecular and chemical triggers associated with arterial remodeling.

Arterial remodeling also has important consequences for the neural regulation of the cardiovascular system because aortic arches are sites for baroreceptors. Readers interested in the phylogenetic development of baroreceptors are referred to reviews by Jones and Milsom (1982) and Van Vliet and West (1994).

Mammalian cardiac development

The cardiovascular changes that occur in mammals during and shortly after birth are described in detail in Chapter 16. A major consequence of these changes is that the left ventricle of the neonate performs significantly more flow and pressure work than it does in the fetus. This section and the one that follows are concerned with two important cellular changes that occur during neonatal life that are likely related to the increased workload of the neonatal heart. The switch from hyperplastic to hypertrophic cardiac growth illustrates how the study of cardiac cell division in nonmammalian species could contribute significantly to our understanding of mammalian cardiac development.

The switch from hyperplastic growth to hypertrophic growth

The cardiac hyperplasia characteristic of fetal life contrasts markedly with the type of cardiac growth postpartum. Except for the first few weeks of neonatal life in certain mammals, postpartum myocyte growth is hypertrophic (Dowell & McManus, 1978; Lewis, Kelly, & Goldspink, 1984; Goldspink, Lewis, & Merry, 1986). As the heart increases in size, myocytes can double their length and increase their width 10-fold. Also, the ratio of cardiac DNA to protein decreases even though myocytes become polyploid and nonmyocytes undergo hyperplasia. One often tragic consequence of this switch to hypertrophic cardiac growth is that myocyte damage or death cannot be repaired in adult humans through myocyte replication. Although manipulation of the developmental switch that stops myocyte division could have important clinical benefits, the molecular nature of this switch in mammals is unknown. However, lower-vertebrate models may prove useful in discovering this switch.

In lower vertebrates cardiac hyperplasia apparently persists throughout most of the adult life. Their hearts, therefore, can be considered as evolutionarily arrested in a hyperplastic growth mode. Although the studies are scant and indirect, the evidence for hyperplastic cardiac growth in fish and amphibians is perhaps the most convincing. For example, adult fish and amphibian myocytes are typically much narrower than mammalian cells (Page & Niedergerke, 1972; Karttunen & Tirri, 1986; Farrell et al., 1988). Perhaps more convincing is the observation that for a 100-fold increase in cardiac mass in rainbow trout, myocyte length increased only 2.5 times, myocyte width increased by less than 50%, and the DNA-to-protein ratio tended to decrease slightly. Therefore, comparative studies of myocyte growth could prove useful in discovering the regulatory process(es) involved in the switch to hypertrophic cardiac growth. An advantage of a fish model for such studies is that the fish heart responds well to relevant experimental manipulations; cold exposure and testosterone treatment can both nearly double ventricular size (Farrell, 1987; Franklin & Davie, 1992). Far more is known about cardiac growth in mammals (see Chapters 3–7), but, unlike in fish, cardiac enlargement in adult mammals is usually associated with a pathology. For example, pressure overload brought on by systemic hypertension results in concentric cardiac hypertrophy in which wall thickness increases owing to augmentation of myocyte diameter. In contrast, volume overload results in eccentric hypertrophy owing to increases in both myocyte length and diameter. Interestingly, dilated, failing adult hearts reexpress the early neonatal or fetal gene program (Gerdes & Capasso, 1995).

The switch from extracellular to intracellular calcium handling

A second significant cellular change during neonatal life is the expression of a protein channel associated with the cardiac sarcoplasmic reticulum (SR) (see also Chapter 2). The SR calcium channel is critical to normal excitation–contraction (EC) coupling in adult mammals because the majority of activator calcium (i.e., that used for binding to the regulatory site on troponin C) passes through this channel, even though a transsarcolemmal calcium influx triggers the SR calcium release. When the ligand ryanodine binds to the receptor site on the SR calcium channel (hence the reference to the SR calcium channel as the ryanodine receptor), the channel becomes locked in a closed state and cardiac contractility decreases. This is not the case in utero. Vornanen (1992) found that the fraction of SR activator calcium increased from 6% in newborn rats to 33% in 11-day-old weanlings and to 87% in adult rats. The developmental switch for the maturation of the SR awaits discovery.

The heart of most adult lower vertebrates, in contrast to adult mammals, remains relatively ryanodine insensitive (Tibbits, Moyes, & Hove-Madsen, 1992; Keen et al., 1994), resembling the situation in fetal mammals. One known exception is the skipjack tuna, in which ryanodine is known to reduce isometric tension of atrial tissue (Keen et al., 1992). These fish also have heart rates and blood pressures more similar to mammals than other fish, pointing to a potential role of these factors in the expression of the ryanodine receptor. Therefore, phylogenetic comparisons have the potential of providing clues to the mystery of the developmental expression of the ryanodine receptor.

Forming a divided heart

All vertebrates show a common cardiac developmental pattern during the earliest stages of development. The heart starts as a straight tube that later bends into an S shape, and the walls of the tube ultimately differentiate into the various chambers of the adult heart (see also Chapter 10). In the case of mammals, the embryonic heart tube becomes completely divided to form the right and left sides of the adult heart. (Birds have a fenestrated atrial septum.) As a result, there is an in-series arrangement of the pulmonary and systemic circulations. At the other extreme, adult fishes retain the tubular arrangement of the heart; blood passes sequentially through the four heart chambers (a sinus venosus, an atrium, a ventricle, and either a conus or a bulbus arteriosus). Again, the respiratory and systemic circulations are in series, but there is no cardiac pump between them as in mammals. In Dipnoan fishes, the pulmonary and the systemic circulations lie in parallel with each other, but lungfish can separate oxygenated pulmonary venous return from the deoxygenated systemic venous return to a significant degree because the atrium is partially septated and flow through the ventricle is nonturbulent and there is a spiral fold in the bulbus cordis. Among the tetrapods (Figure 9.5), amphibians move further from the embryonic heart tube during development by forming two atria, muscular ridges in the ventricle, and a spiral valve in the single outflow tract from the ventricle (see Chapter 13). The ventricle of noncrocodilian reptiles, although not morphologically divided, develops more specializations (see Chapter 14). The muscular ridges in the walls of the ventricle become so well developed that they form three distinct cava (cavum pulmonale, cavum venosum, and cavum arteriosum) within the single ventricle. Again, the pulmonary and systemic venous returns can be kept separate to a large degree within the ventricle and leave the ventricle via two separate outflow vessels (Figure 9.5). Crocodilian reptiles, like birds and mammals, have a right and a left ventricle separated by a common muscular septum (Figure 9.5).

Clearly there have been several successful evolutionary approaches to the degree of division of the embryonic heart, even though there is an overall evolutionary progression toward the complete developmental division of the heart, as exemplified by crocodilians, birds, and mammals. Yet, despite these sorts of substantive structural rearrangements, we still remain uncertain as to what factors control the division of the embryonic heart tube in

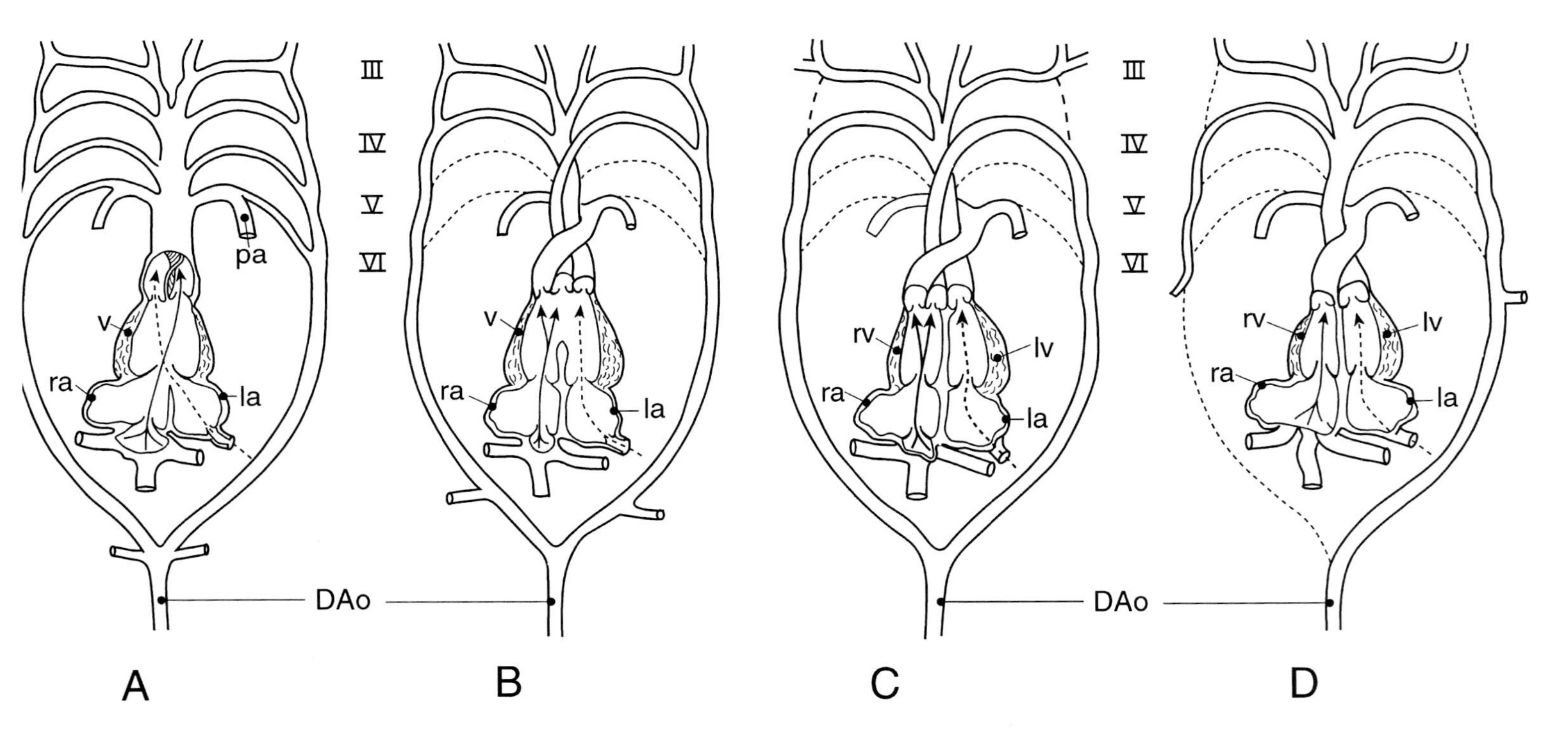

Figure 9.5 Highly schematic diagrams showing dorsal views of the heart and main arterial plan in tetrapods: (*A*) an amphibian, (*B*) a noncrocodilian reptile, (*C*) a crocodile, (*D*) a mammal. Deviations from the general plan for aortic arches are indicated by broken lines, and corresponding aortic arches have identical roman numerals. The basic patterns for the flow of pulmonary venous blood and systemic venous blood through the heart chambers are indicated by the solid and broken arrows, respectively. The evolutionary pattern toward a greater intracardiac anatomical separation of the pulmonary and systemic venous blood is evident (DAo = dorsal aorta, la = left atrium, lv = left ventricle, pa = pulmonary artery, ra = right atrium, rv = right ventricle, v = ventricle). (Redrawn and adapted from Romer, 1962.)

mammals. Likewise, the factors that allow the formation of a partial atrial septum in lungfishes but not in water-breathing fishes and the formation of a complete ventricular septum in crocodilian reptiles but not in other reptiles remain unknown. Perhaps answers to these sorts of questions would lead to a more complete understanding of the factors involved in the embryological division of the mammalian heart.

The coronary circulation

The heart has its own metabolic requirements. In adult birds and mammals the coronary circulation is the principal metabolic supply route. During early mammalian development, prior to the formation of the coronary circulation, the cardiac blood supply is the luminal (venous) blood being pumped through the heart chambers. Subsequent development of the coronary circulation occurs as the embryonic heart increases in size and workload and develops a compact myocardium. There is an evolutionary parallel to this ontogenetic development of coronary circulation.

The cardiac blood supply in most other vertebrates is analogous to that found during early cardiac development in mammals; that is, the heart is composed primarily of trabecular endocardium supplied by venous luminal blood. Certain fishes (Tota, 1983; Farrell & Jones, 1992) and reptiles (MacKinnon & Heatwole, 1981; Farrell, Gamperl, & Francis, in press) have relatively well developed coronary circulations. Even so, in all species except for the skipjack tuna, the trabecular endocardium with its venous luminal blood supply predominates over the compact myocardium and its coronary blood supply. There has been extensive writing on the evolution of the coronary circulation among vertebrates (e.g., Grant & Regnier, 1926; Halpern & May, 1958), but it has yet to be clearly established what selection pressure(s) and cellular signals favor (1) the development of a coronary circulation and (2) the partial or full retention of the trabecular endocardium. Both environmental hypoxia and high cardiac work levels have been suggested as important in the evolution of the coronary circulation in fishes (Davie & Farrell, 1991). Perhaps if this information was known, we might better understand the developmental switch in mammals to compact myocardium and a coronary circulation.

Future directions

The last portion of this chapter focused on a few specific examples of how a comparative phylogenetic approach might prove insightful in unlocking some of the mysteries of the development of cardiovascular systems. There are two others of a more general nature that are worth mentioning. Foremost, research into the linkages (if indeed they exist) between the homeotic gene family and cardiovascular development would be particularly valuable. Morphological diversity among arthropods and tetrapods has largely involved regulatory changes in the expression of conserved arrays of homeotic genes and the interactions between homeotic proteins and the genes they regulate. We need to know if, how, and under what conditions these regulatory changes affect cardiovascular systems. A comparative phylogenetic approach would be valuable for such studies because of the many different cardiovascular patterns found among extant organisms. Lastly, we know very little at the comparative level about how alterations in environmental conditions affect cardiovascular development through alterations to gene expression.

10

Morphogenesis of vertebrate hearts

JOSÉ M. ICARDO

The origin of the heart and early morphogenesis

The origin of the heart has been classically studied in the developing embryo by producing prospective maps of the early blastoderm (DeHaan, 1965; Rosenquist & DeHaan, 1966). Recently, fate mapping has traced the origin of the heart back to the primitive-streak stages (Garcia-Martinez & Schoenwolf, 1993). Precardiac cells ingress through the primitive streak and migrate toward the anterior embryonic pole to form a loose but cohesive epithelial sheet. Then, precardiac cells condense in crescent-shaped areas of the splanchnic mesoderm. The formation of the precardiac mesoderm is the first morphological indication of the heart anlage. The anlage is formed of premyocardial cells, preendocardial cells, and associated extracellular material (Figure 10.1). Later fusion of the paired primordium under the endodermal foregut results in formation of the primitive tubular heart (Figures 10.1–10.2). This process has been reviewed in Manasek et al., 1986; Icardo, 1988; and Icardo, Fernandez-Teran, and Ojeda, 1990.

The mechanisms that regulate condensation of the precardiac cells to form the cardiogenic crescent are poorly understood. The use of cell adhesion molecules (CAMs) has provided some information on this matter. N-cadherin, a calcium-dependent CAM that shows an even distribution on the surface of the early mesodermal cells, changes to an apical localization as the mesoderm splits into the parietal and splanchnic layers, the pericardial coelom forms, and mesodermal cells condense into the cardiogenic crescent (Linask, 1992). Similarly, integrin (a cell surface receptor for fibronectin) redistributes to a basal localization, and Na^+, K^+-ATPase activity becomes restricted to the lateral surfaces of the premyocytes (Linask, 1992). The development of these cell surface domains appears to play a fundamental role in the patterning of the splanchnic mesoderm as well as in the histotypical organization of the precardiac mesoderm. Changes in the expression of several CAMs also appear to be involved in the control of histogenesis at later developmental stages (Baldwin & Buck, 1994).

The origin of the endocardial and myocardial cells is still under investigation. It has tacitly been assumed that the two cell types originate from a common precursor within the precardiac mesoderm (Icardo, 1984). This contention is strongly supported by the fact that, in the chick, groups of both preendocardial and premyocardial cells express cardiac myosin (Tokuyasu & Maher, 1987a; Han et al., 1992), N-cadherin (later restricted to myocardium) (Linask & Lash, 1993), and QH1 (an antigen specific for endothelial cells) (Linask & Lash, 1993). A shared lineage for endocardial and myocardial cells is also suggested by clonal cell analysis of the early blastula in the zebra fish (R. Lee et al., 1994). On the contrary, a recent developmental analysis using replication-defective retroviruses indicates that the precardiac

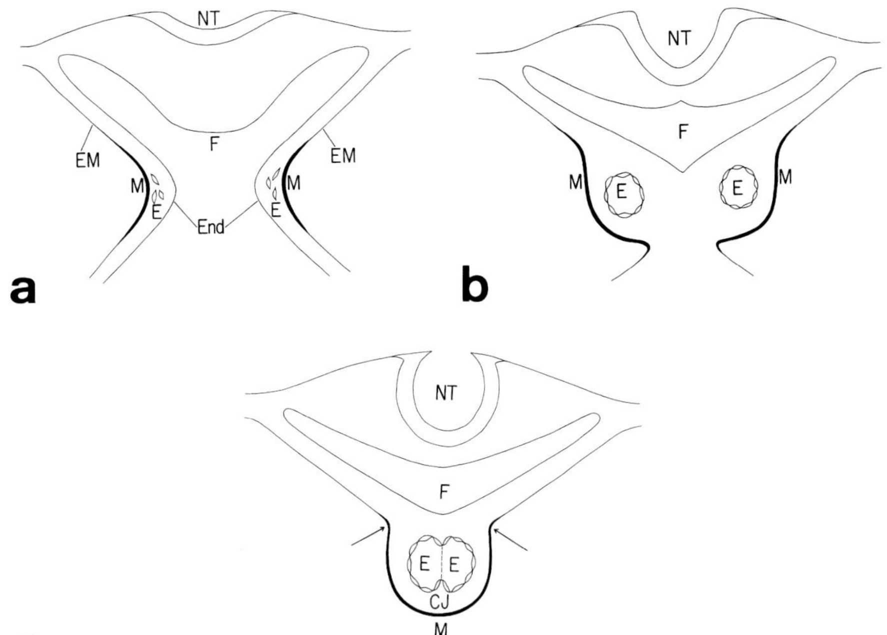

Figure 10.1 Diagrams illustrating the fusion process of the cardiac primordia in vertebrates. (*a*) The prospective myocardium (M) appears as a bilateral thickening of the splanchnic mesoderm (EM). The first endocardial cells (E) appear between the endoderm (End) and the myocardium. (*b*) The endocardium (E) has organized into tubes. The myocardium bends toward the embryonic midline and envelops the endocardial tubes. (*c*) The myocardium and the endocardium have fused. Dorsal to the developing endocardium, the right and left mesodermal layers are still unfused. Arrows indicate the dorsal mesocardium (CJ = cardiac jelly, F = foregut, NT = neural tube). (From Icardo, 1984).

mesoderm is formed by unipotential stem cells that exclusively give rise to cardiomyocytes (Mikawa et al., 1992). The reasons for this discrepancy are unknown. It has been suggested that preendocardial cells may not be targets for integration and expression of the retroviral genome (Jaffredo et al., 1993), although other explanations have been advanced (R. Lee et al., 1994).

Segregation and migration of the preendocardial cells occurs concomitantly with down-regulation of myosin (Tokuyasu & Maher, 1987a) and of N-cadherin (Linask & Lash, 1993). These cells also up-regulate synthesis of extracellular matrix proteins such as fibronectin (FN) and tenascin (Icardo & Manasek, 1991) and start to express several early markers for cells of endothelial lineage (Baldwin & Buck, 1994). Once preendocardial cells sort out, they appear to migrate as isolated cells between the precardiac mesoderm and the pharyngeal endoderm, probably by using their own FN (Icardo & Manasek, 1983). Preendocardial cells soon aggregate into tubes that later become fused cranially to the aortic arches and caudally to the vitelline veins (Icardo, 1984; Manasek et al., 1986).

Formation of the tubular heart involves the midline migration of the paired precardiac mesoderm. This movement is passive, as the endoderm carries the precardiac mesoderm, but the precardiac cells also move actively over the endoderm (Rosenquist & DeHaan, 1966). An early study showed the presence of FN at the endodermal–mesodermal interface (Icardo & Manasek, 1983). It was postulated that this FN was produced by the endoderm

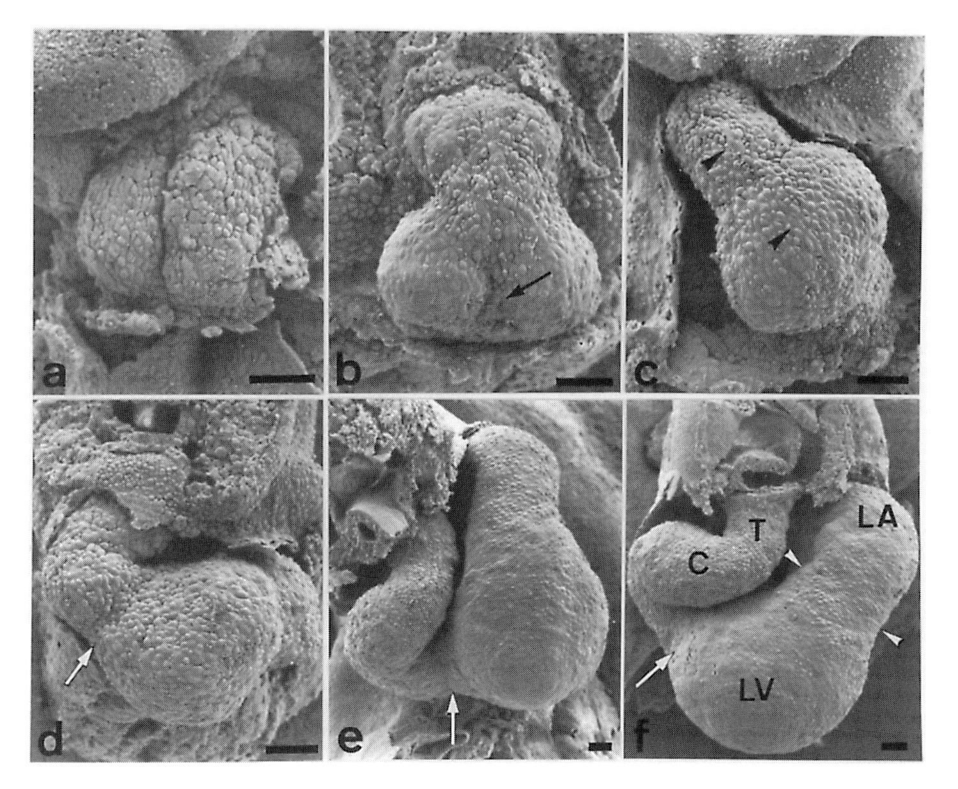

Figure 10.2 Scanning electron microscope composite showing the early development of the mouse heart (Swiss albino mice, 8.5–9.5 days postconception). (*a*) The two cardiac primordia have met at the embryonic midline. (*b*) Fusion of the two primordia results in formation of a single tubular heart. Fusion (*arrow*) is still in progress. (*c*) The heart tube is rotating to the right side of the embryo. Arrowheads indicate remnants of the fusion process. (*d*) With further bending and rotation, the heart is thrown into a loop. (*e*) At the end of the loop, the heart adopts a U shape. (*f*) After looping, the heart progressively adopts an adult configuration. C = conus, LA = left atrium, LV = left ventricle, T = truncus. Arrowheads indicate the atrioventricular canal. Arrows in *d–f* indicate the interventricular sulcus. Bar = 50 μm.

and that endodermal FN could play a complex role, serving as a template for the migration of the precardiac cells and providing morphogenetic information and extracellular cues for the histological organization of the myocardial tissue. Later, a gradient of FN was described in the precardiac areas, suggesting that premyocardial cells would migrate following haptotactic cues (Linask & Lash, 1988). The inhibition of heart formation after treatment of the precardiac areas with anti-FN antibodies appeared to reinforce the role of FN in directing the migration of the precardiac mesoderm (Linask & Lash, 1988). However, the existence of a gradient could not be confirmed by other authors (Icardo & Manasek, 1983; Drake et al., 1990). Also, it is not known whether the injection of the anti-FN antibodies (1) perturbs the cues necessary for directed migration, (2) merely disturbs the organization of a hospitable substratum for migration, or (3) interferes with the histological organization of the precardiac mesoderm or with the process of myocardial differentiation. On the other hand, in situ hybridization techniques have shown that most of the FN at the endodermal–mesodermal interface is synthesized by the precardiac mesoderm itself (Baldwin, Suzuki,

& Solursh, 1990) and not by the endoderm. Therefore, precardiac cells may not depend on exogenous FN as a guidance cue for their movements. Other migratory mechanisms, such as chemotaxis, have been ruled out (Easton, Bellairs, & Lash, 1990).

Heart induction

Induction is an established developmental phenomenon whereby a tissue causes the expression of specific traits in a responding tissue. Induction, however, is not a direct cause-and-effect process. The inductive response usually occurs in several steps, which together specify cell fate. Also, induction is not limited to changes in gene expression but involves changes in cell motility and behavior, in electrophysiological properties, and in cell cycle activity (Jacobson & Sater, 1988).

Although heart induction has been intensively studied over the years (Jacobson & Sater, 1988; Sater & Jacobson, 1990b), we do not yet know precise details of the inductive interactions necessary for the mesoderm to develop into heart tissue. Heart induction appears to occur in at least two steps. First, interaction with the dorsal lip of the blastopore (or the equivalent Hensen's node) appears to be necessary for mesodermal cells to acquire heart competence (Jacobson & Sater, 1988; Sater & Jacobson, 1990b). Permissive interactions necessary for development of this competence do not depend on physical contact between the intervening tissues and may be mediated by soluble substances (the "inducer" of classical embryology). Oncoproteins, activins, and several other members of the growth factor superfamilies, acting alone or in combination with other factors, are among the candidates for mesodermal inducers (Jacobson & Sater, 1988; Jessell & Melton, 1992). Furthermore, it appears that signals produced by the responding cells themselves may also be necessary. Once permissive interactions have been established, committed mesodermal cells can be detected by fate mapping (see above), and, if cultured, they will form beating cardiac tissue. However, these cells will not make a true heart. Further tissue interactions, the so-called formative influences, are necessary for the heart to acquire final differentiative properties, including development of shape.

Secondary induction requires physical contact between tissues (directly or through interposed extracellular matrix) and must be maintained throughout the interaction. The endoderm is the tissue layer most frequently implicated in secondary heart induction. However, it appears that the three germ layers (including the mesoderm itself) are directly implicated in the regulation of heart specification. First, the ectoderm and some of its derivatives, such as the neural plate, are inhibitors of heart differentiation and reduce the extent of the heart morphogenetic field (Jacobson & Sater, 1988). Spatial restriction of the heart field is also achieved by noncardiac differentiation in peripheral areas of the heart field (Sater & Jacobson, 1990a). The ectoderm is also involved in early inductive signaling (Frasch, 1995), provides a substratum for the migration of the early mesoderm, and seems to play another important role in heart development. The extracellular matrix associated with the ectodermal basal surface appears to endow the paired heart primordia with right–left specifications (Yost, 1992). On the other hand, the precardiac mesoderm may be able to self-control the extent of the cardiogenic area through the production of still unidentified activator and inhibitor substances (Smith & Armstrong, 1993). Furthermore, cell–cell interactions within the precardiac mesoderm also appear to have a significant influence on cardiac differentiation. In vitro, early committed mesodermal cells will attain a certain degree of differentiation only if cultured at high cell densities (Gonzalez-Sanchez & Bader, 1990).

The endoderm plays a fundamental role by instructing the precardiac mesoderm to maintain its differentiative process (Jacobson & Sater, 1988; Sater & Jacobson, 1990b). However, it is not known exactly how these differentiative instructions are passed on to the mesoderm. Most probably, the endodermal stimulus, similar to what has been shown in other processes of induction, does not need to be very specific (Icardo & Manasek, 1991). Peptide growth factors (GFs) have recently been identified as inductive signaling molecules during cardiac differentiation. Transforming growth factor-β1 (TGF-β1), platelet-derived GF (Muslin & Williams, 1991), insulin-like GF, and activin A (Yamazaki & Hirakow, 1994) increase the frequency of heart formation in explants of prospective cardiac mesoderm. On the other hand, basic fibroblast GF (bFGF) decreases this tendency in axolotls (Muslin & Williams, 1991). In avians, however, bFGF exerts a positive control on proliferation and differentiation of the cardiogenic cells. These effects appear to depend upon the presence of adequate cell surface receptors (Sugi et al., 1995). Also, cells in the precardiac mesoderm express TGF-β1 and TGF-β2 (Akhurst et al., 1990), suggesting that these GFs may be involved in cardiac differentiation directly or through an increase of the production of extracellular materials. Another GF, activin A, has been implicated directly in the expression of myosin heavy chain α-isoform, a marker for cardiogenic differentiation, in animal pole cells of *Xenopus* (Logan & Mohun, 1993). Thus, autocrine and paracrine roles have been suggested for GFs in both heart differentiation and proliferation. However, functional studies (the use of antibodies and antisense probes to neutralize the activities of specific GFs, gene knockout, etc.) are needed to elucidate these roles completely.

Matrix molecules have also been implicated in the differentiation of the precardiac mesoderm. Inhibition of the synthesis of collagen induces, in a dose-dependent manner, a drastic depression of myofibrillogenesis, suppresses heartbeat (Wiens, Sullins, & Spooner, 1984), and results in down-regulation of cardiac-specific genes (Fisher & Periasamy, 1994). Collagen type IV and laminin have also been implicated in morphogenetic events within the precardiac areas (Drake et al., 1990). The possible role of FN has already been mentioned.

It should be stressed that the reports just cited do not prove a direct role for any extracellular component alone in the differentiation of the precardiac mesoderm. Extracellular matrix molecules are highly interactive, and it is difficult to see how a single component could influence differentiative events (Icardo & Manasek, 1983). On the other hand, some of the effects observed after experimental modifications of the extracellular matrix may be related to physical disruption of cell–matrix interactions, with subsequent loss of the histotypical organization (Fisher & Periasamy, 1994). The expression of phenotypic traits in vitro has been shown to depend upon the exchange of information between the cardiac myocytes and the extracellular matrix (Simpson et al., 1994). Furthermore, the fact that many GFs remain bound to the extracellular matrix after secretion makes determination of primary and secondary events even more complicated.

Heart differentiation

The genetic cascade of events that leads to heart specification is poorly understood. Insights into this matter are provided elsewhere in this volume (Chapter 1). More substantial information is available on the temporal and spatial expression of the cardiac-specific proteins. A summary of the findings for the early chick heart is presented in Table 10.1. None of the cardiac-specific proteins are detected until the cardiac crescent forms and the heart primordium is readily identified under the light microscope (about Hamburger–Hamilton stage 7 in the chick embryo). Do these cells need to be grouped into the cardiogenic cres-

Table 10.1. *Time and site of appearance of cardiac-specific proteins in the early embryonic chick heart*

Protein	Time of appearance[a]	Site of appearance	References
Sarcomeric myosin heavy chain	Stages 7/8	Paired primordium	Han et al., 1992
α-actin	Stage 8	Paired primordium	Wiens & Spooner, 1983
Titin, actin, ventricular myosin	Stages 8/9	Paired primordium	Tokuyasu & Maher, 1987a
Ventricle-specific myosin heavy chain	Stage 8+	Paired primordium	Yutzey, Rhee, & Bader, 1994
Cardiac α-actin	Stage 9	Early ventricle	Ruzicka & Schwartz, 1988
Atrial-specific myosin heavy chain	Stage 9+	Paired (atrial) primordium	Yutzey, Rhee, & Bader, 1994
Smooth muscle α-actin	Stages 9/10	Early heart tube (including nonfused parts)	Ruzicka & Schwartz, 1988
α-actinin	Stage 10–	Early heart tube	Tokuyasu & Maher, 1987b
Cardiac and skeletal troponin isoforms	Stage 10	Early heart tube	Toyota & Shimada, 1981

[a]Stages according to Hamburger and Hamilton, 1951.
Source: Icardo, 1996, with modifications.

cent (i.e., the establishment of a specific architectural organization) to express specific genetic traits? Their developmental concurrence strongly suggests that this may be the case. Hence, formation of the cardiac crescent should be considered an important step within the cardiac differentiation program. Similar patterns of cardiac protein accumulation have been observed in several mammals, although some species differences do appear to exist (Lyons, 1994). The sources listed in Table 10.1 also show that the accumulation of cardiac-specific proteins is progressive, that these proteins soon aggregate into cross-striated patterns, and that differentiation follows a craniocaudal gradient as the precardiac mesoderm fuses caudally and development of the heart tube is completed.

Not only does differentiation of the heart follow a craniocaudal gradient, but each heart chamber also receives specific differentiative instructions, facts recognized a century ago. Pickering (1893) showed that each chamber of the early heart tube has an inherent pulsation rate and that functional development of the different chambers occurs craniocaudally. These functional properties are maintained in cultures of excised heart fragments (Barry, 1942), include development of specific action currents and action potentials, and appear to depend on the surface density and activities of a number of ion channels and pumps (DeHaan, Fujii, & Satin, 1990). The type and number of these intercellular channels vary considerably during development and between species (Wiens et al., 1995).

It should be stressed that these differentiative properties are temporally and spatially determined. In the chick, when excised fragments of the precardiac mesoderm are made to change positions with each other between stages 5 and 7, a normal heart with normal beating develops. If the exchange is made after stage 7, the resulting hearts are abnormal, cardiac bifida may occur, and, in some cases, there are anomalies of the heartbeat (Orts-Llorca & Jimenez-Collado, 1967; Satin, Fujii, & DeHaan, 1988). Similar results are obtained in amphibians (Copenhaver, 1926). What these experiments mean is that, after stage 7, some properties are irreversibly determined spatially and temporally, and thus exchanged fragments behave according to the site of origin (Satin, Fujii, & DeHaan, 1988). In fact, although myosin is detected in the heart primordium very early, ventricular- and atrial-specific myosin heavy-chain isoforms cannot be detected in the chick until stages 8+ and 9+, respectively (Yutzey, Rhee, & Bader, 1994). A similar process of chamber restriction seems to occur in fish (Stainier, Lee, & Fishman, 1993) and mice (O'Brien, Lee, & Chien, 1993; Lyons, 1994). Contact with the endoderm appears to be necessary for the establishment of chamber-specific properties, but the mechanisms are not known. It may simply depend on how long a specific region of the heart has been in contact with the endoderm. The acquisition of specific differentiative properties has been shown to depend on the degree and length of exposure of the responding tissue to growth factors (Jessell & Melton, 1992). Another possibility is the existence of subsets of cells due to changes in the properties of the primary inducer.

Differentiation and morphogenesis

The exact relationship between cell determination and the acquisition of shape is still an unresolved issue. In vivo, the acquisition of differentiated properties occurs concomitantly with development of organ shape. However, morphogenesis does not follow differentiation linearly. Indeed, there is more than simple biochemical differentiation in heart development (Icardo, 1988). A number of experiments repeated over the years have shown that cultures of isolated precardiac mesoderm result not only in formation of beating tissue but in the expression of differentiative properties as well (Icardo, 1988; Icardo & Manasek, 1991). Similarly, cardiac differentiation is obtained in cultures of pluripotent embryonic stem cells (Maltsev et al., 1994). However, heart formation does not occur. Embryonic myocardial tissue growing in the anterior eye chamber of rats develops histological characteristics and pacemaker action potentials similar to those found in adult rats (Tucker, Snider, & Woods, 1988). However, this culture system does not maintain the patency of the heart lumen, and there is no valve development.

The dissociation between cellular differentiation and heart morphogenesis can be illustrated with several other examples. The addition of retinoic acid to chicken hearts at stages 4–9 (Hamburger & Hamilton, 1951) results in expansion of the area of expression of atrial-specific myosin (Yutzey, Rhee, & Bader, 1994). However, this expansion does not convert the rest of the heart tube into an atrial chamber. Indeed, myocardial cells expressing atrial myosin are observed in a still morphologically recognizable ventricle. Mouse myocardial cells (Miner, Miller, & Wold, 1992) can be made to express MyoD, a transcription factor involved in skeletal muscle gene regulation. However, ectopic expression of skeletal muscle–specific (regulatory and structural) genes does not transform mouse myocardial cells into multinucleated myotubes.

Most likely, the history of a given cell cannot be dissociated from its immediate environment. At the organ level, it is the adequate integration of the different components of the heart that ultimately results in development of heart shape and function. Obviously, this includes not only the expression of a set of specific genes but electrical maturation and the acquisition of shape as well (Icardo, 1988).

The cardiac loop

The single heart tube formed in the embryonic midline is not a complete heart. This primitive heart appears to represent the future trabeculated part of the right ventricle only. Continued fusion of the paired anlage in a caudal direction brings about merging of the future trabeculated part of the left ventricle, the atrium, and the sinus venosus, which are thus progressively incorporated into the heart tube. As caudal fusion of the paired primordia progresses, the heart tube bends, rotates to the right, and is thrown into a loop (Figure 10.2). It is at the end of looping that the different parts of the heart can be clearly recognized and that regional divisions are established (Icardo, 1984; Manasek et al., 1986). Evidence suggests that the conotruncal portion of the heart is incorporated into the developing tube at later stages, cranial to the initial site of fusion of the paired primordia (Icardo 1984, 1996). Furthermore, the truncus ("outflow tract," "truncus arteriosus") does not seem to appear until stage 15 in the chick (Garcia-Pelaez & Arteaga, 1993). Indeed, compared with the rest of the heart tube, the truncus presents certain developmental differences that may be related to both the origin and the fate of cells in this area of the heart.

The myocardium of the truncus appears to be formed by continued incorporation of tissue from the splanchnic mesoderm at the arterial end of the heart (Virágh & Challice, 1973). A different origin from the rest of the heart is also suggested by the fact that the truncus never expresses the ventricle-specific myosin light chain 2, which is readily expressed in the conus (O'Brien, Lee, & Chien, 1993). Also, the truncus presents a number of singularities with respect to the expression of several protein isoforms (Ruzicka & Schwartz, 1988; O'Brien, Lee, & Chien, 1993). On the other hand, the endocardium of the outflow tract has a double origin (Noden, 1991). It originates from the heart tube endocardium and from the progeny of cells that migrate away from the cardiogenic plate to mix with mesodermal cells in the head region. In addition, a unique feature of the truncus is that it becomes populated by cells of neural crest origin. These cells participate actively in the septation of this region, in the development of the aortic arches, and in cardiac innervation (Kirby, 1988b, 1993). The developmental fate of the conotruncal region may also help to explain these differences. While the conus is being incorporated into the ventricles, the truncus is being transformed into the proximal part of the two great arteries (for reviews, see Icardo 1984, 1996; Manasek et al., 1986).

The cardiac loop is the first morphological asymmetry to occur in the early embryo, and as such, it has been the subject of research interest for decades. However, the mechanisms implicated in looping are still under discussion. Also, it has not been sufficiently emphasized that looping involves two morphogenetic components: looping itself and the direction of looping. The two components can be separated (Figure 10.3) and appear to be governed by different mechanisms.

Loop formation

What are the mechanisms involved in the formation of the cardiac loop? Isolated and grafted early tubular hearts are able to generate their own bend (Manning & McLachlan, 1990). Hence, most of the requirements for looping appear to be contained within the heart itself. The looping heart is a hollow structure organized into layers. There is an outer layer (myocardium) and an inner layer (endocardium). The space between the two tissue layers is occupied by a broad expanse of extracellular material called cardiac jelly. The endocardium is the tissue layer located at the interface between the circulating blood and the rest of the heart. Early endocardial cells show specific morphologies at specific places and times and appear to be implicated in a number of morphogenetic events during heart development (Icardo, Fernandez-Teran, & Ojeda, 1990). During looping endocardial cells are oriented

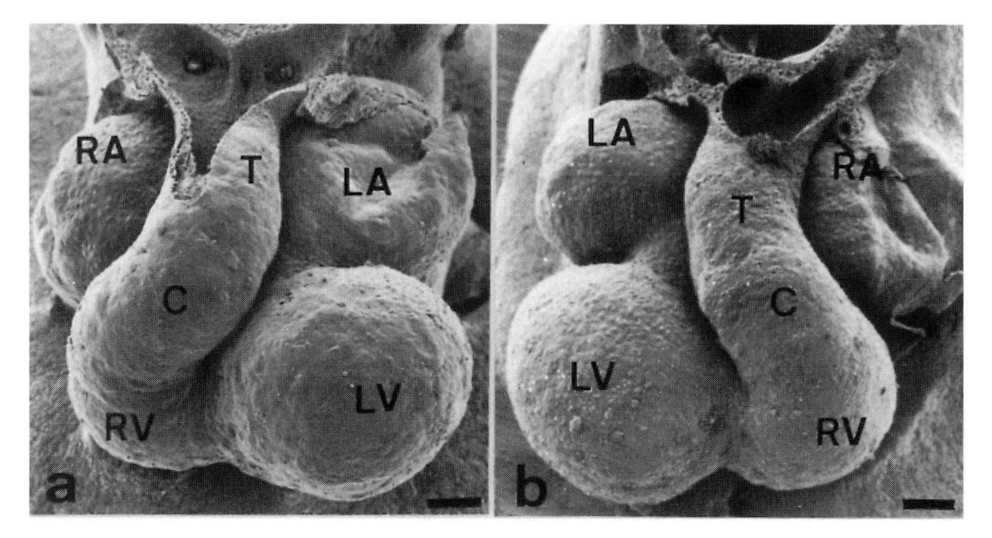

Figure 10.3 Scanning electron microscope micrographs of two iv/iv mouse hearts. The heart on the left (*a*) shows a D-loop. The heart on the right (*b*) shows an inverted, L-loop. Both hearts are otherwise morphologically normal and show a mirror image. The truncoaortic sac and the origin of the aortic arches have been exposed in the two cases (C = conus, LA = left atrium, LV = left ventricle, RA = right atrium, RV = right ventricle, T = truncus). Bar = 100 µm. (From Icardo, Arrechedera, & Colvee, 1995, Fig. 3a.)

in a dorsoventral direction (Icardo, Hurle, & Ojeda, 1982). However, this may be secondary to the rapid circumferential expansion of the heart. The endocardium does not appear to have the structural integrity required to produce looping. The myocardium contains, throughout looping, a single population of developing myocytes. These cells progressively accumulate myofibrils in their cytoplasm, undergo changes in the biosynthetic pattern of specific protein isotypes, and acquire electrical characteristics typical of the mature heart. In addition, cardiac myocytes engage in the synthesis of extracellular material, undergo DNA synthesis and mitosis, and phagocytose cellular debris (for reviews, see Manasek et al., 1986; Icardo, Fernandez-Teran, & Ojeda, 1990; Icardo & Manasek, 1991). However, none of the different myocardial cell activities tested appears to be required for loop formation (Manasek et al., 1984b; Icardo, 1988, 1996).

The layer of cardiac jelly makes up the bulk of the wall thickness. The cardiac jelly is a complex extracellular matrix that undergoes compositional and structural changes during looping (for reviews, see Manasek et al., 1986; Icardo & Manasek, 1991; and Chapter 5 of this volume). It is also essential to heart function. Cardiac jelly increases the speed of narrowing of the heart lumen during systole, facilitates blood output, and, in the absence of anatomic valves, acts as a primitive valvular system that prevents blood regurgitation during the heart contractile cycle (Manasek et al., 1986).

It has been suggested that the force necessary to induce shape changes during looping could be generated by the extracellular matrix (Manasek et al., 1984a, 1984b). Because it is rich in hyaluronic acid and other glycoaminoglycans, the cardiac jelly is able to trap water and ions and become a highly hydrated macromolecular complex. This creates an internal osmotic pressure high enough to, at least, make the heart inflate (Manasek et al., 1984a, 1984b). Since osmotic pressure is uniform throughout the heart tube (Manasek et al., 1986), the heart will inflate uniformly, but it will not necessarily bend. Control of bending may be achieved by a higher stiffness of the cardiac jelly in the area of the dorsal mesocardium where differential distribution of wall stress occurs along the heart tube. Rotation requires the presence of some anisotropic structure to make the heart deform differentially. It has been proposed that wall anisotropy is created by the myocardium, resulting in asymmetric bending (Manasek et al., 1984a).

Myocardial fibrillogenesis is temporally linked to looping. If myofibrillogenesis is disrupted by cytochalasin B, looping is prevented (Manasek et al., 1984a; Itasaki et al., 1991). Given this relationship, it was suggested that myofibrils should exhibit suitable anisotropy along the heart tube. The overall geometry of the filamentous system of the cardiac cells is consistent with the proposal that the myocardium has regions with different compliances and may drive the process of heart rotation during looping (Manasek et al., 1984a). Furthermore, the presence of regional differences in actin bundle orientation during looping correlates well with the presence of lines of deforming stress in the myocardium (Itasaki et al., 1991). Therefore, the myocardium could regulate the rotation of local regions of the heart tube. Proof of this hypothesis would require a direct relationship between alterations in myocyte function and graded changes in looping.

In essence, Manasek's model of heart morphogenesis involves a complex set of cell activities – from the expression of specific genes and myocardial cytodifferentiation to the secretion of extracellular materials and hydration of the matrix – that together result in organ shape. Although the possible role of extracardiac factors in looping has not been considered, this model is a testable hypothesis where many of the different parameters involved can be experimentally manipulated. Nevertheless, experimental testing has resulted in inconclusive and/or contradictory results. The role of the cardiac jelly in supporting heart shape and looping is still unclear. The chick embryonic heart collapses and totally loses its shape after the injection of proteolytic enzymes (Nakamura & Manasek, 1981). Similarly, the heart of *Xenopus* does not loop when proteoglycan synthesis is inhibited (Yost, 1992). These results speak in favor of an important role for the cardiac jelly in supporting early heart morphogenesis. However, when whole rat embryos are cultured in the presence of hyaluronidase, the hearts loop despite the disappearance of the cardiac jelly space (Baldwin & Solursh, 1989). Thus, the presence of hyaluronic acid is not essential to loop formation. At the same time, the relation between myofibrillogenesis and looping is still under discussion. Addition of colchicine to early developing hearts, both in vivo and in vitro, results in considerable disruption of myofibrillogenesis (J. M. Icardo & J. L. Ojeda, unpublished data). Under these conditions, the loops are abnormal, but the hearts still loop (Icardo & Ojeda, 1984).

Direction of looping

The second morphogenetic component of looping is the direction of rotation. Rotation appears to be intrinsic to the heart primordium. Furthermore, the consistent direction of looping indicates the presence of some kind of preexisting (intrinsic) asymmetry. This is clearly illustrated by experiments in which a complete cardiac bifida is formed. When fusion of the paired anlage is prevented, two independent beating hearts develop, one on each side of the embryonic midline (Figure 10.4). In these cases, the left heart shows a relatively normal D-loop, whereas the right heart shows an inverted L-loop (Copenhaver, 1926; DeHaan, 1965). The same results are obtained when the heart is allowed to develop after removal of one side of the cardiogenic area, and similar evidence is obtained from the

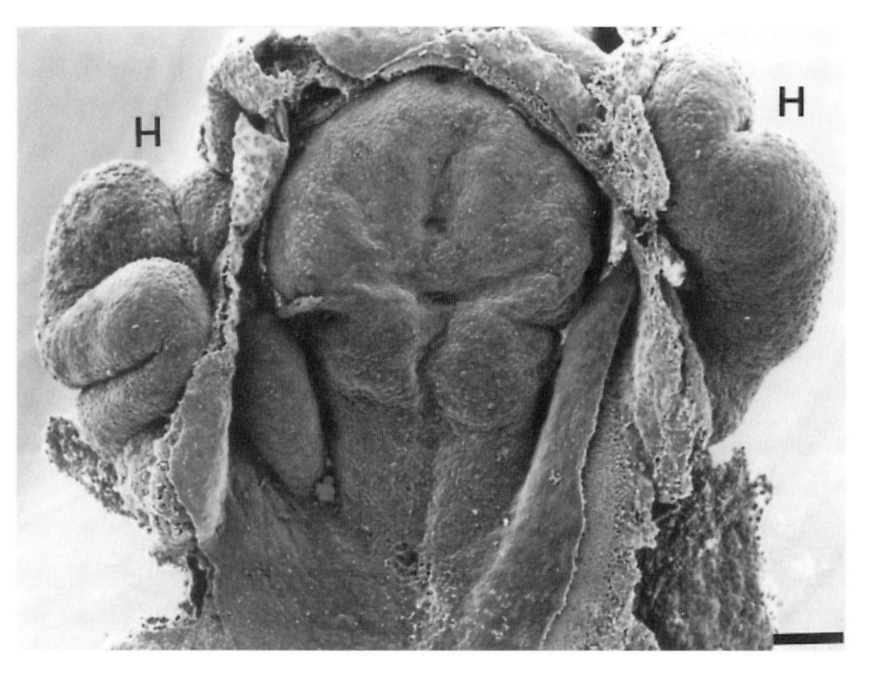

Figure 10.4 Scanning electron microscope micrograph of a double-hearted chick embryo. Fusion of the paired anlage has been prevented by cutting through the midtissues of the anterior intestinal porta. Two independent beating hearts (H) have formed, one on each side of the embryonic midline. Bar = 100 μm. (From Icardo, Fernandez-Teran, & Ojeda, 1990.)

study of conjoined human twins. The left twin shows normal situs while the right twin shows visceral inversion, including an inverted loop (Seo, Shin, & Chi, 1985). These experiments indicate that each cardiac primordium has different positional information engraved. However, the presence of intrinsic differences between the two primordia that may be able to push the emerging tube to the right has not been substantiated by experimental work (Stalsberg, 1970; Icardo, 1988).

To examine the developmental potencies of the right and left cardiac areas, homograft experiments have been performed in amphibians (Copenhaver, 1926) and chicks. In the chick, double-right- and double-left-sided hearts have been produced by transplanting the right or the left cardiac area into a host. When this procedure is performed at stages 4–5, an inverted heart is obtained in a small number of cases (Salazar del Rio, 1974; Hoyle, Brown, & Wolpert, 1992). Otherwise, the hearts develop normally. When these experiments are performed at stage 6 and later, an increase of close to 50% in the number of inverted hearts occurs. Also, these embryos show an elevated number of cardiac bifida and heart abnormalities (Salazar del Rio, 1974; Hoyle, Brown, & Wolpert, 1992). These results indicate that, at stages 4–5, the direction of the curvature (the sidedness of the loop) is not fixed. After stage 5, however, changes in the intrinsic properties of the precardiac mesoderm occur. These changes correlate chronologically with the acquisition by the myocardial cells of fixed developmental properties (see previous discussion). Thus, specific differentiative properties and right/left information are apparently acquired and/or expressed at about the same time.

How is this patterning information acquired? When the extracellular matrix associated with the basal surface of the ectoderm is perturbed at the stages of precardiac cell migration, right/left information is suppressed and heart situs develops randomly (Yost, 1992). These experiments in *Xenopus* implicate heparin sulfate proteoglycans in the establishment of handedness (Yost, 1992). However, many different experiments have induced inversion of the heart loop. In many cases, the specific modifications of the extracellular matrix are not easy to define. Therefore, one has to ask whether there is a single mechanism sensitive to a number of manipulations, or whether there is a cascade of events, each step within the cascade being sensitive to external modifications. Looping is likely a very complex event regulated at multiple levels of organization, with a delicate balance between all the intervening factors (Stalsberg, 1970). These different factors may compensate for each other.

Finally, it should be stressed that, whatever the ultimate mechanisms of looping are, the direction of the cardiac loop is genetically controlled. Furthermore, loop direction is inevitably linked to the establishment of body asymmetry. This is clearly indicated by the analysis of several loss-of-function mutations (Layton, 1978; Singh et al., 1991) that result in situs inversus in 50% of the mice. Furthermore, genetic control appears to be exerted at several different levels within the genome. A recently described insertional mutation (Yokoyama et al., 1993) produced situs inversus in 100% of the subjects. Why such a complex system of signals is necessary to control both body asymmetry and the direction of the cardiac loop is unknown. In this regard, several genes have been reported to be expressed asymmetrically in the chicken heart field during gastrulation (Levin et al., 1995; Lowe et al., 1996). The proteins coded by these genes, which act in sequence, appear to be involved in determination of the heart situs. It is hoped that new data will shed light on the complex series of events involved in formation of the cardiac loop.

Unanswered questions and some critiques

I have tried in this chapter to summarize our knowledge of the early development of the heart, reviewing from my own morphologic perspective old facts and new developments. The techniques of classic embryology and the use of modern, sophisticated tools have yielded an enormous amount of information. The feeling though is that we are still asking the same unanswered questions. How is heart induction established? How do the inducing molecules work? Which molecular pathways are involved in cardiac differentiation? Why does the heart always loop to the right? Which mechanisms control looping?

Perhaps it is the right time to make some critical comments about the way heart development is being perceived. This criticism is not intended, in any way, to be personal. If so, it should start with myself as a volunteer in the heart field. The limitations of classic embryology have been made clear by its own inability to answer questions in a definitive way. Researchers do sometimes duplicate experiments, trying to find more definitive answers. However, translating well-known facts into modern biological language does not improve our knowledge overall. On the other hand, molecular biology seems to have, at this point in time, fallen short of expectations, despite the fact that the newest and most exciting findings come from this area.

Have we not done enough? Or is it just that things are much more complex than anticipated? Probably both. For example, we expected that the cascade of gene activity that results in heart differentiation in relatively simple organisms, like flies, would work similarly in more complex organisms. However, it does not. Functional redundancy appears to work as a safety mechanism. However, another question must be posed: Do similar genes perform the same function in different organisms? On the other hand, are we asking the right questions in the proper way?

Despite all these doubts, the importance of several active fields of study should be underscored. The organization of cells into tissues and organs is the most genuine morphological demonstration of the expression of the cell genetic programs. In fact, many molecular markers do not appear until after tissue organization begins to be established. The study of the molecules involved in cell recognition and adhesion will help us to identify the signals that modulate cell behavior during differentiation and morphogenesis. The mechanisms that control cell death (and cell survival), the role of growth factors, and the dynamics of the signals established between the cells and their immediate microenvironment are promising areas of study also. Finally, it is not unrealistic to anticipate that the deciphering of the genetic cascade involved in heart differentiation will provide us with some of the answers to questions we have been asking for decades.

11

Invertebrate cardiovascular development

BRIAN R. McMAHON, GEORGE B. BOURNE, AND
KA HOU CHU

Introduction

Despite the long history of descriptive invertebrate embryology, there have been few studies specifically on the functional development of the heart or circulatory system. Some mention of the initiation of heartbeat or of some ultrastructural details of heart muscle information have been included en passant in a number of studies, but hardly any studies specifically address the physiology of the developing heart.

Even for adult invertebrates cardiovascular physiology is poorly understood (McMahon, Smith, & Wilkens, 1997). What information we do have is sketchy. For instance, we know a great deal about the physiology of the cardiac control system in crustacean arthropods (Cooke, 1988) but much less about the generation of heartbeat in any other invertebrate group. For these reasons this chapter is limited to a protostome group, including the annelids, arthropods, and molluscs, and a deuterostome group, including the echinoderms, hemichordates, and urochordates, for which there is (1) a modicum of knowledge concerning cardiovascular development and (2) sufficient knowledge of the physiology of the adult cardiovascular system to allow some interpretation of developmental changes. Since there is great diversity in the structure of invertebrate circulatory systems, this chapter is organized by taxonomic group rather than by process. To facilitate comprehension of the evolutionary relationship between the groups covered and the protovertebrates a simplified evolutionary tree is given in Figure 11.1.

Annelida

The structure of adult annelid blood vascular systems has been summarized by Hanson (1949) and Martin (1980); the latter author also examined the phylogenetic and ontogenetic origin of the system, particularly the heart. The blood vascular system of the polychaetes and oligochaetes typically consists of dorsal and ventral longitudinal blood vessels connected by commissural vessels usually around the alimentary tract. Both dorsal and ventral vessels are contractile and have valves. The blood flows anteriorly in the dorsal vessel, which is the main blood-collecting vessel, and posteriorly in the ventral vessel, which serves as the major distributing vessel. In each segment the ventral vessel gives rise to segmental vessels and capillaries. Certain commissural vessels in the anterior segments of oligochaetes that are conspicuously contractile have been referred to as "tubular hearts," "lateral hearts," or "pseudohearts." Tubular hearts are also present in many polychaetes,

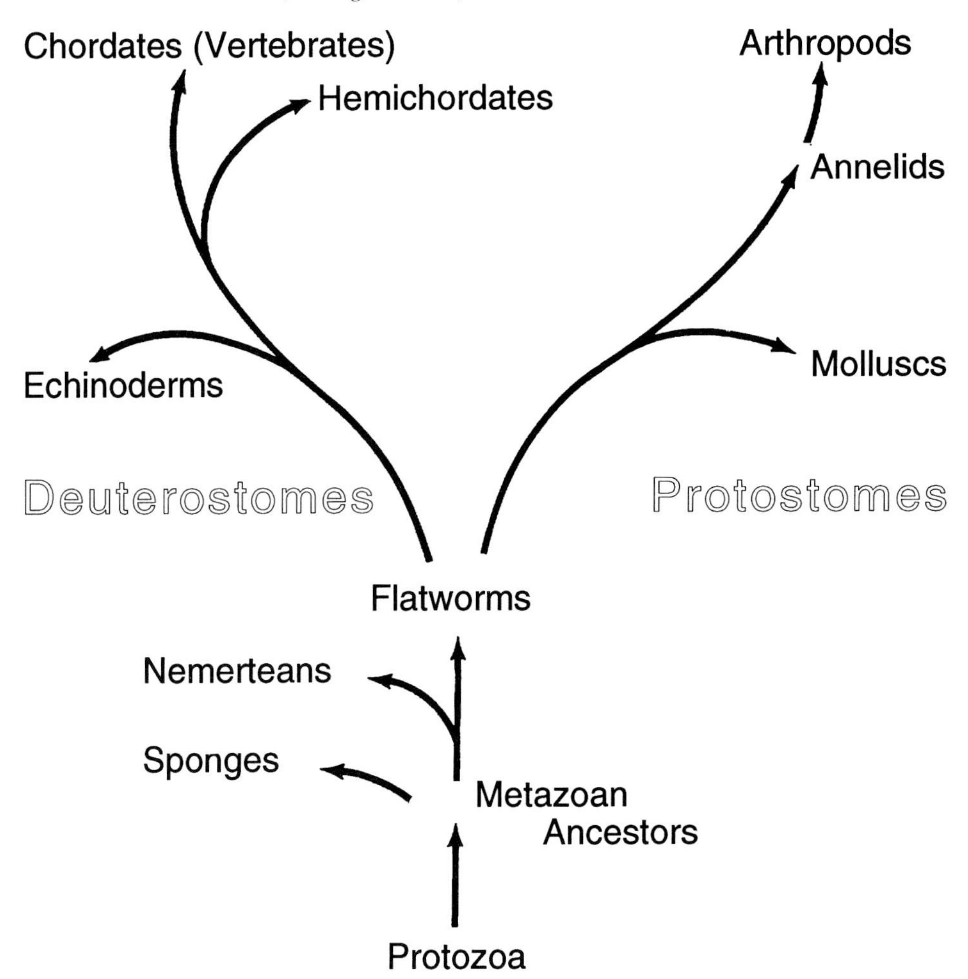

Figure 11.1 Taxonomic relationships between major invertebrate groups and the protovertebrates.

especially in those species with anterior gills or large anterior parapodia (e.g., in *Flabelliderma;* see Spies, 1973). In species without a heart, such as *Nereis virens* and *Nephtys californensis,* most of the blood vessels are contractile (Nicoll, 1954; Clark, 1956).

The earlier literature on the development of the blood vascular system in polychaetes and oligochaetes has been reviewed by Anderson (1966, 1973). The system forms de novo in the site of the former blastocoel. In many cases, a sinus first develops by separation of the inner walls of the somites from the gut epithelium, and the longitudinal vessels are formed by the lateral walls of the sinus coming together above the gut. In other cases, the vessels form directly by separation of the apposed walls of the mesentery (Figure 11.2). The commissural vessels develop by partial separation of the epithelia of the septa between the somites. A modern ultrastructural study on the development of the blood vascular system in the polychaete *Sabellaria cementarium* has been provided by Smith (1986). The system begins to form in the metatrochophore, concomitant with the onset of segmentation. This observation supports the suggestion by Ruppert and Carle (1983) that the adaptive signifi-

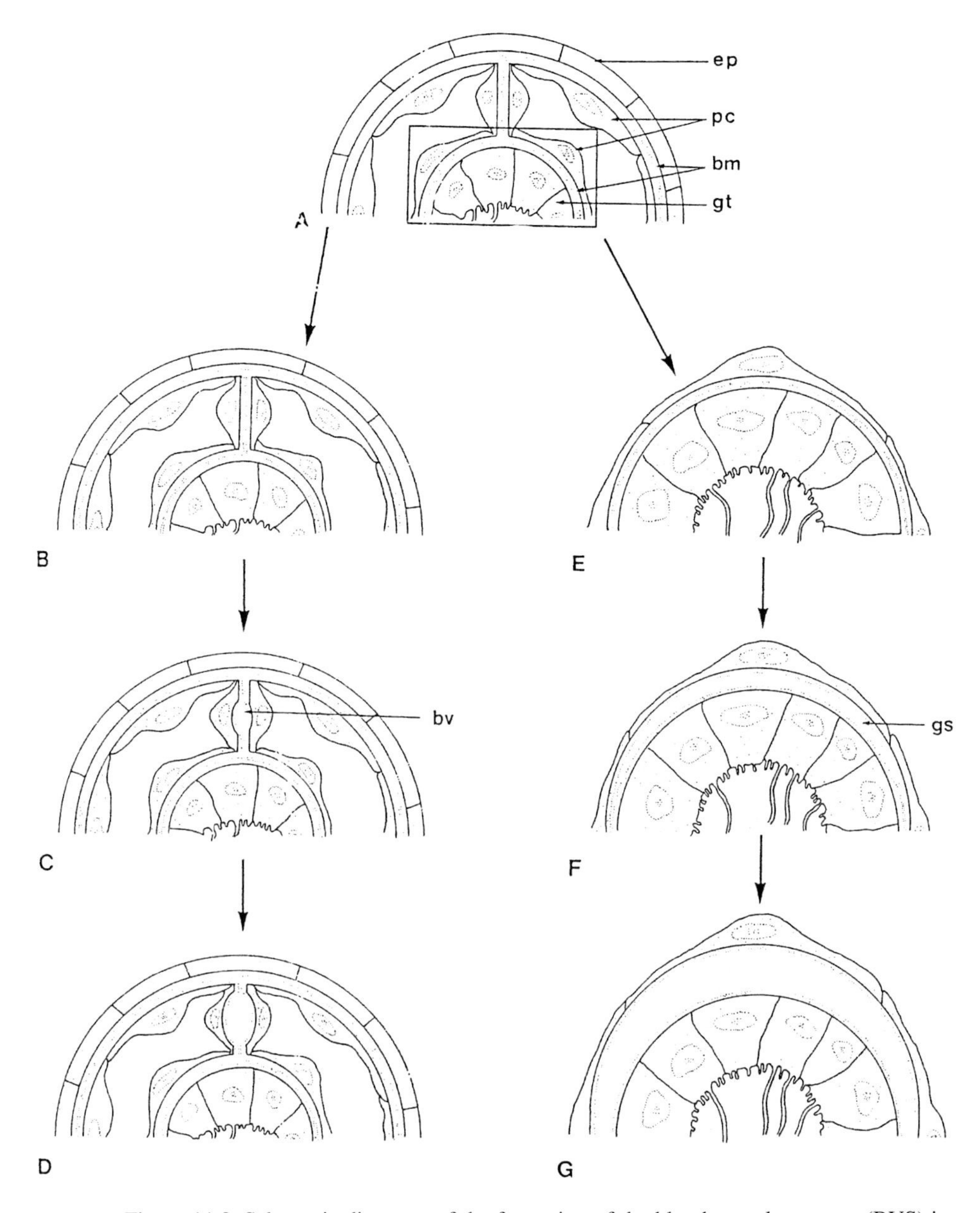

Figure 11.2 Schematic diagrams of the formation of the blood vascular system (BVS) in an annelid. (bm = basal ECM, bv = blood vessel, ep = epidermis, gs = gut sinus, gt = gut, pc = peritoneal cell). (*A*) Half of a cross section of a hypothetical annelid prior to BVS formation showing basal extracellular matrices (ECMs) of the adjacent cell layers lying in direct apposition. (*B–D*) Blood vessel formation; a blood vessel forms by the separation of apposing basal ECMs of adjacent peritoneal cells. (*E–G*) Blood sinus formation; a blood sinus forms by the separation of apposing basal ECMs of peritoneal and gut epithelial cells. (From Smith, 1986.)

cance of the blood vascular system was to provide transport across septal bulkheads between regionally restricted exchange sites in a segmented animal. The major blood vessels and sinuses of the system, formed de novo by the separation of basal extracellular matrices, are fully developed in the nectochaeta, the stage competent to metamorphose. The system in the nectochaeta consists of supraesophageal, lateral esophageal, dorsal, and ventral vessels and of dorsal and ventral sinuses (Smith & Chia, 1985). Because vascular blood fluid was observed, it is likely that the system is functional in the nectochaeta. A similar blood vascular system has been observed in the larvae of a number of polychaete species (Wilson, 1932; Segrove, 1941). In *Scoloplos armiger,* the dorsal and ventral vessels form during embryonic development, but the first pair of lateral hearts do not develop until the embryos hatch to burrowing preadults (Anderson, 1959).

In leeches (Hirudinea) the original blood vascular system is reduced or absent and the coelom assumes an important role in circulation (Martin, 1980). The rhynchobdellid leeches have remnants of a closed blood vascular system typical of oligochaetes, but a series of coelomic channels serves as a supplementary circulatory system. The major vessels of the original blood vascular system generally lie within the new coelomic channels, but the two systems are not connected. In the more modern arhynchobdellid leeches, the blood vascular system has disappeared completely, and a complex network of coelomic vessels has become the sole circulatory system (Mann, 1961). Embryologically these coelomic vessels are derived from cavities generated in the anterior somites during mesodermal development (for reviews, see Mann, 1961; Fernández, Téllez, & Olea, 1992). In *Hirudo medicinalis,* there are four major longitudinal vessels: dorsal, ventral, and paired laterals that are connected to one another at both ends of the animals (Stent, Thompson, & Calabrese, 1979). Blood flows anteriorly in the lateral vessels and posteriorly in the dorsal and ventral vessels. The lateral vessels, referred to as "lateral hearts" or "heart tubes," provide the major propulsive force, distributing blood to the body organs via lateral abdominal sinuses. The myogenic rhythm of the lateral hearts is modulated by both neural and humoral controls, as reviewed by Stent, Thompson, and Calabrese (1979) and, more recently, by Calabrese and Arbas (1989).

Arthropoda

The basic arrangement of the heart and blood vascular system of the arthropods differs greatly from that of the vertebrates and the annelid closed systems just described, and thus a brief outline is presented here. All arthropods have an "open" circulatory system in which fluid circulates within a hemocoelomic cavity and, at some point, bathes tissues directly. This fluid serves the functions of both blood and lymph and is thus termed "hemolymph." In most arthropods hemolymph is circulated within the hemocoel by a dorsally located muscular heart. The ventricle of the heart is suspended within a reservoir chamber (pericardial cavity) by elastic or musculoelastic elements termed "alary ligaments." Muscular contraction forces hemolymph out of the heart and stretches the alary ligaments. The potential energy stored in the ligaments allows the heart to return to presystolic volume as hemolymph refills the heart via ostial valves in the ventricular wall.

In the ancestral arthropod vascular system ostial valves, alary ligaments, and distribution vessels were all arranged segmentally along a tubular heart extending the length of the body. This design strongly supports the contention that the blood vascular system originated as a means of transporting fluids across segmental partitions (Ruppert & Carle, 1983).

Although the contractile region originally spanned the whole length of the body, reduction in the length and number of segments involved in muscular pumping has occurred during evolution of most arthropod groups and is especially marked in the decapod crustaceans.

The complexity of the vascular distribution system also varies greatly within the Arthropoda, ranging from a single arterial vessel (insects) to the extremely complex systems seen in shrimp, crabs, and lobsters. In these decapod crustaceans, the heart is condensed into a muscular ventricle in the posterior thorax, but the original pattern may be seen in a variety of crustaceans in which either the posterior or the anterior aorta still contains muscle (Martin, Hose, & Corzine, 1989; J. L. Wilkens, personal communication), representing a remnant of the original contractile dorsal vessel. The lateral arteries (and lateral segmental arteries, where present) are the descendants of the original segmental delivery vessels.

Horseshoe crabs (Merostomatidae)

Horseshoe crabs are probably the most ancient living arthropods. Horseshoe crabs have the long, tubular, segmentally organized heart described earlier plus a very complex distribution system (Redmond, Jorgensen, & Bourne, 1972).

Unfortunately, functional development of the heart of this group has received minimal attention. Carlson and Meek (1908) demonstrated that the heart of *Limulus polyphemus* develops from a fusion of the mesoblastic walls of the coelomic cavity at about the 20th day of development and becomes active at about the 22nd day. At this stage the heart is a tube formed of a single layer of muscle cells, which may be syncytial, and is suspended in the pericardial cavity as described earlier. Development starts in the posterior of the embryo and proceeds forward.

The hearts of many, but not all, adult arthropods are excited neurogenically via a cardiac ganglion located in the heart wall. Carlson and Meek (1908) pointed out that no organized nervous tissue (i.e., cardiac ganglion) was present at day 22, when heartbeat began. At day 28, however, organized neural development was observed in the middle of the heart, from whence it spread posteriorly and, to a lesser extent, anteriorly. This foreshadows the situation seen in the adult, but the typical adult ganglionic form was not evident at day 33, the latest embryonic stage tested. Carlson and Meek (1908) concluded that the heart is initially excited myogenically but that pacemaker activity is gradually taken over by the developing cardiac ganglion. These authors also suggested that basic myogenicity may also persist into the adult stage, initiating a controversy that remains unresolved today (Watson & Groome, 1989). Interestingly, Yamagishe and Hirose (1992; H. Yamagishe, personal communication) proposed that the heart of the isopod *Ligia exotica* was also initially myogenic, with the neurogenic condition developing in later embryonic or early juvenile life. These authors also appear to suggest that innervation of the heart muscle from the central nervous system precedes the development of the cardiac ganglion.

Crustacea

Artemia. The brine shrimp *Artemia* (Anostraca, Branchiopoda) has one of the simplest circulatory systems seen in the Crustacea. In the adult the heart runs almost the full length of the body, with ostial valves and alary ligaments arranged segmentally, and is of the tubular type (Figure 11.3 bottom), although Ocklund et al. (1982) suggested that the tube is incomplete in the thoracic segments (Figure 11.3 top), with the dorsal surface formed by the body

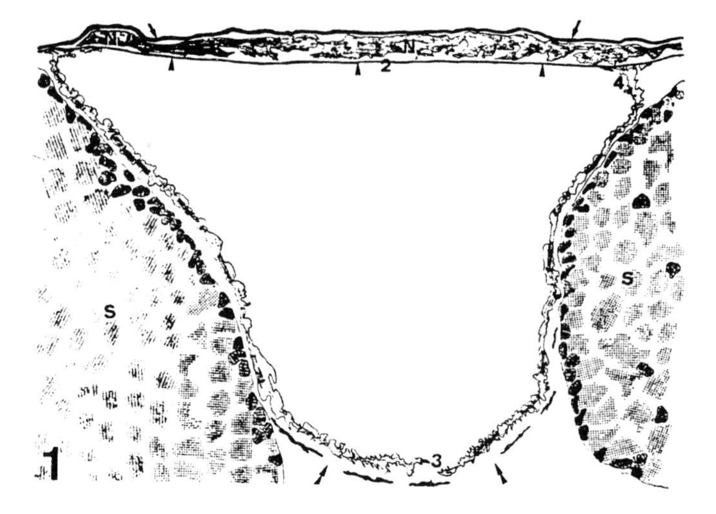

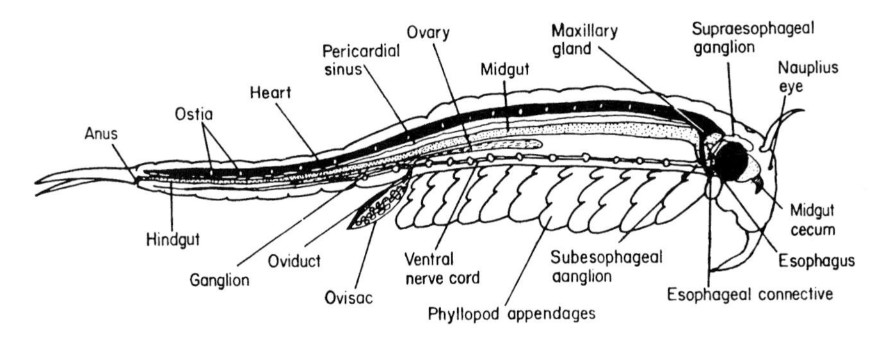

Figure 11.3 Diagram (*bottom*) of the circulatory system of *Artemia.* (From McLaughlin, 1983.) Electron micrograph (*top*) of thoracic region of *Artemia* showing cardiac structure. (From Ocklund et al., 1982.)

wall and the ventral surface formed from connective tissue. There are no distribution vessels at any developmental stage.

Development in *Artemia* is associated with several free-living naupliar and juvenile stages (instars). Spicer (1994) observed that heartbeat initially occurs in the late naupliar instars and associated these first heart pulsations with the differentiation of the first thoracic segment. In subsequent development *Artemia* expresses additional new body segments caudally until the adult number is reached. New heart "chambers," each with a pair of ostia, etc., are added with each body segment. After this stage is complete, the larvae grow by elongation of individual segments. Heart length increases similarly, presumably by elongation of the cardiac muscle cells. Note that heart differentiation here proceeds caudally, in contrast with the anteriorly directed differentiation reported for *Limulus* (Carlson & Meek, 1908).

Decapod crustaceans. Crustacean embryology was reviewed by Shiino (1968), who included some details of heart development in the lobster *Panulirus,* a macruran decapod crustacean. In this species a dorsal aorta is formed from dorsal mesodermal bands that spread out along the dorsal body wall and join to form a tube. In the anterior region the heart is formed by a similar process, but here the dorsal mesodermal bands are widely separated by a connective-tissue membrane, which forms the lateral walls of the heart and the floor of the pericardial sinus. There are obvious parallels between this early decapod organization and that seen in the adult heart of *Artemia.* In decapod crustaceans the heart is usually a condensed, highly muscular region of the dorsal vessel with only two to three pairs of ostial valves and an extremely complex hemolymph distribution system.

Development in the shrimp *Metapenaeus ensis* is characterized by six naupliar, three protozoeal, and three mysid larval instars. Heartbeat is not observed until the later stages of the last nauplius instar, and at first the rate is slow and irregular (McMahon, Chu, & Mak, 1995). These first heart contractions are seen as (1) food reserves become exhausted and (2) activity increases. It is interesting to note that contractions of the posterior aorta and/or intestine commence simultaneously with the first heartbeats.

Heart rate. Aspects of the development of heart function are beginning to be studied for a few crustaceans, including the anostracan branchiopod *A. franciscana* and two decapod crustaceans, the shrimp *Metapenaeus ensis* and the crayfish *Procambarus clarkii.*

In *M. ensis* heart rate is sporadic at initiation in late nauplius 6 but becomes regular and increases in frequency over the last few hours of this instar (Figure 11.4). On transition to

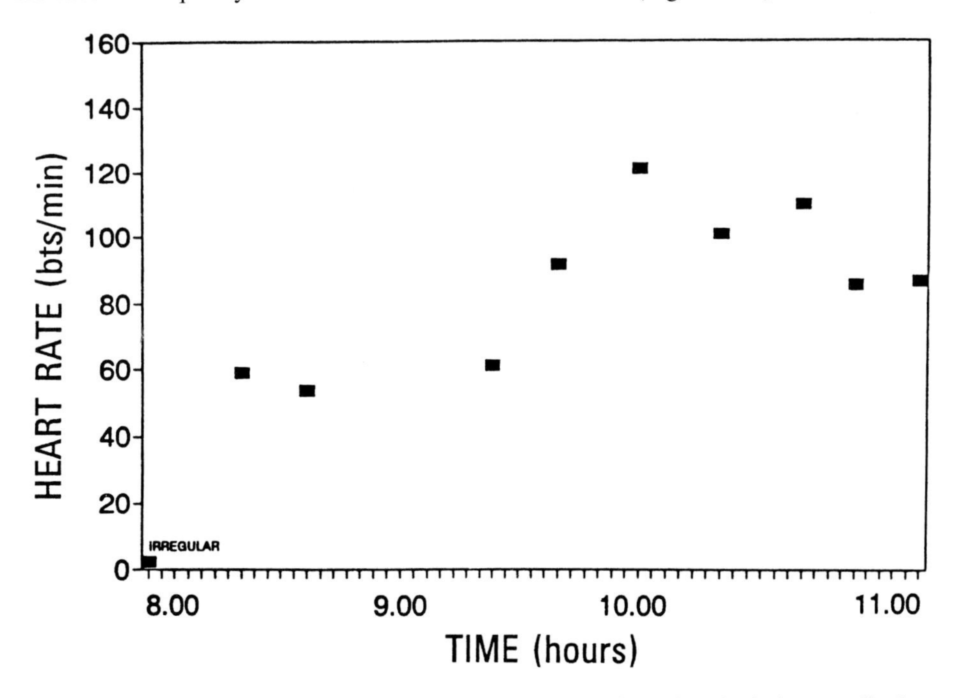

Figure 11.4 Heart rate changes during the initiation of heartbeat in the last naupliar instar of *Metapenaeus ensis* at 25°C.

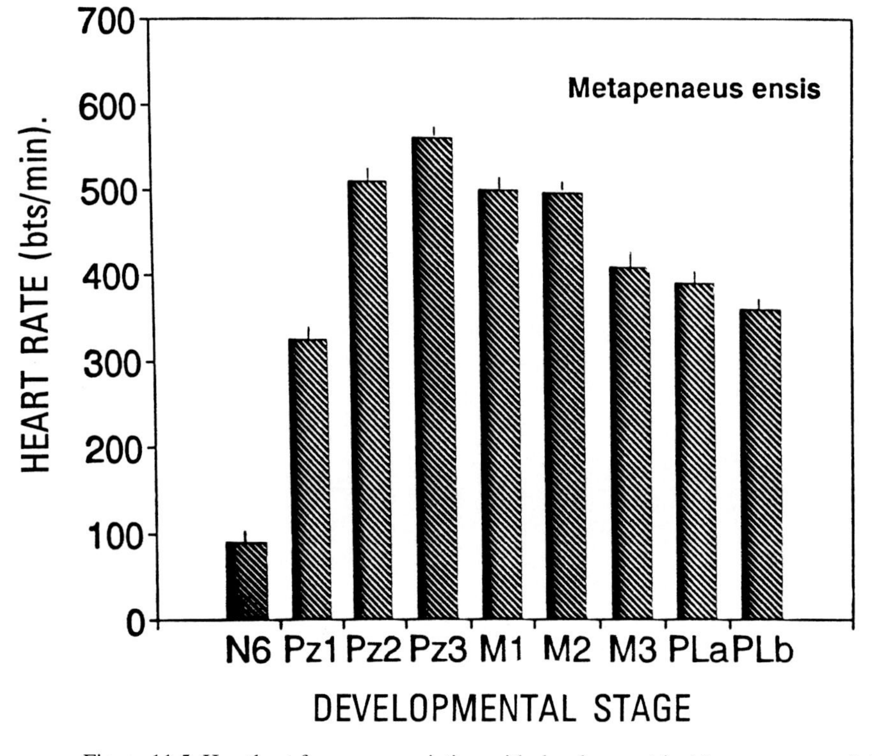

Figure 11.5 Heartbeat frequency variation with development in *Metapenaeus ensis* larval instars at 25°C (N6 = last naupliar instar, PZ1–3 = protozoeal instars, M1–3 = mysid instars, PLa = pelagic, PLb = benthic postlarvae).

the first protozoeal instar, heart rate increases dramatically (Figure 11.5; McMahon, Chu, & Mak, 1995). This may possibly be associated with the continuous-locomotion characteristic of these larvae. Heart rate peaks at the third protozoeal stage and declines throughout the rest of development. Spicer (1994) similarly reported slow and irregular heartbeat commencing at about the third naupliar larval stage in *A. franciscana* and noted rate increase early in development followed by a decrease in later stages. Spicer (1994) noted that the increase was associated with the period of (differential) growth of the heart and the period of decrease was associated with the subsequent period of simple heart elongation. This reasoning cannot be extended directly to the decapod crustacean, where incremental increase in the length of the heart does not occur. It seems more likely that the initial increases seen result from changes in the neuronal and neurohormonal control systems.

Developmental change in heart rate has also been studied in embryos of the crayfish *Procambarus clarkii* (A. Wojciechowski & B. R. McMahon, in preparation). The results are difficult to correlate directly with those for the other two crustaceans since this crayfish exhibits a high degree of direct development, with few free-living larval stages. Nonetheless, the pattern of development is clearly similar. The heart starts beating irregularly at about 17–18 days (20°C) and then increases rapidly over a 1–2 day period before reaching a plateau, where heart rate increases only very slowly until hatching (eclosion), after about 8 more days. After eclosion, heart rate declines slowly (Figure 11.6).

A comparison of heart rate data from crustaceans of widely differing body mass (Maynard, 1960) confirmed that the general relationship that heart rate decreases with increase

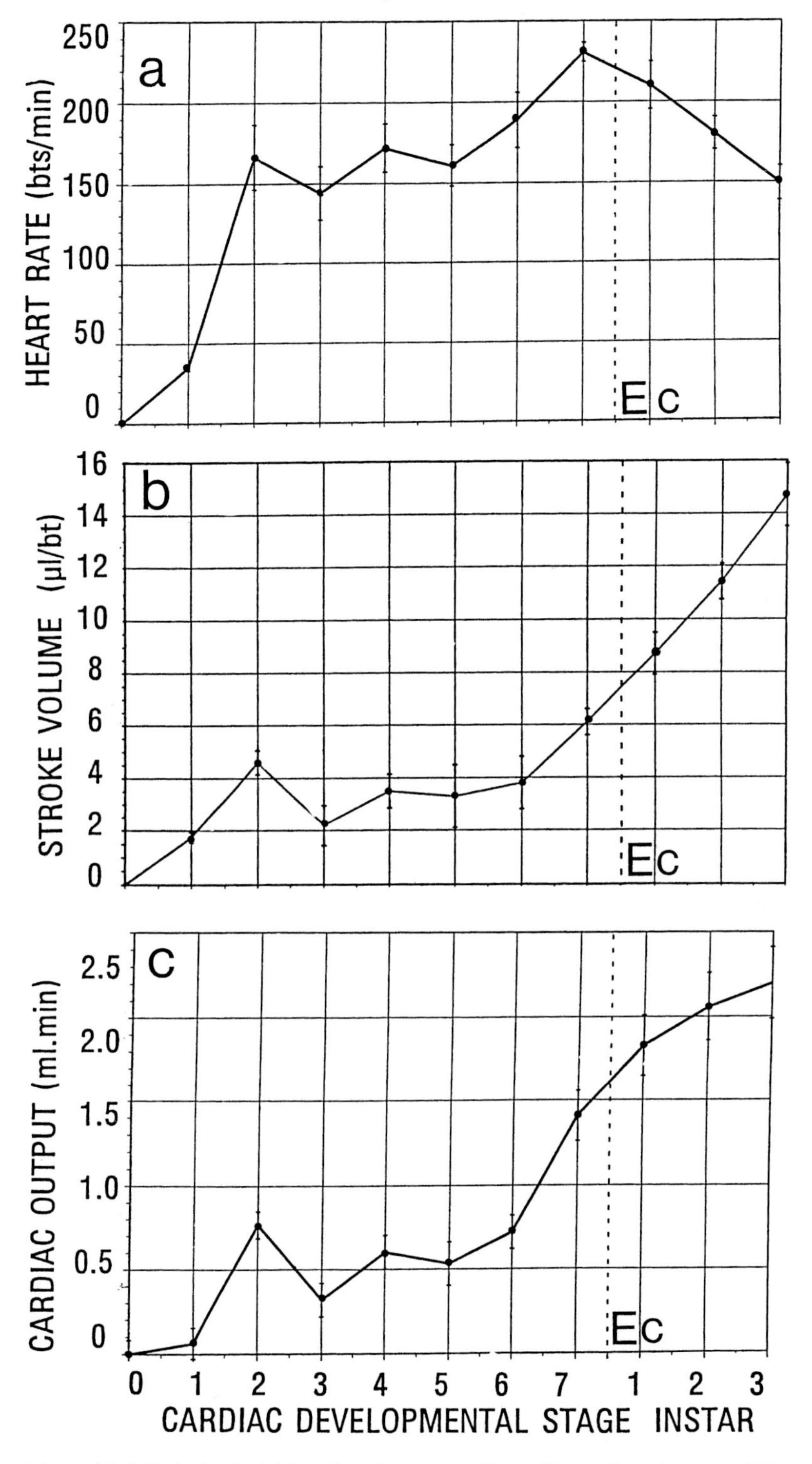

Figure 11.6 Variation in (*a*) heartbeat frequency, (*b*) cardiac stroke volume, and (*c*) cardiac output during direct development in embryos of *Procambarus clarkii* at 20°C.

in body mass also applies to crustaceans (see also Chapter 13 for discussion of this phenomenon). The relationship between heart rate and development revealed here for crustacean larval stages (Figures 11.4–11.6) is more complex because it involves a period when heart rate increases along with body size as well as a period when little change in heart rate occurs.

Stroke volume and cardiac output. Change in stroke volume during development has been measured in both *Artemia* (R. Bourne & B. R. McMahon, unpublished data) and *Procambarus* (A. Wojciechowski and B. R. McMahon, in preparation). In *P. clarkii* stroke volume generally followed a time course similar to that of heart rate (Figure 11.6) and increased rapidly during early heart development but then reached a plateau that lasted throughout most of the remaining embryonic development. Following eclosion, stroke volume increased markedly, in contrast to the decrease in heart rate observed at this time. Cardiac output, calculated from the measurements of heart rate and stroke volume (Figures 11.6), generally increased throughout development. The time course closely followed that for stroke volume, suggesting that change in stroke volume is the more potent factor affecting increase in cardiac output with development.

The plateau phase seen in all cardiac indices through most of the development of the embryonic heart is somewhat puzzling because it indicates that heart function does not increase significantly over the last 40–50% of embryonic development. Our expectation was that cardiac output would increase over this period to fuel the metabolism of the increasing mass of the embryo. This reasoning, however, assumes a single important role for the developing circulation: delivery of O_2 and nutrient to tissues. In fact, there is at least one other. Initially, the circulatory system serves the function of removing food reserves from the yolk mass, included in the egg. As the yolk is consumed, circulation through the vitelline circuit may decrease as circulation through the embryonic mass increases. These factors may explain why cardiac output does not increase appreciably until soon before eclosion, after which the animals are able to move in search of external food sources.

Video microscopic analysis of *Artemia* hearts also indicates that stroke volume and cardiac output increase with the increase in body size (R. Bourne & B. R. McMahon, unpublished data). Although there are some differences in the pattern, it is clear that increasing stroke volume is the major determinant of the increased cardiac output occurring throughout development.

Effects of variation in temperature. Measurements of heart rate on animals developing at a range of temperatures have also been reported. Spicer (1994) showed that heart rate increased significantly with temperature for the latter part of development in *Artemia.* R. Bourne and B. R. McMahon (unpublished data) showed an increase over the range 10–30°C for *Artemia,* but Q_{10} was low (1.3–1.4) throughout an extended developmental range. The latter study also assessed changes in Q_{10} for stroke volume. These were much more dramatic, ranging from 2 at the 20–30°C range to 5 at the 10–20°C range. Change in Q_{10} for cardiac output closely approximated ratios for stroke volume throughout. In all cases Q_{10} was larger for the 10–20°C range than for 20–30°C. This is probably associated with the very slow growth of *Artemia* at 10°C.

Hemoglobin (Hb) is found in nauplii of *Artemia* about 2 hours after hatching. This initial Hb is a type II. Hb III appears next in mid-development. A third Hb (Hb I) was found only at least 7 days after hatching, when functional gills also appear (Heip et al., 1978). Both Hb II and III disappear following larval life, but Hb I remains as the adult variant. It is interesting to note in a respiratory context that the adult Hb I has the lowest O_2 affinity; that is, the high-affinity forms are associated with the larval stages while heart and gill systems are still developing. A similar situation is seen in vertebrate embryos, where high-affinity Hb's are associated with limited oxygen availability.

Insecta

In the Insecta, as in other uniramous arthropods, the heart retains the "primitive" arthropod design, with segmentally arranged ostia and alary suspension. In the cockroach (Orthoptera) lateral vessels are still present, but in the majority of insects these are absent and only a single anterior aorta supplies the cephalic area. A general pattern for the embryology of the insect circulatory system is described by Woodring (1985). Briefly, the hemocoel is produced as a segmental series of lateral coelomic pouches opening into the epineural space. The lateral mesoderm then extends dorsally and eventually fuses in the dorsal midline to form the dorsal vessel (heart), dorsal diaphragm, alary muscles, etc. Generally, the heart starts to pump as soon as it is formed, often long before hatching. In later development, heartbeat tends to slow with each larval instar, as has been described for crustaceans (Woodring, 1985). A. Smits, W. W. Burggren, and D. Oliveros (personal communication) have recorded heart rates in larval, pupal, and adult *Manduca sexta.* They also report heart rate changes during development, but, unlike the situation described for crustaceans, heart rate increases with development (i.e., is faster in adult than in larval animals). Interestingly, the pupal heart rate is slower than that of larvae, perhaps associated with the reduced motility of this stage. Insect hearts are capable of reversing the direction of pumping, and changes in heart beat pattern are reported in development (A. Smits et al., personal communication). In larval *M. sexta* heart rate is regular, but it becomes highly irregular in the pupal stage. Just after emergence of the adult, heart rate quickens and a complex repeating pattern is seen in which heart rate cycles between periods of acardia, followed by periods of moderate-speed (<30–45 beats per minute) forward pumping, followed by periods of slow (10–30 beats per minute) reversed pumping, and ending with a period of high-speed (45–70 beats per minute) forward pumping before acardia returns. The reasons for this pattern are not understood, but it may serve the needs of wing inflation and other complex processes associated with emersion. No simple relationship between heart rate and mass can be seen in this insect (Figure 11.7). No change in heart rate occurs with increase in size in larvae or pupae, and adults have a higher heart rate but are substantially smaller in mass. As with crustaceans, heart rate increases with temperature in all stages.

Mollusca

As with most comparisons between vertebrates and the invertebrate phyla, the functional aspects of both the developing and the adult circulatory systems in the Mollusca are less well understood. Bourne, Redmond, and Jorgensen (1990) provide a summary of the functional properties of the adult circulatory system. With the exception of the cephalopods, the molluscan circulatory system is described as an open one that generally operates at low pressures and slow heart rates. In addition to its transport functions, this system often provides hydraulic ones such as helping extend various body parts. In contrast, the cephalopod circulatory system meets the needs of a more active lifestyle by operating at higher pressures and heart rates (Bourne, Redmond, & Jorgensen 1990).

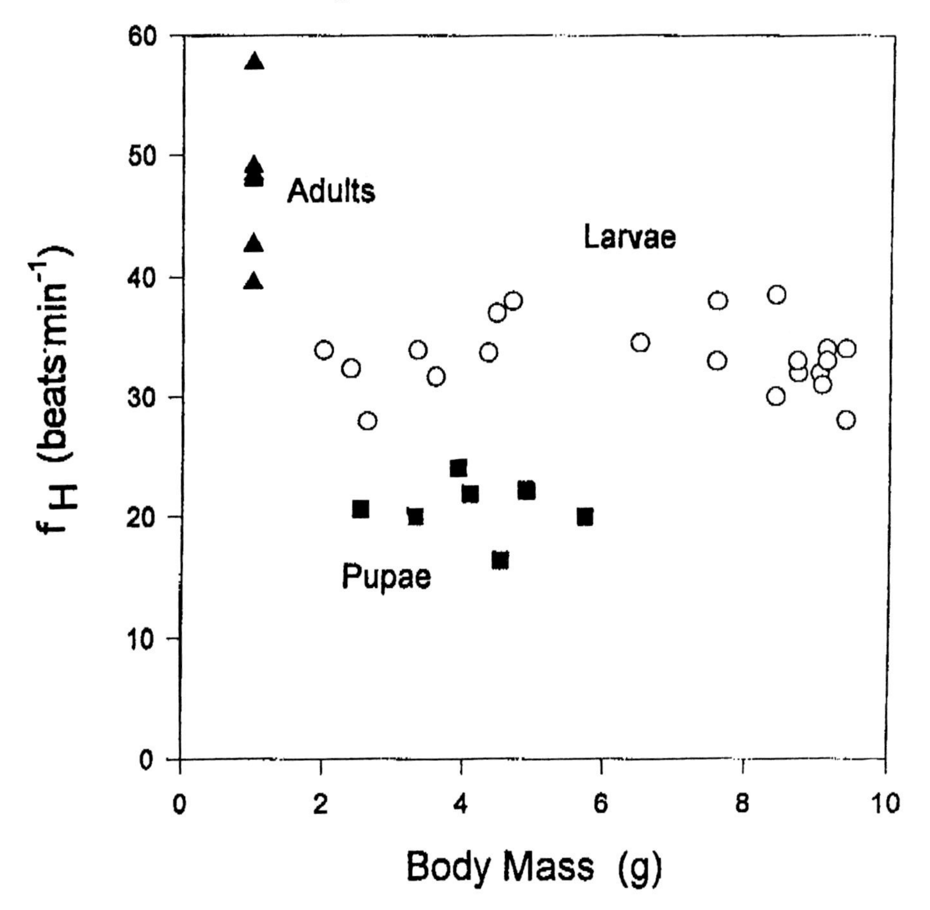

Figure 11.7 Heart rate of larvae, pupae, and adult *Manduca sexta.* (Supplied by A. Smits et al., personal communication.)

The literature covering the descriptive aspects of early molluscan development extends back into the middle of the last century; Raven (1966) offers the most extensive review of these earlier works. Even after Raven's review much of the accumulated literature focuses on the descriptive aspects of individual species (e.g., Damen & Dictus, 1994). However, there is a growing body of information covering experimental embryology mostly concerning mosaic versus regulative development (Atkinson, 1971; McCain & Cather, 1989; Avila et al., 1994). Most of this information deals with the major classes Gastropoda, Cephalopoda, and Bivalvia. With the exception of the cephalopods, molluscan development is typically spiralian. Thus, the remainder of this discussion separates the cephalopods from the rest of the phylum, which is represented by an examination of gastropod development.

Gastropoda

Since this review is primarily concerned with organogenesis, there will be a heavy emphasis on those gastropods that have direct development. In such forms, embryogenesis and larval changes occur within capsules and thus are more amenable to observation and experimental manipulation.

In prosobranch veligers (e.g., *Crepidula, Littorina*) there is an ectomesodermal vesicle behind the velum that exhibits rhythmic contractions (Raven, 1966). During torsion it moves from the right side in front of the mantle fold toward the dorsal side, is overgrown by the mantle, and eventually resides in the mantle cavity (Raven, 1966). Typically, the heart is a pallial complex organ that arises from a common group of mesodermal cells. The other pallial organs associated with the heart include the pericardium, kidney(s), and often the gonadal complex (Raven, 1966). Many authors treat development of the molluscan circulatory, coelomic, and renal systems together (Demian & Yousif, 1973). The pericardial cavity usually appears first, as a cleft in the cell mass (Raven, 1966). The heart tube then appears between the inner pericardium and the gut. Later, constrictions separate the ventricle from the auricle(s). The heart tube connects to blood cavities that appear in the mesoderm at the base of the gill rudiments (Raven, 1966). The blood vessels arise separately from the heart as cavities between mesodermal cells, and as they differentiate, they make secondary connections to the heart (Wasserloos, 1911).

The presence of a polar lobe is a common feature of spiralian development. Atkinson (1971) demonstrated that in *Llyanassa* (*Nassarius*) the heart is one of the so-called lobe-dependent tissues. Following fertilization, morphogenetic determinants that regulate the differentiation of the heart as well as other organs are localized in the polar lobe cytoplasm. Loss of the polar lobe cytoplasm leads to impaired development; such embryos fail to develop the heart and the other structures controlled by the polar lobe.

Functional information on the developing gastropod heart is often collected incidental to other studies. In *Crepidula,* Werner (1955) observed that in the extended veliger, the heart pulsated at 60 beats per minute, but when retracted, the heart rate dropped considerably. Werner (1955) also suggested that the larval heart may have a role in protracting the veliger from the shell. Conrad et al. (1994) produced the most recent functional information while examining the nature of polar lobe determinants. They found that the heart rate of developing *llyanassa* is between 60 and 100 beats per minute. Using heart rate as an index, they suggested that either simulated microgravity or microtubule distribution may affect heart development.

Cephalopoda

Despite certain homologies with the body plan of other molluscs, cephalopods lack most traces of spiral cleavage (Boletzky, 1987a). Instead, cephalopods undergo meroblastic discoidal cleavage, which has an expressed bilateral symmetry from the beginning (Raven, 1966). An additional characteristic of cephalopod eggs is the much greater amount of yolk.

The outer yolk sac appears during early embryogenesis and, with the exception of some squid with small eggs, plays a dominant role in hemolymph circulation throughout organogenesis (Boletzky, 1987b). As the yolk sac is completed, a blood space appears between the yolk sac wall and the yolk syncytium (Boletzky, 1987b). Muscle cells differentiating from scattered mesodermal cells begin to contract periodically, and eventually a coordinated peristaltic wave travels over the entire yolk sac (Boletzky, 1987b). These waves drive blood from the yolk sac sinus to the embryo proper through a broad ventral sinus; the blood returns from the embryo via cephalic and posterior sinuses to the dorsal aspect of the yolk sac (Boletzky, 1987b). Later in development, those embryonic sinuses become part of the venous system. In most forms, the circulation of the outer yolk sac remains lacunar, but in *Nautilus* an elaboration of blood vessels occurs (Arnold & Carlson, 1986).

The cephalopod systemic heart makes its appearance well after the yolk sac circulation is established, that is, toward the end of the early phase of organogenesis (Boletzky, 1987b). At this time, the systemic heart becomes distinct as paired rudiments; these fuse and a small lumen appears (Boletzky, 1987b). About the same time a lumen appears on either side in the mesodermal masses beneath the gill primordia; these are the branchial heart rudiments, and they begin to pulsate as soon as their lumens open (Boletzky, 1987b). However, the systemic heart becomes active only when its junctions to the branchial hearts become established. Coordination of beat between branchial and systemic hearts is achieved gradually as the venous system is more firmly established (Boletzky, 1987b). Angiogenesis then proceeds to establish fully the adult circulatory pattern. Initially, cephalic and posterior aortae develop from the systemic heart, with both developing the secondary and tertiary branching (Boletzky, 1987b). Usually, the yolk sac circulation continues to play a role until hatching (Boletzky, 1987b).

Cephalopod development exhibits a convergent trend with vertebrate developing circulation: the formation of the yolk sac circulation at early stages of embryogenesis. However, the central circulatory pumps of cephalopods, unlike vertebrates, do not appear to play a prominent role in moving blood between the yolk sac and the embryo proper. Those structures reach full functional capacity in cephalopods at a much later stage in development.

Echinodermata

In the early development of echinoderms, the coelom is usually subdivided into the perivisceral coelom, which surrounds the major body organs, and the tubular coelomic complexes, which include the water vascular system and the perihemal system (Hyman, 1955). The detailed development of the coelom varies in different classes (see reviews in Giese, Pearse, & Pearse, 1991), but basically the perivisceral coelom derives from the somatocoel, and the water vascular system and perihemal system form from the axohydrocoel. The water vascular system consists of a ring canal encircling the oral area and radial canals extending through the arms to supply each pair of tube feet. The perihemal system, also called the hyponeural sinus system because of its closer relation to the nervous system (Hyman, 1955), consists of sinuses parallel to canals of the water vascular system. Also present is a hemal system, which is referred to as a blood lacunar system (Hyman, 1955) instead of a blood vascular system because the channels are not definite vessels with endothelial walls. The hemal fluid is gelatinous in appearance and may consist only of a ground substance containing fibrillar material and cells.

Lawrence (1987) reviewed the structure of the four vascular systems and their possible circulatory roles. Although the perivisceral coelomic fluid appears to be the principal circulatory medium, there is increasing evidence to support the role of the hemal system in nutrient transport, particularly in the translocation of nutrients absorbed from the gut (Broertjes et al., 1980a; Broertjes et al., 1980b; Ferguson, 1985; Baird & Thistle, 1986) to the gonad for vitellogenesis (Beijnink, Walker, & Voogt, 1984; Broertjes, De Waard, & Voogt, 1984; Ferguson, 1984; Voogt, Broertjes, & Oudejans, 1985, Byrne, 1988; Eckelbarger & Young, 1992). Because of the gelatinous nature of the hemal fluid, it is likely that the hemal system serves as a temporary repository of nutrients (Grimmer & Holland, 1979; Byrne 1988, 1994). The holothuroids (sea cucumbers) – particularly certain members of Aspidochirotida such as *Holothuria* and *Stichopus* – have the most elaborate hemal systems in echinoderms, with dorsal and ventral vessels associated with the intestine (Cuénot, 1948; Hyman, 1955).

Studies in several species show that the dorsal vessel contracts at 4–12 beats per minute (for a review, see Binyon, 1972). Herreid, LaRussa, & DeFesi, (1976) found 120–150 single-chambered hearts in the region of the upper small intestine of *Stichopus moebii,* which presumably pump fluid through the hemal system in the intestine. Besides the apparent function of the hemal system in nutrient transport, it has been suggested that the vessels associated with the respiratory trees play a role in gaseous exchange (Herreid, LaRussa, & DeFesi, 1976; Herreid, DeFesi, & LaRussa, 1977), but an excretory function has also been proposed (Cuénot, 1948; Hyman, 1955). Herreid, DeFesi, & LaRussa (1977) suggested that the vascular follicles in the pulmonary–intestinal circulation may be involved in production and destruction of coelomocytes. The products from coelomocyte destruction may be eliminated by the respiratory tree (Lawrence, 1987).

Although it is now generally agreed that the echinoderm hemal system plays a role in conveying nutrients for the developing gonads, the function of the axial gland (which together with the axial sinus of the perihemal system makes up the axial complex) remains enigmatic, and many functions (including circulatory, excretory, genital, vestigial, pigment producing, glandular, and immunological) have been proposed (Binyon, 1972). Recent studies of the axial complex have been focused on its possible role in excretion and immune responses. Study of phagocytotic activity in the echinoid axial complex suggests an excretory function for the complex, presumably involving the removal of phagocytes through the rectum (Bachmann, Pohla, & Goldschmid, 1980). Studies of the axial complex of *Asterias rubens* by Leclerc and his colleagues indicated that antibody-like factors are produced (Brillouet et al., 1984; Delmotte et al., 1986) and T- and B-like cells as well as macrophage-like phagocytic cells can be identified in the complex (Anteunis et al., 1985; Leclerc et al., 1986; Bajelan & Leclerc, 1990; Leclerc et al., 1993; Leclerc, Maitre, & Contrepois, 1994) leading to the speculation that it is a primitive immune organ. However, Larson and Bayne (1994) failed to show that the cells in the axial complex of *Strongylocentrotus purpuratus* respond to antigenic challenge in a specific inducible manner, supporting the idea that the formation of antibodies is essentially a vertebrate or chordate feature. Further studies are necessary to resolve this controversial issue.

Another line of studies on the axial complex based on electron microscopy showed that podocytes, which are often found in excretory organs associated with ultrafiltration (Ruppert & Smith, 1988), occur in the axial gland of asteroids and echinoids (Bargmann & von Heln, 1968; Welsh & Rehkämper, 1987), indicating the function of the gland as a filtration surface. Ruppert and Balser (1986) also found podocytes in the pore canal–hydropore complex (which gives rise to the water vascular system) of asteroid larvae. They demonstrated unidirectional fluid transport from the complex, suggesting its role as a functional nephridium. A spherical pulsating structure (6–12 beats per minute) was identified adjacent to the complex, which may act as a muscular pump for the nephridium. Similar pulsating structures, believed to represent the dorsal sac or madreporic vesicle, have been observed in echinoderm larvae (Bury, 1896; Gemmill, 1914, 1915, 1919; Narasimhamurti, 1932). An increase in pulsation rate during larval development was noted in *Asterias rubens* and *Echinus milaris* (Narasimhamurti, 1932). In the starfish *Poronia pulvillus,* however, the pulsation rate in the larva was 9–10 beats per minute but slowed to 6 beats per minute prior to metamorphosis (Gemmill, 1915). Pulsating structures, contracting at about 6 beats per minute, have also been reported in adult asteroids and echinoids and are sometimes referred to as hearts (Narasimhamurti, 1932; Boolootian & Campbell, 1964, 1966; Ruppert's unpublished observation in *Asterias forbsii* described in Ruppert & Barnes, 1994). Ruppert and Balser (1986) speculated that the axial complex of adult echinoderms performs a similar excretory function as the pore canal–hydropore complex in the larva (Figure 11.8). Physiological studies are necessary to ascertain the function of the pulsating structures, and the axial complex in general, in echinoderms. It is interesting to note that in holothuroids, in which the axial gland is absent, the elaborate vessels associated with the intestine and respiratory tree might perform the excretory function discussed above.

Hemichordata

Enteropneusts have a well-defined circulatory system consisting of blood vessels, sinuses, and a heart–kidney, also called the proboscis complex (Hyman, 1959). There are two main longitudinal blood vessels: a dorsal vessel, in which the blood flows anteriorly, and a ventral vessel, in which it flows posteriorly. The paired perihemal coeloms, which are extensions of the metacoels (trunk coelom) to the collar region at both sides of the dorsal vessel, exhibit peristaltic contractions to pump blood in the vessel (Balser & Ruppert, 1990). Blood in the dorsal vessel enters a central sinus, often called the heart, located in the proboscis between the stomochord (buccal diverticulum) and a contractile structure called the pericardium (heart vesicle). The blood in the heart moves into the glomerulus, which comprises fingerlike folds of the proboscis peritoneum against the heart wall. The circulatory system of pterobranchs is less well developed but the heart–kidney is present. The occurrence of podocytes in the glomerulus has been reported in both enteropneusts (Wilke, 1971) and pterobranchs (Dilly, Welsch, & Rehkämper, 1986). Balser and Ruppert (1990) demonstrated fluid transport from the dorsal vessel to the protocoel (proboscis coelom) by injecting dye and iron dextran, suggesting that blood is filtered by the glomerulus, producing primary urine in the protocoel to be released through the proboscis pore. The authors believe that the heart–kidney of hemichordates and the axial complex of echinoderms represent homologous structures, both with a primarily excretory function. The anatomical similarities between the heart–kidney of hemichordates and the axial complex in asteroids have been noted since the end of the last century, and the two have been believed to be homologous structures (Bury, 1896; Gemmill, 1914).

As early as 1894, T. H. Morgan reported a pulsating vesicle in the tornaria larva of the enteropneust *Balanoglossus.* This structure was regarded as homologous to the pulsating structure identified in echinoderm larvae (Gemmill, 1915, 1919; Narasimhamurti, 1932). Ruppert and Balser (1986) have also found a spherical pulsating vesicle next to the hydropore of the larva of *Schizocardium brasiliense.* They provided evidence that the pore canal–hydropore complex of the larva functions as a nephridium in the same way as the complex found in bipinnaria larvae of asteroids (see previous section). In both larvae the pulsating vesicle is believed to serve as a muscular pump for filtration of blood across the wall of the pore canal, producing urine for release at the hydropore (Figure 11.8). Thus the pore canal–hydropore in the tornaria larva performs the same excretory function as its adult derivative – the heart–kidney (Balser & Ruppert, 1990).

Urochordata

The circulatory system of urochordates consists of many large sinuses associated with body organs and connected by blood vessels. The heart is a tubular structure (curved and U-shaped in some species) located at the base of the digestive loop. Embryologically, the heart first appears as an undifferentiated vessel derived from the mesoderm, which invaginates

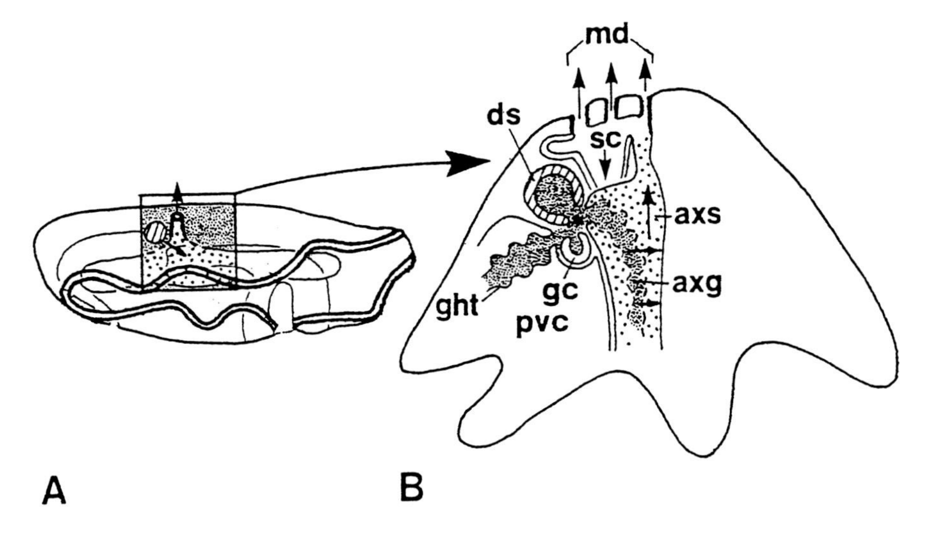

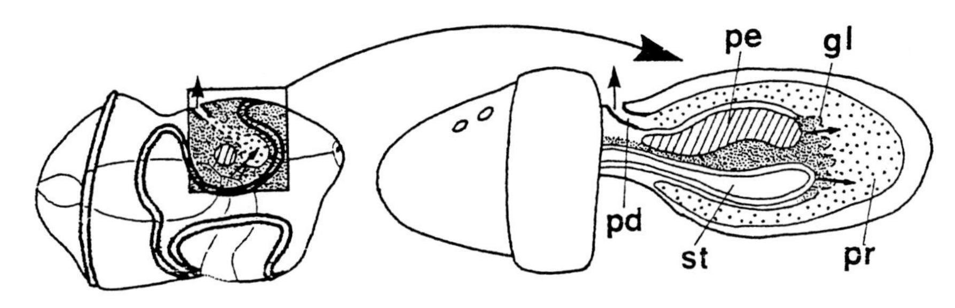

Figure 11.8 A hypothetical model of the development of the excretory organ in asteroids and hemichordates. (*A*) Lateral view of the bipinnaria larva of an asteroid showing the pore canal–hydropore complex. (*B*) Disproportionately enlarged apical portion of the axial complex in an adult asteroid. (*C*) Lateral view of the tornaria larva of a hemichordate showing the pore canal–hydropore complex. (*D*) Disproportionately enlarged proboscis region of an adult hemichordate. Small arrows show the direction of filtration; larger arrows, the direction of ciliary beats; cross-hatching, pulsatile vesicles in larvae and their adult derivatives; dashed lines, coelomic epithelia composed of podocytes; loose stippling, anterior larval coeloms and their adult derivatives; fine stippling, blastocoel of larvae and their adult derivatives. axg = axial gland, axs = axial sinus, ds = madreporic vesicle (dorsal sac), gc = genital coelom, ght = gastric hemal tuft, gl = glomerulus, md = madreporite, pd = proboscis duct, pe = pericardium, pr = proboscis coelom, pvc = perivisceral coelom, sc = stone canal, st = stomochord. (From Ruppert & Balser, 1986.)

along its long axis to form a tube within a tube (Randall & Davie, 1980). The outer tube encloses the fluid-filled pericardium, which is completely sealed and may represent the vestige of the coelom (Weiss, Goldman, & Morad, 1976). The inner tube forms the heart, which remains attached to the pericardium along its length by the raphe. Of the three classes of urochordates (Ascidacea, Thaliacea, and Larvacea), the functioning of the heart has been studied only for ascidians (sea squirts), particularly some of the solitary species (for reviews, see Krijgsman, 1956; Goodbody, 1974; Randall & Davie, 1980). A remarkable feature of the heart is the reversal of heartbeat, which is found in all urochordates (compare the previous discussion on insects). Heartbeat takes the form of a peristaltic wave passing from one end of the heart to the other, and the wave changes in direction at regular intervals of a few minutes. In compound ascidians, the circulatory systems of the individuals are connected. These hearts operate in series but beat independently of each other (Mukai, Sugimoto, & Taneda, 1978). Heart reversal takes place independently in different zooids. Jones (1971) reported that the heart begins to beat and reverses in the late embryo. The physiological mechanisms controlling heartbeat reversal and its functional significance are far from understood.

Concluding remarks

It is evident from the preceding account that we have very little information on the developmental physiology of the cardiovascular systems for any invertebrate group. This is unfortunate because within the invertebrates we see a much wider range of structures used to generate vascular flow than in the vertebrates, and study of these must increase our understanding of fluid flow and control mechanisms generally. The lack of information may be particularly unfortunate within the Annelida because they exhibit a very wide range of blood vascular systems. Study of physiological development of the Annelida could well provide information pertinent to circulatory evolution in the higher protostome groups. Within the arthropods, there has been some recent interest in the ontogeny of crustacean "open" circulatory systems, but we have little physiological information about other arthropod groups. What little we know about the formation and functional characteristics of the molluscan circulatory system indicates two distinct directions, one shown by the cephalopods and the other displayed throughout the rest of the phylum, but much additional work is needed to clarify the basis for these different developmental paths.

There is an even greater dearth of information about cardiovascular development within the deuterostome groups thought to be related to protochordates and the vertebrate evolutionary line. In the echinoderms there are four possible vascular systems, but the extent to which these function in transport is still obscure in either adult or larval forms. The single blood vascular system of the hemichordates may show some similarity with the echinoderm axial complex, but further work is needed to strongly interrelate circulatory structures within these groups and those of the cephalochordates leading on into the vertebrate line.

In short, study of the development of invertebrate circulatory systems has fallen behind that of other physiological systems, especially the nervous system. We feel that there are many significant advances to be made in this area and hope that this account will stimulate further interest in this field.

12

Piscine cardiovascular development

PETER J. ROMBOUGH

Introduction

Fish offer a number of distinct advantages as experimental animals in which to study the development of the vertebrate cardiovascular system. Development is external, and the embryo is transparent in most species, making microscopic examination and manipulation of the living embryo relatively easy. Many common aquarium species are easy to rear, produce large numbers of eggs, and have short generation times. These advantages have led to a surge in interest in recent years in using fish, in particular the zebra fish *Danio (Brachydanio) rerio,* as an animal model for studying the development of vertebrate organ systems (Balter, 1995). One of the more promising techniques to be applied to the study of the cardiovascular system involves induced mutation and subsequent isolation of genes acting at various stages in the developmental cascade (see Chapter 1). Profound mutations, such as the failure of the heart to form, are readily obvious and require little knowledge of normal morphogenetic events. However, identification of subtler effects, particularly functional changes, requires that the investigator have a good knowledge of normal development. Unfortunately, such information is not always readily available. The goal of this chapter is to provide an overview of current knowledge regarding piscine cardiovascular development that will assist investigators in both the planning and the interpretation of their experiments. Where possible, I have attempted to place developmental events in the context of eventual adult function.

In examining the role of the cardiovascular system during development, it is important to recognize that embryos and larvae are not simply small adults. Fish occupy different niches during development than they do as juveniles and adults and are thus subject to different selective pressures. For example, size-dependent processes such as diffusion and the relative importance of viscous, as opposed to inertial, forces are likely to be much more important during early life. One would expect such forces to lead to differences in both structure and function. On the other hand, definitive adult structures arise from those in embryos and larvae. One should not be surprised, therefore, if there is a good deal of congruence in function as well as in structure among the various life stages.

Morphogenesis

The embryonic development of fishes has been the subject of scientific study for more than 150 years (e.g., Vogt, 1842). A preponderance of the studies have dealt with salmonids, but reasonably detailed descriptions of major morphological events are available for at least several dozen other species. Most studies of fish development have been conducted for taxonomic purposes (i.e., species identification). However, the beating heart and the flow of blood through the vasculature are so conspicuous that few investigators have failed to comment, if only in passing, on the major morphological changes that take place in the circulatory system during the course of development. Some of the literature dealing with the morphogenesis of the cardiovascular system has been incorporated into general reviews (e.g., Mossman, 1948; Tavolga, 1949; Mahon & Hoar, 1956; Vernier, 1969; Ballard, 1973a) and texts (e.g. Ballard, 1964; Nelsen, 1953; Rugh, 1956; Torrey & Feduccia, 1979). Unfortunately, a lot of useful information remains scattered in the primary literature.

The heart

The heart is the first recognizable component of the circulatory system to develop. It forms from the fusion of two lateral masses of splanchnic mesoderm in the head region (Hisaoka & Firtlit, 1960; Laale, 1984). The time of formation depends on species and temperature (temperature–time relationships in developing fish usually are expressed in terms of accumulated thermal units, ATU [°C × days], calculated by summing average daily temperatures). In rainbow trout, the paired endocardial rudiments have merged and are visible as a simple tube near the midline ventral to the otic anlage prior to completion of epiboly (≈80 ATU postactivation; stage B12 [B = stage as defined by Ballard, 1973a]). The heart begins to beat weakly at the tail bud stage (≈110 ATU; stage B15a). By about 120–140 ATU (stage B16 or V21 [V = stage as defined by Vernier, 1969]), the heart beats strongly and blood flows through the first set of aortic arches. In rainbow trout, the heart initially has a dorsal–ventral orientation. With the onset of circulation, the heart shifts to a more anterior–posterior orientation and assumes an S-shaped curve. The boundary between the atrium and the ventricle becomes defined by a constriction around 140 ATU (stage B17). The heart continues to twist so that by hatch (≈340 ATU) the main axis of the atrium is dorsal to that of the ventricle. The heart assumes its definitive orientation with the ventricle slightly to the rear and below the atrium by the time larvae are about halfway through yolk absorption. The exact sequence of events and morphology of the embryonic heart vary somewhat among species. General descriptions of how the heart forms in various species can be found in the references listed in Table 12.1.

The early studies of heart development were primarily descriptive. In recent years, interest has shifted to trying to understand the mechanisms leading to the differentiation of the heart. Ballard (1973b, 1982, 1986) and Ballard and Ginsburg (1980) used classic cell-staining techniques to construct rough fate maps for a number of species that show the regions of the blastula involved in the formation of the heart and associated blood vessels. Modern single-cell injection techniques have been used to track cells specifically destined to give rise to the endocardium (R. K. K. Lee et al., 1994) and the myocardium (Stainier, Lee, & Fishman, 1993) of zebra fish. Laale (1984) was able to isolate cardiac progenitor cells from early zebra fish embryos and have them differentiate in tissue culture. Teratogens have been used to study how the rudimentary heart acquires its anteroposterior polarity (Stainier & Fishman, 1992). Most recently, mutagens have been used to generate cardiac mutants in zebra fish in hopes of understanding the genetic basis of heart formation (Fishman & Stainier, 1994; Solnica-Krezel, Schier, & Driever, 1994; see also Chapter 1, this volume).

The myocardium initially forms a simple wall of relatively uniform thickness with abundant cardiac jelly between adjacent myoblasts. Cardiac jelly apparently acts as an interme-

Table 12.1. *References that describe the morphogenesis of the fish heart*

Class/superorder	Genus and species	References
Agnatha	Various species	Percy & Potter, 1986, 1991
Chondrichthyes	*Raja*-sp.	Weber, 1908
	Squalus acanthias	Scammon, 1911
Osteichthyes		
Chondrostei	*Polyodon spathula*	Ballard & Needham, 1964
	Acipenser sp.	Ballard & Ginsburg, 1980
Holostei	*Amia calva*	Ballard, 1986
Teleostei	*Salmo gairdneri*	Vernier, 1969; Ballard, 1973a
	Salvelinus fontinalis	Ballard, 1973a
	Oncorhynchus keta	Mahon & Hoar, 1956
	Coregonus clupeaformis	Price, 1934a, 1934b
	Alosa sapadissima	Senior, 1909
	Catostomus commersoni	Ballard, 1982
	Esox lucius	Markiewicz, 1960
	Fundulus heteroclitus	Armstrong & Child, 1965; Stockard, 1915
	Brachydanio rerio	Hisaoka & Firtlit, 1960; Chen & Fishman, 1996
	Platypoecilus maculatus	Tavolga, 1949

diary in transmitting force from the myocardium to the endocardium. The amount of cardiac jelly gradually diminishes as the myocardial wall thickens and the need for such an intermediary decreases. Trabeculae begin to form once differentiation of the heart chambers is complete; the precise stage probably varies depending on the species. Ventricular trabeculae were present in 130 mg (50 days postspawning) but not 10 mg (27 days postspawning) embryos of the little skate *Raja erinacea* (Pelster & Bemis, 1991). Trabeculae are well developed in rainbow trout larvae by the end of yolk absorption (Kitoh & Oguri, 1985). It has not been documented when they first appear. Maturation of the myocardium in elasmobranch embryos appears to proceed in an asynchronous fashion. Zummo (1983) reported both poorly differentiated and well-differentiated myoblasts in the same region of the ventricle of embryos of the shark *Scyllium stellare.* If one assumes that rainbow trout are representative, it appears that the myocardium consists entirely of spongiosum until near the end of the larval period even in those species that have a compactum as adults (not all fish develop a compact layer; see Farrell & Jones, 1992). Kitoh and Oguri (1985) reported that a peripheral layer of compact myocardium became apparent only near the end of yolk absorption in rainbow trout. The compactum first appeared in the region of the ventricle adjacent to the bulbus arteriosus and gradually extended posteriorly toward the apex. The trout were well into the juvenile phase (4.2–4.6 cm body length) before the compactum covered the whole ventricle.

Simple diffusion of oxygen from the pericardial cavity and the lumen of the heart appears to be sufficient to meet the demands of the myocardium during embryonic and larval development, obviating the need for a coronary circulation. This appears to be true even in those species that have coronary vessels as adults (see Tota, 1983, and Farrell & Jones, 1992, for reviews of the coronary circulation in fish). Rainbow trout, for example, do not develop a coronary circulation until long after metamorphosis (Kitoh & Oguri, 1985). Zummo (1983) reported that the endocardium of *Scyllium stellare* embryos was highly porous at the microscopic level and suggested that seepage of blood through these pores was sufficient to supply oxygen to the myocardium. In lampreys, as in other fish, a layer of endocardium forms early in embryonic development. However, unlike in other fish, the endocardium breaks down early in larval life, after which the myocardium is bathed directly in luminal blood (Percy & Potter, 1991). Percy and Potter (1991) suggested that this was an adaptation to compensate for the lack of a coronary circulation.

The parietal pericardium of fish larvae, as in higher vertebrates but not adult fish, forms a flexible, thin-walled sac. However, unlike in higher vertebrates, the pericardial cavity is filled with a relatively large volume of fluid. Pericardial fluid appears to be secreted by the epicardium in elasmobranch embryos (Zummo, 1983). In agnathan larvae, the pericardial cavity is contiguous with the perivisceral cavity (Percy & Potter, 1991). In some teleost species, the visceral pericardium is highly vascular, and it has been suggested that it may function as an accessory respiratory structure (Turner, 1940; Cunningham & Balon, 1986a).

General circulation

The basic plan of the circulatory system is established relatively early. In rainbow trout, the heart begins to beat shortly after blastopore closure (120 ATU; stage V20). Blood begins to circulate soon thereafter (140 ATU; stage V21) following the simple circuit: heart → aortic arches → dorsal aorta → caudal vein → anal vein → subintestinal vein → vitelline vein → heart (Figure 12.1A). Posterior cardinal veins form at about 160 ATU (stage V22), providing an alternative route for the return of venous blood to the heart (Figure 12.1B). The anal vein disappears around 170 ATU (stage V23), forcing blood from the caudal region to flow back to the heart via the posterior cardinals (Figure 12.1C). At about the same time a branch from the dorsal aorta gives rise to the intestinal artery, which connects to the subintestinal

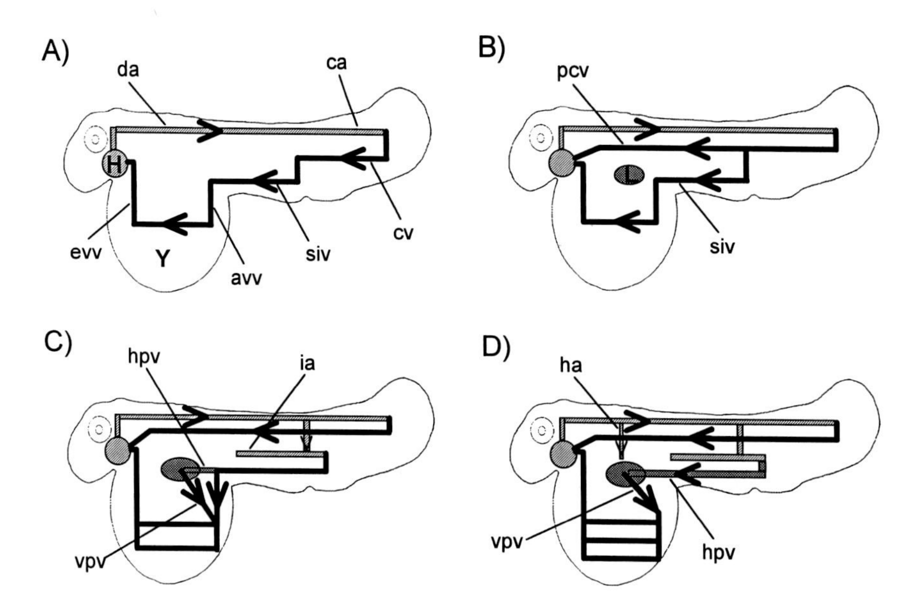

Figure 12.1 Development of the embryonic circulation in rainbow trout (av = anal vein, avv = afferent vitelline vein, ca = caudal artery, cv = caudal vein, da = dorsal aorta, evv = efferent vitelline vein, H = heart, ha = hepatic artery, hpv = hepatic portal vein, ia = intestinal artery, pcv = posterior cardinal vein, siv = subintestinal vein, vpv = vitelline portal vein, Y = yolk sac). (Stages and drawings are after Vernier, 1969.)

vein via a capillary network. The subintestinal vein bifurcates; the ventral branch retains its connection to the vitelline vein, and the anterior branch connects to the developing liver. Portal veins emerge from the liver and contribute to the expanding vitelline plexus. The ventral branch of the subintestinal vein gradually regresses so that by 180 ATU (stage V24) all the blood entering the vitelline plexus passes through the liver (Figure 12.1D). This pattern varies somewhat among teleost species. For example, in the walleye, *Stizostedion vitreum,* the dorsal aorta initially extends only to the posterior limit of the yolk, where it connects via a short intestinal loop directly to the subintestinal–vitelline vein on the surface of the yolk (McElman & Balon, 1979). In the live-bearing platyfish *Platypoecilus maculatus,* the caudal–subintestinal and cardinal veins differentiate at about the same time (Tavolga, 1949), whereas in the northern logperch, *Percina caprodes semifasciata,* the subintestinal–vitelline circulation persists as the major vessel supplying the surface of yolk sac until the yolk is almost completely absorbed (Paine & Balon, 1984). In elasmobranchs, the blood passes directly from the dorsal aorta into the vitelline plexus, bypassing the systemic circulation (Torrey & Feduccia, 1979).

Regional circulation

Skin. Studies of the morphogenesis of the peripheral circulation have focused, for the most part, on those regions involved in respiratory gas exchange, namely, the skin and the gills. Gills typically form relatively late in development, and until then, the skin is the major site of gas exchange. The morphology of the skin of embryos and larvae makes it well suited for the exchange of gases (Rombough, 1988b). The relative surface area of the skin is much greater in embryos and larvae than it is in older fish (e.g., the allometric mass exponent for relative skin area in larval chinook salmon is about –0.7; Rombough & Moroz, 1990). The skin of young fish also tends to be considerably thinner than it is later in life. For example, the skin of a newly hatched Atlantic salmon (≈20 μm) is less than one-third the thickness of the skin of juvenile fish (≈70 μm) (Wells, 1993). Although the skin as a whole does not appear to be particularly well vascularized (Wells, 1993), selected regions of the skin often display high capillary densities. Capillary density usually is greatest on the surface of the yolk sac. The fins, particularly the pectoral fins, also tend to be relatively well vascularized. A few species display extensive vascular networks in other locations (e.g., the pericardium in *Adinia xenica;* Cunningham & Balon, 1985).

Attention has focused on the role of the vitelline circulation in respiratory gas exchange. This is in part because of its high capillary density and in part because of the direction of blood flow. Blood flows through the vitelline circulation in a roughly posterior to anterior direction. The larvae of many species tend to orient themselves so that they face into water currents. This allows them to hold station but it also means that the flow of water past the larvae is in the opposite direction to the flow of blood in the vitelline capillaries. Some species have been shown to actively generate a backward (anterior to posterior) flow of water using their pectoral fins. Liem (1981) suggested that this countercurrent flow of blood and water facilitates gas exchange across the surface of the yolk sac in hypoxic environments.

Species vary considerably in the extent to which the yolk sac is vascularized (Figure 12.2). Cunningham and Balon (1985) estimated that capillaries covered 40–50% of the yolk surface at their maximum extent in the diamond killifish, *Adinia xenica.* At the same stage of development, embryos of the northern logperch, *Percina caprodes semifasciata,* possess only a single vitelline blood vessel (Paine & Balon, 1984). Balon and coworkers (Balon,

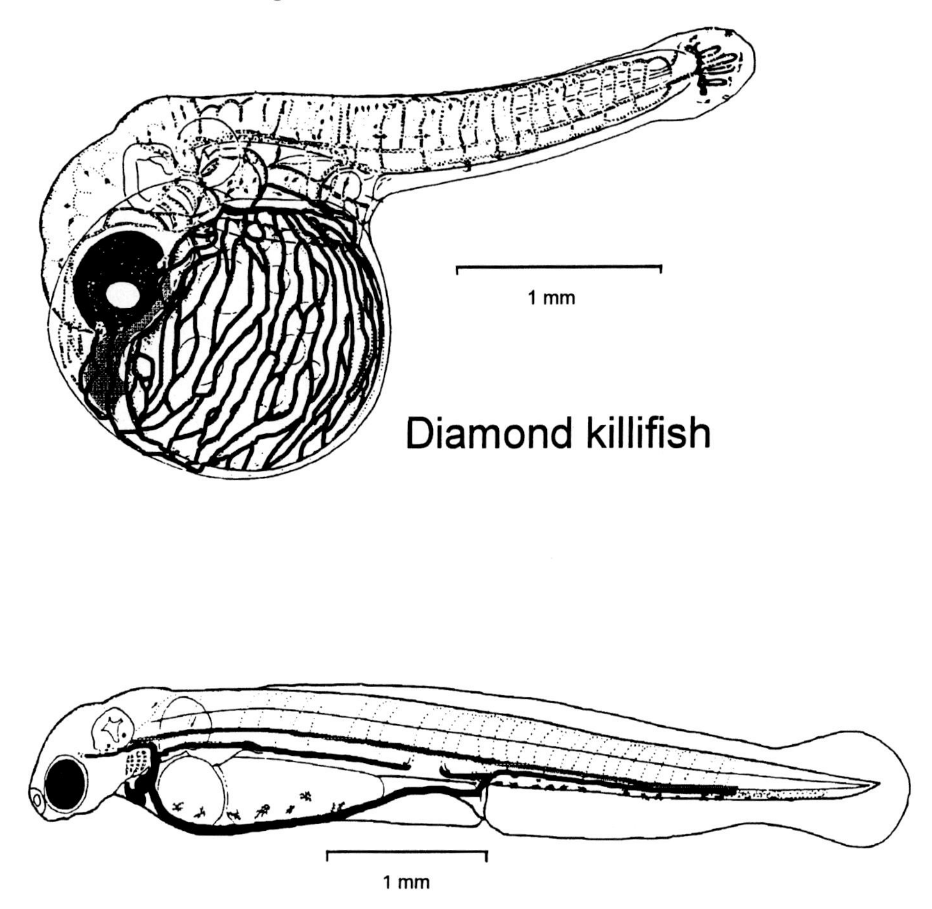

Figure 12.2 Species differ in the degree to which the yolk sac is vascularized. The drawings of the diamond killifish (*Adinia xenica*) and northern logperch (*Percina caprodes semifasciata*) at similar stages of development are from Cunningham and Balon, 1985, and Paine and Balon, 1984, respectively.

1975, 1980; McElman & Balon, 1979, 1980; McElman, 1983; Paine & Balon, 1984; Cunningham & Balon, 1985, 1986a) suggest that such differences are a reflection of the amount of oxygen in the environment. They speculate that species that spawn in hypoxic waters will tend to have a more developed vitelline circulation than species that spawn in well-oxygenated waters. This hypothesis assumes that the gas exchange capacity of the skin is directly related to capillary density. At first glance this assumption appears reasonable, but recent partitioning studies (Rombough & Ure, 1991; Wells, 1993) suggest that the relationship between exchange capacity and capillary density may not be as simple as is generally believed (see the section on the physiology of the peripheral circulation).

Gills. The development of the gill vasculature has been reviewed by Hughes (1984) and Hughes and Morgan (1973). The literature dealing specifically with the ontogeny of the gill vasculature (2 studies of elasmobranchs, 1 of chondrosteans, and about 12 of teleosts; see Hughes & Morgan, 1973, and Rombough, 1988b, for specific citations) is rather sparse considering the importance of the gill and the vast number of studies that have looked at adult structure and function.

The formation of the aortic arches is a prerequisite for blood circulation. According to Vernier (1969), blood begins to circulate in rainbow trout after the first two aortic arches have formed (140 ATU; stage V21). The direct connection of the second (hyoid) arch with the dorsal aorta is only transitory, and it soon fuses with the dorsal portion of arch I (mandibular). All six aortic arches are formed by about 180 ATU (stage V24; Figure 12.3A). The mandibular and hyoid arches fuse ventrally and anastomose with the efferent branchial artery of arch III (Figure 12.3B, C). The connection of the mandibular arch with the ventral aorta degenerates so that by about mid–yolk absorption the remnants of arches I and II receive only oxygenated blood (Figure 12.3D). The remnants of arch I supply blood to the pseudobranch, eye, and brain, whereas the remnant of arch II supplies blood to the operculum. According to Morgan (1974a), afferent and efferent branchial arteries begin to differentiate in rainbow trout at about 280 ATU (stage V28). The afferent branchial artery forms as a branch from the ventral end of the primary gill arch and grows dorsally. Paired lymph vessels form at about the same time. Blood loops form in the developing filament folds, connecting the presumptive afferent and efferent branchial arteries. These loops eventually differentiate into the afferent and efferent filamental arteries. The primary arch becomes occluded near the base of the gill, producing complete separation of afferent and efferent branchial arteries 20–30 ATU before hatch (embryos hatch at ≈340 ATU). The afferent branchial artery becomes surrounded by mesenchyme a few days before hatch and

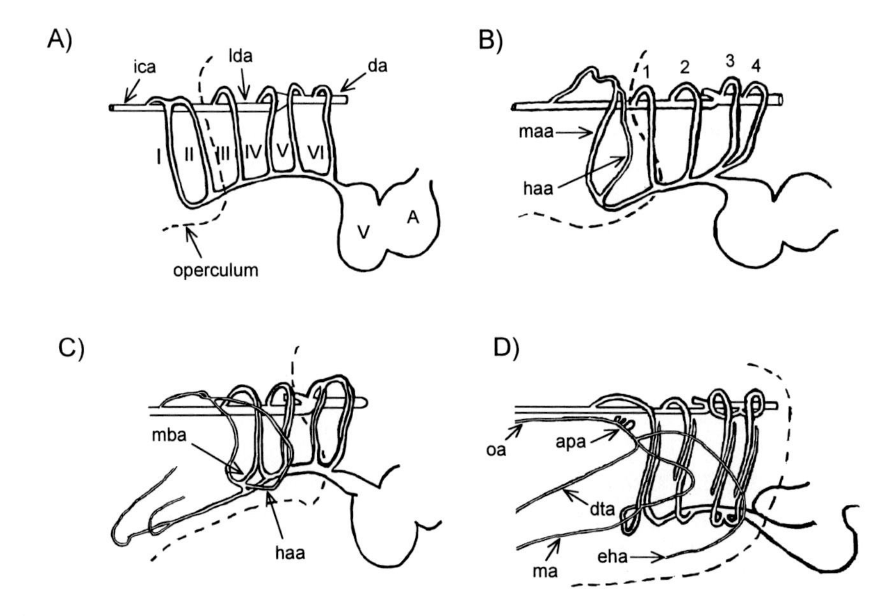

Figure 12.3 Development of the branchial circulation in rainbow trout (A = atrium, apa = afferent pseudobranchal artery, da = dorsal aorta, dta = dentary artery, eha = efferent hyoid (opercular) artery, haa = hyoid aortic arch, ica = internal carotid artery, lda = lateral dorsal artery, ma = mandibular artery, maa = mandibular aortic arch, mba = mandibular branchial anastomoses, oa = ophthalmic artery, p = pseudobranch, V = ventricle). (Stages and drawings are after Vernier, 1969.)

by 360 ATU posthatch (≈700 ATU postfertilization) is more or less completely differentiated. The efferent branchial artery, on the other hand, remains relatively undifferentiated at 360 ATU posthatch. The gill apparently becomes innervated shortly after hatch (Morgan, 1974a).

Secondary lamellae begin to form a few days before hatch in rainbow trout (Morgan, 1974b). The first indication of the presumptive lamellae is an accumulation of mesenchyme between the afferent and efferent filamental arteries. The mesenchyme cells form a column adjacent to the central filamental sinus, and a marginal blood channel opens up connecting the afferent and efferent filamental arteries. The mesenchyme cells separate, leaving spaces through which the blood can flow, and begin to differentiate into pillar cells. Near the end of yolk absorption, lamellar blood channels have formed a vascular sieve very similar in appearance to that seen in adult fish (Solewski, 1949).

The stage at which the gill vasculature develops varies considerably among species. In rainbow trout, both filaments and lamellae are present at hatch (Morgan, 1974b). Chinook salmon, *Oncorhynchus tshawytscha,* have filaments at hatch, but lamellae do not form until between 3 and 10 days posthatch at 10°C (Rombough & Moroz, 1990). Filaments first appear in walleye incubated at 20°C about 2 days posthatch; lamellae appear about 6 days posthatch (P. J. Rombough, unpublished data). At temperatures of 4–7°C, halibut larvae are 47 days old before filaments form and about 60 days old by the time lamellae appear (Pittman, Skiftesvik, & Berg, 1990). Filaments and lamellae do not form simultaneously on all four arches. In walleye, arch III is the first to differentiate, followed by arches II, I, and finally IV (P. J. Rombough, unpublished data).

The surface area of the gills available for respiratory gas exchange expands rapidly once the gills begin to form (DeSilva, 1974; McDonald & McMahon, 1977; Oikawa & Itazawa, 1985; Hughes & al-Kadhomiy, 1988; Rombough & Moroz, 1990). Mass exponents typically are in the range 1.5–3.5, but values as high as 7.1 have been reported (Oikawa & Itazawa, 1985) Rombough and Moroz (1990) noted that in some species the initial period of rapid gill expansion is relatively brief and that the use of mass exponents can be misleading. They suggested that in such cases it might be better to view the early expansion in gill surface area in terms of saltatory development (Balon, 1981). Blood–water diffusion distances decrease during larval development in obligate water-breathers (Morgan, 1974b; Wells 1993). Mean lamellar diffusion distances in rainbow trout decreased from 11.1 μm at hatch to values typical of juveniles (6.8 μm) by 71 days posthatch (Morgan, 1974b). In contrast, mean lamellar diffusion distances of bimodal breathers increase as development proceeds (Mishra & Singh, 1979; Prasad & Prasad, 1984). The increase in lamellar thickness in bimodal breathers is probably an adaptation to prevent the delicate lamellae from being damaged while collapsed when the fish is out of water.

Blood. The stage at which hemoglobin-containing erythrocytes first appear in the circulation is highly variable. Benzidine staining indicated that hemoglobin was present in white sucker embryos almost as soon as the blood began to circulate (McElman & Balon, 1980). The dolphinfish *Coryphaena hippurus* (Benetti & Martinez, 1993) and the halibut (Pittman, Skiftesvik, & Berg, 1990), on the other hand, do not produce significant amounts of hemoglobin until metamorphosis. In species that produce hemoglobin early in development, there appear to be two primary sites of hematopoiesis: one in the intermediate mesoderm near the heart and the other in the blood islands in the yolk sac (Tavolga, 1949; al-Adhami & Kunz, 1976, 1977; Cunningham & Balon, 1985; Ballard, 1986). The sites of hematopoiesis shift near hatch to centers in the spleen and kidney (Vernidub, 1966; al-Adhami & Kunz, 1976). Embryonic and larval hemoglobins generally display a higher oxygen affinity (Bird, Lutz, & Potter, 1976; Macey & Potter, 1982, Iuchi, 1973; Weber & Hartvig, 1984; King, 1994), a reduced Bohr effect (Iuchi, 1973), and greater cooperativity (Weber & Hartvig, 1984) than juvenile and adult hemoglobins. These differences generally are thought to be adaptations to facilitate oxygen uptake in oxygen-poor environments such as interstices of gravel, the egg cases of elasmobranchs, and the ovary/uterus of viviparous species (Rombough, 1988b). However, high oxygen affinity in itself is not sufficient to ensure survival under hypoxic conditions. Galloway et al. (1987) reported that ammocoetes of the lamprey *Geotria australis* were less tolerant of hypoxia than ammocoetes of other lampreys even though the hemoglobin of *G. australis* had a significantly higher oxygen affinity.

Physiology

It is only in the past decade or so that technology has improved to the point where it is possible for investigators to collect the basic physiological data (e.g., blood pressure, electrical recordings of heartbeat, blood P_{O_2}) required to understand cardiovascular function in animals the size of fish embryos and larvae (Burggren & Fritsche, 1995). We are beginning to get some idea of the role played by the circulatory system in respiratory gas exchange (see reviews by Blaxter, 1988; Rombough, 1988b; Burggren & Pinder, 1991), but our knowledge of other aspects of cardiovascular function is fragmentary at best.

Central hemodynamics

When the heart first begins to beat, contractions are weak and irregular. Contractions quickly become regular and strong enough to propel the blood through the aortic arches and systemic circulation. In teleost fishes, the rate of contraction increases progressively throughout most of embryogenesis, reaching a maximum shortly before hatch (Figure 12.4; McElman & Balon, 1979, 1980; Laale, 1984; Klinkhardt, Straganov, & Pavlov, 1987). Heart rate typically increases about 2.5-fold at constant temperature. The faster heart rates of older embryos seem to be due to an increase in spontaneous electrical activity rather than the result of extrinsic modulation. Laale (1984) reported a pattern and magnitude of increase in the rate of spontaneous contraction by explanted rudiments of the zebra fish heart that are almost identical to those seen in intact embryos. Heart rate declines after hatch, but the stage at which this occurs appears to vary among species. In walleye (*Stizostedion vitreum*) larvae, heart rate does not begin to decline until some time after the end of yolk absorption (Figure 12.4; McElman & Balon, 1979). In contrast, the heart rates of both rainbow trout (Holeton, 1971) and arctic char (*Salvelinus alpinus*) (McDonald & McMahon, 1977) begin to decrease shortly after hatch. By the time rainbow trout larvae are halfway through yolk absorption, their heart rate has fallen to about two-thirds its peak value (Holeton, 1971). The heart rate of elasmobranchs follows a pattern similar to that seen in teleosts (i.e., a fairly sharp rise to a maximum, followed by a more gradual decrease), but the timing of events appears to be different. In the little skate (*Raja erinacea*), for example, heart rate reaches it maximum value (68 beats per minute) about 84 days before hatch (Pelster & Bemis, 1991). It is not clear whether this earlier peak is intrinsic to elasmobranchs

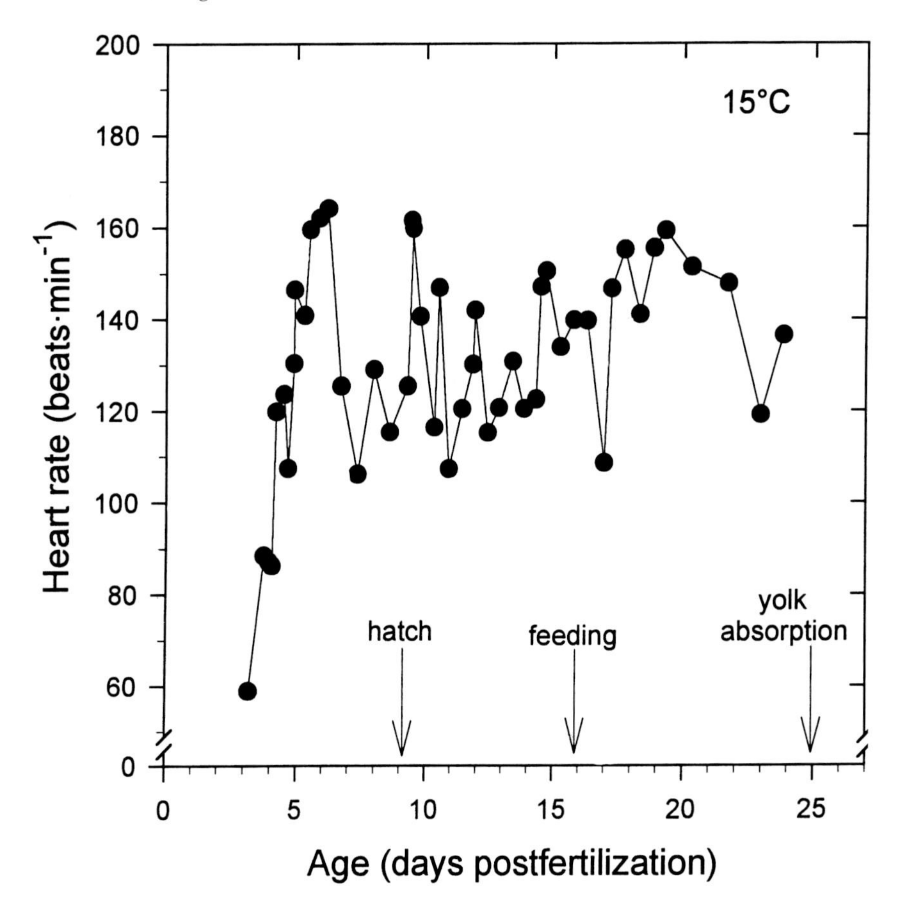

Figure 12.4 Ontogenetic changes in the heart rate of walleye embryos and yolk sac larvae (data from McElman & Balon, 1979).

or simply reflects the fact that elasmobranchs generally are much larger than teleosts at a similar stage of development. Little skate embryos, for example, weigh more than 4 g at hatch, whereas the weight of most newly hatched teleosts is in the range 10–100 mg.

At their peak, heart rates are much higher during early life than they are in juvenile and adult fish. The peak heart rate of walleye embryos is ≈160 beats per minute at 15°C (McElman & Balon, 1979). The rate for Atlantic salmon embryos just before hatch at 8°C is ≈110 beats per minute (Klinkhardt, Straganov, & Pavlov, 1987), and that of rainbow trout larvae 2–5 days posthatch at 10°C is 65–75 (Holeton, 1971). For comparison, the resting heart rate of a 1 kg adult rainbow trout at 10°C is about 38 beats per minute (Kiceniuk & Jones, 1977). Some reports of embryonic and larval heart rates seem unrealistically high. For example, Balon (1980) reported heart rates of more than 120 beats per minute for arctic char larvae at 4.4°C. This is more than 2.5 times the value (45 beats per minute) reported by McDonald and McMahon (1977) for similar-aged arctic char held at 6.5°C and much higher than most reports for larvae of other species at similar temperatures. Experience in my lab suggests that it is easy to become confused when trying to determine heart rate by visual observation through a microscope. In particular, observers often end up counting atrial and ventricular contractions as separate beats. A relatively easy way to circumvent this difficulty is to use video microscopy to record beats and to make counts while running the tape in slow motion. An alternative suggested by Burggren and Fritsche (1995) is to use an impedance converter to record the electrical activity generated by the heart.

Contraction of the embryonic heart is initiated by pacemaker cells. In brown trout, *Salmo trutta,* these cells are located initially in the curve of the heart linking the presumptive ventricle and atrium (Grodzinski, 1954). The primary pacemaker gradually shifts in a caudal and dorsal direction to take up position in the atrium near the opening of the duct of Cuvier by late embryogenesis. This region appears to be homologous with the sinoatrial node of adult fish. Grodzinski (1954) noted two additional regions, one near the junction of the ventricle and atrium (homologous with the atrioventricular node?) and one near the cranial end of the ventricle, that displayed pacemaker activity in older trout embryos. These two regions are normally subordinate to the sinoatrial pacemaker, and their activity becomes evident only if the sinoatrial node is excised. Laale (1984) suggested that it was the right cardiac rudiment of zebra fish that gave rise to the sinoatrial pacemaker. This conclusion was based on the observation that if the two cardiac rudiments were separated and grown in tissue culture, the right rudiment eventually beat faster than the left rudiment (Laale, 1984).

The heart begins to pump blood in what appears to be an efficient manner well before cardiac valves develop. In the halibut *Hippoglossus hippoglossus,* blood begins to circulate early in embryonic development, but the semilunar valves between the ventricle and bulbus are not formed until about 45 days after hatch at incubation temperatures of 4–7°C (Pittman, Skiftesvik, & Berg, 1990). In the little skate, blood circulation starts prior to 27 days postspawning, but the atrioventricular and conal valves are not evident until about 50 days postspawning (Pelster & Bemis, 1991). Developing fish are able to maintain a unidirectional flow of blood in spite of not having functional valves because their hearts function more like a peristaltic pump than like the chambered pump one sees in adult fish. Slow-motion videos of young rainbow trout larvae obtained in my lab indicate that contraction of the heart begins in the atrium and progresses along the ventricle in a posterior to anterior direction as a peristaltic wave. The region of the heart immediately behind the crest of this wave remains in a contracted state long enough to prevent retrograde blood flow. By the time the atrial end of the ventricle begins to relax, the atrium has reentered systole. Examination of videotapes suggests that sustained contraction prevents backflow at the bulbal end of the trout ventricle as well as at the atrial end. Backflow appears to be prevented in much the same manner in elasmobranch embryos. Pelster and Bemis (1991) noted that the conus arteriosus remained occluded during ventricular diastole in embryos of the little skate. They interpreted this as diastolic collapse, but it is difficult to distinguish diastolic collapse following contraction from sustained contraction on the basis of simple visual observations, so it is probably premature to say which alternative is correct.

The pressure generated by the embryonic heart is initially quite low. Pelster and Bemis (1991) reported a ventricular systolic pressure of only about 0.13 kPa (1.0 mm Hg) for a 10 mg little skate embryo (Figure 12.5). Systolic pressure rose relatively rapidly as the embryos grew, however, so that by the time the skate embryos were ready to hatch, their systolic pressure (≈1.8 kPa) was not much less than that found in adult elasmobranchs (Pelster & Bemis, 1991). Systolic pressure in skate embryos was highly correlated with body mass, with a mass exponent of about 0.45 ($M^{0.45}$). Major rearrangements of the peripheral circulation such as the development and then regression of external gills had no apparent effect on systolic blood pressure. Systolic pressures in developing teleosts appear to be similar in magnitude to those recorded in skate embryos of roughly the same size. Pelster and

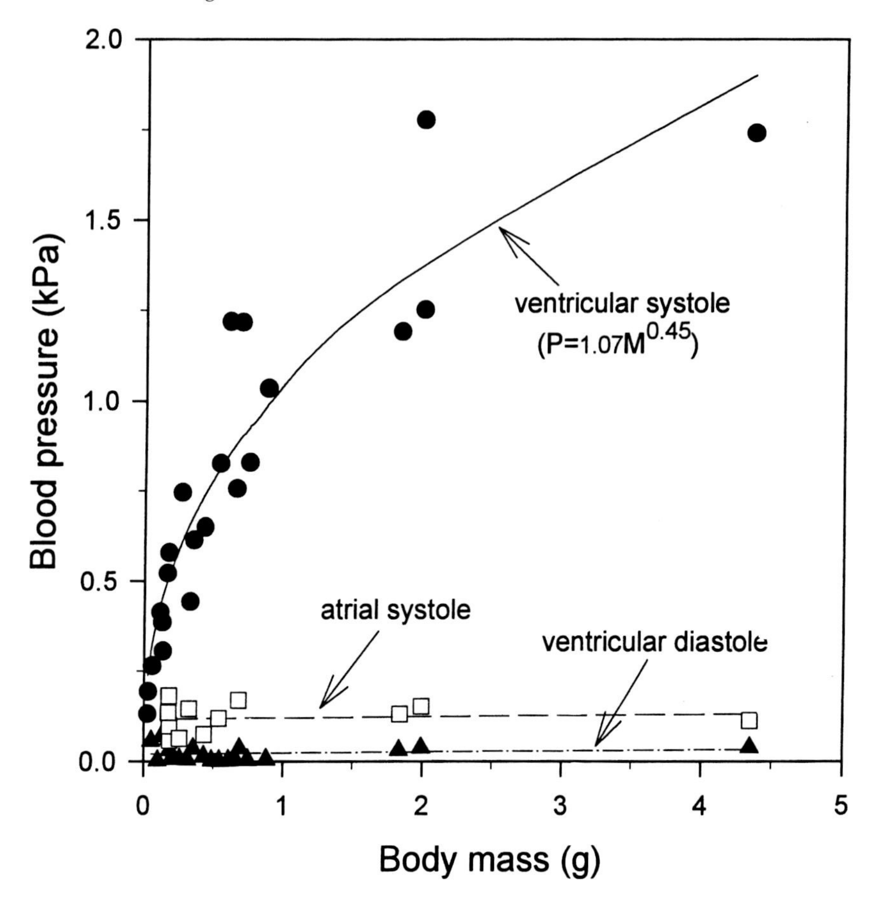

Figure 12.5 Ontogenetic changes in ventricular and atrial blood pressure in embryos of the little skate (data from Pelster & Bemis, 1991).

Burggren (1996) recorded ventricular pressures of about 0.5 kPa in 96-nour zebra fish larvae. Preliminary measurements indicate that the systolic pressure of newly hatched Atlantic salmon larvae also is about 0.5 kPa (P. J. Rombough & A. W. Pinder, unpublished data).

Ventricular pressure in embryos, unlike in adults, probably does not go subambient during diastole. Pelster and Bemis (1991) reported a minimum diastolic pressure in little skate embryos of about +0.03 kPa. Unlike systolic pressure, there was little change in diastolic pressure during the embryonic development of the little skate. The reason diastolic pressure remains positive is not clear but may be related to the fact that the pericardium is compliant in embryos and, therefore, is not able to assist in generating aspiratory pressures. Alternatively, the myocardium simply may not be strong enough in embryos to generate significant elastic recoil. As in older fish, atrial contraction appears to be solely responsible for ventricular filling in embryos. However, the rise in ventricular pressure due to atrial contraction is small. Pelster and Bemis (1991) reported that atrial contraction increased ventricular pressure only from 0.03 to 0.08 kPa in skate embryos. Atrial pressure, unlike diastolic ventricular pressure, changed little during development. Maximum conal and ventral aortic pressures in skate embryos were similar to maximum ventricular pressures (Pelster

& Bemis, 1991). The major function of the conus appears to be to prolong the period of high arterial pressure rather than to supplement the pressure generated by the ventricle. The net result is that a positive ventral aortic pressure is maintained throughout ventricular diastole in the little skate. Pelster and Bemis (1991) report ventral aortic systolic and diastolic pressures of 0.7 and 0.5 kPa, respectively, for a 0.74 mg embryo. In contrast, ventral aortic pressure appears to fall to very close to zero during diastole in teleost embryos. Video recordings of red blood cells moving through the branchial arches of newly hatched rainbow trout show the blood coming to a complete halt between beats (P. J. Rombough, unpublished observations). This is not what one would expect if positive ventral aortic pressure was maintained throughout diastole.

We know relatively little about when and how the heart is brought under central control. It appears that in fish, as in amphibians (Orlando & Pinder, 1995), the heart is innervated well before it is subject to physiological control (Balashov & Soltitskij, 1991). In the golden mullet *Liza auratus,* adrenergic receptors develop and become functional before cholinergic receptors (Balashov & Soltitskij, 1991; Balashov, Soltitskii, & Makukhina, 1991), but whether this is true in other species has not been established. Nechaev, Labas, and Denisor, (1992) examined the stages at which the heart of the blue acara cichlid *Aequidens pulcher* became responsive to various catecholamines. They found that heart rate increased in response to first dopamine and then norepinephrine during late embryogenesis but that epinephrine was not effective until after hatch. The stage at which the heart comes under nervous control appears to vary depending on species. As just mentioned, the heart of the blue acara cichlid responded to various catecholamines prior to hatch (Nechaev, Labas, & Denisov, 1992). Balashov, Soltitskii, and Makukhina (1991) reported that the heart of the golden mullet was not subject to sympathetic control until 5–6 days after hatch. The hearts of some cyprinodontids (e.g., *Fundulus heteroclitus*) come under vagal control during late embryogenesis (Armstrong, 1931; Armstrong & Child, 1965; Cunningham & Balon, 1985, 1986b). In contrast, golden mullet remain insensitive to acetylcholine for at least the first 10 days of larval life (Balashov, Soltitskii, & Makukhina, 1991). The golden mullet is probably more representative of fish generally. The early development of vagal control in the cyprinodontids appears to be correlated with their rather unusual ability to enter diapause if ambient conditions are unsuitable for hatching.

There are no published accounts of how cardiac output changes during the course of fish development. In amphibians, mass-specific output increases during larval development (Hou, 1992; Orlando & Pinder, 1995; Hou & Burggren, 1995b) but whether this holds for fish larvae remains to be established.

Peripheral circulation

Studies of the peripheral circulation have focused on respiratory function, particularly on the uptake of oxygen across the skin and gills. In most species, the skin probably remains the major site of gas exchange well into larval life. Partitioning studies indicate that even in species such as chinook (Rombough & Ure, 1991) and Atlantic salmon (Wells, 1993) that hatch with relatively well developed gills, the skin is the major site of gas exchange until midway through yolk absorption (Figure 12.6A). As noted previously, capillaries are not evenly distributed over the skin surface. At first glance, it appears reasonable to expect the efficiency of gas exchange to be greater in the well-vascularized regions. Partitioning studies indicate, however, that there is little correlation between capillary density and the effi-

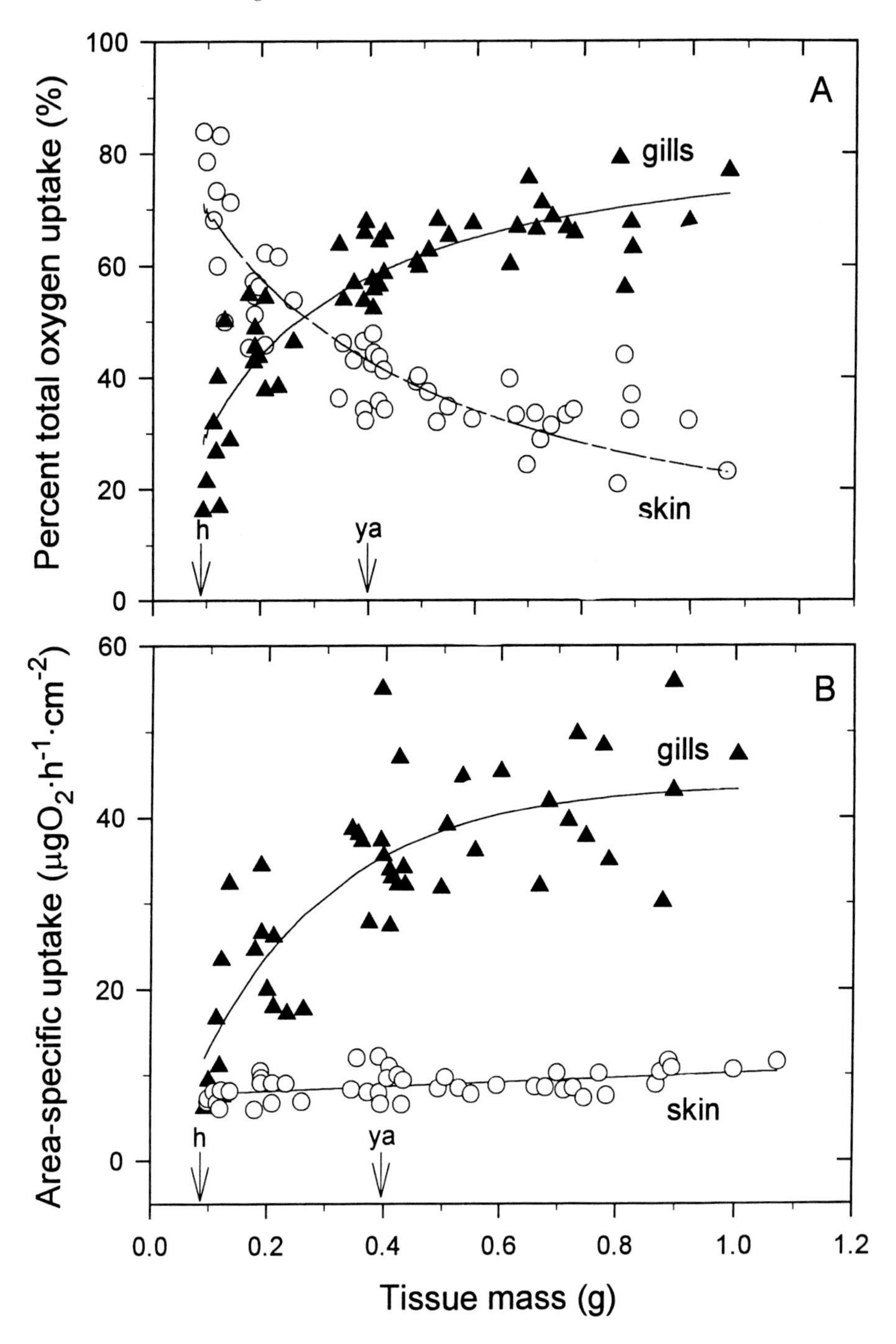

Figure 12.6 (*A*) Ontogenetic changes in relative importance of the skin and gills as sites of oxygen uptake in young chinook salmon. (*B*) Changes in the relative efficiency of oxygen uptake across the skin and gills of young chinook salmon (h = tissue mass at hatch, ya = tissue mass at yolk absorption). (Data from Rombough & Ure, 1991.)

ciency of cutaneous gas exchange. Wells (1993) noted that although capillary density was much higher in the yolk sac than in the head and trunk of Atlantic salmon, oxygen uptake per unit area was virtually identical (27 ng $O_2 \cdot h^{-1} \cdot mm^{-2}$ vs. 25 ng $O_2 \cdot h^{-1} \cdot mmm^{-2}$). In both chinook salmon (Figure 12.6B; Rombough & Ure, 1991) and Atlantic salmon (Wells, 1993), the area-specific uptake of oxygen across the skin is essentially the same at the end of yolk absorption as it is at hatch. If capillary density was a critical factor in determining the efficiency of gas exchange, one would expect area-specific uptake to be greater at hatch, when the vitelline circulation is at its maximum extent, than at the end of yolk absorption, by which time the vitelline circulation has been largely internalized.

The presence of a relatively thick unstirred layer of water immediately adjacent to the skin surface appears to be the reason capillary density has so little effect on the efficiency of cutaneous gas exchange. In still water, the diffusive boundary layer surrounding rainbow trout larvae is more than 1000 μm thick (Figure 12.7). The thickness of the boundary layer decreases as water velocity increases, but even at a relatively high velocity of 2 $cm \cdot s^{-1}$, the diffusive boundary layer is still about 250 μm thick (Pinder & Feder, 1990). (The water velocities salmon or trout larvae are likely to encounter in nature are in the range 0.03–0.2 $cm \cdot s^{-1}$; for comparison, the flow rate through the gills in resting juvenile rainbow trout is about 2.0 $cm \cdot s^{-1}$ [Randall, Lin, & Wright, 1991].) The "pipes in a wall" model developed by Malvin (1988) predicts that at diffusion distances of this magnitude (250–1000 μm), differences in capillary density are unlikely to have much of an impact. The presence of a thick

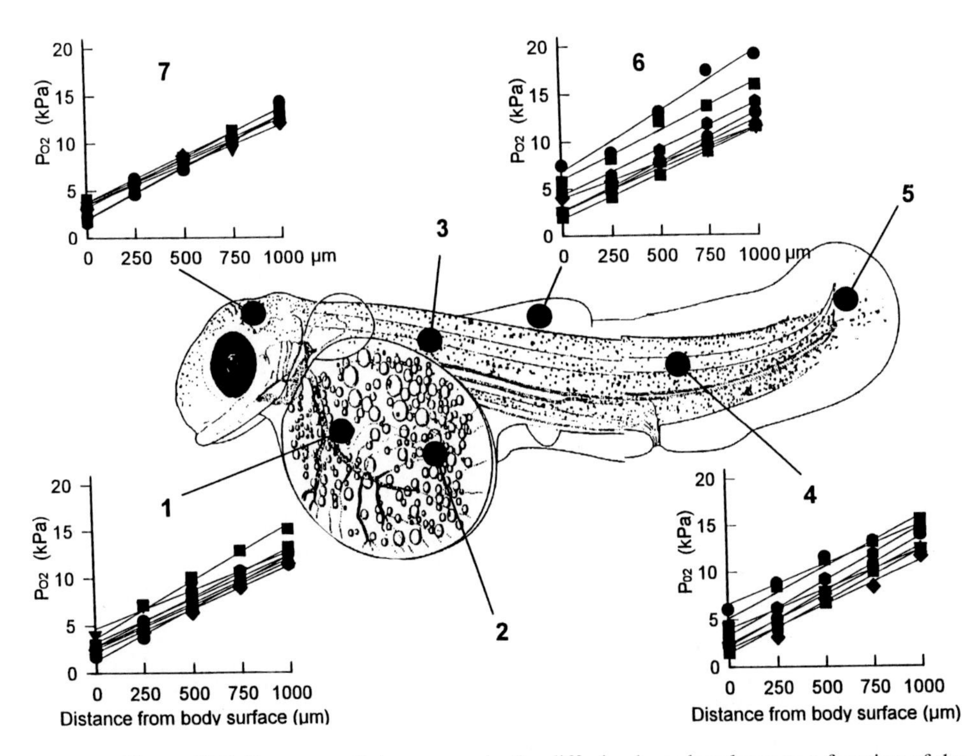

Figure 12.7 Oxygen partial pressures in the diffusive boundary layer as a function of the distance from the skin surface at various locations on the body of rainbow trout larvae. Measurements were made in still water. Lines represent different individuals (P. J. Rombough, unpublished data).

diffusive boundary layer also explains why the diffusive conductance of the skin of salmonids remains relatively stable during larval development in spite of about a 50% increase in skin thickness. If one assumes that the boundary layer is about 250 μm thick, an increase in the thickness of the skin from 20 to 30 μm will result in only about a 3% increase in total diffusion distance.

Although the yolk sac does not appear to be a particularly efficient gas exchanger, this may not be the case for some of the other well-vascularized regions of the skin. The pectoral fins, in particular, would seem to be well suited for gas exchange. Many species, including salmonids, have relatively large, well-vascularized fins that they move rapidly back and forth. The oscillatory frequency of the pectoral fins can be as high as $150 \cdot min^{-1}$ (Peterson, 1975). Because of edge effects, these rapid oscillatory movements are likely to be more effective at breaking up the boundary layer than an externally generated flow of the same velocity passing over a stationary surface such as the yolk sac.

The diffusing capacity of the gills, unlike the skin, increases significantly during larval development (Rombough & Ure, 1991; Wells, 1993). The area-specific uptake of oxygen across the gills of chinook salmon at hatch is about the same as that across the skin (Figure 12.6B, Rombough & Ure, 1991). However, by the end of yolk absorption, the area-specific uptake of the gills is about 3.5 times that of the skin. Most of this increase in efficiency appears to be due to the reduction in the effective thickness of the boundary layer that occurs when the water irrigating the gill is forced to flow through the interlamellar channels. The maximum length of the diffusive pathway in the gill is one-half the interlamellar distance, which in the case of chinook salmon larvae is about 12 μm. The thickness of the diffusive boundary layer adjacent to a free surface such as the skin at typical gill water velocities would be of the order of 100–250 μm.

The partial pressures of oxygen in the blood of cutaneously respiring rainbow trout larvae are much lower than in juvenile fish (Figure 12.8) Values of intravascular P_{O_2} in the vitelline vein, efferent branchial artery, and subintestinal vein recorded using micro-oxygen electrodes ranged from 0.4 to 1.7 kPa in still water and from 2.4 to 6.3 kPa in moving water (Rombough, 1992). The presence of the diffusive boundary layer is the main reason P_{O_2} values were so low. The boundary layer accounted for 95% of the total resistance to oxygen flux in still water and 82% of the total resistance in moving water. Unlike in juvenile fish, oxygen levels appear to be essentially the same throughout the circulatory system in young larvae. Again, this seems to be a reflection of the overwhelming impact of the diffusive boundary layer (it also provides support for the notion that the yolk sac is not a particularly efficient gas exchanger).

Pelster and Bemis (1992) examined the role of the external gill filaments in embryonic gas exchange in the little skate. External gill filaments form about 25–30 days postspawning. The structure of the filaments is relatively simple, with each containing a single afferent and efferent blood vessel. Filaments grow to a maximum length of about 1 cm by 70–75 days postspawning and then regress. The external filaments are completely resorbed and replaced by internal gills by 90–95 days postspawning. The volume and velocity of blood flow through the filaments parallel the expansion and regression of the filaments. Blood velocities recorded using slow-motion video ranged from 0.1 to 0.7 $mm \cdot s^{-1}$. The external gills appear to be perfusion limited and capable of supplying up to 50% of the total oxygen requirement of the embryo (this calculation assumes that movement of the filaments is effective in breaking up the diffusive boundary layer). Both peak systolic and peak pulse pressures decreased along the length of the filament. Vascular resistance decreased slightly during the expansion of the filaments. There was a sharp increase in vascular resistance as

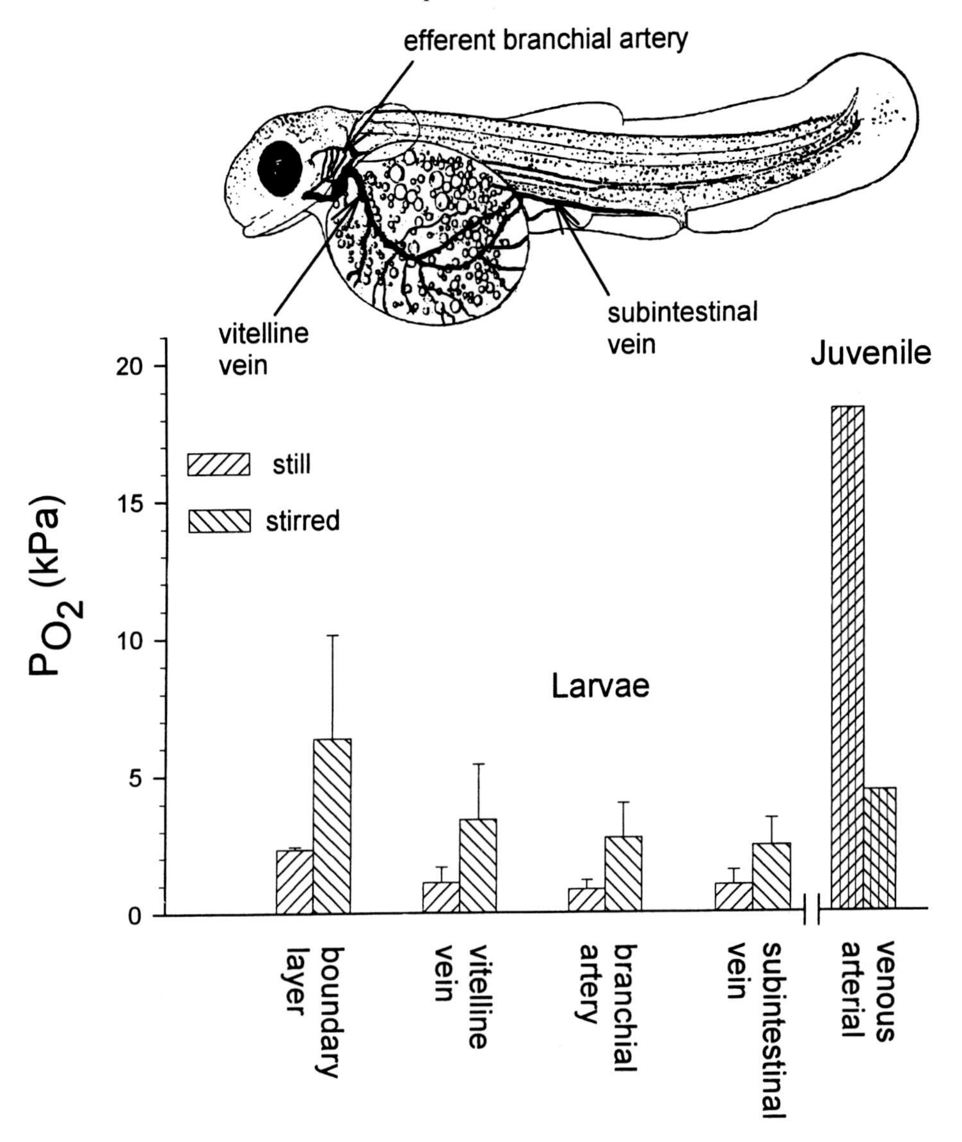

Figure 12.8 Oxygen partial pressures in the boundary layer immediately adjacent to the yolk sac surface and at various locations in the circulation of rainbow trout larvae in still and vigorously stirred water. Typical values of arterial and venous P_{O_2} for juvenile rainbow trout are shown for comparison (data from Rombough, 1992).

the filaments were resorbed. Changes in peripheral resistance had no effect on central pressures (Pelster & Bemis, 1991).

Adaptive responses

Organisms are faced with the task of balancing energy supplies and demands. In vertebrates, the supply and removal of respiratory gases are frequently the limiting step in the chain of events that ultimately provides the energy to power cellular metabolism. Adult fish respond rapidly to changes in supply–demand relationships for respiratory gases by altering the activities of their respiratory and cardiovascular systems in a coordinated fashion (Burleson, Smatresk, & Milsom, 1992; Taylor, 1992). The alterations in the cardiovascular system appear to be driven by changes in respiratory demand, leading to the perception that, at least over the short term, the circulatory system is the "slave" of the respiratory system.

Fish face more pronounced changes in supply–demand relationships for respiratory gases during the course of embryonic and larval development than they do as adults (Rombough, 1988b). The rate of oxygen consumption increases by 2–3 orders of magnitude over a relatively short period as a result of tissue growth. Temperature appears to have a more pronounced effect on metabolic rate during early life. The Q_{10} for metabolism is about 3 in larvae but only about 2 in juveniles and adults. Embryos and larvae are frequently subject to hypoxia either because of reductions in the ambient oxygen level of the microenvironment surrounding the organism (e.g., perivitelline fluid, diffusive boundary layer) or because of physiological hypoxia induced by a combination of activity and limited capacity to make ventilatory adjustments.

The linkage between the respiratory and cardiovascular systems appears to be looser during development than it is later in life. Although both metabolic and heart rate increase during embryonic development, they do not seem tightly linked. In Atlantic salmon, for example, metabolic rate more than doubles during early to mid-organogenesis, but there is very little change in heart rate (Figure 12.9; Klinkhardt, Straganov, & Pavlov, 1987). Temperature appears to have a greater effect on metabolism than on heart rate during early life. Metabolic Q_{10}s during late embryogenesis are typically about 3.0 (Rombough, 1988b; the metabolic Q_{10} for Atlantic salmon embryos based on the data in Figure 12.9 is 3.5). Fisher (1942) reported Q_{10}s of 2.2 and 2.4 for embryonic heart rate in brook trout (*Salvelinus fontinalis*) and Atlantic salmon, respectively (the Q_{10} for the heart rate of Atlantic salmon based on the data in Figure 12.9 is 2.6). The critical indicator of whether the respiratory and cardiovascular systems are closely linked, however, is not heart rate but cardiac output. Cardiac output is the product of heart rate and stroke volume. Unfortunately, we have no information on how stroke volume changes during development or in response to temperature so it is impossible to know for sure how tightly the two systems are linked. Studies of amphibian larvae suggest the link is relatively loose. In *Xenopus laevis* larvae, for example, the mass exponent for cardiac output is about 1.2 but that for metabolic rate is only about 0.8 (Orlando & Pinder, 1995). One would expect mass exponents to be about the same if cardiac output and metabolic rate were closely linked.

The cardiovascular response of many fish larvae to hypoxia is the opposite of that seen in juveniles and adults. Hypoxia elicits reflex bradycardia in juvenile salmonids (Smith & Jones, 1978). The response in salmonid embryos and larvae is moderate tachycardia (Fisher, 1942; Holeton, 1971; McDonald & McMahon, 1977). The heart rate of 1-day-old rainbow trout was about 15% greater at a P_{O_2} of 30 mm Hg than at normoxia (Holeton, 1971). A similar response was seen in 8-day-old rainbow trout larvae (Holeton, 1971) and in arctic char larvae exposed to chronic hypoxia (McDonald & McMahon, 1977). A tachycardic response to hypoxia is consistent with pharmacological observations (Balashov & Soltitskij, 1991; Balashov, Soltitskii, & Makukhina, 1991) that the heart comes under sympathetic control well before parasympathetic control. Holeton (1971) suggested that the drop in the heart rate of rainbow trout larvae between 9 and 16 days (90–160 ATU) posthatch might be indicative of the heart coming under vagal control. Unfortunately, we have no information on

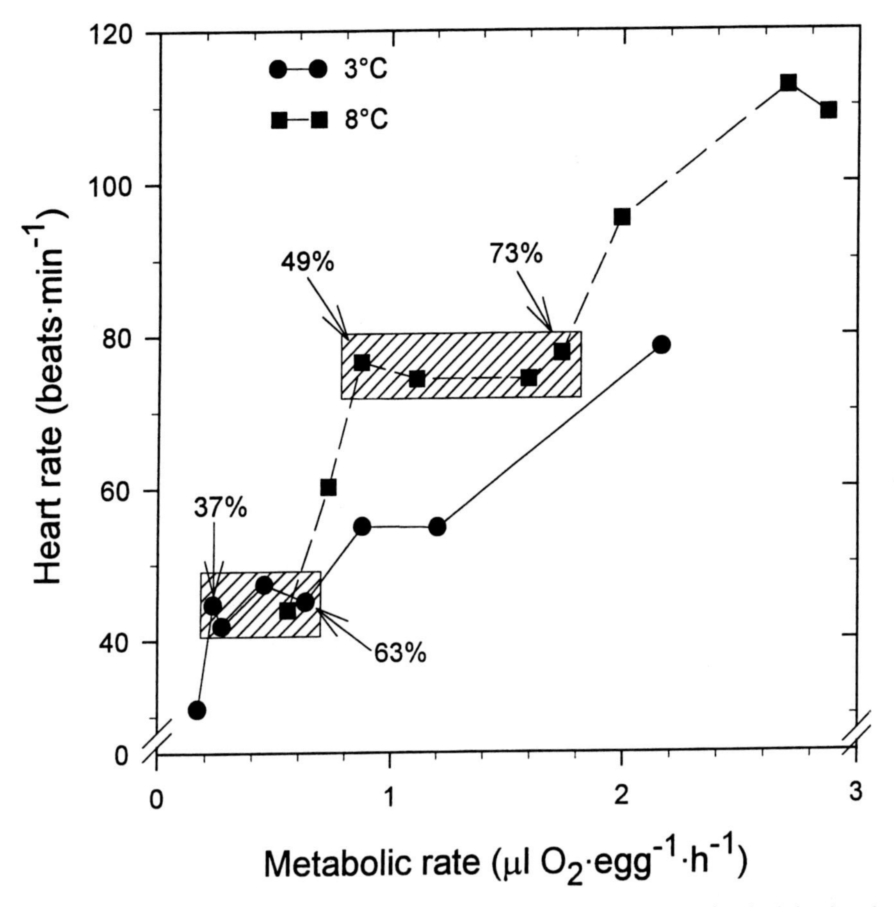

Figure 12.9 Relationship between heart rate and oxygen consumption in Atlantic salmon embryos incubated at 3°C and 8°C. Arrows indicate rates at various stages of development (expressed as percentage of total embryonic period). Note that there was no significant correlation between heart and metabolic rates between 37% and 63% of embryonic development at 3°C and between 49% and 73% of embryonic development at 8°C. (Redrawn using data from Klinkhardt, Straganov, & Pavlov, 1987.)

whether there is a corresponding change in hypoxic response. The heart appears to be extremely tolerant of hypoxia during early life. Fisher (1942) reported that critical oxygen tensions for heart rate (P_{ch}) in salmon and trout embryos ranged from 3.5 to 5.0 mm Hg depending on temperature. The hypoxic tolerance of the embryonic heart is particularly impressive when one considers that Fisher's (1942) estimates for P_{ch} are ambient partial pressures. Based on the difference between intravascular and ambient P_{O_2} values observed in rainbow trout larvae (Rombough 1992), the critical P_{O_2} at the level of the heart is probably about an order of magnitude lower than the reported P_{ch} value (i.e., ≈0.3–0.5 mm Hg). In comparison, the threshold lumenal P_{O_2} for adult fish hearts is about 10 mm Hg (Farrell & Jones, 1992). The response of rainbow trout alevins to carbon monoxide poisoning is similar to that observed during hypoxia, namely, tachycardia (Holeton, 1971). Tachycardia becomes more pronounced in older alevins. The nature of the relationship between cardiac function and activity remains to be elucidated. Heart rate tends to become more variable as larvae grow, but whether this variability is correlated with activity-induced changes in oxygen demand has not been established.

Several lines of evidence indicate that embryos and larvae are less dependent on the circulatory system to supply oxygen to the tissues than are older fish. As just noted, cardiac output does not appear to track metabolic rate as closely during early life. P_{O_2} levels are essentially the same throughout the circulation (Rombough, 1992). There appears to be little correlation between capillary density and the diffusing capacity of the skin (Rombough & Ure, 1990; Wells, 1993). Destroying erythrocytes (Iuchi, 1973) or poisoning their hemoglobin (Holeton, 1971; Pelster & Burggren, 1996) evokes only modest response. It would seem that fish embryos, like most vertebrate embryos (Mellish, Pinder, & Smith, 1994; Burggren & Territo, 1995), are small enough that respiratory demand for oxygen can be met by simple diffusion across the skin. This raises the question of what is the function of the circulatory system. One obvious possibility is nutrient delivery. The nutrient supply for the developing embryo is concentrated in the yolk, and the nutrients must somehow be distributed to the tissues. Diffusion is not a very effective means of delivering nutrients, given the size of the molecules and the fact that many of them require carrier molecules to pass across membranes. The driving force leading to the development of the circulatory system, thus, may be nutrient transport rather than gas exchange. This is not to say that the circulatory system is not a convenient vehicle for gas transport once it is established. The importance of the circulatory system in gas transport increases as the embryo grows and finally becomes essential when the demand for oxygen exceeds the rate at which it can be supplied by simple diffusion.

Concluding remarks

The body of literature dealing specifically with the ontogeny of the cardiovascular system in fish is not large, particularly when compared with the vast literature available for amphibians and birds. As a consequence, our knowledge concerning circulatory development in fish remains rather sketchy.

This is particularly true of functional aspects. For example, we have no information at all for such basic physiological parameters as blood volume, blood oxygen capacity, and cardiac output. There have been only a couple of studies examining the pharmacology of the heart, and we still do not know for certain when the heart comes under central control. We know virtually nothing about control of the peripheral circulation. Our knowledge of the morphogenesis of the circulatory system is somewhat better, but even here there are obvious deficiencies. Much of the information regarding the morphogenesis of the circulation comes from descriptive studies conducted in the first half of this century. We need to reexamine events using more modern techniques. In particular, a better understanding of the microanatomy of the circulation is required before we can fully explore many of the functional aspects of cardiovascular development. Other areas that merit further study are the extent to which fish embryos and larvae are capable of adaptive responses and the nonrespiratory functions of the blood. Even studies that appear to have yielded good information on certain aspects of development could profit from replication. Fish are an extremely diverse group of animals occupying a wide variety of habitats. To date, we have looked at only a handful of the approximately 22,000 extant species. To paraphrase August Krogh, nature already may have conducted the critical experiment that will provide the clue for understanding key events in vertebrate development.

13

Amphibian cardiovascular development

WARREN W. BURGGREN AND REGINA FRITSCHE

Introduction

As research in developmental biology over the decades has moved from classic embryology to studies of the cellular and molecular basis of organogenesis, amphibians have emerged as one of several general models for vertebrate development, joining the chick, the mouse, and more recently, the zebra fish. One of the major reasons why frogs, fishes, or any other group of lower vertebrates can be used as a general model for higher-vertebrate development is that early in development the basic patterns, structures, and physiological processes appear to be largely identical (Figure 13.1).The historically important role of amphibians in the study of cardiovascular development can be attributed to their typically large, semitransparent eggs, their relative ease in rearing, and the ability to grow amphibian cells in culture (Slack, 1991). Biologists studying cardiovascular development have recently added amphibian models to their repertoire of vertebrate cardiovascular systems (e.g., Yost, 1994; Danos & Yost, 1995).

Frequently, however, cardiovascular development in amphibians is studied primarily to model these processes in mammals rather than to understand development in amphibians as a subject of interest in its own right. Consequently, the goals of this chapter are not only to describe the general process of cardiovascular development in amphibians from a "vertebrate model perspective" but also to highlight and celebrate the diversity and the unique events to be found in developing amphibians.

Cardiovascular anatomy of amphibians

The cardiovascular anatomy of mature amphibians is among the most complex of all vertebrates. As is the case for most reptiles (see Chapter 14), the presence of an incompletely divided ventricle provides possibilities for controlled redistribution ("shunt") of oxygenated and deoxygenated blood into the systemic or the gas exchange circuits. The ability of amphibians to respire by cutaneous diffusion as well as by lungs is legendary and is supported by a unique and complicated arrangement of central vessels and valves (Feder & Burggren, 1985). When one then additionally considers a developmental overlay that incorporates an anatomical metamorphosis and a respiratory transition from an aquatic to an aerial mode, then the cardiorespiratory anatomy and physiology of amphibians is indeed elaborate. The following is a brief synopsis; the interested reader is directed to Shelton, 1976; Burggren and Just, 1992; and Burggren, 1995, for reviews.

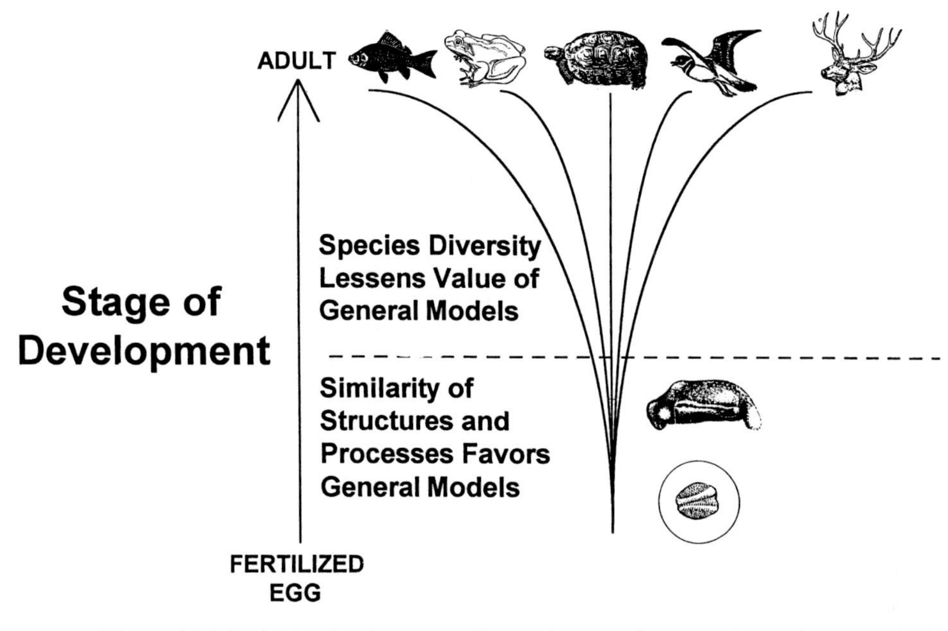

Figure 13.1 Early in development, all vertebrate embryos undergo the same basic processes of organogenesis and differentiation, with similar, if not identical, ontogenetic changes in major organ systems. In these early stages, any one group can be used as a model to gain insight into the development of a structure or process in another group. As development progresses, however, species diversity begins to develop in earnest, and the structures and processes shown by each group become unique and less representative of other, distantly related vertebrates. Not indicated on this diagram is the fact that the horizontal broken line, depicting the transition zone in which general model applicability is lost, may be located at different times in development for different processes.

The mature amphibian circulation: Species diversity

Although "the frog" is often viewed as the typical amphibian, considerable species diversity exists in the cardiovascular anatomy of mature amphibians. There are three orders in the Amphibia – the anurans (frogs and toads), the urodeles (salamanders), and the apodans (caecilians) – each with unique cardiovascular features.

Adult anurans have anatomically distinct left and right atria (Figure 13.2, top). The left atrium receives blood returning from the lungs via paired pulmonary veins. The right atrium receives blood returning from the systemic vascular beds (which contain some highly oxygenated blood returning from the skin; see the later discussion). The single ventricle consists of spongy trabeculate myocardium without coronary vessels. All cardiac output is ejected into a large, elastic conus arteriosus, which divides distally into symmetrical and separate systemic, carotid, and pulmocutaneous arterial arches. Semilunar cusp valves guard the orifice between the ventricle and the conus arteriosus, and that between the conus arteriosus and arterial arches. A single spiral valve, attached along only one edge, runs the length of the conus and helps to distribute primarily oxygenated blood toward the systemic circulation. The pulmocutaneous arch divides distally into a large pulmonary artery and a small cutaneous artery. The cutaneous artery is unique to anuran amphibians and supplies the skin primarily with deoxygenated blood. The skin of anurans thus receives *both* deoxygenated blood via the cutaneous artery and oxygenated blood via vertebral vessels.

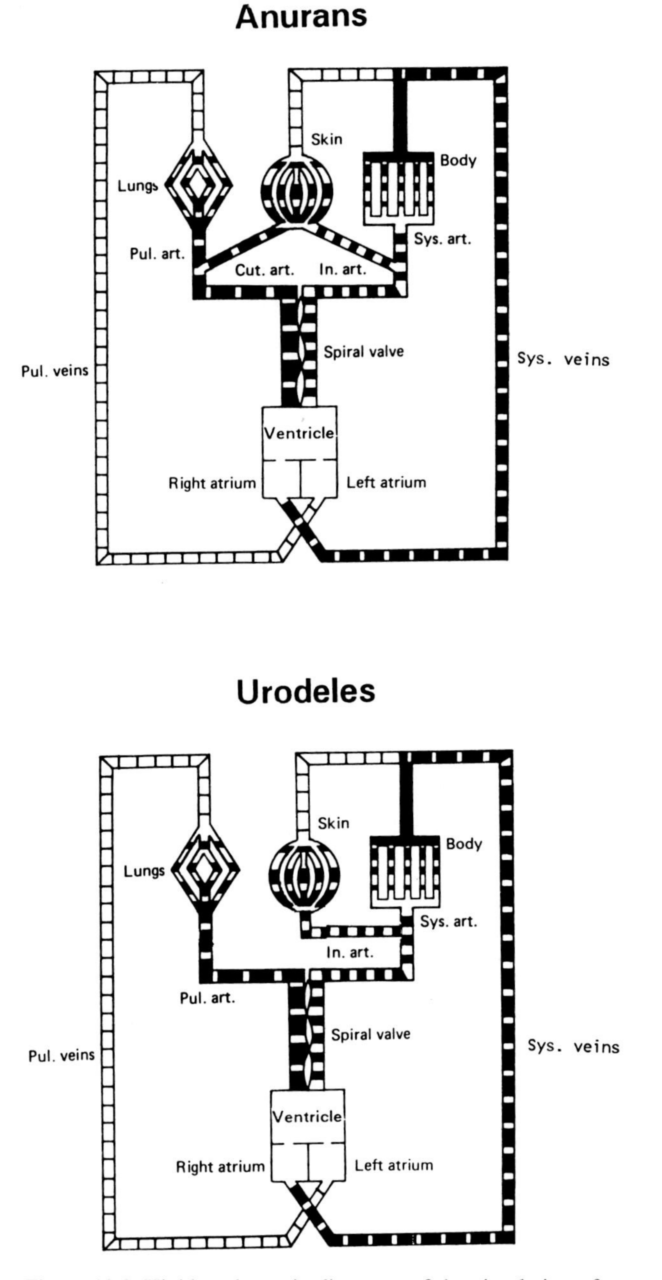

Figure 13.2 Highly schematic diagrams of the circulation of a mature anuran amphibian (*top*) and a mature urodele (*bottom*). Oxygenated blood is depicted as white, and deoxygenated blood is black (art. = artery, Cut. = cutaneous, In. = intervertebral, Pul. = pulmonary, Sys. = systemic). (From Burggren, 1995.)

The cardiovascular anatomy of urodeles (Figure 13.2, bottom) differs from that of anurans in several important ways. Salamanders lack a cutaneous artery, so the skin receives the same blood that is distributed to other systemic tissues, provided by vertebral arteries. Some urodeles – primarily *Chioglossa, Salamandrina, Rhyacotriton,* and the entire family Plethodontidae – completely lack lungs as adults. The interatrial septum is highly reduced or absent, and there is no pulmonary artery. All gas exchange is across the skin in these so-called lungless salamanders (see Burggren, 1989, for references). As adults, several genera of urodeles have an incomplete vertical septum within their single ventricle, including *Siren intermedia* (Owen, 1834; Putnam, 1975), *Necturus maculosus* (Putnam & Dunn, 1978), and *Cryptobranchus alleganiensis* (Putnam & Parkerson, 1985). The hemodynamic implications of this septum, and its possible role in channeling left and right atrial blood through the ventricle, are not known.

Apodans are poorly understood, not only in their cardiovascular anatomy but in almost all aspects of their biology. Notable observations include the fact that the semiterrestrial *Siphonops annulatus* from Brazil has an incompletely separated and fenestrated interatrial septum and lacks a spiral valve (Mendes, 1945), suggesting a decreased separation of oxygenated and deoxygenated blood streams within the heart. In *Typhlonectes compressicauda,* however, there are completely separate atria and a completely divided conus (Toews & MacIntyre, 1978). Still other apodans have significant ventricular septation formed by muscular trabeculae (Putnam & Dunn, 1978).

Developmental changes in the heart and central vessels

The general processes by which the elongated "heart tube" undergoes folding and collapse leading to the generation of a multichambered heart, as described in detail by Icardo in Chapter 10, also hold true for amphibians (Fox, 1984; Danos & Yost, 1995). Thus, at the gross anatomical level there appears to be little that is unique in amphibian embryos from the perspective of atrial and ventricular development, lending credence to the view that early embryos of each vertebrate class can serve as useful models for vertebrate development in general (Figure 13.1).

Nevertheless, during the development of the amphibian cardiovascular system, remarkable changes occur in the central and peripheral circulation associated with larval development and the subsequent metamorphosis from the larva (tadpole) to the adult form. In essence, an amphibian embryo first develops one complete set of vascular structures that serve the larva's existence as a free-living, aquatic entity in part dependent on gills for gas exchange. Then, just as the larva is "settling in" to this aquatic lifestyle, it typically makes a fairly abrupt metamorphic change to a second, often more terrestrial, air-breathing animal that depends heavily on lungs for gas exchange (see Burggren & Infantino, 1994, for a detailed description of the physiology of this respiratory transition). Importantly, each of these distinct stages – embryo, larva, and adult – exhibits some unique features in the organism's central vascular anatomy.

The changes in central vascular anatomy associated with development in a common anuran amphibian (a ranid frog) are shown in Figure 13.3. A prominent feature of the amphibian circulation is the intricately structured spiral valve bracketed by pylangial and synangial valves. These valves form very early, by at least Taylor–Kollros (TK) stage III,[1] and

[1] This stage is equivalent to Gosner stage 28 and Nieuwkoop–Faber (NF) stages 49–50, the period of early limb bud growth. Detailed comparisons of various popular schemes for staging amphibian embryos and larvae are provided in Burggren and Just, 1992.

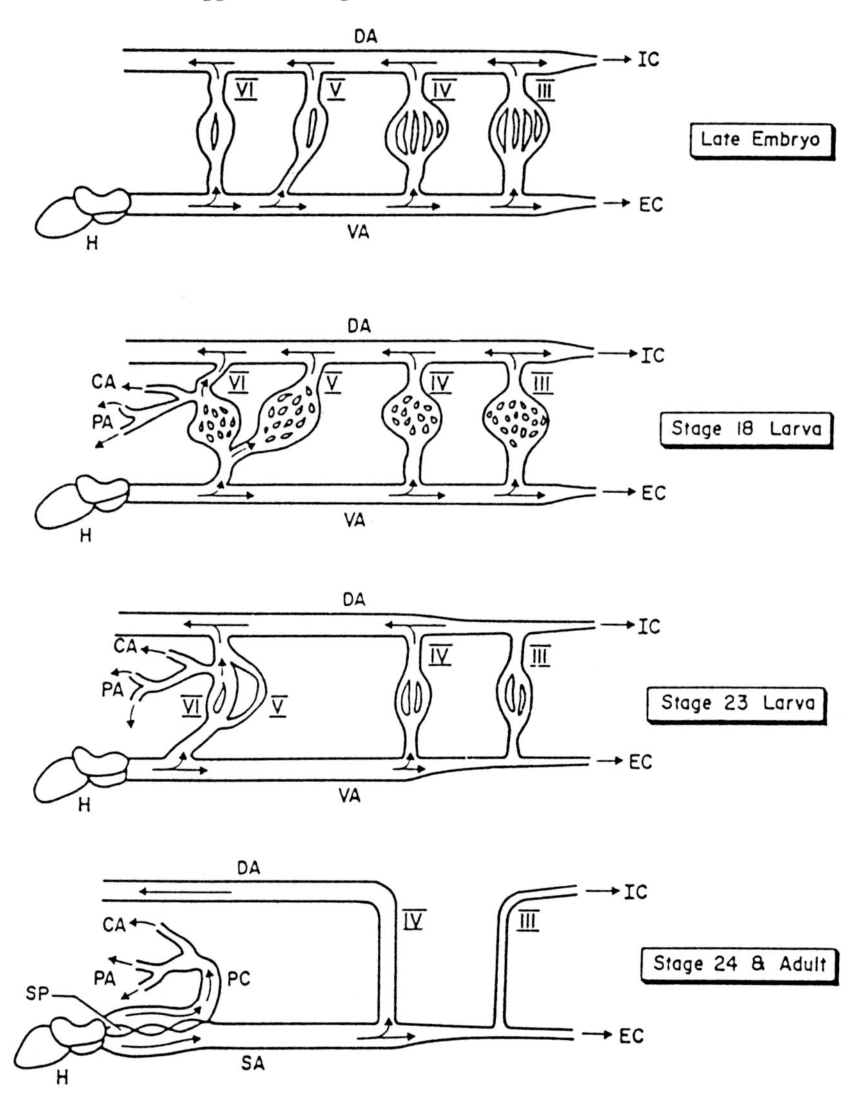

Figure 13.3 Changes in central vascular anatomy during development in an anuran amphibian (a ranid frog) (III–VI = branchial arches, CA = cutaneous artery, DA = dorsal aorta, EC = external carotid artery, H = heart, IC = internal carotid artery, PA = pulmonary artery, PC = pulmocutaneous artery, SA = systemic artery, SP = spiral valve, VA = ventral aorta). (From Burggren, 1984, which provides source references.)

quickly assume a hemodynamic role in *Rana catesbeiana* (Pelster & Burggren, 1991), where they presumably direct relatively oxygen poor blood into the developing pulmocutaneous (gas exchange) circuit. Notice that the well-developed branchial arches, which are vestigial in embryonic reptiles, birds, and mammals, actually enter a phase of great proliferation as internal gills in midlarval development (panel 2, Figure 13.3). These branchial arches are effectively ventilated with water by a branchial pumping mechanism in the midlarval phase and achieve an efficiency of gas exchange comparable to that of adult fishes (Burggren & Just, 1992).

Cardiovascular anatomy reaches its most complex form in the stages prior to metamorphosis (TK stage 18, panel 2, Figure 13.3), when larvae typically are developing functional lungs but still retain internal gills ventilated by buccal movements. At these intermediate stages, gas exchange is partitioned between branchial, pulmonary, and cutaneous sites (see Burggren & Infantino, 1994, for references). Even as it is continuing to develop, the central vascular anatomy must effectively supply and drain *three* distinct gas exchange organs, each with its own characteristic hemodynamic properties (impedance, compliance, etc.). It should be emphasized that, along with this anatomical complexity of the cardiovascular system, comes physiological complexity, particularly in control systems. Receptors that monitor both environmental and internal blood levels of O_2 and CO_2, and possibly pH, are involved in the reflex control of gill and lung ventilation and cardiovascular performance in these intermediate developmental stages (West & Van Vliet, 1992; Burggren & Just, 1992; Burggren & Infantino, 1994).

Metamorphosis from larva to adult is associated with an overall simplification of cardiovascular anatomy (panels 3 and 4 of Figure 13.3; Malvin, 1985; Burggren, 1995). Just as the number of vessels and potential shunt pathways are reduced at hatching or birth in reptiles, birds, and mammals (see Chapters 14–16), the amphibian circulation becomes anatomically simpler following metamorphosis. The branchial arches, so elaborate in mid-larval life, regress from a functional standpoint into mere conduits for blood flow to more distally located structures (transition from panel 3 to panel 4, Figure 13.3). Branchial arch VI becomes modified into the "pulmocutaneous arch," which carries predominantly deoxygenated blood into the pulmonary and cutaneous vascular beds. Branchial arches III–IV become incorporated into the newly forming root of the dorsal aorta, which carries blood to the trunk and external and internal carotid arteries, which in turn carry blood to the head. Not shown in Figure 13.3 is the extremely vascularized "carotid labyrinth," which in adult anurans is located at the base of the bifurcation of the carotid arches. This structure is both highly perfused and richly innervated and has been implicated in both cardiovascular and respiratory control (Smyth, 1939; West & Van Vliet, 1992).

Finally, observations of the interesting anatomical differences of adult anurans, urodeles, and apodans (both within and between the orders) have seldom been followed up from a developmental perspective. Most developmental studies have been confined to anurans, especially *Rana* and *Xenopus*. Study of the development of a partly divided ventricle in some apodans and urodeles, or the degeneration of pulmonary vessels in lungless salamanders, could provide interesting general insights into developmental processes in the cardiovascular system.

Hemodynamic maturation

The advent of microtechniques suitable for investigation of cardiovascular physiology in animals in the milligram range (see review by Burggren & Fritsche, 1995) has led to new, more complete analyses of hemodynamic processes in vertebrate embryos. The developmental physiology of amphibian embryos and larvae has been the focus of research in numerous recent studies. It is fair to say that we now know as much about the early physiological development of the anuran amphibian cardiovascular system as we know about that of any other vertebrate.

Onset of heartbeat and developmental patterns of change

Heart rate is the easiest of all cardiovascular variables to measure, and as a physiological variable it can be informative as to the nature and extent of maturing cardiovascular control. Consequently, the literature on physiological development is replete with measurements of heart rate in embryos, larvae, and fetuses of numerous vertebrate classes (see also numerous other chapters in this book). Unfortunately, two critical variables – body temperature and exact stage of development – have not always been carefully controlled. As a result, it is a daunting task to provide broad characterizations of heart rates and their patterns of change during development, not just for amphibians but for any vertebrate.

As in all other vertebrates, the heart begins to beat extremely early in amphibian development, presumably to distribute nutrients to the growing tissues and to provide a pulsatile flow for angiogenesis (see below). The forming heart of anurans begins to beat as it undergoes S-folding (Nieuwkoop & Faber, 1967) and within a few hours establishes a steady rhythm. In fact, the heart of anuran embryos initially beats like a metronome, with very limited heart rate fluctuation. External perturbations such as temperature fluctuation and hypoxia alter heart rate but appear to do so by directly affecting cardiac muscle rather than through any extrinsic reflex mechanisms (Fritsche & Burggren, 1996).

The initial heart rate of anuran embryos and larvae seems to be highly species specific, as evident in Figure 13.4. As the embryo increases in mass, heart rate may initially sharply decrease (*Rana catesbeiana*) or sharply increase (*Eleutherodactylus coqui, Xenopus laevis*). In *R. catesbeiana,* almost all of the fall in heart rate during subsequent larval development can be explained on the basis of changes in body mass, which increases from 40 mg at hatching up to 400 g as an adult (W. Burggren, unpublished data). Simply put, larger animals tend to have lower heart rates, both inter- and intraspecifically (Schmidt-Nielsen, 1984). Although a similar relationship exists for *X. laevis* (Orlando & Pinder, 1995), scaling effects, or so-called allometry, cannot explain all aspects of heart rate change during anuran development. In *E. coqui* and *X. laevis,* for example, both heart rate and body mass increase sharply early in development, suggesting that organogenesis – not allometry – dominates heart rate relationships in the embryo and early larva. Put another way, the development of new tissue (organogenesis) requires an increase in cardiac output, and this increasing demand for blood flow, fulfilled in part by an increase in heart rate, overrides allometric influences on heart rate. However, later in larval development and in adults the decrease in heart rate shown by *E. coqui* and *X. laevis* can be fairly accurately predicted from allometric relationships between heart rate and body mass established for interspecific comparisons of adult vertebrates. Thus, changes in resting heart rate in developing amphibians appear to be dominated by one or both of two interacting, and in some cases antagonistic, influences: organogenesis and allometry. Incidentally, the respiratory transition to air-breathing, which demands considerable rearrangement of cardiovascular anatomy, is correlated with changes in heart rate (multiple points AB on Figure 13.4), but whether there is a true causal relationship between the onset of air-breathing and changes in heart rate and other aspects of cardiac performance remains to be determined for anurans.

Control over heart performance is evident early in developing anurans. Figure 13.5 shows the ontogenetic changes in heart rate regulation in *R. catesbeiana,* compiled from studies by Burggren and Doyle (1986a, 1986b), Pelster and Burggren (1991), Pelster et al., (1993), and Kimmel (1992). By TK stage III, both cholinergic inhibitory and adrenergic excitatory nerve fibers influence heart rate, but there is little evidence of central reflex regulation of the heart. After this point in development, however, there is increasing evidence

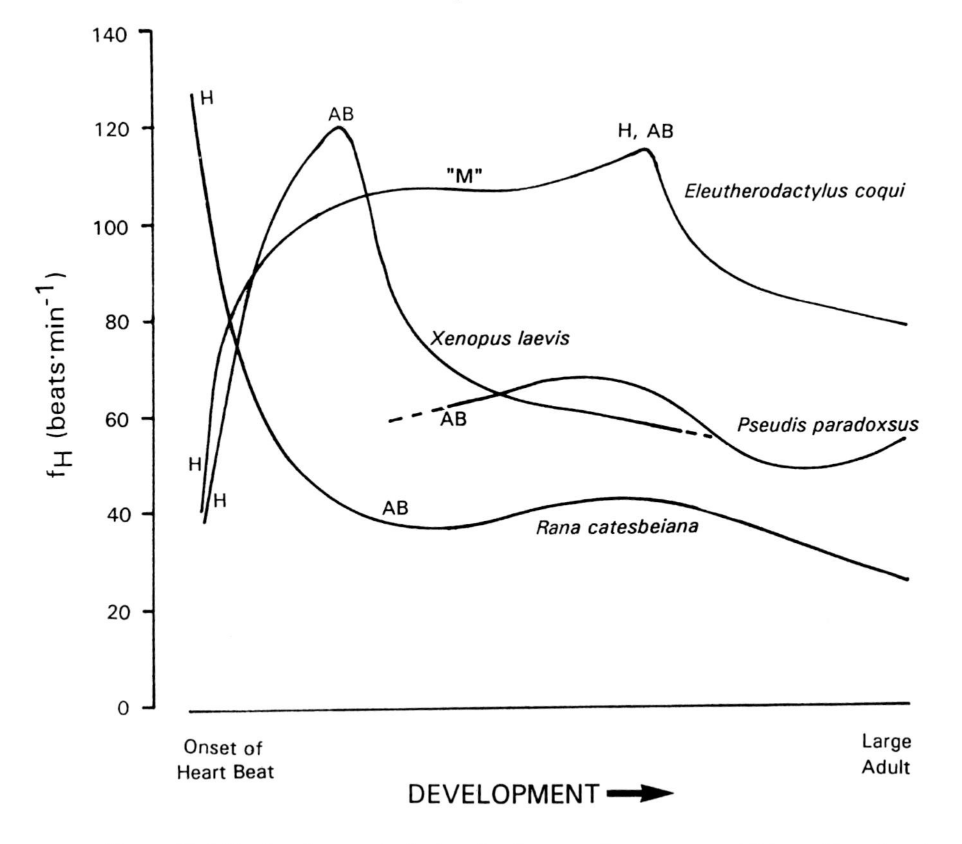

Figure 13.4 Resting heart rate during development in anuran amphibians. Note that the abscissa, development, has been "normalized" such that the origin is the first point at which heart rate can be measured, and the far right of the graph is heart rate in the largest adults (H = hatching, AB = onset of air-breathing, M = metamorphosis). (From Burggren, 1995, which provides sources for the data on each species.)

that cardiac reflexes associated with changes in gill/lung ventilation, exercise, or hypoxic exposure are affected by changes in vagal and/or sympathetic tone.

Blood pressure generation

Blood pressure increases manyfold during development in vertebrate embryos and larvae. The advent of commercially available servonull micropressure systems has allowed blood pressure measurements in the anuran amphibians *X. laevis* and *R. catesbeiana* at early embryonic stages weighing as little as 1 mg, with subsequent measurements made during embryonic and larval development (Pelster & Burggren, 1991; Hou & Burggren, 1995a; Fritsche & Burggren, 1996). In *Xenopus* and *Rana* both systolic and diastolic blood pressure increase with increasing body mass during embryonic and early larval development. The linear relationship between blood pressure and log body mass in developing amphibian embryos and larvae (Figure 13.6) shows a correlation coefficient of .6–.7, suggesting that allometry alone cannot adequately explain all of the increase in blood pressure during development. The increase in blood pressure in vertebrate embryos occurs concomitant

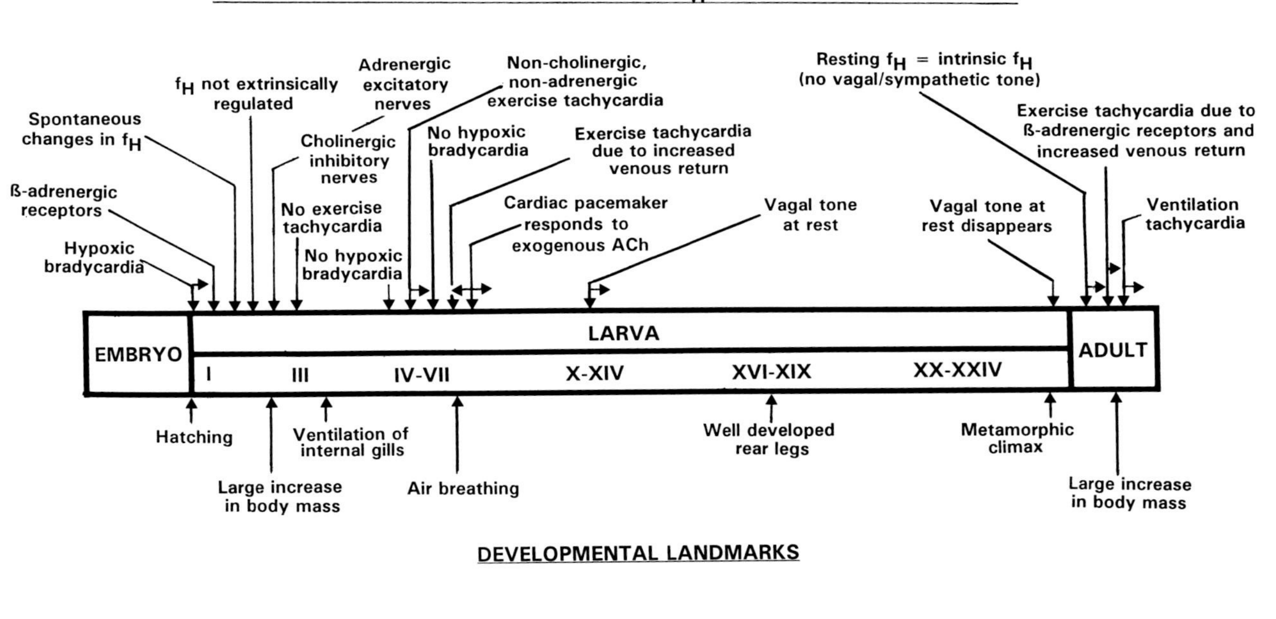

Figure 13.5 Developmental changes in heart regulation in the bullfrog, *Rana catesbeiana*. An individual animal's life span is indicated by the horizontal bar. Major developmental landmarks and the appearance, modification, and/or disappearance of cardiac regulatory mechanisms are shown. Vertical arrows represent a single observation of the indicated event. A vertical arrow combined with a horizontal arrow indicates the onset of a continuing process. (After Burggren, 1995.)

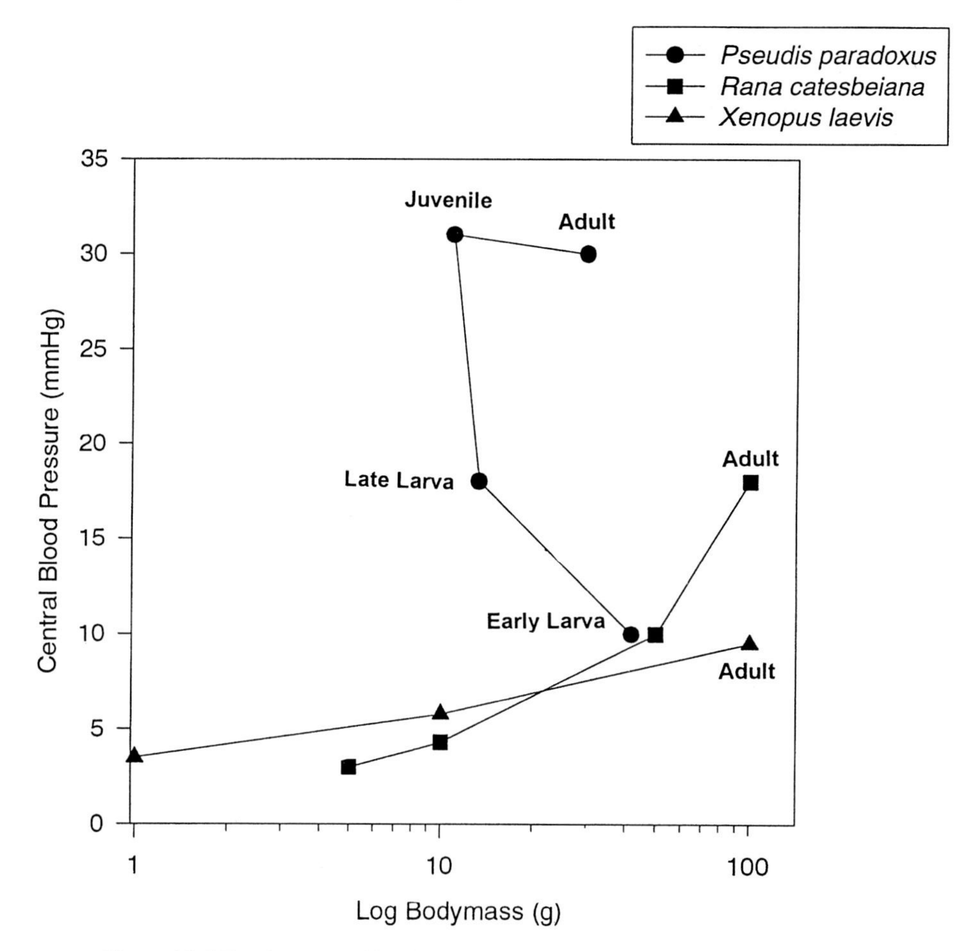

Figure 13.6 Resting mean blood pressure during development in three different amphibians. Note the decrease in body weight during the main part of development in *Pseudis*. (data compiled by P. Territo, unpublished).

with a large increase in cardiac output and a decline in peripheral resistance. The reasons for the decrease in resistance during development could be explained by viscosity changes, addition of parallel vascular beds, increased intrinsic diameter of resistance vessels, and development of humeral and/or neural vascular control systems reducing vascular resistance (Zahka et al., 1989). Of note, the positive correlation between body mass and blood pressure evident in developing anurans contrasts with the independence of body mass and blood pressure interspecifically among adult vertebrates (Schmidt-Nielsen, 1984).

The trend for blood pressure to rise progressively during development does have interesting exceptions among the anurans. In the paradoxical frog, *Pseudis paradoxus,* blood pressure stays relatively stable as body mass of the larva *decreases* in the later stages of larval development (Burggren et al., 1992). Blood pressure then sharply increases during metamorphosis in the absence of any body mass change, and then remains unchanged as the juveniles grow to adult body mass (Figure 13.6). Clearly, there are other factors more important than changes in body mass regulating blood pressure during development.

Intraventricular and central arterial blood pressure measurements in larval *Xenopus* and *R. catesbeiana* indicate that functional valves separate the ventricle, conus arteriosus, and truncus arteriosus early in the embryonic period (Pelster & Burggren, 1991; Hou & Burggren, 1995a). Diastolic pressures measured in the central regions of the circulation show distinct regional differences. There is no backflow from the truncus to the conus and from the conus to the ventricle during diastole or systole. In early larval *Xenopus* the blood pressure waveform from the truncus arteriosus shows two distinct peaks (Figure 13.7). The first peak corresponds to contraction of the ventricle and the second peak to contraction of the conus arteriosus. In older larvae the second peak is not as evident, though the temporal separation of the peaks (reflecting conduction velocity) does not change. Apparently, the importance of conus contraction is greater in young larvae than in older larvae of *Xenopus* (Hou & Burggren, 1995a). The role of the conus arteriosus in actively pumping blood also decreases during development in bullfrog larvae (Pelster & Burggren, 1991). This phenomenon is seasonal, however, with a strong conal contraction persisting primarily in fall–winter animals. Arterial systolic pressure is also much higher in spring–summer animals (10 mm Hg at TK stage III) than in fall–winter animals (4 mm Hg at TK stage III), when measured at the same temperature (≅22°C). The underlying physiological mechanisms for this seasonal change in blood pressure and site of pressure generation in bullfrog larvae is unclear. It should be noted that bullfrogs can spend up to 3 years in the larval state before metamorphosing, so these seasonal changes observed in the lab may have interesting but as yet unknown ramifications for survival in the wild.

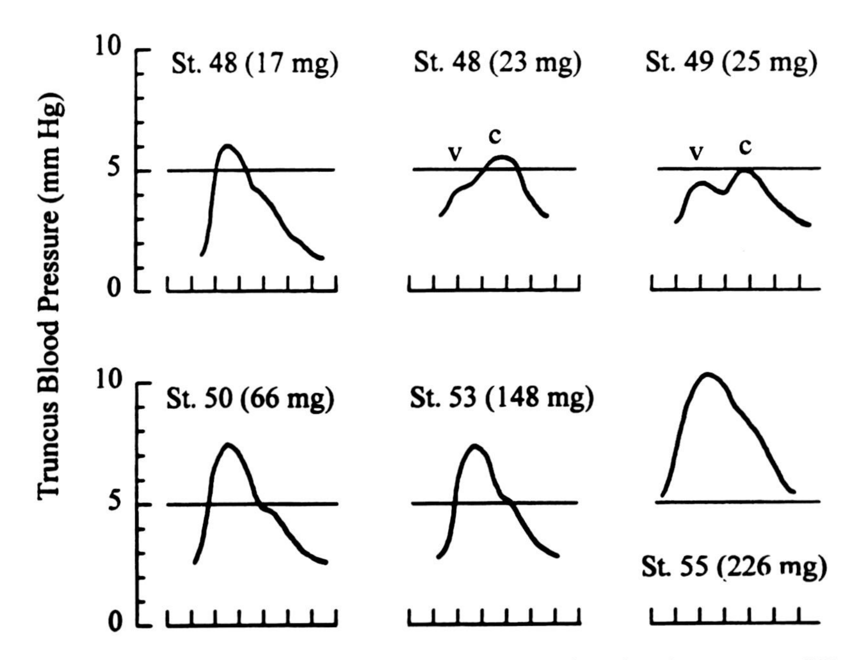

Figure 13.7 Recordings of truncus arteriosus pressure from larval *Xenopus* at different Nieuwkoop–Faber stages of development. Especially around early larval stages 48–49, the truncus arteriosus pressure signal is biphasic, with the ventricle (v) contraction producing the first peak and the conus (c) contraction producing the second. Later in development, conus contraction contributes less to the overall pressure waveform in the truncus arteriosus.

Little is known about the onset of blood pressure regulation during development in anuran amphibians. Certainly, receptors and/or receptor-mediated responses that could reasonably be involved in regulating both the heart and the peripheral vasculature occur by TK stage II in *Rana* (Kimmel, 1990, 1992; Protas & Leontieva, 1992; Pelster et al., 1993) and by NF stages 45–49 (the period of early hind limb bud growth) in *Xenopus* (Bride, 1975). The pharmacology of the vasculature of larval *Ambystoma tigrinum* has also been extensively characterized, revealing the ability to actively regulate blood flow and blood distribution early in larval development (for references, see Malvin, 1985; Burggren & Just, 1992). Major changes in heart rate with no change in stroke volume, which would cause associated changes in blood pressure in a simple compliant mechanical system, produce little or no change in arterial blood pressure in *R. catesbeiana* and *P. paradoxus,* suggesting the presence of mechanisms that regulate blood pressure.

Studies specifically designed to elucidate the ontogeny of blood pressure regulation in anurans during embryonic and larval development will be well worth the effort.

Cardiac output and blood flow generation

Cardiac output has been determined in the embryos and larvae of only a few amphibian species. Video microscopic techniques have been used in *X. laevis* (Hou & Burggren, 1995b; Orlando & Pinder, 1995; Fritsche & Burggren, 1996), and pulsed Doppler techniques have been used in the larger larvae of *Ambystoma tigrinum* (Hoyt, Eldridge, & Wood, 1984). In both species cardiac output increases with increasing body mass. In *Xenopus,* for example, cardiac output is about 0.1 $\mu l \cdot min^{-1} \cdot mg^{-1}$ at 10 mg body mass, rising to 0.6 $\mu l \cdot min^{-1} \cdot mg^{-1}$ at 1000 mg body mass (Hou & Burggren, 1995b; Fritsche & Burggren, 1996; Figure 13.8). The increase in cardiac output in *Xenopus* is due primarily to an increase in stroke volume, since after an initial increase, heart rate decreases during development in this species. This increase in stroke volume may result from an increase in the size and number of myocardial cells as well as an increase in the percentage of contractile protein within individual myocardial cells during development. In adults, changes in cardiac output are closely correlated with changes in oxygen consumption, and cardiac output is considered a limiting factor for oxygen consumption (Schmidt-Nielsen, 1984). In larval *Xenopus,* on the other hand, as illustrated in Figure 13.8, cardiac output increases faster than body mass (Hou & Burggren, 1995b; Orlando & Pinder, 1995; Fritsche & Burggren, 1996), whereas oxygen consumption increases slower than body mass (Hastings & Burggren, 1995; Territo & Burggren, in press). Apparently, some factor other than cardiac output per se limits oxygen consumption at rest during larval development of *Xenopus.*

Evidence for intrinsic control of cardiac output is limited but significant (see also Chapter 7). Large beat-to-beat variations in stroke volume are associated with spontaneous fluctuations in heart rate in larval *R. catesbeiana* as early as stage III. Increased preload induced by venous volume loading leads to an increase in heart rate in TK stage V–VIII larvae (and adults) with pharmacologically blocked cholinergic and adrenergic receptors (Burggren & Doyle, 1986b). Both of these observations suggest a functioning Frank–Starling mechanism relatively early in development, as has been suggested for chick embryos as early as Hamburger–Hamilton stage 24 (see Chapter 7). Extrinsic mechanisms (neural/hormonal) affecting stroke volume in developing amphibians have not been well categorized but presumably occur with a time course similar to those that control heart rate (Figure 13.5).

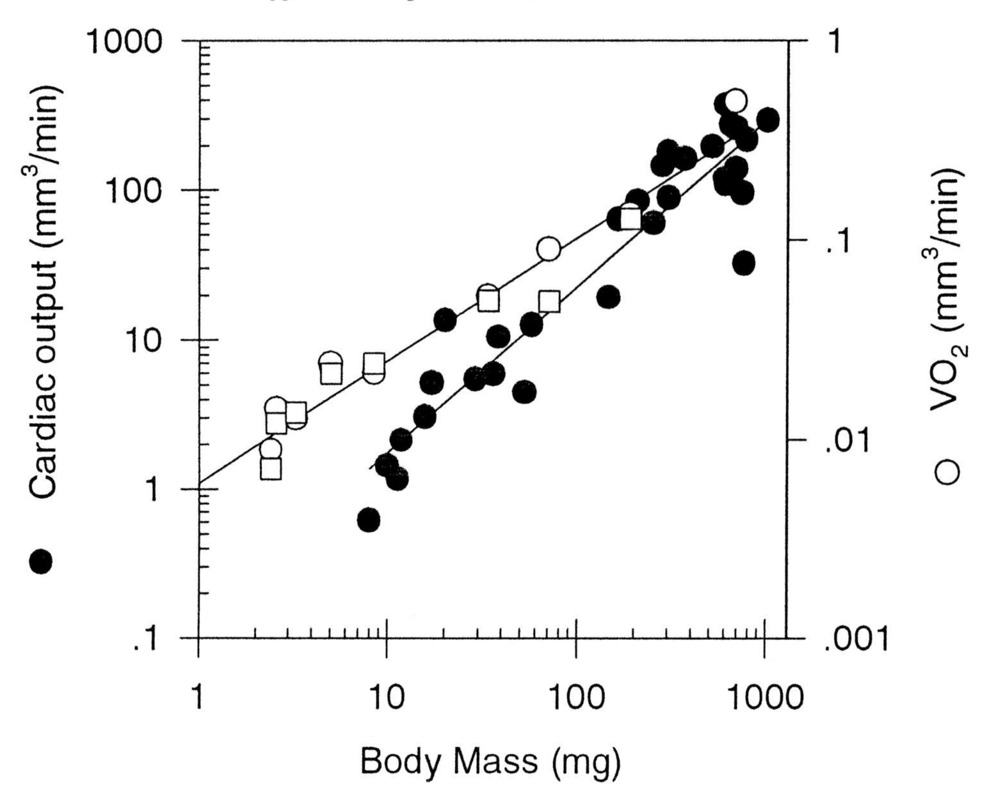

Figure 13.8 Developmental changes in cardiac output and oxygen consumption for larval *Xenopus*. Cardiac output increases faster than body mass (b = 1.11, r = .89), whereas oxygen consumption increases slower than body mass (b = 0.62, r = .97), implying that cardiac output is not a limiting factor for oxygen consumption.

Control of the peripheral circulation

Studies on the development of peripheral vascular control in amphibians are few. Pharmacological studies on in vitro and in situ perfused pulmonary and branchial vascular beds of ambystomid larvae revealed a cholinergic vasoconstriction and an adrenergic vasodilation of the branchial vasculature (Malvin, 1985). Catecholamines have no effect on pulmonary vessels, whereas acetylcholine causes a pulmonary vasoconstriction. Changes in the local environment of the blood vessel, such as changes in CO_2 or pH, alter blood flow in the branchial and pulmonary vessels of *A. tigrinum* (Malvin, 1985). Studies on in situ perfused vascular beds in larvae of *R. catesbeiana* have shown that catecholamines dilate the branchial vasculature in larval stages as early as TK stage III (Kimmel 1990, 1992). Intra-arterial injections of catecholamines in young larvae of *X. laevis* (from NF stage 47) increase systemic blood pressure without any significant change in heart rate and cardiac output, suggesting that the peripheral vessels respond very early to catecholamines in this species (Fritsche & Burggren, 1996; Figure 13.9). Exposure of these animals to acute hypoxia results in an increased vascular resistance early in development, whereas no change in peripheral resistance is found later in development (Fritsche & Burggren, 1996). These findings suggest that the peripheral vessels are affected by both neuronal/humoral and local mechanisms very early in development.

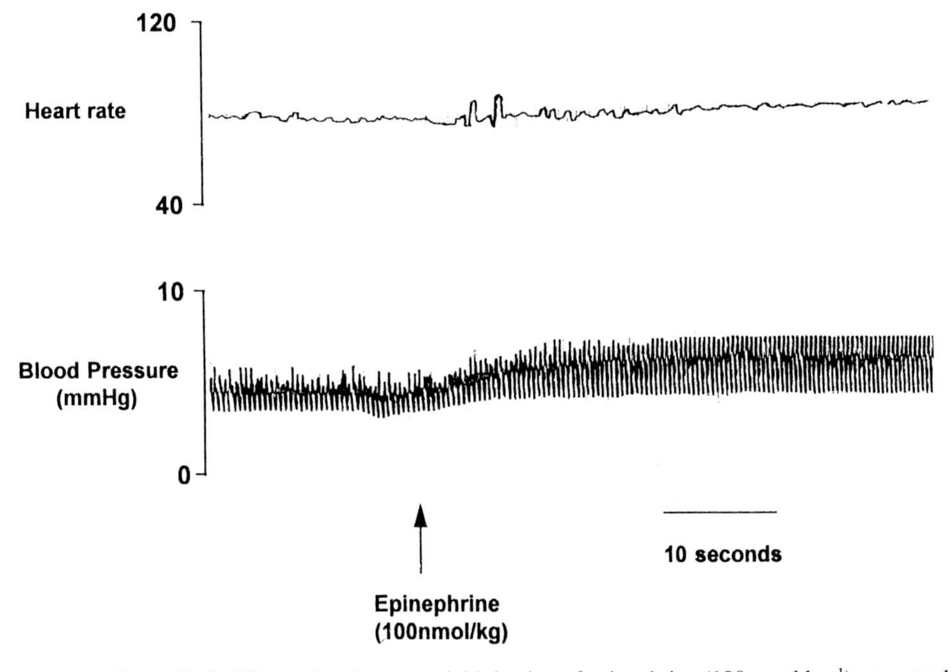

Figure 13.9 Effects of an intra-arterial injection of epinephrine (100 nmol·kg^{-1}) on central blood pressure and heart rate in a larval *Xenopus* weighing 10 mg. The injection was made via a glass microelectrode (tip size = 5–10 μm) inserted directly into the vessel.

The ontogeny of vascular control systems (neuronal, humoral, intrinsic) is practically unexplored in amphibians, as evident from the scant description just given. Key questions remain to be answered: When during development do efferent autonomic nerves appear on the blood vessels? When do the nerves contain neurotransmitters? Which neurotransmitters develop first? When are the different receptors expressed? When are the different control systems "turned on"? Answering these questions will be important in understanding cardiovascular development and the potential developing amphibians have for regulating cardiovascular functions in response to environmental alterations.

Diffusive versus convective oxygen delivery in early amphibian embryos: Do vertebrate embryos need hearts?

Underlying our understanding of developmental changes in cardiovascular anatomy and physiology in amphibians is the common assumption that the onset of heart function is closely matched to the embryo's need to receive nutrients and especially oxygen by convection in addition to diffusion. This long-standing hypothesis, which we call "synchronotropy" after the synchronous onset of heartbeat and the need for convective oxygen and nutrient supply (Burggren & Territo, 1995; Territo and Burggren, in press), has, surprisingly, not been rigorously tested until very recently. Experiments on anuran and urodele amphibians (Mellish, Pinder, & Smith, 1994; Burggren & Territo, 1995; Territo & Burggren, 1995; Mellish, Smith, & Pinder, in press), zebra fish (Pelster & Burggren, 1996), and chick embryos (Cirotto & Arangi, 1989) all provide compelling evidence that the embryonic heart begins to beat *prior* to the absolute need for convective blood supply.

These observations lead to an alternative hypothesis we have termed "prosynchronotropy." A variety of techniques have been applied to embryos in these studies to perturb Hb-O_2 transport (carbon monoxide exposure, destruction of red cells by phenylhydrazine) or to prevent blood flow altogether (cardiac mutants, surgical intervention). Typically, neither oxygen consumption nor inward oxygen conductance is greatly altered, if affected at all. Perhaps most convincing are the experiments of Mellish, Smith, and Pinder (in press) in which the heart primordia were completely removed in stage 19 axolotls (*Ambystoma mexicanum*). The embryos continued to develop and at stage 39, oxygen consumption and conductance were not significantly different from controls.

If the heart does not begin to pump blood to support aerobic processes, then what determines the timing of initial heart function? To put it another way, why does the heart begin to function "early" from an oxycentric perspective? Numerous possibilities have been discussed (Mellish, Pinder, & Smith, 1994; Burggren & Territo, 1995), including the need for metabolic substrates other than oxygen and the maintenance of ionic and osmotic balance. The most likely reason for the onset of heartbeat well before diffusive oxygen supply is outstripped by oxygen demand may be the need to provide a fluid with oscillating pressure and flow to assist angiogenesis. A pressurized circulation could act directly by forcing open newly developing vessels. However, it could also act less directly by stretching tracts of endothelial cells forming presumptive vessels. Endothelial cells grow more rapidly when subjected to sheer and stress, and growth factor secretion may similarly respond to mechanical deformation at the tips of sprouting vessels during angiogenesis (see Risau, 1991b; Davies & Tripathi, 1993, and Chapter 4, this volume, for references).

Developmental plasticity of the amphibian cardiovascular and respiratory systems

The developmental trajectory of any animal represents a combination of a genetic blueprint determined at fertilization and environmental variables (both internal and external) that may prolong, accelerate, or otherwise modify developmental processes. In vertebrates that gestate within a protective uterus (mammals) or that are enclosed in an egg brooded by a parent (most birds, a few reptiles and amphibians), the embryonic environment may be quite stable in terms of oxygen, carbon dioxide, pH, temperature, and nutrient availability. However, vertebrates such as amphibians and fishes typically lay eggs that are much more exposed to the environment, and consequently, developmental processes in these embryos can routinely be subjected to far greater natural perturbations than those in birds or mammals (see Chapters 19 and 20). Figure 13.10 illustrates schematically how environmental disturbances during key periods in development (so-called critical windows) can produce modified adult phenotypes as a result of abnormal developmental trajectories. Some abnormal developmental trajectories are unalterable once embarked on by a tissue, organ, or whole animal, even if the original disturbing environmental stimulus is removed. Other trajectories can be overridden by removal of the environmental stimulus. Experiments on anuran amphibian larvae exposed to chronic hypoxia suggest that there is considerable developmental plasticity in the face of environmental change and uphold the concept of both reversible and nonreversible changes in developmental trajectories. For example, profound changes in respiratory surface architecture (e.g., lung volume, index of lung surface area) facilitate gas exchange during environmental hypoxia in the developing bullfrog larva

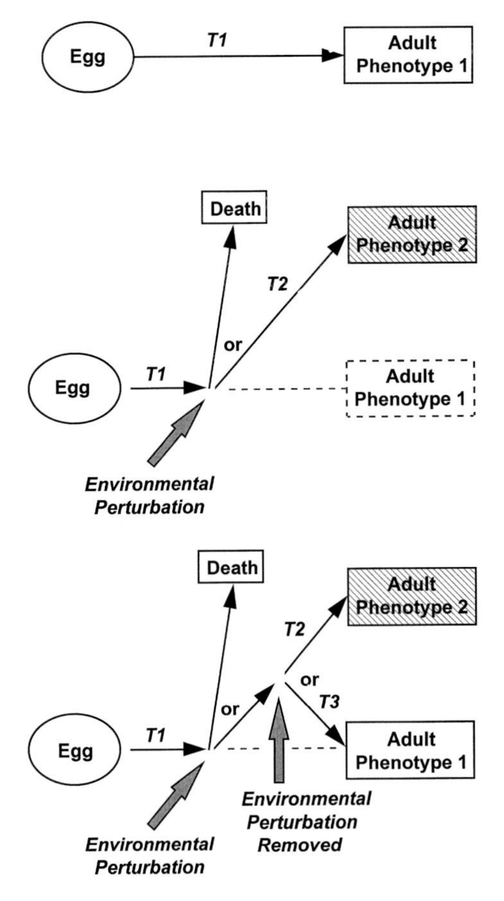

Figure 13.10 The interrelated concepts of developmental plasticity and developmental trajectories are depicted in this highly schematic drawing. In all three cases, a fertilized egg is assumed to develop into an adult showing a characteristic phenotype by way of a developmental trajectory dictated by a complex interaction of genes and environment. (*Top panel*) Trajectory 1 (T1) shows the normal developmental pathway taken in the absence of any perturbing environmental factors. (*Middle panel*) At the arrow, a perturbing environmental stimulus occurs (e.g., hypoxia, temperature change, pharmacological factors) and causes development to proceed along a new developmental trajectory (T2), ultimately leading to a new phenotype. If the perturbing stimulus is severe, mortality will occur. (*Bottom panel*) If the perturbing environmental stimulus occurs but is later removed (second arrow), then one of two events may occur. Development may continue along trajectory T2 even in the absence of further environmental disruption, leading to the formation of the new adult phenotype. Alternatively, trajectory T3 can be followed, which results in a return to trajectory T1 and eventual production of the normal phenotype. Examples of all three possibilities are to be found among studies on amphibian cardiorespiratory systems.

(Burggren & Mwalukoma, 1983; Pinder & Burggren, 1983). Importantly, some of these changes (e.g., lung volume) are partially or fully reversed upon return to normoxia either before or after metamorphosis (Pinder, 1985). Developmental plasticity in amphibians is a double-edged sword that not only can result in misdirected developmental trajectories induced by environment and leading to abnormal phenotypes but also can allow full or partial repair of environmentally induced misdirection of trajectories.

A perspective on future research

Research on cardiovascular development in amphibians has burgeoned in the past decade, in part because of attention focused on this class by comparative physiologists but also because of the interest of developmental biologists who are realizing the demonstrated utility of amphibians as general models for early cardiovascular development. How the basic physiological variables (heart rate, stroke volume, cardiac output, blood pressure, peripheral resistance) change during amphibian (particularly anuran) cardiovascular development is beginning to be understood. However, investigations of amphibian cardiovascular development must begin to move from this necessary, early descriptive phase toward studies that define basic mechanisms. The perturbation of normal systems – by "gene knockout," surgical intervention, or environmental manipulation – will likely reveal insights into amphibian development similar to those emerging for other vertebrate classes.

Importantly, the great species diversity of amphibians can be exploited to provide additional knowledge about cardiovascular development. For example, amphibians may prove especially useful in studies of ventricular septation and valve formation. The former is more rudimentary and the latter more extensive than in reptiles, birds, or mammals.

A final area on which the interests of both amphibian developmental biologists and general cardiovascular modelers may converge is the issue of developmental plasticity. The ability of amphibian embryos to return to their genetically dictated developmental trajectories after changes induced by acute or chronic environmental perturbation, and the whole issue of developmental plasticity in general, will likely prove to be a fascinating area for future research using developing amphibians.

14

Reptilian cardiovascular development

STEPHEN J. WARBURTON

Introduction

Little is known of cardiovascular development in reptiles, despite their important place in evolutionary history. Three main factors have contributed to the neglect of these animals in developmental research. First, reptiles were not commonly bred in captivity until recently, and embryos were available only by retrieval from the field. This approach to obtaining specimens leads to problems of inadequate supplies of material and problems in timing of developmental events, especially regarding fertilization dates and thermal incubation histories. Second, unlike avian embryos, the chorioallantois of reptiles adheres tightly to the overlying eggshell membrane within the first few days of development, making invasive procedures difficult without incurring excessive blood loss. Third, reptiles have extended development times relative to birds and most mammals, 450 days in the most extreme case, which is a time course to tax the most patient of investigators. The large array of developmental pathways available to reptiles suggests there may not be a standard reptile developmental timetable. Not only is there a huge range of developmental times, but there are also many types of eggshells (e.g., low and high permeability) and different reproductive modes (i.e., viviparous, ovoviviparous, and oviparous).

Recently, however there has been a resurgent interest in reptile embryos and their physiology. Indeed, one of the "drawbacks" of studying reptilian embryos, the long incubation period, may prove a boon to some types of developmental investigations. For those interested in the timing of developmental events, the 21-day incubation period of the domestic fowl (even less for many fish and amphibians) limits the resolution with which certain events, especially those at the initiation of circulation, can be timed. Animals with long incubation periods such as the American alligator, which develops to hatching in about 3 months, provide an expanded time scale to evaluate the chronology of developmental events.

Because reptiles reproduce by both egg-laying and live-bearing, they may well be the ideal system in which to study the evolution of the placenta and the evolution of mammalian-style reproduction. Closely related species exist wherein one species is egg-laying and a close relative is live-bearing, and the differences in their embryonic physiology and anatomy indicate adaptations for live-bearing versus egg-laying. Thus, the many modes of development that exist in reptiles provide clues about the evolution of the mammalian placenta. Given the knowledge that the study of reptilian cardiovascular development may confer, it seems imperative to continue to gather more information about early anatomy and physiology in this diverse group.

Structural development of the circulatory system

Reptilian hearts and circulatory systems present an array of structural diversity, from the three-chambered hearts of lizards and snakes to the fully separated four-chambered hearts of the crocodilians. Briefly, the ventricular chambers of most reptiles (excluding the crocodilians) are anatomically contiguous; however, both the presence of muscular ridges within the ventricle and streamlined flow separation permit almost complete separation of arterialized from dearterialized blood within the heart. The separation of flow is not fixed, however, and both left-to-right and right-to-left shunting occur during specific physiological states or with pharmacological intervention. The mechanism of shunting is not agreed upon but appears to be a combination of what have been called "washout shunt" and "pressure shunt." Details of these two hypotheses are best acquired from the extensive literature on the subject (e.g., Burggren, 1985b; Heisler & Glass, 1985; Hicks & Malvin, 1992; Comeau & Hicks, 1994).

The anatomy of the great vessels is grossly similar throughout the Reptilia, with a pulmonary artery and left and right aortic arches leaving the heart. In the crocodilians, there is complete septation of the left and right ventricles and this separation has resulted in the left aortic arch being assigned to the right ventricle, alongside the pulmonary artery. A similar anatomical arrangement occurs during avian development (Chapter 15), with the primary difference that crocodilians maintain the left arch as adults. The crocodilian circulation is notable for the presence (in the adult) of a functional connection between the left and right aortic arches, the foramen of Panizza (Figure 14.1). It is therefore possible for crocodilians to achieve an extracardiac right-to-left shunt through the foramen of Panizza if the pressure in the pulmonary circulation exceeds that in the systemic circulation. Further details on the structure and function of crocodilian hearts are available in several papers (e.g., Shelton & Jones, 1991; Jones & Shelton, 1993). The embryonic origin of the foramen of Panizza is unclear and is suggested to appear late in development, although others consider this unlikely (Greil, 1903; Goodrich, 1930). There are no recent studies of the embryology of this unique cardiovascular structure and conclusive data are needed.

Of great interest in the reptiles is the fate of the various aortic arches (Figure 14.2). As with heart shape, reptiles provide substantial variations on the ultimate disposition of the embryonic arches. For example, the fate of the ductus caroticus (or carotid duct) is different in various reptilian orders. The ductus caroticus (sometimes inaccurately referred to as the ductus Botalli) is the segment of the dorsal aorta between the third and fourth aortic arches and when retained in the adult is a connection between an internal carotid artery and a systemic arch. This structure is normally present as a large-diameter patent vessel in adult rhynchocephalians (tuataras, *Sphenodon* spp.), a primitive, geographically restricted, lizardlike animal that constitutes its own order in the Reptilia (Figure 14.3, top left). The ductus caroticus may persist in other reptilian orders as well, although it is apparently only in the Lacertilia (lizards) that it appears with great regularity (O'Donoghue, 1917; Figure 14.3, bottom left). The ductus arteriosus (ductus Botalli), or the complete sixth aortic arch, also remains patent in the tuataras (Figure 14.3, top left) and has been described as ordinarily being patent in adults of some chelonian (turtle) species (O'Donoghue, 1917) but may exist as a common abnormality in others (Burggren, 1976; W. Burggren, personal communication; Figure 14.3, top right). The ductus arteriosus also appears to be commonly retained in Ophidia (snakes; Figure 14.3, bottom right), although it frequently appears only as the ligamentous remnant of the original vessel. The ductus arteriosus in snakes is frequently found bilaterally, despite the fact that most snakes have only a single complete pulmonary artery (and lung).

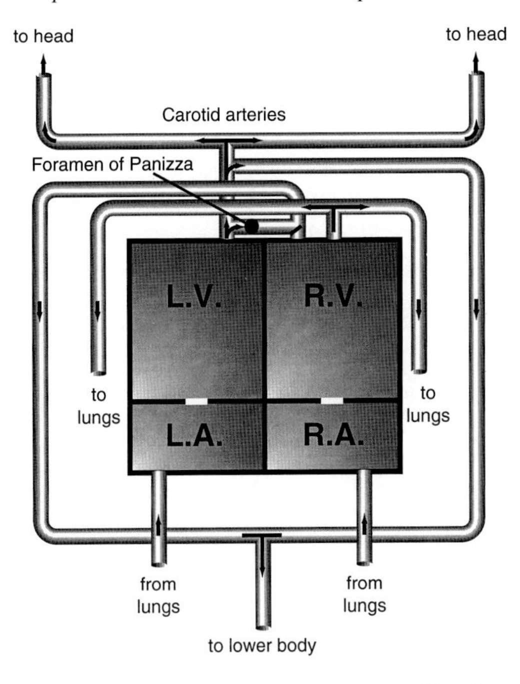

Figure 14.1 Schematic of mature crocodilian heart and great vessels (L.A. = left aortic arch, R.A. = right aortic arch, R.V. = right ventricle, L.V. = left ventricle). Arrows represent direction of blood flow. Note that flow through the foramen of Panizza, which connects the left and right aortic arches, is controlled by a valve that closes either the right ventricle or the foramen but not both.

Body shape may be one factor that determines which arches remain and which are lost in development (O'Donoghue, 1917). In species where there is little change in body proportions during development, both the ductus caroticus and the ductus arteriosus can persist, as in the tuatara. In species where the anterior thorax becomes broadened (e.g., some lizards), the ductus arteriosus is eliminated, and in lizard species that become wide and elongated (varanid lizards), both the ductus arteriosus and the ductus caroticus are eliminated. Snakes, with long necks and highly compressed width, lose the ductus caroticus and maintain the ductus arteriosus. Obviously, more data are needed to verify this conclusion, but it is interesting to speculate that the probability of a persistent arch within any given species may increase as a species' body form approaches that of the tuatara; thus, in species that diverge widely from the "general plan," the frequency of abnormal persistent arches may be lowest.

There is also interesting evidence that the bulbus cordis, which many textbooks describe as being subsumed into the mature ventricle of reptiles, not only persists as a mature structural entity but also exists as an important functional entity. In the embryo, the bulbus cordis is the last chamber of the heart to beat, during the progression of contraction from the sinus venosus to the bulbus cordis. In adult turtles, the bulbus apparently retains this property despite being almost entirely incorporated into the pulmonary outflow tract. Cinematographic evidence indicates that the thickened ring that represents the remnants of the bulbus cordis is actively contractile and that contraction is delayed relative to the bulk of the ventricle (March, 1961). Contraction of this sphincterlike ring at the base of the pulmonary artery may alter flow patterns and distribution of blood into adult pulmonary and systemic

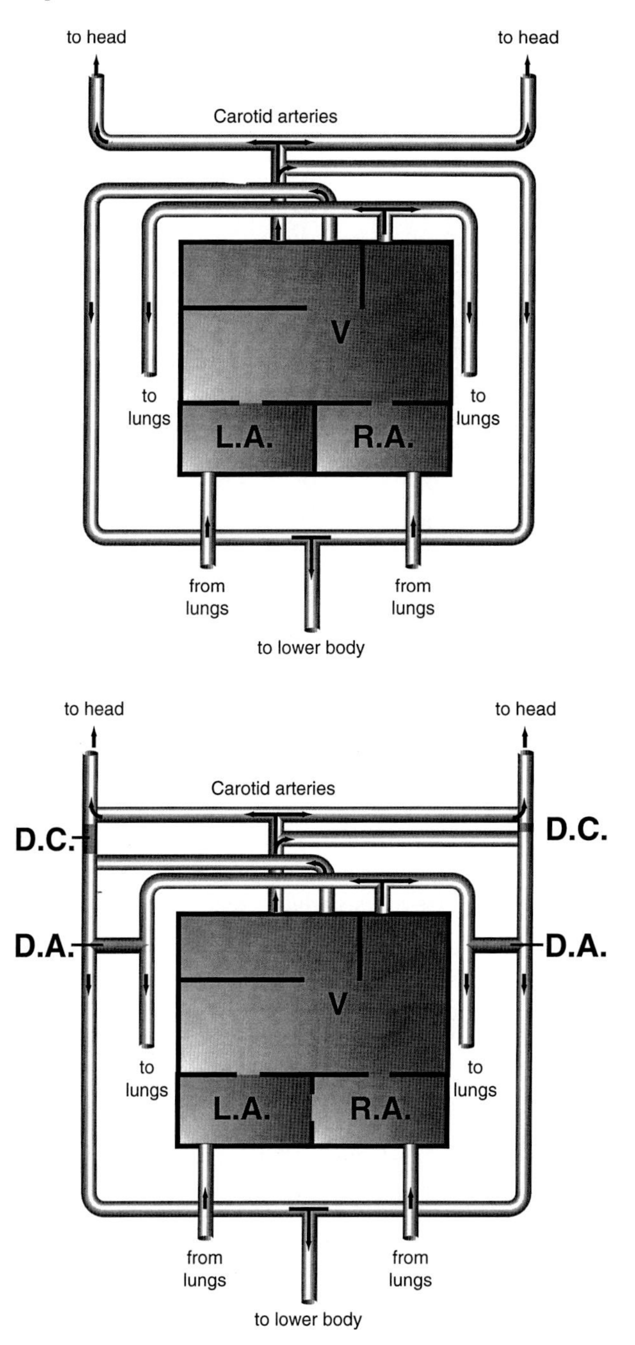

Figure 14.2 (*Top*) Schematic of a "typical" adult reptile heart and great vessels. (*Bottom*) Schematic of a "typical" embryonic reptile heart and great vessels. Darker areas represent vascular segments that may or may not be lost in the adult (D.A. = ductus arteriosus, D.C. = ductus caroticus, L.A. = left atrium, R.A. = right atrium, V = ventricle).

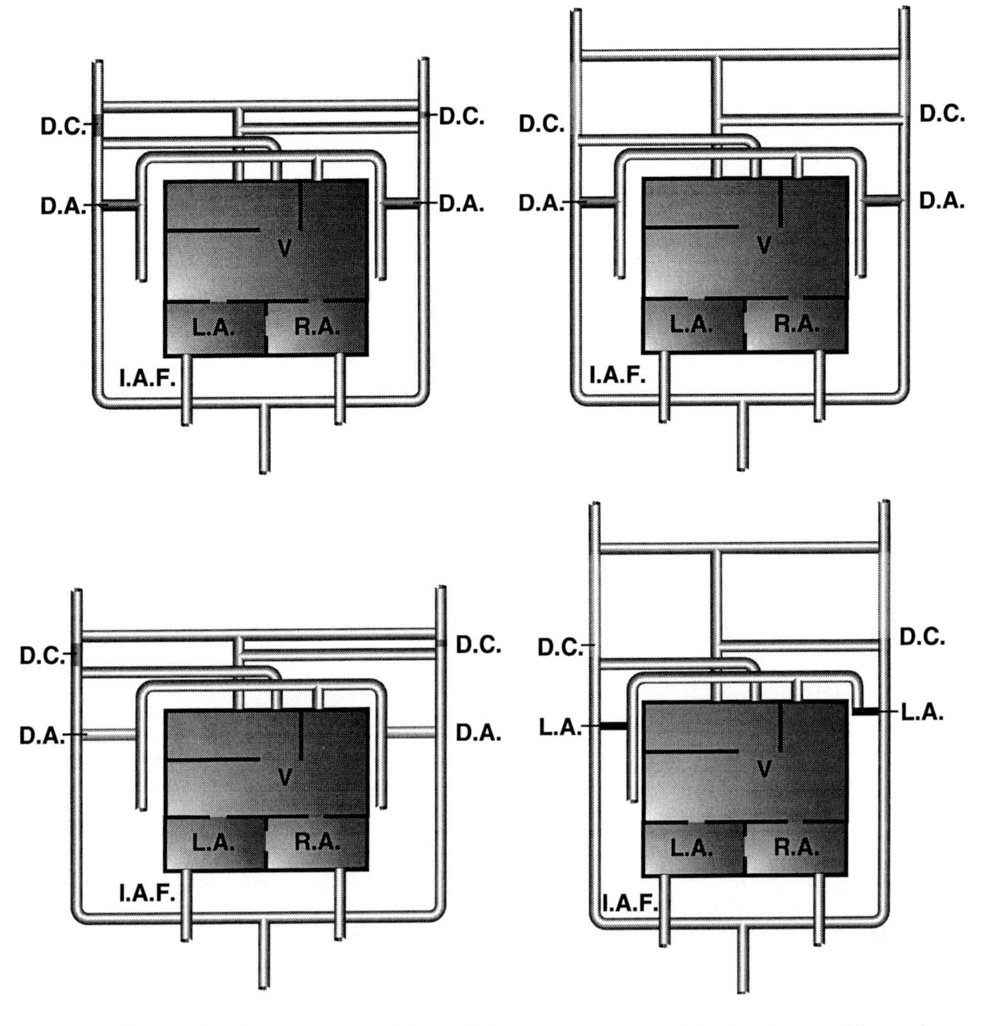

Figure 14.3 Representation of branchial arches lost or maintained in four reptilian orders. Aspect ratio of diagram represents modification of "ideal" proportions. See text for discussion. (*Top left*) Heart and great vessels of an adult tuatara with "ideal" proportions and a ductus caroticus (D.A.) and ductus arteriosus (D.A.). (*Top right*) Circulatory system of an adult turtle with ductus arteriosus. (*Bottom left*) Heart and great vessels of an adult lizard with "wide" proportions and patent ductus caroticus. (*Bottom right*) Circulatory system of an adult ophidian with "long and narrow" proportions. In this animal, the ductus arteriosus has been replaced by the ligamentum arteriosum (L.A.). (I.A.F. = interatrial foramina.)

circuits (Woodbury & Robertson, 1942; Burggren, 1977). Thus, the study of function in the embryonic bulbus cordis may shed valuable light on the function of the adult heart and the control of the blood distribution into pulmonary and systemic circuits.

Finally, although reptilian embryos do not possess a foramen ovale, the interatrial septum is pierced by secondary perforations that probably permit blood flow between the atria, similar to the fenestrated atrial septum in birds. These perforations apparently fuse closed upon the initiation of pulmonary ventilation (Goodrich, 1930). We know nothing of the mechanisms that close the fenestrations.

Extraembryonic vessels

Like birds, reptiles that develop in eggs exchange gases via a chorioallantois (or by vitelline vessels early in development). In birds, a decrease in the gradient for oxygen exchange (i.e., decreased ambient oxygen) leads to increased chorioallantoic vascularization. Thus, a lesser degree of chorioallantoic vascularization in reptilian embryos compared to avian embryos may be expected, given the much lower mass-specific metabolism of reptiles. Birchard and Reiber (1993) found that quail embryos have a twofold higher vascular density than similar-sized turtle embryos. Prehatching quails, however, have a fivefold higher oxygen consumption just prior to hatching relative to prehatching turtles. This implies there are other differences between the two species, such as diffusion distances, that account for the quail embryo's ability to take up oxygen five times faster while having only twofold greater vascular density. The difference may also be due to greater oxygen extraction in the quail, although measurements of arterial and venous oxygen pressures in another reptile embryo (crocodile) reveal values very similar to those found in avian embryos (Grigg, Wells, & Beard, 1993). Differences in oxygen affinity could also partially account for the disparity, but there is little difference in P_{50} between blood from crocodile embryos and blood from avian embryos. Baumann and Meuer (1992) determined P_{50} in 6-day chick embryonic blood to be 43.8 mm Hg, and this compares well with the P_{50} of embryonic crocodile blood, which initially has very low affinity (P_{50} = 52 mm Hg at 30°C) but attains a higher affinity late in development (P_{50} = 30 mm Hg at 60% of incubation) (Grigg, Wells, & Beard, 1993). Model analysis of gas exchange in a shunted system like the chorioallantois reveals that gas exchange depends on the ratio of cardiac output directed to the gas exchange organ versus that to the body proper (e.g., Tazawa & Johansen, 1987). It may be that reptilian embryos deviate from an optimal point and that this degree of shunting differentiates avian embryos from reptilian embryos.

The study of reptilian extraembryonic membrane development is further enriched by the existence of viviparous, truly placental species. Viviparity has evolved only in the Squamata (snakes and lizards), for reasons that are unclear. Crocodilians, chelonians, and tuataras are solely oviparous. There is evidence that reptilian placentation evolved separately many times rather than arising from a single ancestor (Blackburn, 1992). This lack of a common origin has resulted in several different types of placental structures because of the differing embryological origin of the membranes (Stewart & Blackburn, 1988). Thus, the study of viviparous and oviparous reproductive modes in similar species will provide important information on the evolution of gas and nutrient exchange during the transition from egg-laying to live-bearing.

Erythrocyte and hemoglobin function

Blood, an essential component of the cardiovascular system, has only recently been studied in embryonic reptiles. In most mammals, a significantly higher oxygen affinity is associated with the embryonic or fetal form of hemoglobin relative to adult hemoglobin, presumably to augment oxygen transfer from maternal blood to embryonic blood. It has been proposed that the high-affinity hemoglobin of mammalian embryos was probably inherited from oviparous ancestors (Manwell, 1958; Metcalfe, Dhindsa, & Novy, 1972; Grigg & Harlow, 1981). Reptiles developing in eggshells, especially those in underground nests, may encounter hypoxic (as well as hypercapnic) conditions during development, and a high-affinity hemoglobin could therefore enhance oxygen uptake. In viviparous reptiles, there does not appear to be a unique embryonic hemoglobin; rather, embryonic hemoglobin is allosterically modulated into a high-affinity hemoglobin by decreased levels of nucleotide triphosphates (Ingermann, 1992). The one exception known to date is the rattlesnake (*Crotalus viridis oreganus*), in which a distinct embryonic hemoglobin has been identified (Ragsdale & Ingermann, 1991). Oviparous estuarine crocodiles (*Crocodylus porosus*) and green sea turtles (*Chelonia mydas*) have distinct embryonic hemoglobins (Isaacks, Harkness, & Witham, 1978; Grigg, Wells, & Beard, 1993). Perhaps surprisingly, in both these species, despite a surge of 2,3-diphosphoglycerate (2,3-DPG) late in development, there appears to be no resultant allosteric modification of hemoglobin affinity. The lack of hemoglobin sensitivity to 2,3-DPG in some embryos suggests that a different explanation for the function of 2,3-DPG in early development may exist (Grigg, Wells, & Beard, 1993).

Hemoglobin function in adult crocodilians is modified primarily by bicarbonate (Bauer & Jelkmann, 1977; Brittain & Wells, 1991), unlike embryonic crocodilian hemoglobin, which, in *C. porosus,* is modified by the organic phosphate ATP (Grigg, Wells, & Beard, 1993). Thus, oxygen delivery in this species seems to be evolutionarily conservative in embryos and becomes divergent only in the adult form. Finally, in some viviparous snake species, pregnancy causes maternal hemoglobin to shift to a lower affinity (Ingermann, Berner, & Ragsdale, 1991; Ragsdale et al., 1993), again, presumably to enhance oxygen transfer. The mechanism of decreased hemoglobin oxygen affinity is allosteric modification induced by an increase in erythrocyte nucleotide triphosphates (Ragsdale et al., 1993). Interestingly, these same authors demonstrated that the decrease in affinity can be induced both in nonpregnant females and in males by progesterone implants.

Cardiovascular responses to hypoxia

As with many species, reptiles have several possible adaptive responses to hypoxic challenge: structural, physiological, and behavioral. Structural changes include morphological changes to decrease diffusion distances, increased ventricular size, and increased hematocrit. Physiological acclimation to hypoxia includes increased cardiac output by increased heart rate, altered peripheral resistance to alter blood flow patterns (e.g., altered shunt fractions), and increased anaerobic metabolism. In embryos, behavioral responses are not possible. At present we have only a few glimpses into the suite of responses exhibited by reptilian embryos.

Structural responses

Two studies examined hypoxia-induced structural changes in reptilian embryos, with somewhat different results. Turtles (*Pseudemys nelsoni*) incubated in 10% oxygen were smaller at hatching than normoxic controls but had larger hearts, both in absolute weight (2.8 and 3.2 mg dry mass, respectively) and in relative dry weight (0.18% and 0.25% of dry yolk-free body mass, respectively) (Kam, 1993). Alligator embryos (*Alligator mississippiensis*) incubated in 17% oxygen had significantly smaller hearts than normoxic controls (0.14 and 0.22 g); however, relative heart weights did not differ (0.29% and 0.27% of body mass including yolk; Warburton, Hastings, & Wang, 1995). The more severe hypoxia experienced by the turtle embryos may account for these differences. Whether the larger hearts were due to increased cardiac output or increased peripheral resistance is unknown. In both studies, hematocrit was higher in hypoxic-incubated reptiles, and this increased viscosity may also have imposed more load on the developing heart. Structural acclimation of the chorioallantoic membrane to hypoxia has been studied only in *Pseudemys nelsoni,* in which chronic hypoxia caused increased vascular length density ($mm \cdot mm^{-2}$), which became significant late in development (approximately 85% of incubation), although neither relative surface area ($mm^2 \cdot mm^{-2}$) nor vascular density index (grid intersections$\cdot mm^{-1}$) was affected (Kam, 1993).

Physiological responses

Reptilian embryos have a ductus arteriosus and the equivalent of a foramen ovale, and therefore, it is assumed that they shunt most of the cardiac output away from the nonventilating lungs, as avian and mammalian embryos do. It is important to recognize that this assumption remains unproved. Prior to lung ventilation, acute hypoxic vasoconstriction of pulmonary vasculature is probably irrelevant to embryonic shunt patterns (assuming minimal embryonic lung perfusion), whereas vasomotor responses of the chorioallantois to acute hypoxia are unknown. These vessels may react like pulmonary vessels and constrict when exposed to hypoxia, they may react like systemic vessels and vasodilate, or, of course, they may also be entirely unresponsive to hypoxia. Hypoxic vasoconstriction of these vessels might cause increased afterload on the heart or, with compensatory vasodilation, may result in redistribution of blood flow to the systemic circulation with no increase in afterload. Hypoxic vasodilation may result in decreased afterload or, in conjunction with systemic vasoconstriction, in increased flow (relative to systemic flow) to the chorioallantois. The embryonic cardiovascular response to hypoxic stress is therefore difficult to predict and should be investigated.

The myriad of developmental paths suggests there may also be a variety of physiological responses, different in degree, timing, or even existence. Heart rate responses to hypoxia in embryos of two reptile species, American alligators (*A. mississippiensis*) and kingsnakes (*Lampropeltus geletus*), have been studied to begin to elucidate species differences. These two species were chosen because they have different degrees of eggshell permeability, and thus perhaps different response patterns to hypoxia. Alligators have a rigid shell with a relatively high resistance to water vapor (ca. 1 mg $H_2O \cdot h^{-1}$ to a dry environment) and presumably to other gases as well, whereas kingsnakes have a shell that is relatively permeable (ca. 270 mg $H_2O \cdot h^{-1}$). Upon exposure to a hypoxic environment, alligator embryos 40% through incubation exhibit bradycardia proportional to the degree of hypoxia (Figure 14.4A; Warburton, Hastings, & Wang, 1995). This response is in contrast to adult hypoxic tachycardia as seen in hatchling alligators, the typical response of an air-breathing vertebrate. The response of the kingsnake embryos is quite different. When exposed to low ambient oxygen concentrations, kingsnake embryos halfway through incubation exhibit a prompt and reversible tachycardia (Figure 14.4B). The reason for these different responses is unknown but may relate to different degrees of autonomic development or other unknown factors.

Reptilian development and insights into vertebrate embryonic cardiovascular function

The study of the unique cardiovascular designs of reptiles provides more than filling in additional cells in the matrix of all possible developmental patterns. Some of the cardio-

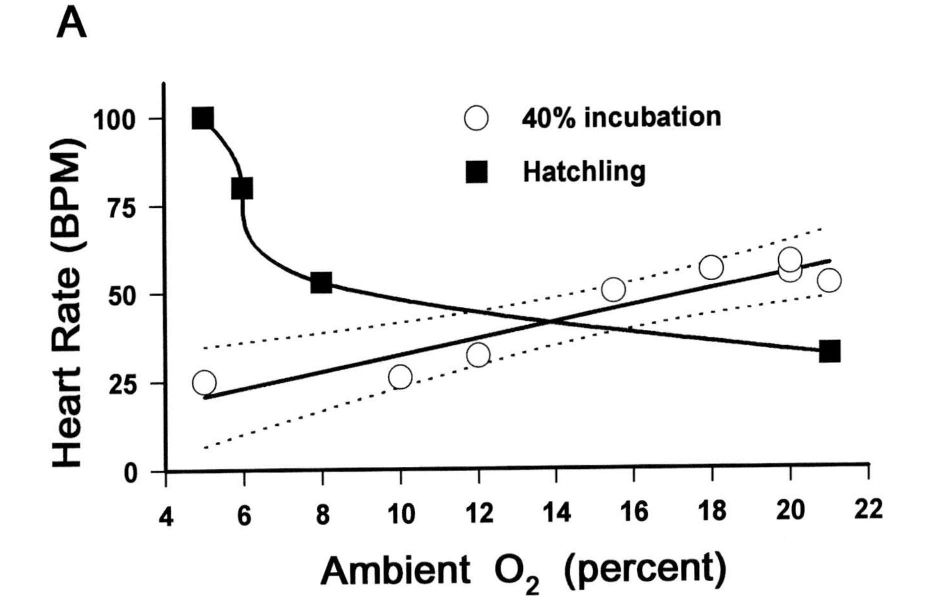

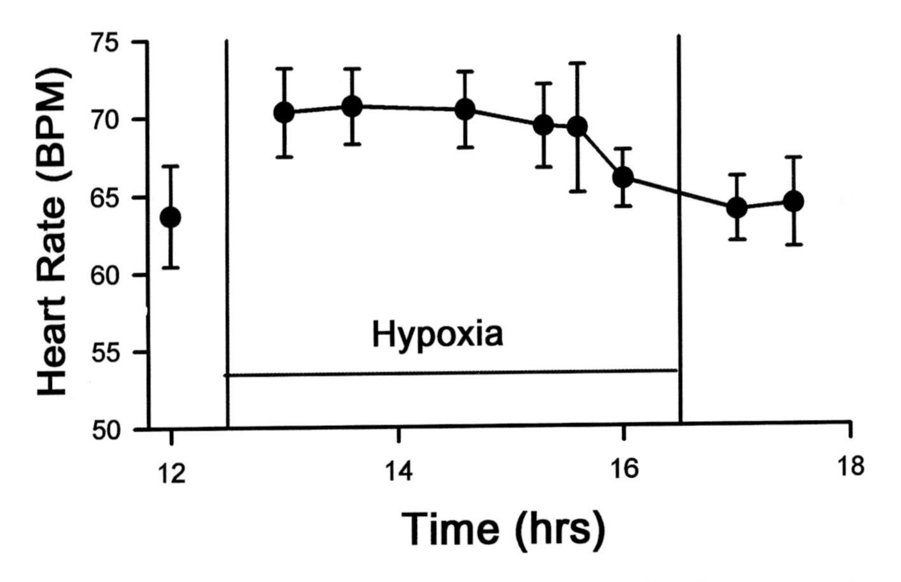

Figure 14.4 (*A*) Heart rate response of embryonic and juvenile alligators to acute increasing hypoxia. Embryo data are fitted to linear regression with 95% confidence intervals. Hatchling data are from one representative animal. (*B*) Heart rate response of kingsnake embryos to 12% oxygen for 2 hours. Symbols are mean ± SD, $n = 6$. (S. Warburton, unpublished data.)

vascular structures that exist in reptiles may cause us to reconsider some of the paradigms we accept about mammalian development. As discussed, reptilian embryos possess a ductus arteriosus and the equivalent of a foramen ovale. In mammalian and avian development, these structures are presumed to allow deoxygenated blood to bypass the pulmonary circuit but yet permit a system that can quickly change to the fully separated adult form at the time of birth or hatching. Reptile embryology makes this viewpoint less tenable because these species also possess, even as adults, intra- and extracardiac shunt pathways through which almost complete right-to-left shunting can occur. Thus, the question is raised, why do animals that already possess functioning pulmonary bypass circuits have additional bypass circuits as embryos?

Despite the diversity that exists within the cardiac structure of reptiles, all members of the class have one characteristic in common. All reptiles are capable, as adults, of achieving large right-to-left central vascular shunts, and these shunts are under physiological control. For example, adult turtles are capable of varying intracardiac shunt within the partially septated ventricle (Figure 14.3, top right). Thus, it would not be surprising to find that embryonic turtles likewise use this pathway to shunt blood away from nonventilated lungs. This seems not to be the case, however, as turtles also have the same shunt pathways we associate with mammalian and avian embryos, the ductus arteriosus and foramen ovale. Thus, given their "built-in" ability to shunt blood, the role of the embryo's second set of shunt pathways clearly deserves further investigation.

A similar situation exists in crocodilians because although they have a truly four-chambered heart, crocodilians are still capable of extracardiac right-to-left shunting via a patent left aortic arch arising from the right ventricle. In response to increased pulmonary vascular resistance, right ventricular pressure increases until it exceeds systemic pressure, then the valve of the left aortic arch opens to decompress the right ventricle, resulting in the shunting of dearterialized blood (Figure 14.1). Note, however, that the carotid arteries arise from the right aortic arch and thereby preferentially receive oxygenated blood. Again, embryonic crocodilians have a patent ductus arteriosus and foramen ovale. To date, we know nothing of the relative contributions of each shunt pathway (i.e., adult vs. embryonic) during embryonic development.

In summary, reptilian embryogenesis can follow many different pathways via viviparous or oviparous development and lead to complex three-chambered or four-chambered hearts. Elucidation of cardiovascular development in reptiles will play an important role in allowing us to delineate the evolutionary forces leading to the mammalian and avian embryonic forms. The many paths reptiles follow in development allow us to begin to determine which aspects of vertebrate cardiovascular development are universal and which are specific for a particular taxonomic group.

15

Avian cardiovascular development

HIROSHI TAZAWA AND PING-CHUN LUCY HOU

Introduction

There are more than 8550 avian species, and egg size ranges from not more than 1 g to about 1500 g. Depending on embryonic maturity at hatching, birds are grouped into two classes: altricial (embryos hatch with naked bodies, with closed eyes, and without the abilities of locomotion and thermoregulation) and precocial (hatchlings can walk, swim, or dive shortly after hatching). Despite the wide variety of avian eggs, however, study of avian embryonic cardiovascular systems has been limited primarily to domesticated birds, with emphasis on the chick (*Gallus domesticus*) embryo.

An incubation temperature of about 38°C is adequate for the growth of the chick embryo, which hatches on day 21 of incubation. As the embryo grows, the cardiovascular system develops rapidly. Because of the restricted locomotion of embryos within the confines of the eggshell, rapid development of the cardiovascular system, and easy acquisition of fertile eggs year-round, the chicken egg provides an excellent experimental animal model for the study of the development of cardiac and vascular regulation (see Chapters 7, 8, 10, and 19). Also, during the last half of incubation, it becomes possible to measure some physiological variables from developing embryos without impeding gas exchange through the eggshell.

This chapter describes (1) the development of chick embryos, especially their hearts, (2) developmental patterns of circulatory parameters during the period of rapid embryonic growth prior to pipping, and (3) a model analysis of cardiovascular shunts of a late embryo.

The developmental time of chick embryos varies, probably depending upon strains, fresh egg mass, shell porosity, and slight differences in incubation temperature (Romanoff, 1960a; Tullett & Burton, 1982; Zhang & Whittow, 1992). Thus, embryo mass and morphological developmental stage attained may vary among embryos measured on the same incubation day. A staging table based on morphological development of the embryos is available (Hamburger & Hamilton, 1951), but the determination of stages is not possible without sacrificing the embryos. Incubation days are described in almost all chronic studies of individual embryos and even in many acute experiments for comparison with chronic results. Therefore, in this chapter, the developmental patterns of circulatory variables are depicted against the incubation days to summarize the reported values. In these experiments, incubation temperature was not necessarily fixed at 38°C, which might have caused the results of the studies to differ.

Embryogenesis and development of the chick heart

Embryogenesis in bird eggs begins at the moment of fertilization; development of the embryo proceeds in the oviduct. It may be influenced by oviductal transit time and the body temperature of the hen. Soon after laying, embryonic development can be stopped or maintained by changing ambient temperature. Storage of newly laid fertile eggs below "physiological zero" temperature induces dormancy of the embryo; incubation at a temperature higher than 38°C accelerates embryonic development. By changing ambient temperature, chick embryos at specific developmental stages can be obtained. However, hatching occurs at an incubation temperature between 35.5°C and 39.5°C, and constant incubation temperature outside this range kills embryos (Romanoff, 1960a).

During the first few days at optimal incubation temperature, the albumen that initially surrounds the yolk loses water, becomes heavy, and sinks toward the lower end of the egg. Eventually, the developing embryo floats to the top of the yolk and lies on its left side, just below the shell membranes. Early embryos with yolk sacs are easily observed through a window opened in the eggshell. The yolk sac is well vascularized in its medial region, the area vasculosa. For early embryos, the area vasculosa is a gas exchanger when it contacts the inner shell membrane. Because of the easy access to the embryo and area vasculosa, fenestrate egg preparations have long been used to investigate the cardiovascular system of the early embryo. Beginning on day 5 of incubation, the chorioallantois protrudes from the embryo and begins to cover the inner shell membrane. It spreads over the whole inner shell membrane and by day 12 envelops all the contents of the egg, including the embryo. Its outer surface is well vascularized, which replaces the area vasculosa as the second respiratory organ.

Embryo development is divided into prenatal and paranatal stages (Visschedijk, 1968; Rahn, Ar, & Paganelli, 1979). During the first 19 days or so of incubation (prenatal stage), the embryo exchanges respiratory gases with atmospheric air by a diffusive process through the porous eggshell and the area vasculosa or chorioallantoic membrane. Toward the end of the prenatal stage, the yolk diminishes in size and the embryo grows rapidly and fills the eggshell almost entirely. The allantoic fluid, which reaches its maximum volume by the end of the second week, also diminishes, so that virtually none remains on day 19. Then, the embryo first penetrates the air cell with its beak and begins to breathe the air (internal pipping) and then cracks (pips) the shell with the egg tooth (external pipping). The period from internal pipping to hatching is the paranatal period and lasts about 1–1.5 days in the domestic fowl. During the paranatal period, the respiratory function of the chorioallantois is replaced by that of the lungs. In some sea birds, the paranatal stage begins with external pipping and lasts several days (Pettit & Whittow, 1983).

Figure 15.1 shows chick embryo developmental patterns in terms of daily changes in wet mass (Romanoff, 1967; Van Mierop & Bertuch, 1967; Tazawa, Mikami, & Yoshimoto, 1971; Lemez, 1972; Clark et al., 1986; Haque et al., 1996). Although the range of the reported values is large, all reports agree that the embryo mass increases in a geometric fashion toward the end of the prenatal stage and then the growth rate declines. The developmental pattern during the entire incubation period is expressed by a logistic equation for precocial embryos and by a power function for altricial ones (Hoyt, 1987). However, for the precocial chick embryos (Figure 15.1), the increase in body mass till day 19, when growth rate decreases, is also well expressed by a power function of incubation time. The following equation is derived from the data shown in Figure 15.1:

$$\text{body mass} = 0.24I^{4.02} \qquad (r = .999),$$

where body mass indicates wet body mass of the embryo (in mg) and I indicates incubation time (in days).

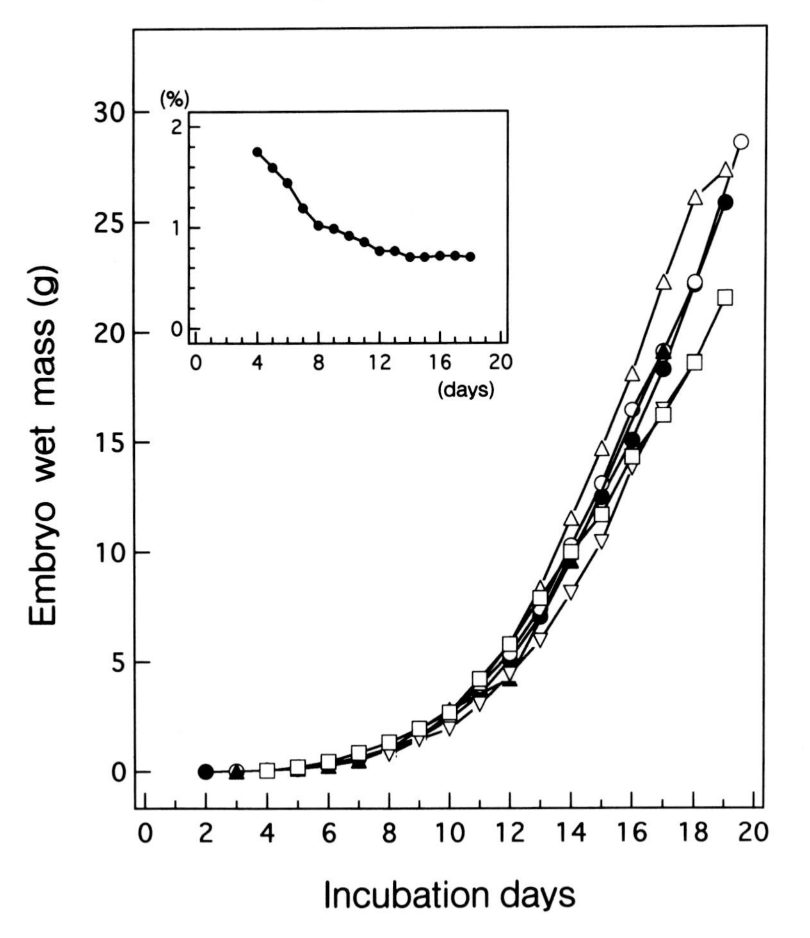

Figure 15.1 Daily changes in wet mass of chick embryos during the prenatal stage from six reports ● = Romanoff, 1967; ○ = Van Mierop and Bertuch, 1967; ▽ = Tazawa, Mikami, & Yoshimoto, 1971; □ = Lemez, 1972; ▲ = Clark et al., 1986; △ = Haque et al., 1996. Insert: The ratio of heart to embryo mass (in %) during prenatal development (Romanoff, 1967).

Heart mass also increases as a power function of incubation time (from day 4 to day 18; Romanoff, 1967):

$$\text{heart mass} = 12.62I^{3.26} \qquad (r = .999),$$

where heart mass is in micrograms. As incubation and embryo growth proceed, the ratio of heart to embryo mass decreases (see insert in Figure 15.1). Such a relationship is also reported for ventricle and embryo mass during the early period of development; that is, the ratio of ventricular mass to embryo mass decreases with development (Clark et al., 1986; Hu & Clark, 1989). The largest percentage decrease occurs early in development when the heart becomes a four-chambered structure; that is, as a proportion of body mass, the primordial heart is relatively large compared with the more mature heart. The greater relative heart mass very early in development may relate to the circulatory requirements of the embryo, which has initially high vascular resistance and poor vascularization (Clark et al., 1986; see also Chapters 7 and 10, this volume).

The primitive heart is a paired tubular structure that soon fuses to form a single tube. It begins to elongate rapidly in the pericardial cavity containing it and forms a ventricle and the bulbus within 2 days of incubation. The heart begins to beat at about 30 hours of incubation, and the blood begins to circulate after about 40 hours when the intraembryonic circulation connects the vessels of the yolk sac. The limitation imposed upon the growing heart by lack of space and the impact of bloodstreams upon the inner surface of the contorted tube influence the external configuration and internal structures of the heart (Chapter 19). The heart with four chambers is formed by Hamburger–Hamilton stage 35 (days 8–9 of incubation). Out of six initial pairs of aortic arches, the third, the right fourth, and the sixth aortic arches persist as the carotid arteries, the aorta, and the pulmonary arteries, respectively (Romanoff, 1960a). The remaining arches disappear in late embryos. However, development, appearance, and disposition of the aortic arches are not always consistent (Levinsohn et al., 1984). The anatomy and morphology of intra- and extraembryonic circulatory systems change extensively with growth of the embryo.

Developmental pattern of cardiovascular variables

Blood volume

The blood and blood vessels are formed first in the extraembryonic yolk sac before a primitive vascular system and heart develop within the body of the embryo. The yolk sac is a main organ for hematopoiesis during embryonic development. As embryos grow, the volume of circulating blood increases geometrically (Figure 15.2; Yosphe-Purer, Endrich, & Davies, 1953; Barnes & Jensen, 1959; Lemez, 1972; Kind, 1975). The average increase in blood volume shown in Figure 15.2 is well expressed by a power function of incubation day (I):

$$\text{blood volume} = 1.85 I^{2.64} \qquad (r = .996),$$

where blood volume is in microliters. The rate of increase in blood volume with incubation day is not as fast as the growth rate, and the mass (embryo)-specific blood volume decreases with development from about 1 $ml \cdot g^{-1}$ on day 4 to 0.15 $ml \cdot g^{-1}$ on day 18 of incubation (see insert in Figure 15.2). Thus, circulating blood volume is extremely large during the very early period when the primordial heart forms the four-chambered heart. The logarithmic increase in blood volume is linearly related to the logarithmic increase in embryo mass:

$$\text{blood volume} = 0.44 \times (\text{body mass})^{0.67} \qquad (r = .997).$$

Stroke volume and cardiac output

The Frank–Starling mechanism and Bowditch's staircase phenomenon exist in chick embryos as early as 3–7 days (Faber, 1968; Keller, Hu, & Tinney, 1994). In spite of the absence of cardiac valves in these embryos, there is no retrograde flow during systole from the ventricle to the atrium or from the truncus arteriosus to the ventricle; the ventricular stroke volumes are "forward" stroke volumes that depend on the circulatory blood volume and increase with embryonic growth (Faber, Green, & Thornburg, 1974). Taking advantage of the transparency of early embryos and extraembryonic membranes, Faber, Green, and Thornburg (1974) calculated the stroke volume by taking moving pictures of the beating heart. Earlier, Hughes (1949) had incised the sternum and photographed the exposed heart of later embryos to calculate the stroke volume. Figure 15.3 presents measured stroke volume and diastolic filling volume determined for the exposed heart of later embryos.

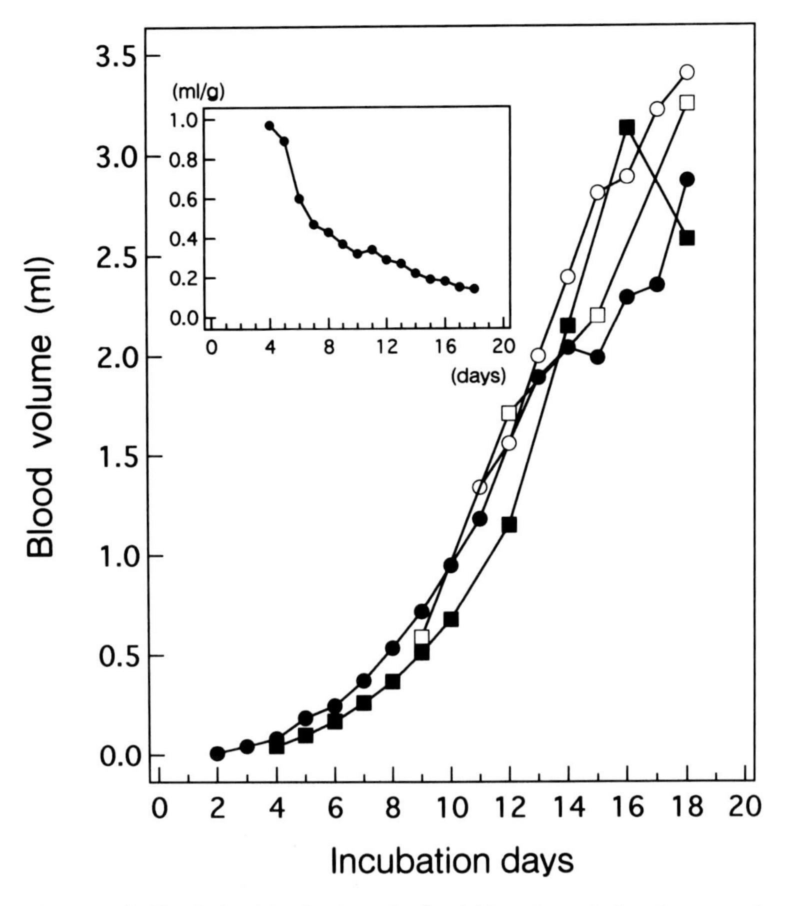

Figure 15.2 Circulating blood volume in the chick embryo during the prenatal stage from four reports. ○ = Yosphe-Purer, Endrich, & Davies, 1953; □ = Barnes & Jensen, 1959; ● = Lemez, 1972; ■ = Kind, 1975. Insert: The ratio of mean blood volume to mean body mass (from Figure 15.1).

Although the embryos were exposed to a considerable degree of trauma, the result indicates that the stroke volume tends to increase in a power function of incubation time even in later prenatal stages.

Another way to estimate the stroke volume of early embryos is to determine mean dorsal aortic blood flow (by a pulsed Doppler velocity meter) and divide it by heart rate (see insert in Figure 15.3; Hu & Clark, 1989). The increase in stroke volume during the period from day 2 to day 6 of incubation is expressed by a power function of incubation day (I):

$$\text{stroke volume} = 2 \times 10^{-3} I^{3.46} \qquad (r = .958),$$

where stroke volume is in microliters.

In early embryos (from day 2 to day 6 of incubation), the mean dorsal aortic blood flow is a measure of cardiac output (Hu & Clark, 1989) (see insert in Figure 15.4). Because the heart rate almost doubles during this period (discussed later), dorsal aortic blood flow increases faster than the stroke volume, as expressed by

$$\text{DAo} = 0.15 I^{4.01} \qquad (r = .974),$$

where DAo is dorsal aortic blood flow (in $\mu l \cdot min^{-1}$).

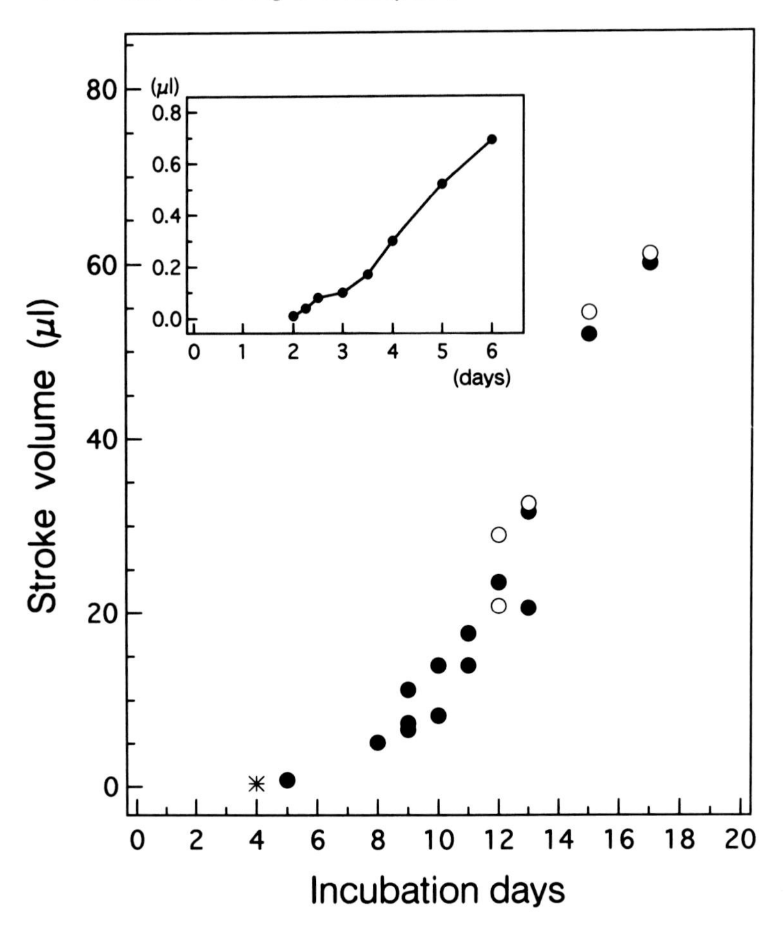

Figure 15.3 Stroke (*unfilled circles*) and diastolic filling (*filled circles*) volumes of the exposed hearts of later chick embryos determined by Hughes (1949). Insert: Stroke volume of early embryos determined from the quotient of mean dorsal aortic blood flow divided by heart rate (Hu & Clark, 1989).

During the same period, embryo mass also increases in a power function of incubation day, and thus the mass-specific aortic blood flow is about 0.5–1.2 $ml \cdot min^{-1} \cdot g^{-1}$. Faber, Green, and Thornburg (1974) calculated the cardiac output of early embryos (3–5 days) by multiplying stroke volume by heart rate and arrived at a mass-specific cardiac output of about 1 $ml \cdot min^{-1} \cdot g^{-1}$. The determination of cardiac output in late embryos is difficult and only a few data are plotted in Figure 15.4. The cardiac outputs calculated from the stroke and diastolic filling volumes shown in Figure 15.3 give mass-specific values ranging from 1.3 $ml \cdot min^{-1} \cdot g^{-1}$ on day 12 to 1 $ml \cdot min^{-1} \cdot g^{-1}$ on day 17 of incubation. The cardiac outputs estimated by model analysis (White, 1974; Rahn, Matalon, & Sotherland, 1985; Tazawa & Takenaka, 1985) yield a mass-specific cardiac output of about 0.9–1.5 $ml \cdot min^{-1} \cdot g^{-1}$ on day 16. Consequently, the cardiac output of chick embryos may increase almost in parallel with embryo mass during the later period of prenatal development. Based on this speculation, the cardiac output of late chick embryos can be calculated by

$$CO = 0.24I^4,$$

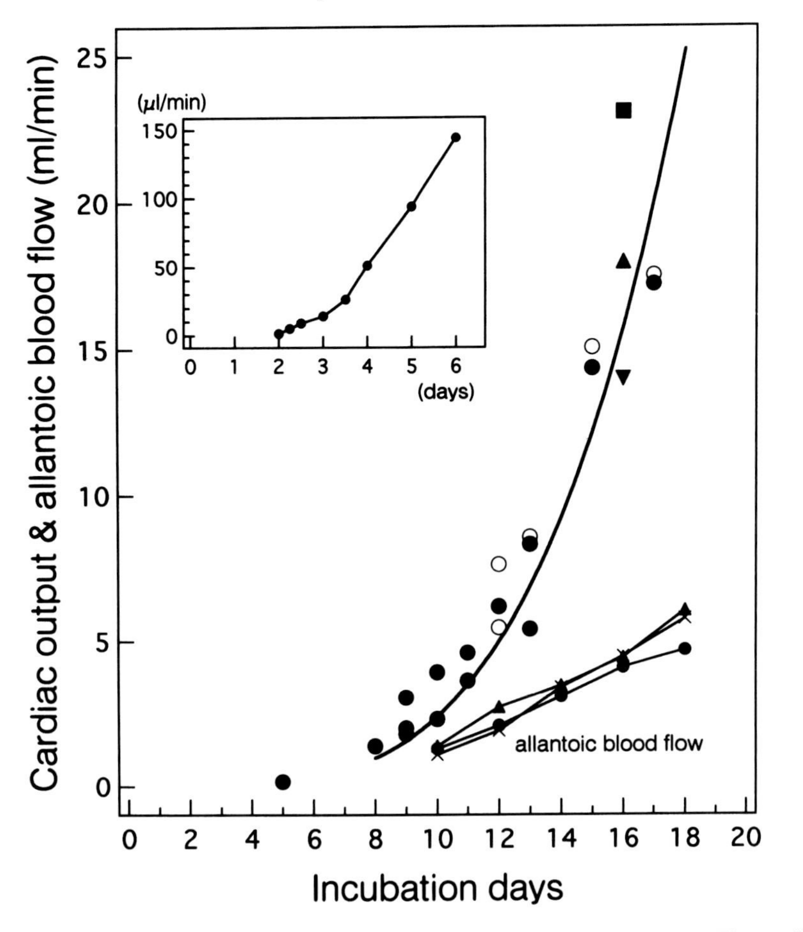

Figure 15.4 The cardiac output and allantoic blood flow of late embryos. The cardiac outputs estimated from the product of heart rate and stroke (*unfilled circles*) and diastolic filling (*filled circles*) volumes (shown in Figure 15.3) are presented by open and closed circles, respectively. Those derived from model analysis are shown by filled squares (Tazawa & Takenaka, 1985), filled up-pointed triangles (White, 1974), and filled down-pointed triangle (Rahn, Matalon, & Sotherland, 1985). Thick solid curve indicates the cardiac output estimated by assuming that the ratio of cardiac output to embryo mass is constant; i.e., 1 $ml \cdot min^{-1} \cdot g^{-1}$. The allantoic blood flow data are from three reports. × = Tazawa & Mochizuki, 1976; ● = Tazawa and Mochizuki, 1977; ▲ = Bissonnette and Metcalfe, 1978. Insert: Mean dorsal aortic blood flow determined by a pulsed Doppler velocity meter (Hu & Clark, 1989).

where CO is cardiac output (in $\mu l \cdot min^{-1}$). The thick solid line drawn in Figure 15.4 presents the cardiac output estimated by the above equation.

In contrast to the determination of cardiac output, allantoic blood flow can be measured without difficulty (Figure 15.4; Tazawa & Mochizuki, 1976, 1977; Bissonnette & Metcalfe, 1978). In early embryos the extraembryonic circulation appears to accommodate more than half of the cardiac output (Faber, Green, & Thornburg, 1974). Allantoic blood flow increases with embryonic development, but the mass-specific value decreases from about 0.5 $ml \cdot min^{-1} \cdot g^{-1}$ on day 10 to about one-half of that value on day 18 (Figure 15.4). Because cardiac output tends to increase in parallel with embryonic growth, the volume fraction of flow to the chorioallantoic gas exchanger decreases as embryos grow. Allantoic blood flow must decrease toward the end of prenatal development with the atrophy of the allantoic artery and finally cease, to prevent hemorrhage during pipping.

Blood pressure

Early measurements of blood pressure using a micropipette technique were made for the vitelline artery in early embryos and for a branch of the allantoic arterial system in older embryos through a window opened in the eggshell (Van Mierop & Bertuch, 1967; Girard, 1973a). A recent technique using a servonull micropressure system (see Burggren & Fritsche, 1995, for details of the technique) made it possible to measure not only the vitelline arterial pressure but also atrial, ventricular, and dorsal aortic pressures in early embryos (Clark et al., 1986; Hu & Clark, 1989; Keller, 1995). Both the systolic and diastolic pressures of the vitelline artery increase with incubation time (see insert in Figure

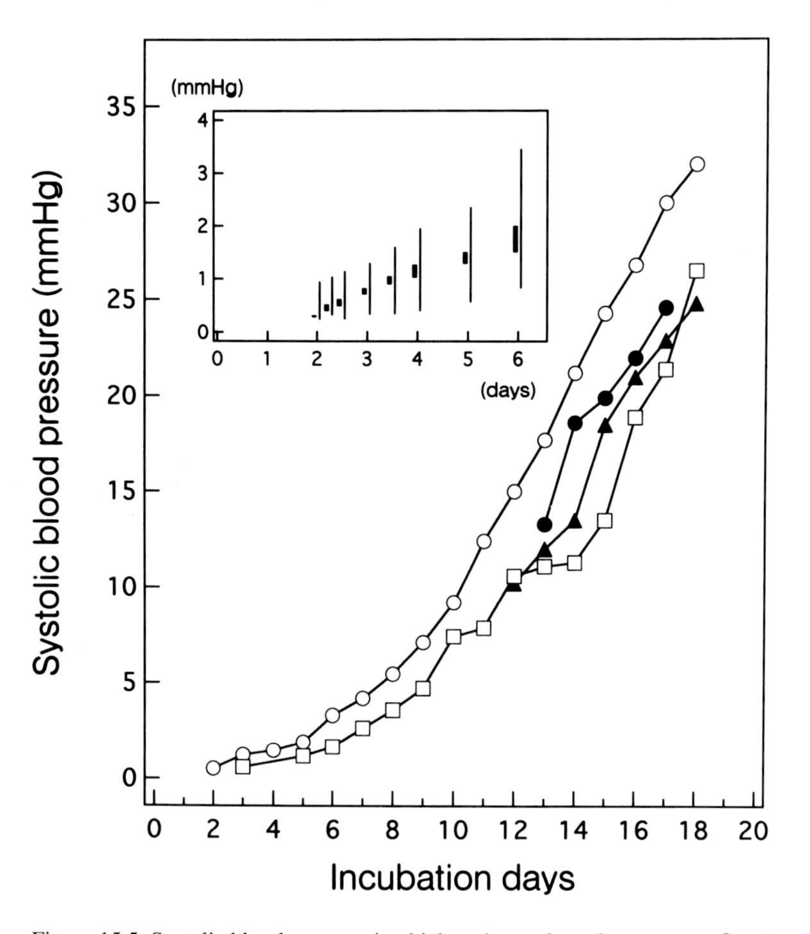

Figure 15.5 Systolic blood pressure in chick embryos from four reports. ○ = Van Mierop & Bertuch, 1967; □ = Girard, 1973a; ● = Tazawa, 1981b; ▲ = Tazawa & Nakagawa, 1985. Insert: Vitelline arterial pressure (*thick line*) and ventricular pressure (*thin line*) (Clark et al., 1986; Hu & Clark, 1989). Each line connects the systolic and diastolic pressures.

15.5). The relationship between vitelline arterial systolic pressure (P_{Vsys}, in mm Hg) and incubation day (I) is expressed by

$$P_{Vsys} = 0.14I^{1.53} \qquad (r = .979).$$

When embryos grow, the allantoic blood vessels floating in the allantoic fluid become thick, so that a needle catheter can be implanted through a small hole opened in the eggshell and the hole closed after insertion of the catheter (Tazawa et al., 1980). This method allows determination of the allantoic blood pressure of late chick embryos while maintaining adequate gas exchange (Tazawa, 1981b; Tazawa & Nakagawa, 1985). Figure 15.5 shows the developmental patterns of arterial systolic pressure determined by the micropipette and needle-catheter techniques (Van Mierop & Bertuch, 1967; Girard, 1973a; Tazawa, 1981b; Tazawa & Nakagawa, 1985). The arterial blood pressure increases geometrically with embryonic development. Van Mierop and Bertuch (1967) found a linear relationship between arterial systolic pressure and embryo mass (and also incubation time) when they are plotted on logarithmic scales (except in early stages of development). The regression equation for the relationship between arterial systolic pressure and embryo mass is expressed by

$$P_{sys} = 5.87 \times (\text{body mass})^{0.54} \qquad (r = .999),$$

where P_{sys} is systolic blood pressure of the vitelline (early embryo) and the allantoic (later embryo) artery (in mm Hg). The relationship with incubation day (I) (from 4.2 to 18 days) is expressed by

$$P_{sys} = 0.054I^{2.24} \qquad (r = .998).$$

Although the measurements by Girard (1973a) were underestimates by at least 10–20% because of damping in the measurement system, the arterial systolic pressure is also expressed by a power function of incubation time (3–18 days):

$$P_{sys} = 0.042I^{2.18} \qquad (r = .992).$$

For the systolic blood pressure of the allantoic artery (P_{Asys}) determined by the needle catheter system (Tazawa & Nakagawa, 1985), the regression equation is expressed by

$$P_{Asys} = 0.040I^{2.25} \qquad (r = .985).$$

The slope is consistent among these three studies and is greater than that of vitelline arterial pressure in early embryos.

The pulse pressure increases with embryonic development, and the ratio of the pulse pressure to systolic blood pressure is larger than that in the adult bird (Van Mierop & Bertuch, 1967; Tazawa & Nakagawa, 1985). The blood pressure of the vitelline artery is already pulsatile on day 2 of incubation (Van Mierop & Bertuch, 1967; Hu & Clark, 1989). Despite the lack of cardiac valves, the configuration of the arterial pressure tracings and the ventricular pressure waves suggest that mechanisms corresponding to arterial and atrioventricular valves are functional as early as day 3 of incubation (see insert in Figure 15.5). For the tubular heart with a peristaltic beat in early embryos, the swelling of the cardiac jelly performs the role of valves (Barry, 1948).

Although the egg must be partially opened before the procedure can be performed, simultaneous measurements of dorsal aortic blood flow and blood pressure show that intravenous injection of isoproterenol (isoprenaline) into day 4 (stage 24) chick embryos causes no change in mean arterial pressure (or in mean atrial pressure and heart rate), but it does cause a decrease in dorsal aortic blood flow and an increase in vascular resistance (Clark, Hu, & Dooley, 1985). It is likely that resistance is regulated before pressure is. Although a similar experiment can also be performed with fenestrate eggs, the microcatheterization technique allows the simultaneous measurements of arterial and venous blood pressures. Adrenergic sensitivity of blood pressure begins on day 6 of incubation, and on day 7 vascular α- and β-receptors can be demonstrated by blocking agents (Girard, 1973b). The implantation of the catheter into both the artery and the vein makes it possible to determine the effect of autonomic agents on the blood pressure of the late embryos without altering gas exchange through the shell or O_2 transport by the blood (Tazawa, Hashimoto, & Doi, 1992). Figure 15.6 shows the simultaneous measurement of allantoic arterial and venous pressures before and after administration of 2 μg adrenaline (Ad) through the venous catheter. In the same manner, day 13–16 chick embryos have been examined for their responses to a parasympathomimetic drug (acetylcholine, ACh), a parasympathetic blocking agent (atropine), sympathomimetic drugs (Ad, noradrenaline [NAd], and isoprenaline), and α- and β-adrenergic blocking agents (phentolamine and propranolol, respectively) (Tazawa, Hashimoto, & Doi, 1992). Although the embryos are sensitive to 0.5–2 μg ACh, preinjection of 1 μg atropine invalidates the effect of ACh. Nevertheless, 50 μg atropine alone does not change arterial blood pressure. The day 13–16 embryos respond to an administration of NAd with increasing arterial blood pressure and to isoprenaline (β-adrenoceptor stimulator) with decreasing arterial blood pressure. Phentolamine decreases, and propranolol increases, the arterial blood pressure, indicating constant α- and β-adrenergic tone in the blood vessels. Both Ad and NAd fail to change the arterial blood pressure after administration of an increased dose of phentolamine (α-receptor blocker). After α-receptors are blocked by phentolamine, subsequent injection of isoprenaline fails to decrease the arterial blood pressure, suggesting that the blood vessels are already relaxed. Antagonistic effects of isoprenaline and propranolol on the arterial blood pressure are also clearly demonstrated. These experiments indicate that in chick embryos the sympathetic receptors already function before the end of the second week of incubation, but the parasympathetic nerve seems not to function until at least day 16.

Heart rate

Although the heart rate (HR) is counted from blood pressure recordings, it can be more simply measured by electrocardiography (ECG) (Cain, Abbott, & Rogallo, 1967; Laughlin,

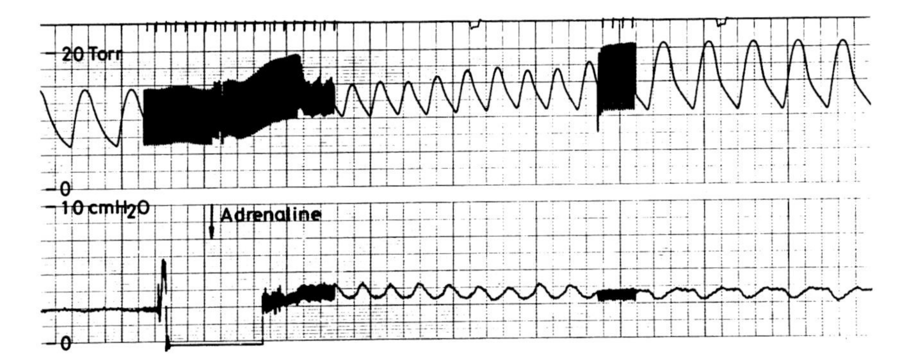

Figure 15.6 Simultaneous measurement of allantoic arterial (*upper tracing*) and venous (*lower tracing*) blood pressures of a 14-day chick embryo. The arrow indicates injection of 2 μg adrenaline in 20 μl saline through the venous catheter. The time marker at the top of the panel indicates 1 s. Both arterial and venous pressures increase within a few seconds of injection, and transient ventricular arrhythmia occurs. The venous pressure becomes pulsatile following adrenergic stimulation.

Lundy, & Tate, 1976; Tazawa & Rahn, 1986), impedance cardiography (ICG) (Haque et al., 1994; Howe, Burggren, & Warburton, 1995; Komoro et al., 1995), ballistocardiography (BCG), and acoustocardiography (ACG). BCG measures ballistic movements of the egg produced by the cardiac contractions of the embryo within the eggshell. The cardiogenic ballistic movements are multidirectional (Sakamoto et al., 1996), and several methods and systems have been developed to determine them: piezoelectric transducer (Cain, Abbott, & Rogallo, 1967) and piezoelectric film (Tazawa et al., 1993), audiocartridge (Tazawa, Suzuki, & Musashi, 1989), laser speckle meter (Tazawa et al., 1989a), laser displacement meter (Hashimoto, Narita, & Tazawa, 1991), and electromagnetic induction coil (Ono et al., in press) measuring systems. ACG detects air pressure changes outside the eggshell produced by mechanical movements of the embryonic heart (Rahn, Poturalski, & Paganelli, 1990) or by pulsatile gas exchange resulting from pulsatile blood flow (Wang, Butler, & Banzett, 1990). Although the mechanism is not clear, the pressure changes originating in cardiac contractions of the embryo can be detected by attaching a condenser microphone airtightly to the eggshell; the signal is called an acoustocardiogram (Rahn, Poturalski, & Paganelli, 1990; Akiyama et al., in press). Both BCG and ACG are noninvasive methods that can determine the HR during the last half of incubation up to hatching (BCG) or pipping (ACG). The HR during the first half of incubation can be measured by ECG & ICG, which are semi-invasive methods.

In contrast to the developmental patterns of blood flows and pressures, which tend to increase in parallel with embryo growth, changes in embryonic HR are not linearly related to embryo mass. Figure 15.7 shows some developmental patterns of embryonic HR in chick embryos (Cain, Abbott, & Rogallo, 1967; Van Mierop & Bertuch, 1967; Girard, 1973a; Laughlin, Lundy, & Tate, 1976; Hu & Clark, 1989; Zahka et al., 1989; Tazawa et al., 1991; Howe, Burggren, & Warburton, 1995). The mean HR increases asymptotically during early development with rapid morphogenesis. Then, HR seems to change relatively little during the last half of incubation. However, the HR increases again during pipping and hatching (Tazawa et al., 1991). The major increase during the external pipping period seems to coincide with an increase in O_2 consumption.

Measurement technique seems to affect HR pattern (Figure 15.7). The HRs determined from the blood pressure waves measured in fenestrate eggs show a depressed pattern (Van Mierop & Bertuch, 1967; Girard, 1973a; Hu & Clark, 1989; Zahka et al., 1989). HR measurements in nonfenestrate eggs by BCG (Tazawa et al., 1991), ECG (Cain, Abbott, & Rogallo, 1967; Laughlin, Lundy, & Tate, 1976), and ICG (Howe, Burggren, & Warburton, 1995) do not require opening the eggshell, and gas exchange is maintained through the eggshell. The HR in this group is routinely higher. If blood pressure is measured after closure of a hole in the eggshell, HR is as high as that determined by BCG, ECG, and ICG (Tazawa and Nakagawa, 1985). The removal of the eggshell, which disturbs adequate gas exchange through the chorioallantoic membrane and eggshell (Tazawa et al., 1988a), may result in underestimation of the HR during the later period of development. Another possibility is that low ambient temperature cools the fenestrate eggs.

The HRs shown in Figure 15.7 are mean values. The instantaneous heart rate (IHR), however, is very variable. Figure 15.8 shows the IHRs of an embryo determined on days 13–17 of incubation by ACG (Akiyama et al., in press). Variability increases with embryonic development, and characteristic, transient bradycardia begins to occur around day 14 or 15 of incubation. In addition, transient tachycardia occurs toward the end of incubation and HR becomes more arrhythmic. There are several possible explanations for this cardiac arrhythmia. There may be intrinsic influences of the pacemaker and conducting system or extrinsic factors such as sympathetic and vagal functions. Atropine injected intravenously

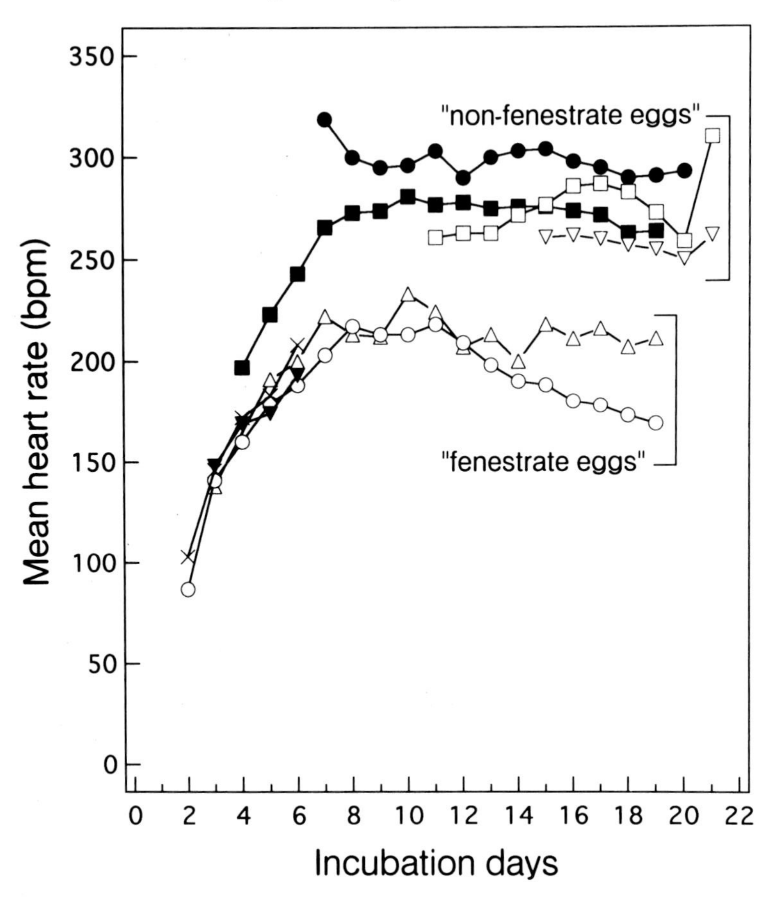

Figure 15.7 Mean heart rate in chick embryos from eight reports ■ = Cain, Abbott, & Rogollo, 1967; ○ = Van Mierop & Bertuch, 1967; △ = Girard, 1973a; ▽ = Laughlin, Lundy, & Tate, 1976; × = Hu & Clark, 1989; ▼ = Zahka et al., 1989; ○ = Tazawa et al., 1991; ● = Howe, Burggren, & Warburton, 1995.

seems to mitigate the bradycardia. Both Ad and NAd increase the IHR of many embryos. The variability of IHR seems unaffected by Ad and NAd. Whereas propranolol (β-adrenergic blocker) exerts a pronounced negative chronotropic effect, phentolamine (α-adrenergic blocker) decreases the IHR of a few embryos among those tested. Both drugs seem not to change the variability. Because these observations were preliminary, further study is needed to elucidate a mechanism of HR variability of developing embryos together with the response of IHR to altered environments.

Model analysis of cardiovascular shunts and cardiac output in the late embryo

Based on anatomical findings, White (1974) illustrated the cardiovascular system of a 16-day chick embryo. Figure 15.9 presents an illustration showing the principal arterial system of a late embryo (Tazawa & Takenaka, 1985). The coronary arteries and right and left brachiocephalic arteries, supplying left ventricular blood to the heart, head, eyes, and thorax

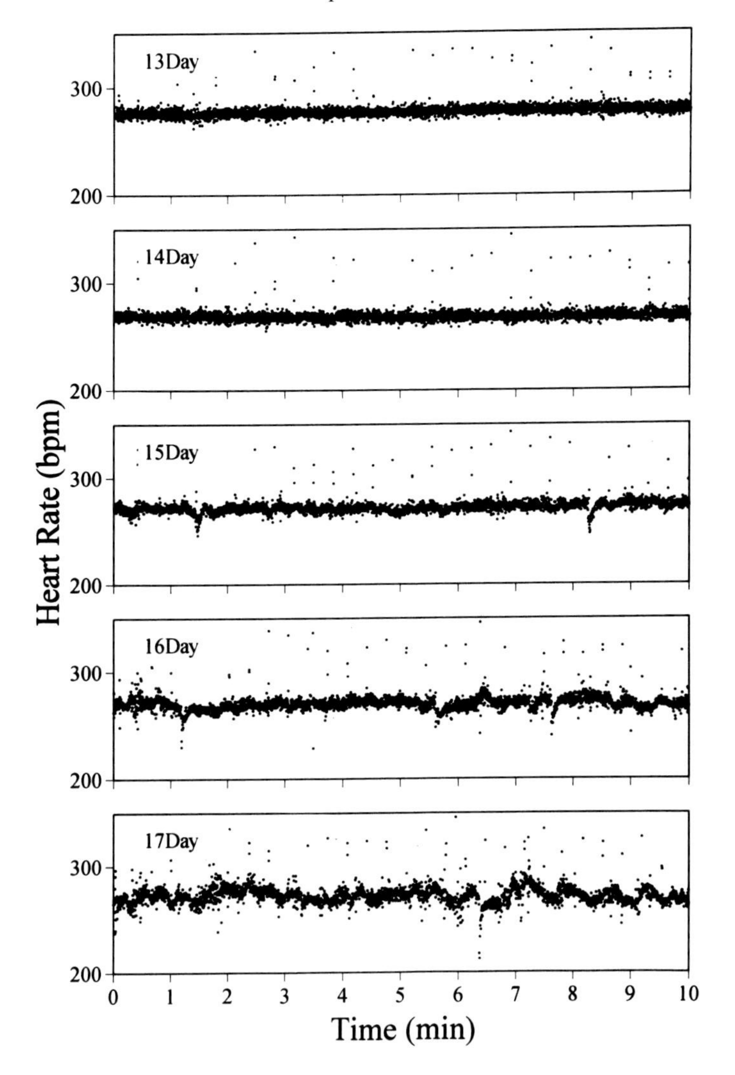

Figure 15.8 Instantaneous heart rate of a chick embryo determined from day 13 to day 17 of incubation. The egg was installed with a condenser microphone on the eggshell on day 12. A 10-min recording of IHR is presented here for each incubation day.

(referred to as the anterior body), ramify from the aortic arch. The right ventricle gives rise to the pulmonary arches, which proceed to the lungs through the pulmonary arteries. The pulmonary arch also communicates with the aortic arch through the ductus arteriosus. The dorsal aorta begins with a junction of the aortic arch and ductus arteriosus. The tissue mass supplied with blood through the branches from the dorsal aorta is referred to as the posterior body. The allantoic arteries bifurcate from the sciatic arteries, supplying blood to the chorioallantoic gas exchanger. The blood supply from the heart is partitioned to the four compartments in the illustrated model (Figure 15.10).

Because the lungs do not function as a gas exchanger until pipping, late embryos have an anatomical shunt of the ductus arteriosus connecting the pulmonary arteries and the dorsal aorta and also an intracardiac shunt between the two atria, the interatrial foramina. Figure

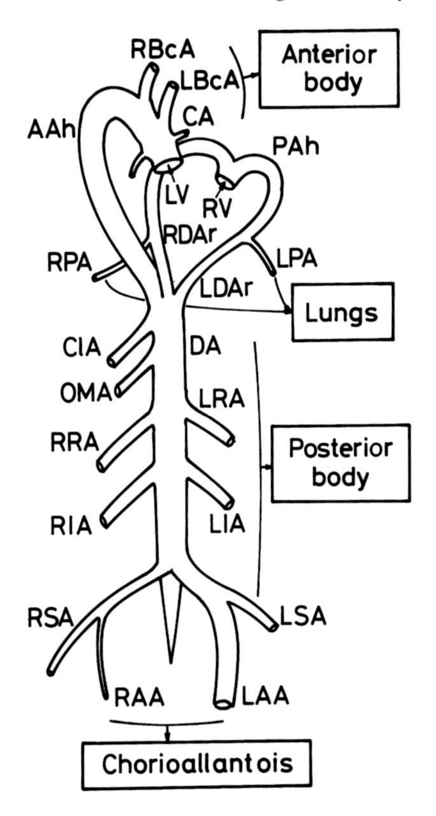

Figure 15.9 The principal arterial system of the late embryo, supplying blood to four compartments: anterior body, posterior body, lungs, and chorioallantoic membrane (AAh = aortic arch, CA = coronary artery, CIA = coeliac artery, DA = dorsal aorta, OMA = omphalomesenteric artery, PAh = pulmonary arch, RAA and LAA = right and left allantoic arteries, RBcA and LBcA = right and left brachiocephalic arteries, RDAr and LDAr = right and left ductus arteriosus, RIA and LIA = right and left iliac arteries, RPA and LPA = right and left pulmonary arteries, RRA and LRA = right and left renal arteries, RSA and LSA = right and left sciatic arteries, RV and LV = right and left ventricles). (Reprinted from Tazawa & Takenaka, 1985, with permission of the publisher.)

15.10 illustrates the intra- and extracardiac shunts in relation to the four heart chambers. Oxygenated blood in the allantoic vein from the chorioallantois returns to the right atrium through the posterior venous channel, which also delivers deoxygenated blood from the posterior body and nutrient-rich blood from the yolk. The aperture of the posterior venous channel faces toward the interatrial foramina, and that of the anterior vena cava is directed toward the right atrioventricular canal (White, 1974). Thus, oxygenated blood from the posterior venous channel preferentially enters the left atrium, and the deoxygenated blood from the anterior body is directed to the right ventricle. A small admixture of these two bloodstreams may occur in the atria. The dotted and broken lines indicate possible intracardiac shunting. The fraction of the posterior venous channel blood returning to the pulmonary circulation is designated by x. Since it includes the oxygenated blood from the gas exchanger that proceeds to the pulmonary arch, x is defined as the fraction of functionally shunted blood. And y is the fraction of the deoxygenated blood returning to the systemic tissues.

The cardiovascular system of the late chick embryo presented in Figure 15.10 is redrawn as a simplified circulatory model in Figure 15.11 (Tazawa & Takenaka, 1985). The chorioal-

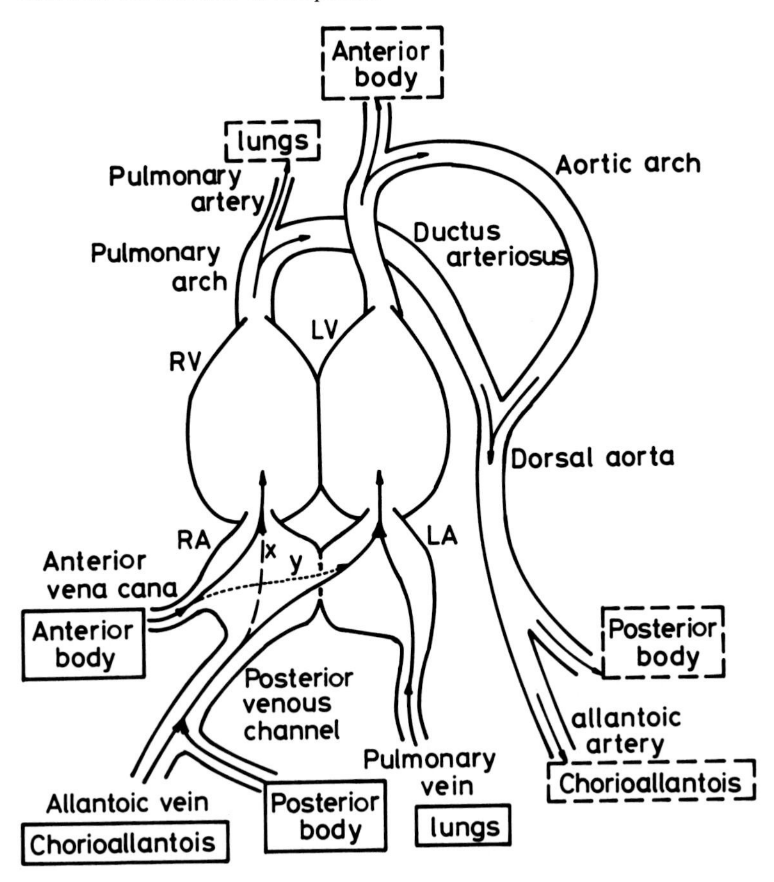

Figure 15.10 The intra- and extracardiac shunts in relation to the four compartments. (Reprinted from Tazawa & Takenaka, 1985, with permission of the publisher.)

lantois takes up oxygen from atmospheric air and consumes some amount of O_2. The remaining O_2 is consumed by the nonfunctional lungs, the anterior body, and the posterior body in fraction (*u, v,* and *w,* respectively). A fraction, *l,* of the right ventricular output to the pulmonary arches proceeds to the pulmonary artery, and thus a remaining fraction $(1 - l)$ is shunted through the ductus arteriosus to the dorsal aorta. The left ventricle delivers its output by a fraction, *m,* to the anterior body. A fraction, *n,* of the dorsal aortic blood flow perfuses the posterior body. Fluid-flow and mass-flow equations are derived for this circulation model by Tazawa and Takenaka (1985). From the equations, the functional intracardiac shunting fraction (*x* and *y*) and blood flow distribution coefficients (*l, m,* and *n*) are determined only by the oxygen content involved in the model and O_2 consumption coefficients (*u, v,* and *w*). The experimental data on blood O_2 content and several assumptions give intracardiac shunting fractions ($x = 37\%$, $y = 6\%$), cardiac outputs ($\dot{Q}_l = 10.4$ ml·min^{-1}, $\dot{Q}_r = 12.7$ ml·min^{-1}), and associated blood flows for a 16-day embryo as shown in Figure 15.11. The venous return to the left atrium from the lungs is about 2 ml·min^{-1}, and the total venous return to the right atrium reaches about 21 ml·min^{-1} (i.e., 4 ml·min^{-1} from the choriollantoic gas exchanger, about 9 ml·min^{-1} from the anterior body, and 8 ml·min^{-1} from the posterior body). However, the right ventricular output is nearly halved by intracardiac shunting. Consequently, both the right and left cardiac outputs become similar in spite of the large difference

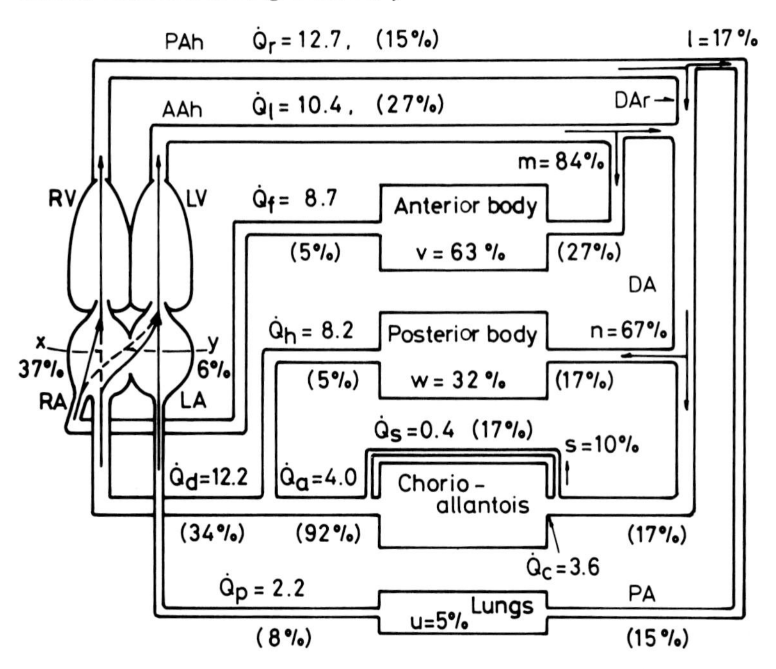

Figure 15.11 The simplified circulatory model of the late embryo ($\dot{Q}_r$ and $\dot{Q}_l$ = right and left ventricular output: $\dot{Q}_p$ = pulmonary blood flow; $\dot{Q}_a$ = allantoic blood flow; $\dot{Q}_f$ and $\dot{Q}_h$ = blood flow of anterior and posterior bodies; $\dot{Q}_s$ = blood flow of allantoic shunt; $\dot{Q}_c$ = blood flow of chorioallantoic capillary; DAr = ductus arteriosus; PA and PV = pulmonary artery and vein; 1, m, and n = blood flow distribution coefficients; u, v, and w = fractions of oxygen consumption taken by the nonfunctional lungs, the anterior body, and the posterior body, respectively). Blood flows are in milliliters per minute. Other abbreviations are the same as in Fig. 15.9. The percentages shown in parentheses indicate blood O_2 saturation. (Reprinted from Tazawa and Takenaka, 1985, with permission of the publisher.)

between the venous returns to the right and left atria. In addition, no less than 80% of the right ventricular output bypasses the lungs through the ductus arteriosus.

Using dye injection and O_2 content determinations, White (1974) suggested that a large part of the oxygenated blood from the chorioallantois flows through the interatrial foramina to the left heart; he estimated the cardiac output to be about 18 ml·min^{-1} in a 16-day embryo. Rahn, Matalon, and Sotherland (1985) calculated a cardiac output of 13 ml·min^{-1} for a 16-day embryo on the basis of a two-compartment model.

Concluding remarks and perspective

Because the gas exchange of avian embryos takes place by molecular diffusion, chicken eggs have been used as an experimental animal model to elucidate the mechanism of diffusive gas exchange through the eggshell. Consequently, the respiratory functions of the blood and the chorioallantoic capillary plexus have been well investigated. In this connection, needle catheterization of allantoic blood vessels has been developed for blood sampling and for a few studies of some circulatory functions conducted with later prenatal chick embryos. Autonomic control of blood pressure and HR of chick embryos is still controversial, and needle catheterization of the allantoic artery and vein will allow the investigation of autonomic control during the last half of prenatal development while maintaining gas exchange through the eggshell. Direct measurement of the allantoic blood flow as has been done in a separate experiment (Tazawa, Lomholt, & Johansen 1985) and simultaneous measurement of allantoic arterial blood pressure will elucidate the pressure–flow relationship of the allantoic circulation in late embryos developing within the eggshell. During the last stages of prenatal development and during the paranatal period, embryos switch their respiration from chorioallantoic diffusive gas exchange to convective and diffusive processes by the lungs. Circulatory functions must change dramatically during this period. Injection of isotope-labeled microspheres through the allantoic venous catheter to the heart will contribute to an investigation of intra- and extracardiac shunts before and during pipping behaviors prior to closure of allantoic circulation.

It is difficult to measure cardiac output and cardiovascular functions of embryos while maintaining adequate gas exchange through the eggshell and chorioallantois. Such experiments give only limited information on the cardiovascular functions. More informative data depend on exposure of the embryo to the atmosphere. Once the embryo is exposed to air, the transparency of early embryos and access to the cardiovascular system through a window opened in the eggshell are obvious experimental advantages, and multiple parameters of cardiovascular functions can be measured. Over the past 15 years, early cardiac morphogenesis in the chick has been studied in the early embryo to provide an experimental animal model to investigate the relationship between cardiovascular function and form. The use of computerized cinematography and a video imaging system, together with the servonull pressure system and pulsed Doppler velocity meter, will further improve the study of cardiovascular physiology of early embryos under the condition of direct exposure to air (e.g., Keller et al., 1991a; Burggren & Fritsche, 1995).

The embryonic HR is a parameter that can be measured noninvasively or semi-invasively in eggs of various sizes. The developmental patterns of embryonic HR have been determined for several species of altricial and precocial birds (Tazawa et al., 1991; Tazawa, Kuroda, & Whittow, 1991; Tazawa & Whittow, 1994; Tazawa, Watanabe, & Burggren, 1994). It seems that individual species have their own averaged developmental patterns, probably relating to metabolic requirements and other unknown factors, although there are some intraspecies differences. The developmental patterns of O_2 consumption are distinctly different between the precocial and altricial species, and HR patterns also differ. Comparison between the developmental patterns of O_2 consumption and HR is evidenced by a developmental pattern of O_2 pulse (i.e., O_2 consumption per single heartbeat). This is another parameter to be determined in the precocial and altricial species for a comparative study. The measurement of embryonic HR and O_2 pulse in both species will give the allometric relationship of HR (and O_2 pulse) with fresh egg mass, showing similarity or difference between the species.

The embryonic HR determined while maintaining gas exchange through the eggshell functions as a marker for alterations in cardiovascular function following environmental stresses, including excessive egg turning and altered ambient temperatures and gases. Because in artificial incubation the eggs must be turned several times a day, egg turning is not an environmental stress, but lack of turning and excessive turning or shaking and vibration may stress developing embryos. Although egg turning and lack of turning are known to influence embryonic HR (Vince, Clarke, & Reader, 1979; Pearson et al., 1996), the influence of excessive turning as an external stress needs to be studied. The embryonic HR changes with alteration in ambient temperature. The measurement of HR of embryos exposed to altered ambient temperatures showed that the chick embryos could survive an

8°C environment without beating hearts for long periods, but they could not survive increased ambient temperatures that led to internal egg temperatures reaching 46–47°C (Tazawa & Rahn, 1986; Tazawa et al., 1992; Ono, Hou, & Tazawa, 1994). A recent preliminary study on HR response to altered O_2 showed that embryonic tolerance to environmental hypoxia differed between developmental stages (Komoro et al., 1995). These investigations measured the embryonic HR as a marker, and additional study will elucidate further embryonic responses to altered ambient temperatures and gases.

Recently, developmental patterns of HR of sibling embryos were measured in altricial birds and it was found that they were similar within clutches and dissimilar between clutches (Tazawa, Watanabe, & Burggren, 1994). Burggren, Tazawa, and Thompson (1994) suggested that measurements of developmental patterns of sibling HR will give a clue to genetic and maternal environmental influences on embryonic physiology. The developmental patterns of embryonic HR were also found to differ between normal embryos and embryos with C locus mutations (Howe, Burggren, & Warburton, 1995). Because the embryonic HR indicates genetic anomalies of developing mutative embryos, further application of HR measurement as a marker is expected to be fruitful in the study of the developmental physiology of avian embryos.

Generally, measurement of a single variable does not yield the entire picture of cardiovascular function, and almost all current studies on the embryonic cardiovascular system use chick embryos. Because the chick is one of the most derived species through artificial selection, its development may be different from other avian species. Investigation based on measurements of multiple variables is needed in other avian species for comparative study of the developing cardiovascular functions. The ontogeny of cardiovascular regulation must be studied to determine not only the onset of regulation but also how the regulation changes with embryonic development. The acute and chronic effects of environmental factors on cardiovascular development remain to be studied in avian embryos.

16

Mammalian cardiovascular development: Physiology and functional reserve of the fetal heart

KENT L. THORNBURG, GEORGE D. GIRAUD, MARK D. RELLER, AND MARK J. MORTON

Functional adaptations of fetal ventricles

Embryonic and fetal circulations

As cardiac myocytes differentiate, they take on the well-described characteristics of striated muscle that underlie the function of the newly formed heart (Faber, Green, & Thornburg, 1974; Keller, Hu, & Tinney 1994). Thus, as shown in previous chapters, the embryonic heart displays many functional features in common with the fully mature heart. The embryonic mammalian heart becomes a four-chambered muscular pump at the completion of the cardiac septation stage and remains so throughout the animal's life. However, from the time of septation, the right and left ventricles operate as parallel pumps in the embryo and fetus to accommodate use of the placenta as the organ for nutrient and gas exchange for the duration of prenatal life.

Because the fetus depends on the placenta for oxygen acquisition, a significant portion of the cardiac output must pass through the placenta for gas exchange. Freshly oxygenated blood must then be distributed to vital organs. The fetal lungs obviously cannot be used for oxygenation before birth, and accordingly, they receive only a small fraction of the cardiac output (Heymann, Creasy, & Rudolph, 1973). The intrauterine circulatory arrangement is similar to the adult circulation but has four unique shunts: (1) placenta, (2) ductus venosus, (3) foramen ovale, and (4) ductus arteriosus (Figure 16.1). Blood from the fetal body is delivered to the placenta via a pair of umbilical arteries. In most species (the guinea pig is an exception), the oxygenated umbilical vein blood enters the fetal inferior vena cava at the site of the liver by passing through the ductus venosus. As caval blood reaches the heart, it is preferentially directed into the left atrium rather than the right by streaming across the foramen ovale, a flap valve in the interatrial septum. Therefore, well-oxygenated placental blood is preferentially shunted to the left ventricle, from which the coronary vessels, brain, and upper body are supplied. Deoxygenated blood returning from the upper body via the superior vena cava enters the right atrium, travels through the right ventricle, and is delivered from the pulmonary artery to the descending aorta via the ductus arteriosus. All four of these shunts normally close soon after birth.

The combined cardiac output (both chambers of the parallel pump combined) of the fetus is about 500 $ml \cdot min^{-1} \cdot kg^{-1}$. In contrast, the resting output of the single series chamber of the adult heart is just under 100 $ml \cdot min^{-1} \cdot kg^{-1}$. In the fetus, about 40% of the cardiac output is directed to the placenta, and 7% and 10% to the lung and heart, respectively. Table 16.1 shows the blood flow distribution to various fetal organs per unit weight.

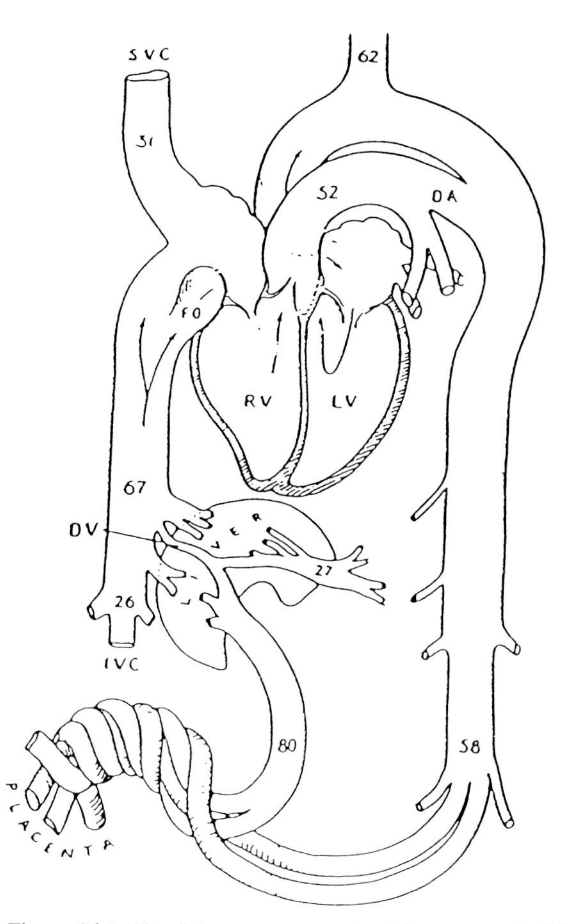

Figure 16.1 Circulatory arrangement of the mature fetal lamb. Numbers indicate mean oxygen saturation (%). Note the higher saturation in the umbilical veins than in the fetal aorta (DA = ductus arteriosus, DV = ductus venosus, FO = foramen ovale, IVC = inferior vena cava, LV = left ventricle, RV = right ventricle, SVC = superior vena cava). (From Born et al., 1954.)

Table 16.1. *Regional blood flow in near-term fetal sheep*

Tissue	Blood flow ($ml \cdot min^{-1} \cdot [100\ g]^{-1}$)
Lung	159
Left ventricle	265
Brain	151
Carcass	15
Adrenal	397
Brown fat	66
Kidney	245
Placenta	138

Source: Teitel, 1988. Reprinted with permission from Thornburg and Morton, 1994.

For the first 70 years of this century, it was believed that the fetal heart ventricles (right and left) developed similarly in shape and volume. Pohlman commented on this arrangement in 1909, as did Sir Geoffrey Dawes in his classic text *Foetal and Neonatal Physiology* (1968). However, up to the time that Dawes's book was written, the individual fetal heart ventricles had been little studied as individual muscle pumps. With the development of the technology required to instrument acutely and chronically the circulatory system in fetal sheep, it became possible to study heart function in unanesthetized fetal sheep. By the 1980s, a tremendous amount of research had been directed toward understanding the mechanical function of the adult heart in many animal models and in humans, and the primary determinants of stroke volume were known (Braunwald, Ross, & Sonnenblick, 1976). These determinants also apply to the fetal heart. More recently, experiments have been designed to determine the separate roles of the two ventricles in controlling cardiac output in the fetus (Anderson et al., 1981; Thornburg & Morton, 1983, 1986; Anderson et al., 1984; Morton, Pinson, & Thornburg, 1987; Reller et al., 1987; Pinson, Morton, & Thornburg, 1987, 1991).

Stroke volume determinants in the mammalian fetus

In the nongrowing adult heart there are three primary determinants of stroke volume (the volume of blood that is ejected from the ventricle each beat). One determinant is *preload.* Preload is the stretching force on the myofibrils per unit cross-sectional area at the end of diastole, just prior to muscle contraction. The increase in stroke volume following increasing preload is known as the Frank–Starling mechanism (Braunwald, Ross, & Sonnenblick, 1976). Mean atrial pressure, end-diastolic ventricular pressure (filling pressure), and end-diastolic ventricular volume are often used as indices of preload. A second determinant of stroke volume is *afterload,* the force per unit cross-sectional area of myocardium during contraction. Stroke volume is depressed by acute increases in afterload such as when arterial pressure is increased. A major determinant of afterload in the fetus is the total impedance of the vascular tree during ejection. Total impedance is the sum of the mean vascular resistance plus the vascular impediment due to pulsatile pressure and flow. A third factor that regulates stroke volume is *contractility,* the intrinsic strength of contraction that is independent of preload or afterload. For fetuses that are rapidly growing, a fourth factor must be considered: *chamber enlargement* by growth. As the heart chamber grows larger, it will eject a larger stroke volume for each beat if the fraction of chamber volume ejected is constant. During fetal life of sheep the heart grows in approximate proportion to the fetal body, about 3% per day.

The study of fetal sheep has been instrumental in advancing knowledge of the function of the prenatal heart and vascular system (Assali, Holm, & Sehgal, 1962; Meschia, Battaglia, & Bruns, 1967; Dawes, 1968; Rudolph & Heymann, 1970; Friedman, 1972; Kirkpatrick, Covell, & Friedman, 1973). The development of techniques that provided for the instrumentation of fetuses in utero allowed investigators to compare the stroke volumes of the right and left ventricles (Heymann, Creasy, & Rudolph, 1973). To everyone's surprise, the right ventricular stroke volume was found to exceed the left ventricular stroke volume by about 30–50% (Heymann, Creasy, & Rudolph, 1973; Anderson et al., 1981). This is very different from the adult, in which stroke volumes are necessarily equal. Identification of the stroke volume difference between the fetal ventricles raised several questions: Could preload (filling pressure) be greater for the right ventricle than for the left? Was it possible for contractility to be preferentially greater for the right ventricle than for the left? Was right ventricular afterload so low that the ejection fraction for the right ventricle exceeded the left? Gilbert (1982) showed that the fetal whole-heart function curve had a reproducible shape and that cardiac output increased steeply with increases in mean atrial pressure until a pressure was reached where it flattened into a plateau. Further studies were designed to investigate the above questions in detail by generating "ventricular function curves" in fetal sheep for each heart ventricle separately (Thornburg & Morton, 1983, 1986). The function curve was then used to describe the relationship of ventricular output at different levels of preload (mean filling pressure).

Studies that used electromagnetic flow sensors on the ascending aorta and pulmonary artery to measure simultaneously output from both ventricles revealed that the function curves for the ventricles were similar to each other in shape (Thornburg & Morton, 1983, 1986; Reller et al., 1987) and similar to the biventricular curves shown by Gilbert (1982). Figure 16.2 shows near-term fetal ventricular stroke volumes plotted as a function of the mean atrial pressure (which was used as an index of preload). This figure demonstrates that right ventricular stroke volume always exceeds left ventricular stroke volume in the fetus at any given filling pressure. Because the experimental design ensured that mean atrial pressures were equal for the two ventricles, preload differences do not explain the right ventricular dominance in stroke volume. Furthermore, because the mean blood pressure in the pulmonary artery is nearly equal to the mean pressure in the ascending aorta, differences in afterload is an unlikely explanation for the large difference between the two ventricular stroke volumes. It is unlikely that the ventricles have different levels of contractility (Friedman, 1972). How then can the stroke volumes be so different if the classic determinants of function do not differ? The remaining possibility is anatomical configuration. Therefore, fetal hearts were arrested in diastole with potassium chloride, and the passive pressure–volume characteristics of each ventricle were determined in vitro to define the anatomic features of the two fetal ventricles (Pinson, Morton, & Thornburg, 1987). In addition, fetal hearts were fixed at their normal resting transmural pressures in order to study their anatom-

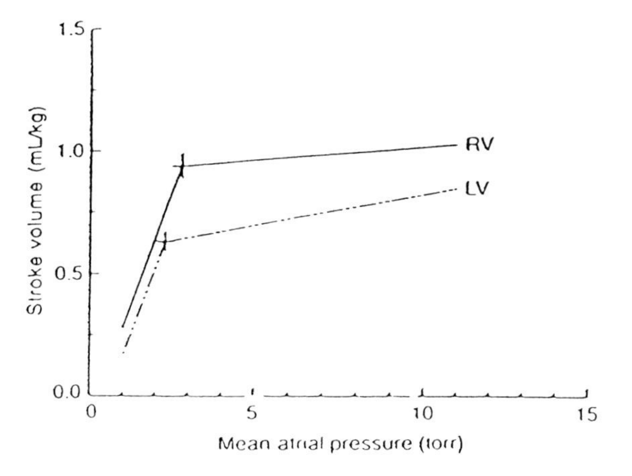

Figure 16.2 Average function curves are shown for the right and left ventricles of the fetal sheep. Both ventricles show a typical shape with rapidly increasing stroke volume at low increases in atrial pressure and a plateau where stroke volumes increase very little with further increases in mean atrial pressure. The break points of the two curves are not significantly different from resting values of mean atrial pressure and mean stroke volume. (From Reller et al., 1987.)

ical differences. Casts were made of the individual ventricles, and their radii of curvature, and wall thicknesses in cross section were determined. Three significant features were found: (1) the end-diastolic volume of the right ventricle is much larger than that of the left ventricle; (2) the radius of curvature of the right ventricle is significantly larger than that of the left ventricle; (3) the wall thicknesses of the two ventricles are approximately equal.

These anatomical findings were very helpful in explaining differences in stroke volume between the two ventricles. It became clear that the right ventricular stroke volume exceeds the left primarily because the right ventricular chamber is larger than the left and ejection fractions are equal. Most important, however, the radius to wall thickness ratios of the two ventricles were found to be quite different. Although the ventricular free wall thicknesses are about the same, the circumferential radius of curvature (r) to wall thickness (h) ratio is nearly twice as great for the right ventricle as it is for the left. The law of Laplace (Equation 1) relates wall stress (S_w) to transmural pressure (P_t) and the radius to wall thickness ratio (r/h).

$$S_w = (P_t/2)(r/h). \qquad (1)$$

Note that wall stress is directly proportional to its r/h. Equation 1 predicts that for a given transmural pressure, an increase in the r/h ratio will result in a proportional increase in wall tension. In the sheep fetus, the r/h ratio was found to be 4.6 for the mid–right ventricle and 2.6 for the left ventricle.

From the apparently high stress in the wall of the fetal right ventricle, one can predict that the right ventricle is at a mechanical disadvantage compared with the left ventricle. Experiments proved this to be true (Reller et al., 1987). Figure 16.3 shows changes in stroke volume for the right and left ventricles, normalized to 100% at the resting level. Pressures in the pulmonary artery and aorta were increased simultaneously by inflating an occluder around the descending aorta. As the pressure was simultaneously increased, the stroke volume of both ventricles decreased as predicted by the biomechanical effects of increasing afterload. For example, as the pressure load to both ventricles is increased by 20 mm Hg (fetal mean arterial pressure is normally 45 mm Hg), stroke volume from the left ventricle decreases by only 10%. However, right ventricular stroke volume decreases more dramatically, by 50%. Thus, the right ventricle is at a severe mechanical disadvantage compared with the left, as indicated by the finding that right ventricular stroke volume drops so much more severely than left stroke volume for equal increases in arterial pressure.

Functional changes with chronic pressure loading

Because the fetal heart is a structurally dynamic tissue, it will respond to chronic changes in pressure load by thickening its wall (similar to adult cardiac hypertrophy). This would lead to a decrease in the radius to wall thickness ratio. Such a response would allow a test of the hypothesis that the r/h ratio correlates inversely with increased performance (decreased pressure sensitivity). After 10 days of pressure loading produced by chronic pulmonary artery constriction, right ventricle wall thickness increased so that the r/h ratio of the right ventricle decreased dramatically from 4.6 to 3.2, approaching the 2.6 ratio of the normal fetal left ventricle (Pinson, Morton, & Thornburg, 1991). Figure 16.4 shows that following pressure loading, the fetal right ventricle, with its decreased r/h ratio, is better able to eject against increases in arterial pressure. In fact, its function is nearly comparable to that of the left ventricle. These data support the hypothesis that the mechanical function

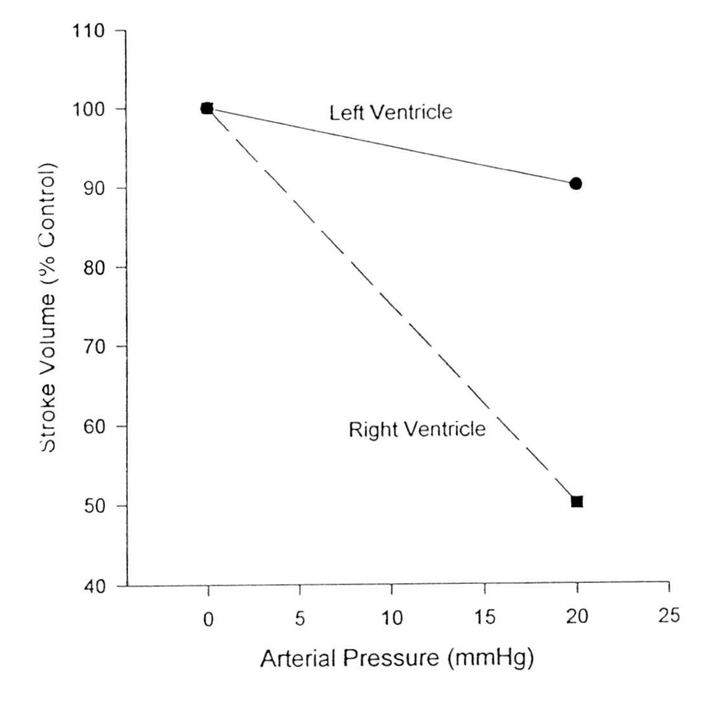

Figure 16.3 Simultaneous average responses of the right and left ventricles to increased arterial pressure ($n = 9$). Stroke volume is expressed as a percentage of a control volume; and arterial pressure, as the increment above control. The linear coefficient for each ventricle was calculated, and the average slope was forced through 100% on the *y*-axis. Lines extended through the range studied. Right ventricular pressure sensitivity (–2.5% ± 1.4% stroke volume·[mm Hg]$^{-1}$) was more than five times the left ventricular pressure sensitivity (0.5% ± 0.7% stroke volume·[mm Hg]$^{-1}$; $P < .001$). (From Reller et al., 1987.)

of the myocardium is partially determined by the anatomical architecture and the resultant radius to wall thickness ratio.

Adaptations of the fetal coronary tree

Coronary blood flow regulation in immature myocardium

The coronary artery is the source of oxygenated blood for the heart. In general, the left ventricular free wall of the sheep is supplied by a left coronary artery, and the right ventricle by a right coronary artery. General hemodynamic principles can be used to understand the basic determinants of coronary flow in the fetus and in the adult. Coronary flow (Q_c) is determined by the mean pressure in the coronary artery (P_a) minus the pressure in the coronary vein (coronary sinus or right atrium, P_{ra}) divided by the mean resistance to flow (R_c). The term $P_a - P_{ra}$ is known as the driving pressure. In equation form,

$$Q_c = (P_a - P_{ra})/R_c. \qquad (2)$$

In the fetus, as in the adult, coronary flow is greatest during diastole because muscular compression retards flow during contraction (Berne & Rubio, 1979).

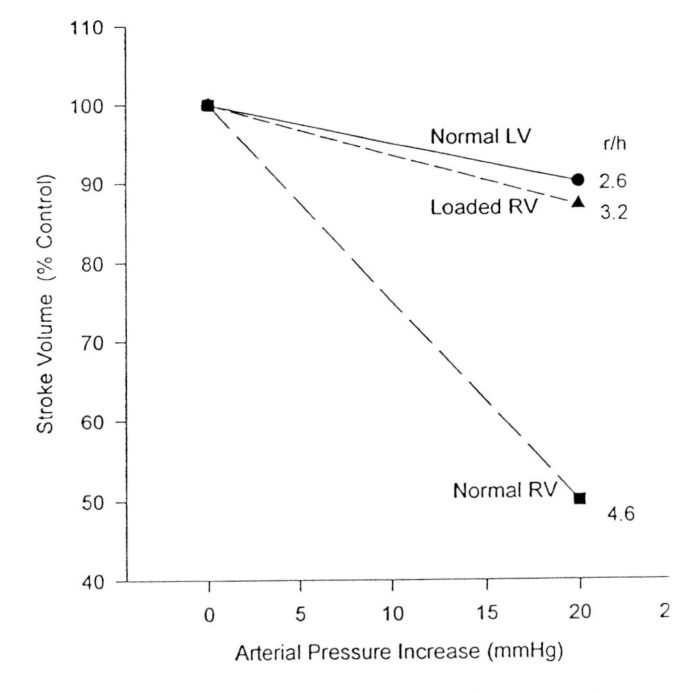

Figure 16.4 Left ventricular and right ventricular stroke volume as percentages of a control volume plotted as a function of increased arterial pressure (control data from Figure 16.3) showing decreased pressure sensitivity following 10 days of chronic loading of the right ventricle. Note that *r/h* (radius to wall thickness ratio), which is normally 4.6 for the right ventricle, was decreased to 3.2 following pressure loading.

It is not surprising that blood flow to the heart muscle is different during fetal life than during adult life. Fetal mean arterial pressure (50 mm Hg) is about half of the adult value. Equation 2 predicts a lower driving pressure and therefore a lower flow. In addition, fetal arterial blood (8 ml $O_2 \cdot dl)^{-1}$ contains only about half of the oxygen content of fully saturated adult blood. Therefore, if fetal and adult heart oxygen requirements were the same (per gram), one would expect that flow to the fetal heart muscle (per gram) would need to be much higher than adult levels to compensate for lower fetal driving pressure and blood oxygen content.

Many studies in adult animals suggest that myocardial blood flow is tightly linked to the metabolic needs of the myocardium. For the adult it has long been known that physiologic alterations in myocardial oxygen requirement, like increased heart rate or contractile work, are accompanied by significant increases in coronary flow via locally controlled vasodilation (autoregulation). Less is known for the fetus. In the 1980s, Fisher, Heymann, and Rudolph (1982a, 1982b) used radioactive microspheres to measure blood flow in the fetal heart. They found that resting fetal blood flows were significantly higher per unit tissue volume than reported adult flows. Resting blood flow for adult dogs (and humans) is about 80–100 $ml \cdot min^{-1} \cdot (100\ g)^{-1}$ and increases to about 400 $ml \cdot min^{-1} \cdot (100\ g)^{-1}$ following chemical dilation at normal arterial pressure. At rest, fetal myocardial flows per 100 g are more than twice adult resting values and increase by four times to about 800 $ml \cdot min^{-1} \cdot (100\ g)^{-1}$ after chemical dilation at fetal arterial pressure.

Oxygen consumption and coronary flow

The differences in anatomic configuration and predicted wall stresses between the right and left ventricular chambers of the fetal sheep heart (Pinson, Morton, & Thornburg, 1987) suggest that they might have different oxygen consumption needs and different coronary blood flows. Wall stress is a known determinant of oxygen consumption. Because fetal right ventricular wall stress and wall mass are higher than those of the left in vitro, the higher wall stress found in the right ventricle predicts that the oxygen consumption on the right side of the heart will be larger than on the left side. Unfortunately, this prediction is difficult to test in fetal sheep. Myocardial oxygen consumption ($M\dot{V}_O2$) can be determined for the left ventricle as the product of left ventricular coronary blood flow and the difference in oxygen content between blood entering the myocardium $[O_2]_A$ and leaving the myocardium $[O_2]_V$:

$$M\dot{V}_{O_2} = \dot{Q}([O_2]_A - [O_2]_V). \qquad (3)$$

But because it is not anatomically possible to sample blood leaving right ventricular muscle, the oxygen content difference across right ventricular myocardium cannot be measured. Thus, oxygen consumption on the right side can only be estimated from other indices.

However, right ventricular myocardial flows can be measured. Fisher, Heymann, and Rudolph (1982a, 1982b) were able to compare myocardial blood flows between the ventricles of near-term fetal sheep. They found that right ventricular blood flow (per gram) is significantly higher than left ventricular flow. This finding fits nicely with the higher predicted wall stress differences between ventricles and with the prediction that oxygen consumption is higher in the right ventricle. Smolich and colleagues (1989) have also shown that right ventricular myocyte cross-sectional areas and capillary densities are larger in the right ventricle than in the left. It appears that the fetal right ventricle is adapted anatomically and physiologically to function with higher wall stress and higher blood flow than is the left ventricle.

Until recently, there was very little information to explain the mechanisms by which the fetal myocardium regulates its perfusion. The extraordinarily high fetal myocardial flows raise questions. For example, with right ventricular blood flow and wall stress being so high compared with their left ventricular counterparts, might not the right ventricle be especially vulnerable to inadequate coronary flow during hypoxic episodes or episodes of high oxygen demand? In keeping with this question, fetal right ventricular function is known to be seriously impaired under intense working conditions, such as with acute pressure loading. As pulmonary artery pressure is increased by inflating an occluder around the pulmonary artery, right ventricular stroke volume decreases. The initial rate of drop in stroke volume is the expected mechanical response to increased afterload (as pressure increases) according to the load–rate of shortening relationship of striated muscle, as evident in Figure 16.3. However, the finding that stroke volume drops dramatically to zero and ventricular function fails at a critical arterial pressure (at about 80 mm Hg, the "toleration point") is not obviously explained by simple muscle mechanics. The toleration point is the maximal pressure the right ventricle can generate and where stroke volume falls as further vessel restriction is applied (Reller et al., 1992b). Once determined, the point is reproducible for any given fetal heart.

Based on these findings, it can be hypothesized that the loss of ventricular function is the result of a coronary flow level inadequate to meet increasing myocardial oxygen demand. To test this hypothesis, Reller and colleagues (1992b) prepared fetal sheep to see whether coronary flow could increase even further (coronary reserve) at the toleration point when the right ventricle was working maximally. Maximum coronary flow reserve was estimated by generating dose response curves for adenosine, a coronary vasodilator, infused into the left atrium.

Table 16.2 compares the mean pulmonary artery pressure at increments of inflation of a pulmonary artery occluder and the mean pressure from the carotid artery. Note that while pulmonary artery pressure increases by about 20 mm Hg, mean pressure in the carotid artery is not altered. Heart rate does not change in these experiments except at the highest pressure. Fetal arterial blood gas pressures and pH did not change during these experiments. The product of heart rate times blood pressure is often used to estimate the workload (and, therefore, oxygen consumption). The rate–pressure product increases for the right ventricle from about 11,000 to 18,000 beats·mm Hg·min^{-1} over the pressure range studied whereas the increases for the left ventricle did not reach statistical significance.

Blood flow to the right ventricle increases dramatically with increases in workload placed upon the right ventricle (Reller et al., 1992b; Figure 16.5). Two important points should be noted regarding these data: (1) right ventricular flow increases in proportion to right ventricular workload, and (2) even though little if any extra workload was given to the left ventricle, left ventricular flow was maintained as a constant portion of right ventricular flow (67%) throughout the procedure. The maximal flow obtained during loading was compared with the maximal flow during dilation with adenosine in order to determine whether all right ventricular flow reserve was "used up" during maximal pressure loading. Whereas right ventricular blood flow can increase approximately three times during loading, right ventricular coronary flow can increase by nearly four times during adenosine administration. Furthermore, when adenosine is administered during pressure loading at the toleration point, the flow reserve is still present.

These data indicate that the sharp drop in right ventricular stroke volume during maximal workload at the toleration point is not likely due to ischemic reduction in myocyte function. Instead, the coronary tree is able to dilate far beyond the level stimulated by this maximal workload. These experiments do not address the question, however, of whether the fetal heart has "access" to the entire reserve under loading conditions (without chemical assistance). The fact that there was no significant decrease of the subendocardial–epicardial flow ratio with increasing arterial pressure, and no increase in right atrial pressure, argues

Table 16.2. *Comparison of hemodynamic variables at baseline and during incremental pulmonary arterial occlusion*

	Mean PAP pressure			
Variable	53	57	63	71
Systolic PAP (mm Hg)	76	84	92	99
RVSV (1.26 ml·kg = 100%)	100%	93%	73%	52%
Systolic CAP (mm Hg)	58	57	60	58
Mean CAP (mm Hg)	46	46	49	46
Right atrial pressure (mm Hg)	4	3	5	3
Heart rate (beats·min^{-1})	141	147	155	182
RV (*RP*)	11,000	13,000	14,000	18,000
LV (*RP*)	8,000	8,000	9,000	11,000

Note: PAP = pulmonary arterial pressure, RVSV = right ventricular stroke volume, CAP = carotid arterial pressure, RV and LV (*RP*) = right ventricular and left ventricular heart rate–pressure product.

Source: Reprinted with permission from Reller et al., 1992b.

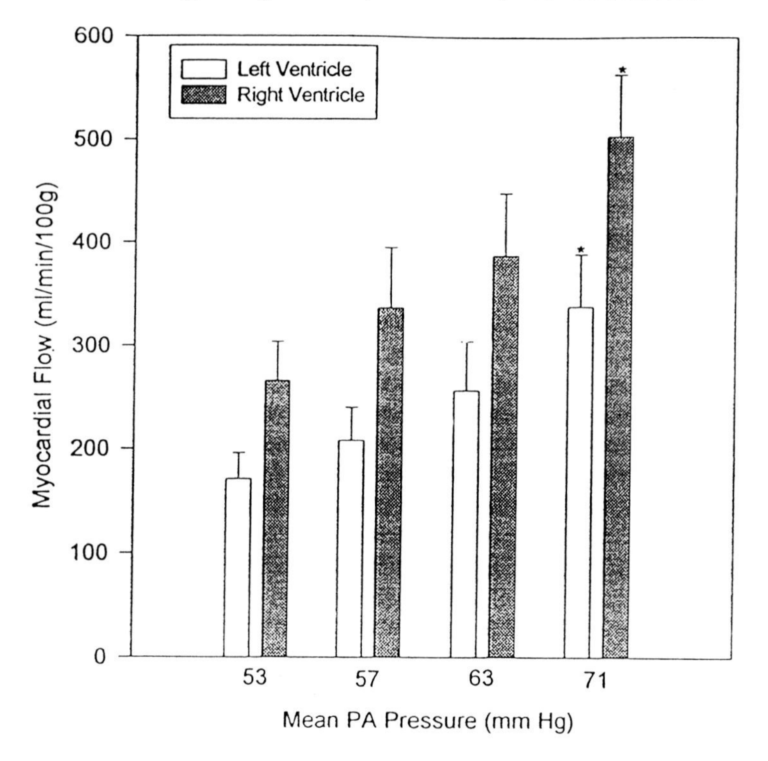

Figure 16.5 Myocardial blood flow in fetal sheep (n = 7) at four workloads indicated by mean pulmonary arterial pressure. The highest pulmonary artery pressure (71 mm Hg) is the average toleration point for the right ventricles studied. Note that left ventricular flow goes up in proportion to right ventricular flow, even though only the right ventricle is made to eject against an increasing pressure.

against the hypothesis that coronary flow constraint is the explanation for the loss of contractile function under acute and severe pressure-loading conditions.

It is important to emphasize that coronary flow is highly dependent on coronary driving pressure (Equation 2). Reller et al. (1992a, 1992b) showed that coronary perfusion pressures were similar under control conditions and during loading or infusion of adenosine. Therefore, changes in coronary driving pressure cannot explain changes in coronary flow in these experiments.

Fetal cardiac blood flow during hypoxemia

The role of hypoxemia in regulating coronary blood flow is one of the most interesting aspects of blood flow regulation in the developing mammalian heart. Fisher, Heymann, and Rudolph (1982a) showed that fetal myocardial oxygen delivery could be maintained under acute hypoxemic conditions for oxygen content reduction up to 50%. Until recently, coronary regulation under conditions of chronic hypoxemia had not been studied in the fetus. However, Reller et al. (1992a) studied coronary blood flow in fetal sheep that were chronically spontaneously hypoxemic. Blood from the carotid arteries of these hypoxemic fetuses had a lower P_{O_2}, higher P_{CO_2}, lower pH, lower oxygen content, and unchanged hematocrit compared with normal counterparts. Gestational ages and weights did not differ between control and hypoxemic fetuses.

One interesting finding of this study was that myocardial blood flow in the hypoxemic fetuses (Figure 16.6, right portion) was almost exactly equal to the maximum flow with adenosine in normal fetuses. One might expect that this enormous blood flow (700 ml·min^{-1}·[100 g]$^{-1}$) was an indication that the myocardium had used up its reserve in response to low oxygen delivery and that no flow reserve would be present. However, a normal flow reserve was found to persist (Figure 16.6). This suggests a massive remodeling of the coronary tree with a dramatic increase in vascular conductance under the stimulus of hypoxemia. Experiments such as this raise another question: is it possible for coronary flow to increase during acute hypoxemic episodes as might occur when human fetuses are carried to high altitude?

To determine the role of acute hypoxemia on coronary flow in the fetus, a condition of acute maternal and fetal hypoxemia was created by allowing the ewe to breathe low-oxygen mixtures of gas. Fetal coronary blood flow was then measured (Reller et al., 1995). Hypoxemia was acute and severe so that P_{O_2} decreased to about 9 mm Hg (normal, 25 mm Hg), which resulted in an arterial oxygen content of 2 ml·dl^{-1} (normal, 8 ml·dl^{-1}). Remarkably, there were no significant changes in coronary perfusion pressure.

To assess the role of the systemic vasodilator nitric oxide (NO; previously known as endothelium-derived relaxing factor) on coronary flow, nitro L-arginine (L-NA), a competitive inhibitor of NO synthase (NOS), was infused into the left atria of fetuses (Reller et al., 1995). In about half of these fetuses, acute hypoxemia was superimposed on the L-NA infusion. Driving pressure was increased during L-NA infusion but not during adenosine infusion or during the hypoxemia period.

However, the surprising result of the Reller et al. (1995) study is the dramatic increase in left ventricular blood flow under conditions of acute hypoxemia, greatly surpassing flow under conditions of maximal chemical dilation. Blood flow to the left ventricle increased

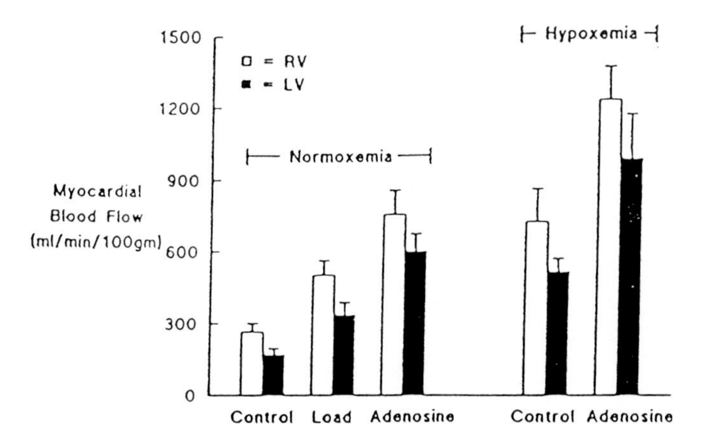

Figure 16.6 Right ventricular and left ventricular myocardial blood flow measured under conditions of normoxemia (normal blood oxygen) and hypoxemia (low blood oxygen). Flows were measured in normoxemic fetuses under control conditions, during pressure loading, at the toleration point, and during adenosine infusion. Myocardial blood flow was also measured in a group of fetuses at control conditions and during adenosine infusion. Maximal myocardial flow with adenosine in the chronically hypoxemic fetuses was significantly greater than any other measured flow. Control hypoxemic blood flow did not differ from maximal myocardial flow in normoxemic fetuses. (From Reller et al., 1992a.)

from about 230 ml·min^{-1}·(100 g)$^{-1}$ to over 1000 ml·min^{-1}·(100 g)$^{-1}$ during hypoxemia. This finding was surprising not only because flows were so high but because there were no previous reports of any physiologic stimulus increasing coronary blood flows above that produced by chemical dilation. These data contradict current dogma that chemical dilation of the coronary tree always surpasses dilation by physiologic stimulation.

In addition to the effect of NO on coronary flow reserves, the relationship between coronary blood flow and arterial blood oxygen content is sensitive to changes in the NO pathway (Reller et al., 1995). For a given oxygen content, myocardial blood flow is substantially reduced when the NOS reaction is rendered nonfunctional by L-NA (Figure 16.7). This indicates that NOS is an important determinant of coronary flow in the fetus. Also striking is the finding that during severe hypoxemia, myocardial flow cannot increase beyond the levels found during adenosine infusion if NOS has been blocked. Further research is needed to determine if NO plays a special role in the enormous flow reserve in the fetal heart during acute hypoxemia.

Summary and future directions

The fetal heart is not merely a miniature version of the adult heart. Its working cells are immature and the architecture of the fetal ventricles is different. During fetal life, the right ventricle is larger than the left and it does most of the work. This fact is important because right ventricular output serves the placenta, where oxygen is acquired. By virtue of its larger radius to wall thickness ratio, the right ventricle is more pressure sensitive than the left ventricle. Therefore, increases in fetal arterial pressure that accompany episodes of hypoxemia or stresses that stimulate epinephrine release will depress right ventricular out-

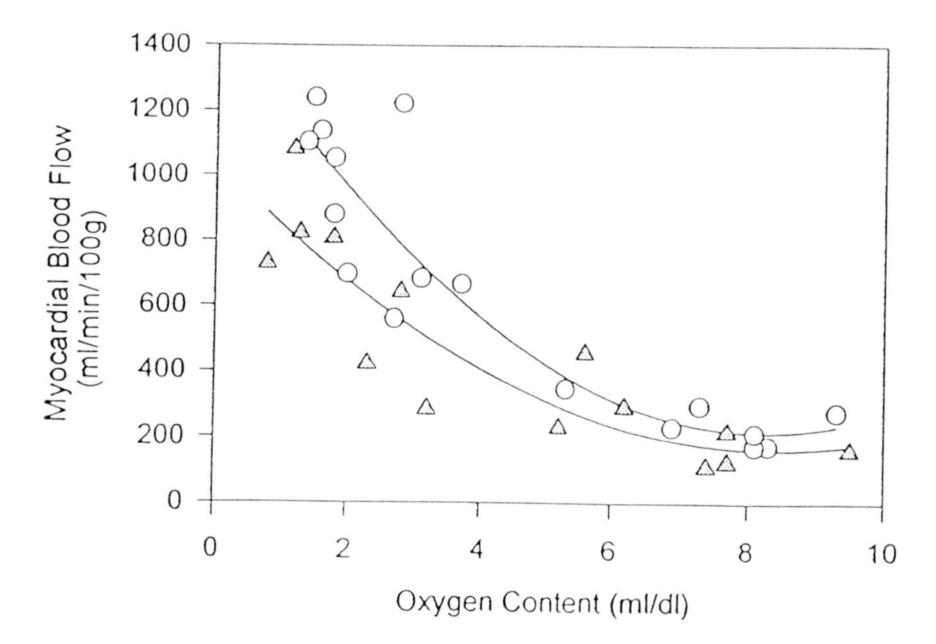

Figure 16.7 Myocardial blood flow shown as a function of oxygen content in arterial blood in the fetal sheep. Circles indicate control conditions. Triangles indicate NOS blockade with L-NA. (From Reller et al., 1995.)

put and fetal oxygenation. The oxygen consumption of the right ventricle is estimated to be larger than the left and its coronary flow is also larger.

Coronary blood flow regulation in the fetus also differs from regulation in the adult myocardium. The fetal myocardium is able to increase blood flow to at least four times control levels when adenosine is infused. However, hypoxemic conditions in the sheep fetus, whether chronic or acute, cause degrees of vasodilation that are enormous (>5 times resting levels) and that are not found in the adult under any circumstances. During hypoxemia, blood flows to the fetal myocardium far exceed $1\ l \cdot min^{-1} \cdot (100\ g)^{-1}$. Cardiac flows higher than chemical dilation levels are not found in adults.

The fetal coronary tree appears to have communication between the ventricles. If the right ventricle is selectively required to eject against a higher pressure, the left ventricle increases its flow in proportion to right ventricular flow even though left ventricular work has not changed. This phenomenon requires further investigation because it does not appear to be present in the adult heart.

The fetus lives in a low-oxygen environment and evidently depends substantially on nitric oxide to maintain blood flow over a wide range of oxygen levels. This mechanism is very powerful in the fetus compared with the adult. The mechanisms that regulate the expression of endothelium-derived vasodilators have not been investigated in the fetus. Just as intriguing is the rapid growth of the coronary vascular tree under conditions of chronic low oxygen. Spontaneously hypoxemic fetuses are able to manufacture a very low resistance arterial tree in just days, without losing the flow reserve of the normal arterial tree. Vascular endothelial growth factors and other angiogenic growth factors are undoubtedly elaborated locally to regulate the development of new vessels. What is unique to the fetal heart is the finding that new vessels at the arteriole level are being constructed and not just capillaries. Investigation in this exciting area of physiology awaits the curious.

Part III

Environment and disease in cardiovascular development

17

Oxygen, temperature, and pH influences on the development of nonmammalian embryos and larvae

BERND PELSTER

Introduction

The plasticity of the cardiovascular system during development in response to environmental perturbations is certainly important, because embryos of lower vertebrates usually are free-living in very early stages and do not develop under the protection of the maternal organism. This means that they usually are exposed to the same or a similar environment as the adults. Alterations of the environment, such as natural fluctuations in temperature, salinity, or oxygen availability, as well as pollution, thus also affect the developing animal. Embryos and larvae may be more sensitive to environmental changes than adults because their organs and organ systems are still in the process of growth and development. In addition, metabolic pathways, pathways of ion regulation, and even pathways for hormonal and neuronal control of organ function may not be fully developed to allow for a coordinated response to changing environmental conditions.

Many studies have looked at the effect of environmental variables like oxygen availability, temperature, and osmolarity on oxygen consumption ($\dot{M}_{O_2}$) or mortality of aquatic larvae. In terms of circulatory physiology, heart rate was the initial variable that was accessible. Recent technical developments like micropressure systems, pulsed Doppler systems, and video imaging techniques have opened up the possibility of gaining insight into variables like blood pressure and blood flow in miniature animals (Keller 1995; Burggren & Fritsche, 1995; Colmorgen & Paul, 1995).

The aim of this chapter is not to present a complete review of the literature available in this field but to excerpt and describe general physiological responses that typically are observed in embryos when facing hypoxia or changes in temperature, osmolarity, or pH.

Hypoxia

Hypoxia and cardiac performance

With the onset of cardiac activity, heart rate initially increases with development, as has been shown for chicken embryos (Romanoff, 1960b; Van Mierop & Bertuch, 1967; Clark & Hu, 1982), little skate (*Raja erinacea*) embryos (Pelster & Bemis, 1991), white suckers (*Catostomus commersoni;* McElman & Balon, 1980), zebra fish (*Danio rerio;* Pelster & Burggren, 1996), the direct-developing neotropical frog *Eleutherodactylus coqui* (Burggren, Infantino, & Townsend 1990), and the African clawed frog *Xenopus laevis* (Hou &

Burggren, 1995a; Fritsche & Burggren, 1996). As an example, blood pressure measurements obtained from *Raja erinacea* embryos during the very first days of cardiac activity (i.e., between 18 and 23 days of development, which in total takes about 150 days) show a slow but very regular pattern of cardiac contractions with slowly increasing frequency and pressure (Figure 17.1). This may be due in part to a shift in the pacemaker unit (Carlson, 1988; Burggren & Warburton, 1994). First, the presumptive ventricle begins to beat, which is the slowest unit in the hierarchy of pacemakers in the adult, followed by the atria and the sinus venosus, which is characterized by the highest frequency in the adult. For the chick embryo, however, Kamino, Hirota, and Fujii (1981) reported that the earliest site of pacemaker activity is the region of the future sinus venosus.

In many species this initial increase is followed by a slow decrease in heart rate until hatching, but in some species heart rate may also remain fairly constant or even increase slightly (Burggren & Warburton, 1994). The early appearance of cholinergic and adrenergic receptors in the heart has been shown for amphibians and birds, but cardiac innervation appears to be functional much later than the receptor response to humoral stimulation (Pappano, 1977; Kimmel, 1992; Protas & Leontieva, 1992).

As adults many lower vertebrates respond to hypoxia with a bradycardia (Jones & Milsom, 1982; Randall, 1982; Burggren & Pinder, 1991). The late appearance of a functional innervation of the heart certainly raises the question of whether larvae are able to modify heart performance in response to environmental "challenges" like hypoxia. The response of the various species to hypoxia appears to be quite variable and, as observed in *Xenopus laevis,* even changes from a tachycardia in very early stages to a bradycardia in later stages (Hou, 1992; Fritsche & Burggren, 1996). At least in some species, the first response of the heart to hypoxia is observed before innervation appears to be functional.

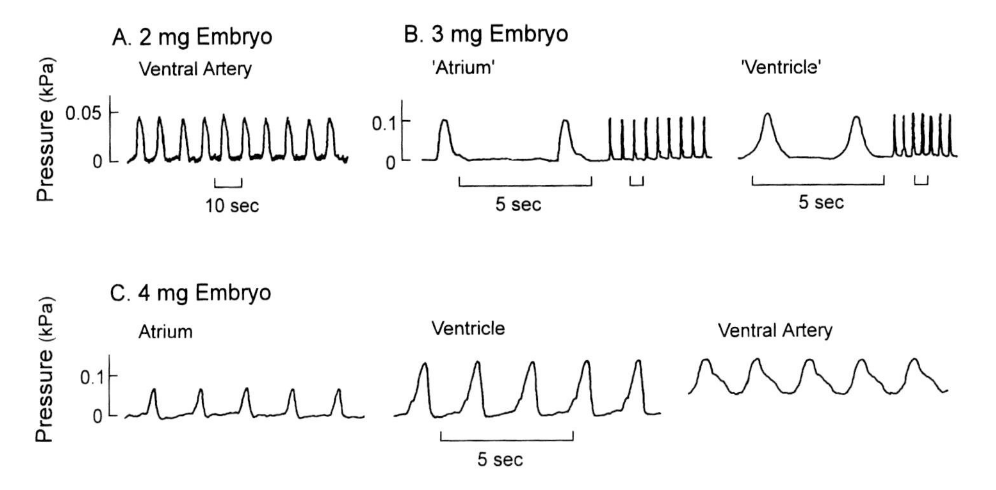

Figure 17.1 Original recordings of blood pressure in the central cardiac system of the little skate (*Raja erinacea*) during the first 5 days of cardiac activity (days 18–23 after fertilization; development is completed after about 150 days). (See Pelster 1991.)

When surfacing was prevented in Gosner stages 25–31, the heart rates of larvae of the frog *Rana berlandieri* decreased by about 12% during hypoxia, and the synchrony of ventilatory activity and heart contractions was lost (Wassersug, Paul, & Feder, 1981). In the same species, Feder (1983) reported an initial increase in heart rate and gill ventilation with decreasing P_{O_2}. In severe hypoxia, however, both parameters decreased. The final decrease in cardiac and ventilatory activity could be a direct effect of hypoxia, as also observed in bullfrog (*Rana catesbeiana*) larvae. Larval *Rana catesbeiana* in hypoxia down to a P_{O_2} of 4 kPa show no bradycardia after Taylor–Kollros (TK) stage IV, whereas early larval stages respond with a severe bradycardia, which appears to be a direct effect of metabolism and not a regulatory effect (Burggren, 1984; Burggren & Doyle, 1986b). The first study analyzing the influence of hypoxia on cardiac output in anuran larvae revealed a decrease in cardiac output only at severe hypoxia (P_{O_2} = 3.6–6 kPa) in larvae just after hatching. In later stages, this decrease, caused by a decrease in heart frequency, disappears (Orlando & Pinder, 1995). As in other studies, this modification in heart rate was attributed to a direct effect of oxygen availability, not to a regulatory response of the central cardiac system. The diving bradycardia observed in larval stages 37 and later of the paradoxical frog (*Pseudis paradoxus*), however, indicates the presence of cardiovascular control mechanisms (Burggren et al., 1992), which may also be responsive to a decrease in oxygen availability.

In larvae of the marine fish *Agonis cataphractus* a decrease in heart rate and ventilation rate was observed below a P_{O_2} of 10.7 kPa (Colmorgen & Paul, 1995). In this study the use of a video imaging system made it necessary to restrain the animals. The buildup of unstirred layers around the skin therefore may have rendered the animals more sensitive to hypoxia. The diffusional boundary layer represents the main resistance (80–90%) to oxygen uptake in larval fish (Rombough, 1992). Larval salmonids 1 day after hatching, on the other hand, respond to hypoxia down to a P_{O_2} of 2.7 kPa with a mild (10%) tachycardia (Holeton 1971; McDonald & McMahon, 1977).

Neotene salamanders (*Ambystoma tigrinum*) show a mild bradycardia when air access is prevented (Heath, 1980). With proceeding development *Ambystoma tigrinum* and *Ambystoma maculatum* embryos become increasingly sensitive to anoxia. In *Ambystoma maculatum* embryos of stages 36–41 (scheme of Bordzilovskaya et al., 1989), heart rate remained more or less constant down to a P_{O_2} of 5.4 kPa; below this, a severe decrease was observed, finally culminating in cardiac arrest (Adolph, 1979). During hypoxia a significant increase in the flow of microspheres to the lung at the expense of the flow to the gill arches indicated a redistribution of blood flow toward the lung to facilitate aerial gas exchange in *Ambystoma tigrinum* (Malvin & Heisler, 1988).

In day 7, 11, and 19 chicken embryos, incubation in an atmosphere of 10% oxygen induces a 10% decrease in blood pressure and heart rate, whereas hyperoxia increases blood pressure and heart rate (Girard, 1973a; Tazawa, 1981a). According to Pappano (1977), the adrenergic and cholinergic limbs of the autonomic cardiac system in the chick become functional at days 12 and 20, respectively. This clearly is later than the first bradycardia is observed in response to hypoxia in the chick (Girard, 1973a).

In lower vertebrates that have been analyzed so far, cholinergic and adrenergic receptors in the heart are functional earlier than the innervation is completed. Thus, humoral control mechanisms may be of importance in modifying cardiac activity in very early larval stages. This could, for example, explain the tachycardia, which is observed in a few species during hypoxia. The reduction in cardiac activity in response to environmental hypoxia displayed by many lower vertebrates must in part, however, be attributed to a metabolic depression.

Hypoxia and ventilation

Even in very early larvae a decrease in oxygen availability usually provokes a change in ventilatory activity. In newly hatched arctic char (*Salvelinus alpinus*) ventilatory activity and the coordination of ventilatory movements increase during hypoxic exposure (McDonald & McMahon, 1977). An increase in swimming activity under hypoxic conditions has also been reported for several fish larvae, including speckled trout (*Salvelinus fontinalis;* Shepard, 1955), smallmouth and largemouth bass (*Micropterus dolomieui* and *Micropterus salmoides;* Spoor, 1977, 1984), and northern anchovy (*Engraulis mordax;* Weihs, 1980).

Similarly, the earliest stages of anurans usually respond to hypoxia with an increase in gill ventilation (West & Burggren, 1982; Feder, 1983; Burggren & Doyle, 1986b; Burggren & Just, 1992). In larval *Rana catesbeiana* the increase in gill ventilation is observed only up to TK stage XIV, whereas even down to TK stage IV an increase in surfacing activity, associated with lung ventilation, was observed. In later stages (after TK XIV) hypoxia will cause a decrease in gill ventilation in order to prevent loss of oxygen to the environmental water (Burggren & Doyle, 1986b). However, the development of the pulmonary circulation and the extension of the pulmonary surface, actually favoring lung ventilation, occur relatively late during development (Burggren, 1984; Burggren & Doyle, 1986b). Nevertheless, an increase in surfacing activity has been reported for many anurans and thus appears to be a general response of anuran larvae to hypoxia (Wassersug & Seibert, 1975; Feder, 1983; West & Burggren, 1983; Burggren & Doyle, 1986b; Orlando & Pinder, 1995). These changes in behavior clearly are related to oxygen uptake and represent the attempts of the animals to access a different oxygen source, to avoid the unfavorable environment, or at least to reduce the thickness of unstirred boundary layers hindering gas exchange. Thus, even very early larvae of lower vertebrates sense fluctuations in oxygen availability. In response, behavioral changes and humoral modifications of cardiac activity are observed; neuronal control of cardiac activity is possible only in later stages.

Hypoxia and hemoglobin function

In mammals hypoxia usually causes an increase in the oxygen transport capacity of the blood, that is, an increase in hematocrit and in hemoglobin concentration. Similar observations have been reported for the turtle *Pseudemys nelsoni* (Kam, 1993), but stimulation of red cell production was not observed in chicken embryos during early development (Baumann & Meuer, 1992).

Another important response to hypoxia by developing lower vertebrates is a modification of the hemoglobin oxygen affinity. Compared with adult hemoglobin, the hemoglobin of viviparous embryonic vertebrates is characterized by a higher oxygen affinity, and this appears to be true for higher, as well as for lower, vertebrates. In viviparous fish and elasmobranchs the higher P_{50} is in part due to a higher intrinsic oxygen affinity of the pigment and in part due to a different concentration of organic phosphates within the red blood cells (Hartvig & Weber, 1984; Weber & Hartvig, 1984; Ingermann, Terwilliger, & Roberts, 1984; Ingermann & Terwilliger, 1984; Korsgaard & Weber, 1989). Although in reptiles the organic phosphate compound appears to be of great importance, a specific embryonic hemoglobin with a higher intrinsic affinity has been reported in the rattlesnake (Ragsdale & Ingermann, 1993; see also Chapter 14, this volume). In birds the higher P_{50} in embryos is brought about by an intrinsic hemoglobin in conjunction with a characteristic pattern of changes in the organic phosphate concentration, including a switch from adenosine triphosphate (ATP) to 2,3-bisphosphoglycerate (2,3-BPG) (Bartlett, 1980; Baumann & Meuer, 1992).

Chronic hypoxia at a P_{O_2} of 10 kPa for 4 weeks produced almost no change in blood P_{50} of larval *Rana catesbeiana,* but in adults a severe decrease was observed (Pinder & Burggren, 1983). This contrasts with the results obtained in birds, in which hypoxia results in a premature appearance of adult erythrocytes with an increase in P_{50} mainly due to a decrease in ATP concentration and a shift in red cell pH (Baumann, 1984; Baumann & Meuer, 1992). The variability in the modification of the affinity of the respiratory pigment in response to hypoxia may well reflect differences in the importance of convective gas transport for the metabolism of the developing embryo or larva. In early stages of small larvae, convective gas transport – although present – may be of no significance for oxygen supply (Pelster & Burggren, 1996). A modification of hemoglobin oxygen affinity during hypoxia in this case would probably not improve oxygen uptake. In larger embryos or larvae, in which convective gas transport is essential for a stable oxygen supply to tissues, hemoglobin oxygen affinity is important for the effectiveness of blood oxygen transport. In those embryos a change in oxygen affinity may stabilize convective oxygen transport in the face of fluctuating oxygen availability.

Hypoxia and metabolism

Oxygen uptake usually decreases in response to hypoxia. In animals that initially are able to keep the rate of oxygen uptake fairly constant or respond with very moderate decline in oxygen uptake ("oxygen regulators"), a critical P_{O_2} (P_c) is defined, below which $\dot{M}_{O_2}$ declines rapidly. Below the critical P_{O_2} anaerobic metabolic pathways significantly contribute to the production of ATP. The critical P_{O_2} of an egg is dependent on the rate of oxygen uptake, the diffusion coefficient, and the geometry of the egg. It thus can be calculated as $P_c = \dot{M}_{O_2} r^2/6D$ (Hamdorf, 1961). After onset of circulation, analysis of critical P_{O_2} becomes more complicated, and the diffusion distance between the blood and the surface must be taken into account. The ability of the embryo or larva to reduce oxygen demand by reducing metabolic activity also will be of importance for the definition of the critical P_{O_2} and may even change with development (Hastings & Burggren, 1995). Nevertheless, the critical P_{O_2} is dependent on $\dot{M}_{O_2}$. An increasing demand for oxygen due to a stimulation of metabolic activity, for example, can be expected to result in an increase in critical P_{O_2}.

In several fish species, eggs have been shown to be quite resistant to hypoxia, but as development proceeds the embryos become more and more sensitive to a decrease in environmental oxygen tension (Hayes, Wilmot, & Livingstone, 1951; Alderdice, Wickett, & Brett, 1958; Rombough, 1986). A detailed analysis of the influence of hypoxia on the metabolism of larval *Xenopus laevis* reveals that early embryos and larvae simply decrease the rate of oxygen uptake during periods of hypoxia and that regulatory capacities are observed only in larvae beyond larval Nieuwkoop–Faber (NF) stages 54–57 (Hastings & Burggren, 1995). In early larval stages the decrease in $\dot{M}_{O_2}$ was not compensated by an increase in lactate production. An increase in whole-body lactate levels could be detected only in larval NF stages 52 and later. The unstirred perivitelline fluid represents the major resistance to gas exchange in the eggs of *Misgurnus fossilis* (Berezovsky et al., 1979), and accordingly, in trout, Rombough (1988a) observed a decrease in the slope of the increase in critical P_{O_2} after the embryos started muscular contractions (Figure 17.2A), which mixed the perivitelline fluid (see also Chapter 12). Similarly, the decrease in critical P_{O_2} after

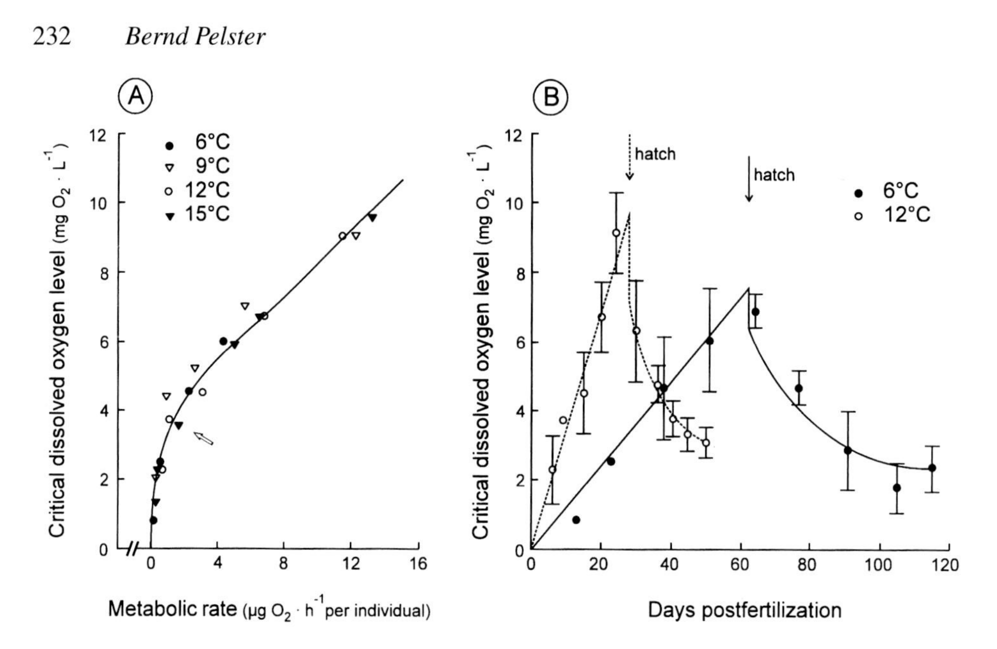

Figure 17.2 (*A*) Relationship between critical dissolved oxygen level and routine metabolic rate at temperatures between 6°C and 15°C for steelhead (*Salmo gairdneri*) embryos. The arrow indicates the onset of muscular contractions of the embryo, which mix the perivitelline fluid. (*B*) Relationship between critical dissolved oxygen level and development for steelhead trout at 6°C. (Modified from Rombough, 1988a, with permission.)

hatching can be explained by a significant reduction in the resistance to gas exchange by eliminating the egg capsule. Following the decrease in critical P_{O_2} after hatching, critical P_{O_2} may either stay fairly constant or increase again (Rombough, 1986, 1988b). These observations clearly underline the importance of external convective gas transport in the developing embryo.

The onset of circulation thus should also decrease the overall resistance to gas exchange and perhaps even increase the resistance of the animals to oxygen deficiency. This contention, however, is not upheld by the current literature, which rather suggests that at the onset of circulation convective gas transport is not the predominating function of blood flow. In fish embryos the heart starts contracting before the red blood cells appear. Exposure to carbon monoxide at levels sufficient to block the oxygen transport function of the hemoglobin has little apparent effect on larval rainbow trout (Holeton, 1971) or even chicken embryos (Cirotto & Arangi, 1989). Recent experiments have demonstrated that the African clawed frog, *Xenopus laevis* (Burggren & Territo, 1995), and also the zebra fish, *Danio rerio* (Pelster & Burggren, 1996), can be raised in an atmosphere containing 2% CO without any obvious physiological impairment of cardiac function until far beyond hatching. In the "cardiac-lethal" mutant of the salamander *Ambystoma tigrinum,* the heart is formed normally, but it never starts beating. Thus, the mutant larvae completely lack a functional circulatory system. Nevertheless, these animals are active and usually survive until a few days after hatching (Justus, 1978; Smith & Armstrong, 1991). Up to Bordzilovskaya et al. stage 39, $\dot{M}_{O_2}$ and G_{O_2} (oxygen conductance) are not greatly reduced in cardiac-lethal mutants (Mellish, Smith, & Pinder, in press).

Collectively, these observations clearly suggest that embryonic and larval gas exchange mainly occurs via the skin. Rombough and Ure (1991) analyzed the contribution of various organs and tissues to gas exchange and found that about 80% of the total oxygen taken up by recently hatched salmonid larvae was transferred through the skin. This strategy is facilitated by the fact that the larval fish swim using red muscle right under the epithelium. These muscles later move to the lateral line (El-Fiky & Wieser, 1988; Burggren & Pinder, 1991). To overcome the problem of the unstirred boundary layers, which represent the major resistance to gas exchange (Rombough, 1992; see also Chapter 12, this volume), some embryos create external circulation with cilia, as has been described for *Rana palustris* egg masses (Burggren, 1985a). Orange chromide (*Etroplus maculatus*) males display active and passive fanning of the eggs, which greatly increases breeding success (Zoran & Ward, 1983). Larval *Monopterus albus* have adopted a special behavior to create a countercurrent flow of water and blood at the body surface (Liem, 1981); countercurrent flow is a very effective arrangement for respiratory gas exchange.

These examples suggest that the circulatory system and blood flow can be of importance for convective gas transport and gas exchange, but there are also examples that show that in small larvae the absence of convective gas transport does not impair $\dot{M}_{O_2}$ and development. Body size probably is a crucial parameter in this context.

Morphological changes induced by hypoxia

Pelagophil fish larvae adapted to well-oxygenated waters usually have no special respiratory organ, or it is poorly developed if present. The blood vessels do not come near the surface of the body, and red blood cells appear only well after the onset of fluid circulation in embryonic blood vessels. In fishes living in poorly oxygenated water, however, as many freshwater fish do, special embryonic respiratory organs are well developed. These organs can be dense nets of blood vessels in the dorsal or ventral fins, respiratory vessels on the gill cover and the yolk sac, or even external gills. Such external branchial structures are present in *Misgurnus fossilis* and in elasmobranch embryos that are confined in an egg case during their development (Balon, 1975; Wourms, 1977). In the African lungfish these external gills can even be retained in the adults (Burggren & Johansen, 1986). In little skate embryos external gill filaments can contribute significantly to the overall embryonic gas exchange (Pelster & Bemis, 1992). It appears quite possible that in these embryos with additional respiratory devices the role of cutaneous gas exchange is not as predominant as in salmonid embryos. The embryonic respiratory organs of fish embryos and larvae are resorbed by the time gills with the secondary lamellae develop, which provides additional evidence of their function in gas exchange.

In larval stages of frogs and salamanders, chronic hypoxia may result in branchial or lung hypertrophy, including a decrease in diffusion distance between blood and respiratory surface as well as an increase in respiratory surface area (Bond, 1960; Guimond & Hutchison, 1976; Burggren & Mwalukoma, 1983; Burggren & Just, 1992). In larval *Rana catesbeiana* chronic hypoxia (10 kPa) results in an increase in the number and size of the internal gill filaments (Burggren & Mwalukoma, 1983). In larvae of the arctic char (*Salvelinus alpinus*) chronic hypoxia results in a decrease in the number of secondary gill lamellae, but growth of the individual lamellae is stimulated (McDonald & McMahon, 1977). Specialized larval structures to support gas exchange must have an intense vascular supply, so their development requires transient modifications of the circulatory system. During the reabsorption of larval structures the capillary system must be reabsorbed as well. Thus, hypoxia or even local metabolic demand may be an adequate stimulus to modify the capillary supply of special organs (see also Chapter 4).

In chicken embryos, hypoxia has indeed been reported to cause increased capillarization of the chorioallantoic membrane (Dusseau & Hutchins, 1988; Hudlicka, Brown, & Egginton, 1992). Using a whole-body perfusion technique, Adair and coworkers (1987) demonstrated that in response to hypoxia structural changes in the vascular system of chick embryos increased blood flow to the tissue at a given perfusion pressure gradient. A decrease in capillary supply to both fast and slow muscles has been observed in tench, although these changes perhaps were influenced by seasonal temperature differences (Egginton & Johnston, 1984). These changes in capillarization may, however, not be a direct effect of hypoxia but a secondary effect resulting from a change in cardiac output. Blood flow appears to be an important parameter in angiogenesis (Hudlicka & Tyler, 1986; Hudlicka, Brown, & Egginton, 1992; see also Chapter 4, this volume).

In larval bullfrogs the capillarization of the skin, which is the main organ for gas exchange at this stage, is almost doubled during chronic hypoxia at a P_{O_2} of 9.3–10.6 kPa (Burggren & Mwalukoma, 1983). This coincides with a reduction in the blood–water diffusion barrier, which also facilitates gas exchange. In bird embryos hyperoxia (P_{O_2} = 60 kPa) causes a significant (22%) increase in the weight of heart muscle; hypoxia induced by covering part of the eggshell did not significantly reduce heart mass (McCutcheon et al., 1982). Hypoxia-incubated (P_{O_2} = 10 kPa) Florida red-bellied turtle (*Pseudemys nelsoni*) embryos showed an increase in heart muscle mass relative to body mass (Kam, 1993).

Generally speaking, hypoxia in embryos and larvae causes a reduced rate of development and a retarded growth rate (Alderdice, Wickett, & Brett, 1958; Hamdorf, 1961; Kinne & Kinne, 1962; Garside, 1966; Gulidov, 1974; Carlson & Siefert, 1974; Siefert, Carlson, & Herman, 1974; McDonald & McMahon, 1977; Adolph, 1979; Brooke & Colby, 1980; McCutcheon et al., 1982; Kam, 1993). Hypoxia may even change organ differentiation, as has been demonstrated for fish (Hamdorf, 1961) and birds (McCutcheon et al., 1982). In birds, for example, hypoxia induced a 19% decrease in liver mass, although heart mass was unaffected (McCutcheon et al., 1982). In terms of successful larval development in water, predation and starvation appear to be the major environmental parameters, and a small decrease in larval growth may result in a substantial increase in the time spent as a larva. Thus, a delay in development caused by hypoxia may, as a secondary effect, significantly increase the risk of predation (Houde, 1987). After transient hypoxia below the threshold for malformation of the embryo, development can proceed with a certain delay after return to normoxia. Increasing degrees of hypoxia result in malformations and finally death of the embryo (Garside, 1959; Spoor, 1977).

Apart from the critical P_{O_2}, the effect of hypoxia is dependent on the developmental stage at which the embryo or larvae is exposed. An interesting observation in this respect is that some fish and amphibian embryos initiate hatching almost immediately when the environmental P_{O_2} is reduced within a certain time window prior to hatching (Alderdice, Wickett, & Brett, 1958; Hamdorf, 1961; Cunningham & Balon, 1986b; Bradford & Seymour, 1988); hypoxia apparently triggers release of the hatching enzyme (Ishida, 1985). In some species, this hypoxia even appears to be essential for the onset of hatching, as described for the frog *Pseudophryne bibroni* (Bradford & Seymour, 1988).

Temperature

Temperature certainly is an important factor determining life on earth and thus also is an important variable for the development of the central cardiac system. The magnitude of the optimal or tolerable temperature range, in which the organism can exist in a steady state, is subject to large interspecies variation and to the previous thermal history of the individual (Rosenthal & Alderdice, 1976; Hutchison & Dupré, 1992). Temperature exerts profound effects on the morphological development of the central cardiac system, as well as on many other tissues.

Oxygen transport and cardiac performance

Almost all lower vertebrates are poikilothermic: their body temperature varies with environmental temperature and is not controlled by means of metabolic activity. Even in birds, clearly homeothermic as adults, the embryo is poikilothermic for most of development (Tazawa et al., 1991). Because the cardiovascular system is a central component of convective gas and nutrient transport, a regulatory change in cardiac performance during temperature changes is to be expected in situations where a regulatory metabolic response is observed. But even in the absence of a primary relationship between temperature and cardiovascular functions, a temperature change will change gas transport of the blood and thus may require a response from the central cardiovascular system. An increase in temperature causes an increase in metabolic activity but decreases the physical solubility of gases in body fluids. Moreover, it usually decreases the oxygen affinity of the respiratory pigment. This facilitates unloading of the hemoglobin in the tissue but, on the other hand, reduces oxygen uptake at the respiratory surface. This may cause a decrease in arterial hemoglobin oxygen saturation and thus require a change in cardiac activity to meet the increase in oxygen demand of the tissues.

As $\dot{M}_{O_2}$ in poikilothermic animals is reduced with decreasing temperature, heart rate falls. Nevertheless, a complex interdependence can exist between these two parameters, as observed in embryos of the neotropical frog *Eleutherodactylus coqui.* Although the Q_{10} for the decrease in heart rate with decreasing temperature in this species was calculated to be about 1.5–2, the Q_{10} for oxygen uptake in the same temperature range varied between 3.5 and 6.5 (Burggren, Infantino, & Townsend, 1990).

In addition to convective gas transport, temperature also influences diffusive gas transport. While physical gas solubility decreases with increasing temperature, the diffusion coefficient increases. As a consequence, Krogh's constant of diffusion increases only by about 1% per °C in aqueous solutions ($Q_{10} = 1.1$; Dejours, 1981). As already pointed out, the Q_{10} of enzymatic reactions, and thus of aerobic metabolism, typically is somewhere around 2–3; consequently, diffusive gas transport cannot keep up with the increasing oxygen demand of the tissue. Therefore, internal P_{O_2} decreases with increasing temperature, and the critical P_{O_2} at which anaerobic metabolic pathways start to dominate in cellular metabolism increases with temperature (Rombough, 1988b). Thus, increasing temperatures exacerbate the effect of hypoxia on fish embryos and larvae (Garside, 1966; Florez, 1972; Siefert, Spoor, & Syrett, 1973; Spoor, 1977, 1984). These considerations clearly provide an explanation for the observation of Bradford (1990) that ovum size is inversely related to prevailing ambient temperature. In an analysis including 55 species of amphibians, average ovum volume at 10°C was about 20 mm^3 and average volume at 30°C was only 1 mm^3. This relationship apparently has also been shown to occur within populations of individual anuran species (Bradford, 1990). The decrease in ovum size provides a better surface-to-volume ratio and thus facilitates gas exchange at higher temperatures.

Metabolic adjustments and cardiac performance

Chicken embryos are poikilothermic during development, so embryo temperature follows environmental temperature, and $\dot{M}_{O_2}$ decreases in proportion to the temperature decrease. This relationship holds for altricial embryos, which require extensive parental care after hatching and in which the ability to control body temperature is achieved only late after hatching, and for most of the development of precocial birds, which require little parental care after hatching. In precocial embryos, like those of chickens or ducks, the first signs of a regulatory response aimed at preventing a decrease in body temperature are detected at the time of hatching or even a few days earlier. Chicken embryos at days 18–19 have been reported to increase heat production as a response to cooling of the egg (Freeman, 1964, 1970; Tazawa & Rahn, 1987; Tazawa et al., 1988b; Kuroda et al., 1990). Whereas in early stages body temperature immediately declines with declining environmental temperature, in these stages body temperature remains constant for about 0.5–1 hour before declining. In later stages, particularly after external pipping, this effect is even more pronounced, as indicated by a significant decrease of Q_{10} from the value of about 2, observed for $\dot{M}_{O_2}$ in early stages with no sign of a homeothermic response (Tazawa et al., 1988b, 1989b).

An increase in metabolic activity due to environmental cooling requires the consumption of additional oxygen and nutrients. The additional supply can be met by an increase in arteriovenous gas partial pressure and concentration gradients and/or by an increase in cardiac output. If we transfer our knowledge from exercise physiology and thermoregulation of adults to the embryonic organism, we have to expect that both parameters will change. Accordingly, a number of studies have reported a linear or almost linear relationship between heart rate and temperature for most of the developmental period, giving no indication of a regulated response (Romanoff, 1960b; Cain, Abbott, & Rogallo, 1967; Girard, 1973a; Clark, 1984; Tazawa & Nakagawa, 1985). Furthermore, Tazawa and Rahn (1986) observed an exponential decrease in heart rate to a new plateau, or steady-state value, in response to a step reduction in temperature, which was quite similar to the results obtained for the decrease in $\dot{M}_{O_2}$ under these conditions (Tazawa & Rahn, 1987). Shortly before pipping, however, changes in heart rate with temperature also indicated incipient homeothermy by a slight decrease in Q_{10}, although these results were not as clear as the changes observed for $\dot{M}_{O_2}$ under these conditions (Tazawa et al., 1992).

Similar results were obtained for blood pressure, which in early embryos also decreases immediately after exposure to lower temperatures (Tazawa & Nakagawa, 1985). In later stages – embryos of 17–18 days – blood pressure may remain constant for about 1 hour or even more before it decreases in response to a sudden decrease in environmental temperature down to 26–28°C. Girard (1973a) also reported a linear decrease in systolic blood pressure in 3-day-old chicken embryos, but in 7-day-old embryos pressure initially increased with a temperature reduction down to about 32°C and then decreased with a further decrease in incubation temperature. Low-temperature exposure also results in a decrease in cardiac output, which is to be expected because of the decrease in heart rate. Nevertheless, the decrease in heart rate can be associated with an increase in filling volume and stroke volume (Casillas et al., 1993). For 16-day chicken embryos the temperature-induced decrease in cardiac output, however, was estimated to exceed the decrease in blood pressure, indicating an increase in peripheral resistance under these conditions (Tazawa & Nakagawa, 1985).

If environmental temperature decreases drastically, coordinated development and survival are endangered. Exposure to 18°C reduces heart rate of chicken embryos initially

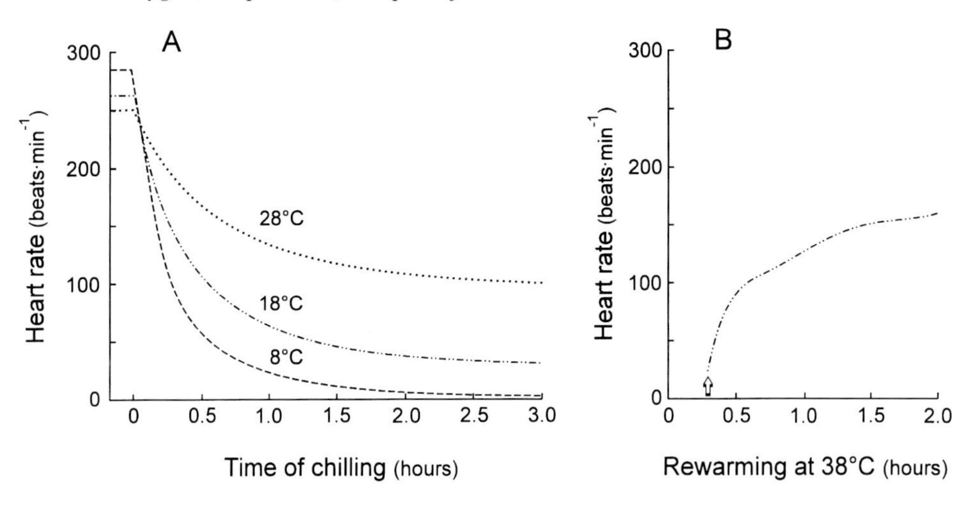

Figure 17.3 (*A*) Decrease in heart rate of 10-day-old chicken embryos exposed to 28°C, 18°C, or 8°C temperature. (*B*) Recovery of cardiac activity during rewarming to 38°C in an embryo that was chilled down to 18°C for 43 hours and underwent cardiac arrest minutes before rewarming. (Modified from Tazawa & Rahn, 1986, with permission.)

down to a plateau, then after several hours the heart rate becomes irregular and finally the heart stops (Tazawa & Rahn, 1986). Subsequent rewarming sometimes results in a recovery of cardiac activity (Figure 17.3), but usually recovery is incomplete even after a day or two. After the onset of irregular cardiac activity, recovery is unlikely. At lower temperatures (8°C), however, heart rate decreases progressively to complete cardiac arrest. The reduction in metabolic activity at 8°C allows survival for several hours, and after rewarming, cardiac activity recovers. In this respect early chicken embryos are more resistant to low temperatures than late embryos; the immature organ can withstand complete cardiac arrest for a longer period than the adult organ can (Tazawa & Rahn, 1986; Tazawa et al., 1992). Exposure to 44°C or 45°C also caused cardiac arrest.

The ability of birds to protect body temperature in the face of changing environmental temperature apparently is achieved only very late in development; early stages clearly are poikilothermic. Lower temperatures cause a decrease in cardiac activity and thus a reduction in oxygen supply to the tissue. Metabolic activity of the tissue, however, also decreases, and thus a temperature decrease does not necessarily induce hypoxemia. Incomplete recovery or failure to recover after transient cooling, however, may indicate a mismatch between tissue metabolism and convective nutrient and oxygen supply, causing cell and tissue damage. Homeothermia develops only at about the time of hatching, and this is about the time when lung respiration starts. It may be that limitations of gas exchange through the eggshell or the limited fuel stores within the egg do not allow for the higher metabolic rate associated with homeothermia throughout embryo development.

Temperature-dependent morphological changes

Higher temperatures increase metabolic activity and cause an accelerated differentiation of tissues and organs but retard the growth of such structures relative to low temperatures. Because of the limitations in oxygen transport encountered at high temperatures (see earlier discussion), eventually oxygen supply becomes limiting and thus weight increment is retarded (Alderdice & Velsen, 1971; Laurence, 1975; Rosenthal & Alderdice, 1976; Hamor & Garside, 1976, 1977). Thus, although growth rate is reduced at low temperatures, the embryos may eventually grow to a larger size in the cold.

Besides this general effect on growth rate, temperature also influences cellular metabolism and tissue capillarization. Lowered temperatures usually result in elevated enzyme activities of oxidative metabolism and in increased capillary density of tissues (Hudlicka, Brown, & Egginton, 1992). Thus, Johnston (1982) reported an increase in the mitochondrial compartment and in capillary supply during cold acclimation in adult crucian carp red and white muscle. Similarly, Garside (1959) observed a higher degree of vascularization in the vitelline circulatory system of embryonic lake trout *Salvelinus namaycush* at low temperatures.

Salinity and pH

Water pH is a crucial parameter affecting the vitality of aquatic animals. When shed in freshwater, pike and salmonid eggs take up water, and the perivitelline compartment is formed (Loeffler, 1971). The permeability of the egg membrane changes with development and typically increases with temperature (Alderdice, 1988). This demonstrates exchange of water between the egg and the environment, and changes in the environmental water therefore are translated to the embryo. There is, however, almost no information available on the effect of pH or salinity on the development of the cardiac system.

The influence of salinity on $\dot{M}_{O_2}$ is highly variable among species. Whereas salinity hardly changes $\dot{M}_{O_2}$ in herring and plaice (Holliday, Blaxter, & Lasker, 1964; Almatar, 1984), striped mullet (*Mugil cephalus*) shows a salinity tolerance curve with a narrow peak at 26–32%. Higher and lower salinities significantly decrease survival rates (Sylvester, Nash, & Emberson, 1975). In *Cyprinodon macularis* rapid changes in salinities cause developmental arrest (Kinne & Kinne, 1962). Salinity tolerance also is dependent on other parameters, including oxygen tension or temperature (Alderdice & Forrester, 1971). Exposure to low pH reduces survival of early life stages, and again the sensitivity to acid exposure varies with species and also with developmental stage (Ingersoll et al., 1990a; DeLonay et al., 1993). Detailed studies mainly on salmonid embryos and larvae reveal that the influence of acid exposure is dependent on a number of other parameters, for example, Ca^{2+} or aluminum concentration in the water (Wood et al., 1990; Ingersoll et al., 1990a; Ingersoll et al., 1990b; Cleveland et al., 1991; DeLonay et al., 1993). Acid exposure finally results in a disturbance of the ion equilibrium between the various cellular compartments. The proper function of the cardiovascular system clearly is dependent on the ion composition of the extracellular and intracellular compartments. The influence of these changes on the development of the central cardiac system, however, is poorly understood.

Conclusion

In spite of the variability in physiological adaptations and responses observed in embryos and larvae of lower vertebrates, several general traits may be detected. To summarize some outstanding examples: eggs and early developmental stages can be quite durable with respect to fluctuations in environmental oxygen concentration and temperature, whereas limitations of respiratory gas exchange render later stages more sensitive; the sensory system to detect hypoxia is present in very early stages, but the innervation of the cardiac system, and thus the structures for neuronal control of the circulatory system, are completed only very late in development; even in typical homeothermic animals like birds, homeothermia is established only late in development, and early embryos are poikilothermic. Thus, the ability of embryos and larvae to cope with fluctuations in environmental oxygen availability, temperature, and pH is not just a reflection of responses and strategies known from adults. Larval stages may be much more sensitive than adults, and the variability in individual responses and the high variability in sensitivity of different species to changing environmental conditions require a very careful and complex analysis that includes many species. Environmental studies to predict, for example, the impact of pollution on an ecosystem cannot be based on adults only but must include larval stages.

18

Modeling gas exchange in embryos, larvae, and fetuses

ALAN W. PINDER

Introduction

This chapter introduces the "ideal" models used to study respiratory gas exchange in vertebrate embryos, fetuses, and larvae. The assumptions, simplifications, and limitations of each model are discussed, along with experimental tests. In some cases, departures from ideal behavior have themselves been modeled, including such phenomena as shunts, heterogeneous perfusion, and diffusion/perfusion inequality.

What is different about early life stages?

Gas exchange in early developmental stages has several unique characteristics and is always very different from that in adults of the same species. There may be no specialized gas exchange organ at all, as in most fish and amphibians, or there may be specialized gas exchange organs that are unique to embryonic or fetal life, as in amniote eggs or placental mammals. Gas exchange requirements increase exponentially during development as the embryo grows or converts yolk into metabolizing tissue; thus, there is never a true "steady-state" gas exchange, and conditions for gas exchange continuously change. Embryos always start development with a small enough body mass, and with a low enough oxygen consumption, to satisfy all gas exchange requirements by simple diffusion. "Simple" diffusion is complicated by a number of extraembryonic structures, however. The egg capsule of fish and amphibian eggs, composed of both fibrous and thick gelatinous layers, may impose a significant resistance to gas exchange because of its thickness. The shells of bird and reptile eggs similarly impose a resistance to diffusion external to the embryo. Eggs are sometimes laid in gelatinous masses (amphibians) or large clutches (reptiles), so that neighboring eggs may impose additional diffusion barriers and also decrease local oxygen concentrations. Aquatic eggs are surrounded by layers of relatively stagnant water, the diffusion boundary layer; terrestrial eggs may be buried in soil; and mammalian embryos are surrounded by another organism; all of which add yet more resistance to diffusive gas exchange.

With growth and cardiovascular development, there is a switch to convective gas transport, with direct diffusion and circulatory gas transport providing parallel routes of exchange and transport during the transition. The stage at which this occurs depends on the size of the embryo and on the metabolic rate. Small ectothermic embryos, such as those of fish and amphibians, may hatch and commence feeding before circulatory oxygen transport is essential. At the other extreme, circulatory oxygen transport becomes necessary very early in mammalian and avian development because of their relatively large size and high metabolic rate.

There are innumerable variations on developmental themes for which gas exchange has not been modeled or studied in detail: the numerous, independently evolved exchange systems in ovoviviparous and viviparous fish, reptiles, and amphibians (e.g., Webb & Brett, 1972a, 1972b; Toews & MacIntyre, 1977; Blackburn, 1992; Wourms, 1994), mouth-brooding and other forms of egg ventilation in fish (Blaxter, 1988), and brooding eggs in cutaneous pouches in some amphibians (Duellman & Trueb, 1986) are just a few examples of little-studied systems. Whatever the specific gas exchange system, the embryo is surrounded by external diffusion barriers and commonly has a less efficient blood flow pattern for gas exchange, so that developing tissues are more hypoxic than in the adult. This may merely reflect limitations of gas exchange, or it may be important to the embryos for reducing the danger of spontaneous and deleterious oxidation of proteins and other cell constituents (Ar & Mover, 1994). Certainly, the relative hypoxia of embryos compared with adults is reflected in the higher O_2 affinities of embryonic and fetal hemoglobins (Ingermann, 1992; Holland, 1994; Weber, 1994).

General concepts in gas exchange modeling

Gas exchange modeling reduces biological complexity to a minimum of functionally important variables to interpret data, explain observed phenomena, and formulate testable hypotheses (Scheid, 1987). Models have been used, in conjunction with experiments, to test which steps are most limiting to the movement of gases at different developmental stages, to explain how changes in ventilation and perfusion affect gas exchange, and to predict how various inhomogeneities affect gas exchange efficiency. Models are useful only when they can be experimentally tested and when predictions from alternative models can be distinguished from each other.

The respiratory gas cascade: Steps in series. Gas exchange is usually broken down into a series of steps, each with an associated conductance (and its inverse, resistance). In adults the usual steps are ventilation, diffusion across the gas exchanger (including diffusion across the tissue barrier, diffusion in blood, chemical reaction rates with hemoglobin, and sometimes diffusion through unstirred layers), and blood flow. The steps are the same for O_2 and CO_2, although in opposite directions and often with different conductances. Ideal gas exchange models appropriate for the structures of adult vertebrates have been developed and periodically reviewed (e.g., Piiper & Scheid, 1982, 1989). Many of the inhomogeneities that decrease gas exchange efficiency have also been modeled and experimentally tested (Wagner, Saltzman, & West, 1974; Piiper, 1993; Powell, 1993).

Because of the different properties of O_2 and CO_2 in aqueous solution and in interactions with red cells (Dejours, 1981), the two gases often show quite different patterns of change with development. Oxygen uptake and transport are most frequently studied, perhaps because oxygen lack is more immediately fatal. Diffusion, often in aqueous solution, is generally more important in embryos than in adults, and CO_2 has a much higher (roughly 20 times) Krogh's diffusion constant than O_2 in water. Thus P_{CO_2} gradients are much smaller, and CO_2 buildup may be less of a problem. Of course, CO_2 exchange is complicated by the reactions of CO_2 with water and thus its involvement in acid–base balance.

Diversity of vertebrate gas exchangers. Although gas exchange in developing vertebrates is very different from that in adults of the same species, especially in the usual lack of regulated ventilation, the concepts and some of the steps are the same. Four main types of gas exchanger in adults have been modeled (Figure 18.1; Piiper & Scheid, 1989): ventilated pool (mammalian lung), infinite pool (or open; amphibian skin), countercurrent (fish gill), and crosscurrent (bird lung). All four models have analogs in developing vertebrates. The embryonic model of gas exchange lacking an adult vertebrate analog is direct diffusion to tissues, with no specialized gas exchanger and no blood transport of gases. Adult vertebrates are too large for direct diffusion to tissues other than the most superficial cutaneous or mucosal surfaces, although many adult invertebrates remain small enough for oxygen delivery by direct diffusion (see also Chapter 11).

Basic definitions and relationships. The terminology of this chapter follows that of Piiper and Scheid (1989). In its simplest form, a respiratory gas exchange model has three variables: rate of gas transfer, $\dot{M}$; conductance, G (or its inverse, resistance, R); and gas partial pressure difference, ΔP:

$$\dot{M} = G \times \Delta P = \Delta P/R.$$

Because at steady state $\dot{M}$ is the same for all steps of gas exchange, the relative conductances are inversely proportional to the relative ΔP across each step ($G_1/G_2 = \Delta P_2/\Delta P_1$; $R_1/R_2 = \Delta P_1/\Delta P_2$), and thus ΔP can be used to analyze the relative conductances of the serial steps of gas exchange.

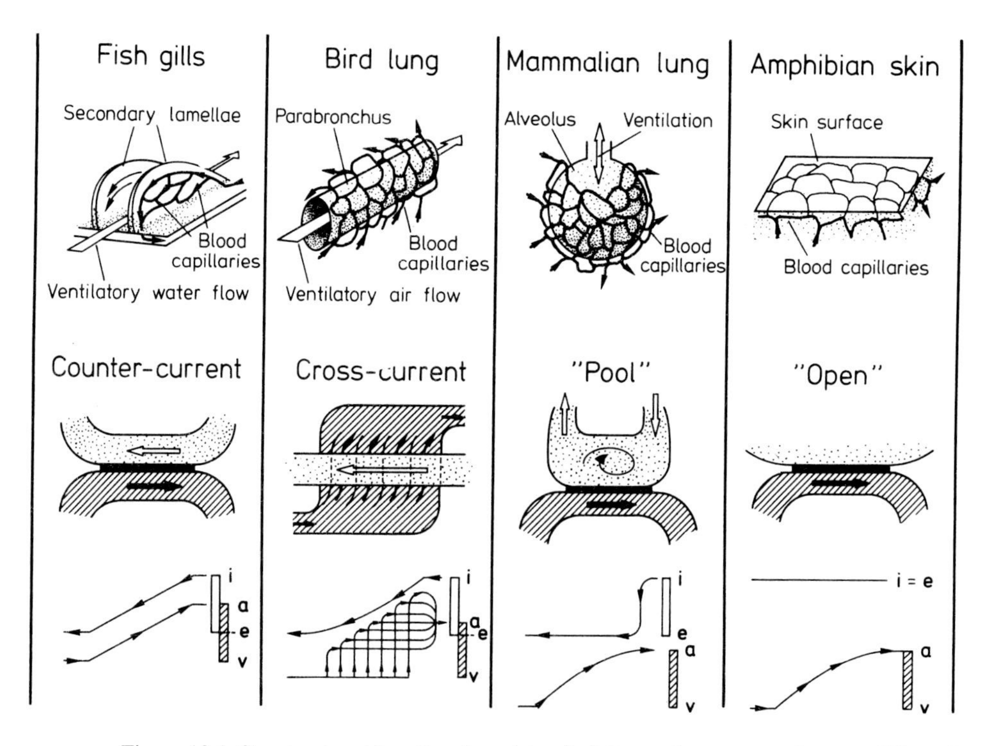

Figure 18.1 Structural and functional models of adult vertebrate gas exchangers (Piiper & Scheid, 1982). Graphs at the bottom show changes in P_{O_2} of blood and respiratory medium flowing through the gas exchanger. *Open bars:* range of P_{O_2} in respiratory medium; *hatched bars:* range of P_{O_2} in blood (i = inspired, e = expired, a = arterial, v = venous).

The diffusive conductance (G_{diff}) or diffusing capacity (D) of a gas exchanger includes diffusion through the tissue barrier (D_m) in series with the resistance of blood and rate of reaction with hemoglobin (Hb):

$$1/G_{diff} = 1/D = 1/D_m + [(1/V_c)(1/\theta)],$$

where V_c is the volume of capillary blood and θ is the reaction rate (Roughton & Forster, 1957). The membrane-diffusing capacity (D_m) is determined by Krogh's diffusion constant for that gas (K, the product of the physical diffusion constant and solubility), the surface area for diffusion (A), and the length of the diffusion path (L), and is the only component of diffusing capacity in the absence of perfusion (e.g., in diffusion across eggshells or jelly capsules):

$$D_m = KA/L.$$

The convective conductance of blood gas transport (G_{perf}) is the product of the rate of blood flow ($\dot{Q}$) and the amount of gas carried per unit blood (β, capacitance, the increment in gas content per increment of partial pressure, $\Delta C/\Delta P$):

$$G_{perf} = \dot{Q}\beta.$$

Because β for both O_2 and CO_2 depends on P (β is the slope of the sigmoid oxygen equilibrium curve), gas exchange models are made more accurate but more complicated by incorporating nonlinear capacitance curves.

A basic measure of the effectiveness of a perfused gas exchanger is the degree to which the blood and external respiratory medium equilibrate across the diffusion barrier, which is determined by the ratio of diffusive conductance to perfusive conductance, $D/\dot{Q}\beta$ (Piiper & Scheid, 1989). If $D/\dot{Q}\beta > 10$ (i.e., a much higher diffusive conductance than perfusive conductance), the blood will almost completely equilibrate with the external water or air. Thus, ΔP between the respiratory medium and arterialized blood will be small, and the rate of gas exchange will be largely determined by the rate of perfusion of the gas exchanger (Figure 18.2A). The gas exchanger would be "perfusion limited" (e.g., mammalian lungs under most circumstances) (Wagner, 1977). In a "diffusion limited" gas exchanger (Figure 18.2B), $D/\dot{Q}\beta$ is low (<0.1), and the blood does not equilibrate with the external medium. Thus, there is a large ΔP between arterialized blood and the external medium (Piiper & Scheid, 1989). An example of a diffusion-limited gas exchanger is amphibian skin (Piiper, Gatz, & Crawford, 1976; Feder & Burggren, 1985; Pinder, Clemens, & Feder, 1991). Some gas exchangers, such as fish gills, are partly limited by both diffusion and perfusion (Scheid, Hook, & Piiper, 1986).

Effects of inhomogeneities in gas exchangers. Various inhomogeneities in the gas exchanger decrease gas exchange efficiency and result in a ΔP between the external medium and blood leaving the exchanger (Figure 18.2C–E; Piiper, 1993; Powell, 1993). Ventilation/perfusion ($\dot{V}/\dot{Q}$) inequality is responsible for most of the small alveolar–arterial P_{O_2} difference in lungs (Wagner, Saltzman, & West, 1974), analogous to $\dot{Q}_{maternal}/\dot{Q}_{fetal}$ inequality in placental gas exchange. In embryos and fetuses, which often lack ventilation of the external medium, $D/\dot{Q}$ or $D/\dot{Q}\beta$ inhomogeneity may significantly decrease gas exchange efficiency (Figure 18.2C). Shunt, the fraction of blood flow to the gas exchanger bypassing the exchange surface, can be regarded as an extreme $\dot{V}/\dot{Q}$ or $D/\dot{Q}$ inequality in which $\dot{V}$ or D is zero (Figure 18.2D). A diffusion shunt may also occur, in which gases diffuse between arteries and veins leading to and from the gas exchanger (Figure 18.2E). Perfusion (and ventilation) is generally modeled to be constant, but in fact it is pulsatile; thus, there may be changes in $D/\dot{Q}$ or $\dot{V}/\dot{Q}$ with each heartbeat. Distinguishing among the vari-

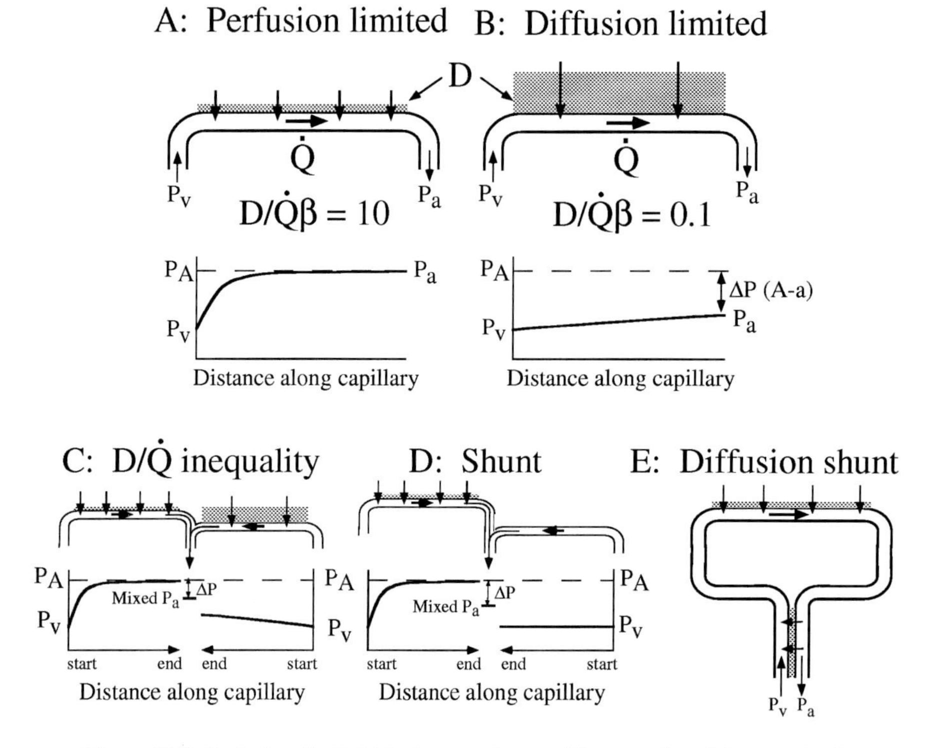

Figure 18.2 Perfusion-limited ideal gas exchanger (diagram *A*) and four mechanisms generating a P_{O_2} difference between respiratory medium (P_A) and arterialized blood (P_a). In each example, mixed venous blood (P_v) flowing through capillaries at rate $\dot{Q}$ exchanges gases across a surface (stippled) with a diffusive conductance, D. In diffusion limitation (diagram *B*), D is much lower than the perfusive conductance ($\dot{Q}\beta$), so that gases do not equilibrate between respiratory medium and blood. In $D/\dot{Q}$(or $D/\dot{Q}\beta$) inhomogeneity (diagram *C*), blood from pathways equilibrated with the external medium (*left*) is mixed with nonequilibrated blood from diffusion-limited pathways (*right*). In true shunt (diagram *D*), equilibrated blood is mixed with blood that bypasses the gas exchanger. In diffusion shunt (diagram *E*), blood equilibrates across the gas exchanger, but some O_2 is lost in countercurrent exchange between supply arterioles and venules. The various mechanisms causing ΔP between blood and the respiratory medium are not mutually exclusive; ΔP may be the result of several mechanisms, including mechanisms not diagrammed here. (After Piiper & Scheid, 1989; Piiper, 1993.)

ous potential causes of lack of equilibration is often very difficult. For example, although $D/\dot{Q}$ inhomogeneity may significantly decrease gas exchanger efficiency (Piiper, 1961), it is almost impossible to measure (Yamaguchi et al., 1993). When the cause of incomplete equilibration cannot be ascertained, the size of an equivalent shunt ("functional shunt") may be calculated.

Gas exchange models for developing vertebrates

Gas exchange in vertebrate development can be categorized into three broad functional patterns: anamniote (fish and amphibians), oviparous amniotes (birds and most reptiles), and viviparous amniotes (mammals, with close parallels in all other vertebrate classes except birds). Embryo size, metabolic rate, embryonic respiratory structures, and respiratory medium differ among the three patterns of development. Different steps appear to be most limiting in each of the three functional patterns, resulting in different questions guiding model development. Therefore, the discussion of specific models of gas exchange has been divided into sections corresponding to the three patterns of development.

Fish and amphibian eggs and larvae

Simple diffusion models: Eggs. Fish and amphibians comprise 25,000–30,000 species adapted to a wide range of environments, and thus there are many variations and exceptions to the general patterns discussed here. Most attention has been focused on oxygen diffusion through the egg capsule external to the embryo, using simple diffusion models. Eggs are generally small compared with amniote eggs, usually 0.5 to about 6 mm in diameter (Duellman & Trueb, 1986; Blaxter, 1988). They are usually aquatic and may be laid singly, in loose masses, or in cohesive masses, buried in gravel or sand, attached to vegetation, or pelagic.

A complete model of gas exchange in a fish egg has not yet been published, although Rombough (1988b) summarizes the steps and what is known about the relative resistances of each step. The diffusion barriers he considered were the external boundary layer of stagnant water, the acellular egg capsule (sometimes called the chorion in fish, although it is not similar to the amniote chorion), and the water-filled perivitelline space between the inside of the capsule and the embryo (Figure 18.3).

Boundary layers of hypoxic and hypercapnic water are found whenever O_2 and CO_2 diffuse between an organism and surrounding fluid. Equations for estimating the thickness and ΔP across boundary layers for spherical eggs are presented by Daykin (1965) and Rombough (1988b). Boundary layer thickness and ΔP increase with decreasing external water velocity, increasing egg size, and increasing $\dot{M}_{O_2}$. At the high end of boundary layer resis-

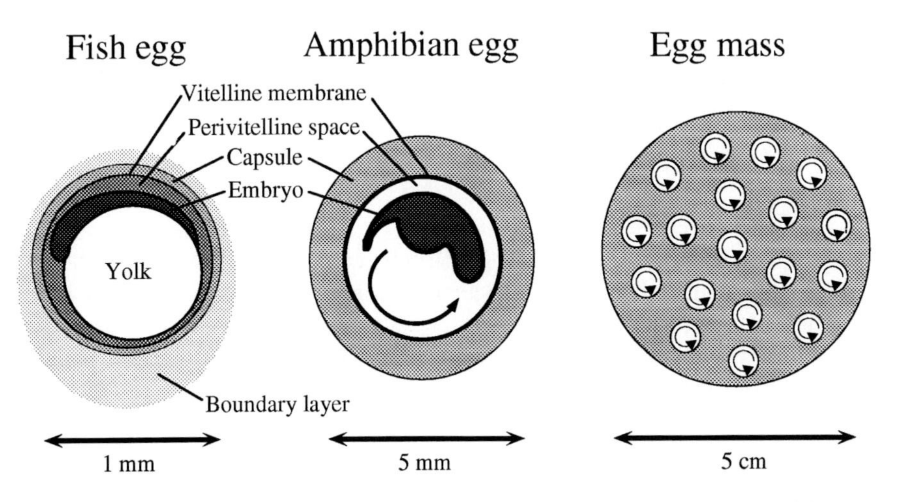

Figure 18.3 Diffusive barriers in a fish egg, amphibian egg, and egg mass. Fish eggs and amphibian eggs are quite similar, although amphibian eggs tend to be larger and have a thicker jelly capsule, and have epithelial cilia that create convection in the perivitelline space. Convection within egg capsules inside egg masses may increase the apparent conductance of the egg mass.

tance, ΔP was estimated to be 52 torr for a 4 mm radius (center to outside of capsule) salmonid egg near hatching at 10°C and 85 cm·h^{-1} flow velocity (Daykin, 1965). For demersal eggs in still water, or pelagic eggs that move with the surrounding water, externally forced water currents may be near zero. There may still be water convection, however, caused by respiratory exchange itself. Hypercapnic, hypoxic water is relatively dense and tends to sink, setting up toroidal circulation around the egg (O'Brien et al., 1978).

Diffusion through the egg capsule depends on the geometry and K_{O_2} of the capsule. Unfortunately, capsule K_{O_2} has not been measured but has instead been calculated using $\dot{M}_{O_2}$, capsule geometry, and ΔP across the capsule derived from the change in the critical P_{O_2} (the P_{O_2} at which $\dot{M}_{O_2}$ becomes oxygen limited) after removal of the capsule (references in Rombough, 1988b). According to these calculations, the capsule is relatively O_2 impermeable, having a K_{O_2} only about 10% of water (in contrast to amphibian egg jelly, which has a K_{O_2} about two-thirds that of water; Burggren, 1985a; Seymour, 1994). The estimate of K_{O_2} in fish egg capsules is therefore uncertain, and there may be differences between species (Rombough, 1988b). Direct measurements of K_{O_2} are necessary before diffusion across egg capsules can be modeled confidently. Of note, the ΔP_{O_2} across the capsule measured with microelectrodes is only a few torr (Berezovsky et al., 1979; Sushko, 1982).

The perivitelline space, if unstirred, is another large diffusion barrier. Although it has the K_{O_2} of water, the perivitelline space can be an order of magnitude thicker than the capsule (Rombough, 1988b). The total conductance within the egg to the embryo can be reduced by movements of the embryo, which start quite early in development, so that gas transport across the perivitelline space is convective rather than diffusive. Later in development, the perivitelline space may be stirred by the pectoral fins as well as by occasional body movements (Rombough, 1988b).

Amphibian eggs are generally larger than fish eggs and usually have a very thick gelatinous capsule that dominates the resistance to gas exchange. In addition, eggs may also be embedded in a mass of jelly. Both single eggs and egg masses have been modeled (e.g., Burggren, 1985a; Seymour, 1994; Pinder & Friet, 1994; Seymour & Bradford, 1995); the most complete review is by Seymour and Bradford (1995).

The diffusing capacity of the capsule is determined by its geometry and Krogh's diffusion constant (K) in jelly:

$$D = K\, 4\pi r_o r_i/(r_o - r_i),$$

thus

$$\dot{M} = [K\, 4\pi r_o r_i/(r_o - r_i)]\ \Delta P,$$

where r_i is the inner capsule radius and r_o is the outer radius (Figure 18.3; Seymour & Bradford, 1995). In this model, D is very sensitive to r_i and less so to r_o and to the thickness of the capsule ($r_o - r_i$), especially with thick capsules. As metabolic rate increases during development, capsule diffusing capacity also increases because r_i increases during development; thus, ΔP does not increase much, and P_{O_2} within the capsule is maintained (Seymour & Roberts, 1991). Although a diffusion boundary layer was not included in this model, it would have the effect of increasing r_o and thus would not greatly affect gas exchange. Although the perivitelline space increases with r_i, amphibian embryos are covered with cilia (unlike most fish embryos), which stir the perivitelline fluid fairly vigorously (Burggren, 1985a). Thus, gas conductance through the perivitelline space is by convection, and the resistance of this step of gas transport is thought to be low (Burggren, 1985a).

Egg masses, which can be 15 cm or more in diameter, have a much higher diffusive resistance than single eggs. In many egg masses there are probably open channels between eggs so that water convection can carry O_2 in and CO_2 out of the egg mass. Strathmann and Chaffee (1984) provide general models of egg masses, including calculations of convection through loosely constructed masses. Convection may be caused by waves, water currents, or temperature or solute gradients in and around the egg mass. Thus, gas transfer in loose egg masses is difficult to model because of complex interactions between convection and diffusion. Obviously, an egg mass with convection can be much larger than one dependent entirely on diffusion. Convection inside egg capsules in an egg mass may increase the effective diffusive conductance of the egg mass (Figure 18.3; Burggren, 1985a).

Egg masses dependent on diffusion have high resistance and large ΔP and may be anoxic in the center (Seymour & Roberts, 1991; Pinder & Friet, 1994). Burggren (1985a) measured relatively small gradients in the "firm" egg masses of *Rana palustris,* but there may have been channels for water convection. The model for diffusion into an egg mass differs from that of a single egg because O_2 consumption is distributed throughout the egg mass. Seymour and Roberts (1991) calculated a "global" P_{O_2} profile through an egg mass by calculating the ΔP_{O_2} and $\dot{M}_{O_2}$ of finite spherical shells, starting from the outside of the mass and assuming homogeneous distribution of $\dot{M}_{O_2}$. They then superimposed the "local" profiles calculated for individual eggs on the global profile (Figure 18.4). In the species they studied (*Limnodynastes tasmaniensis*) diffusion was sufficient to provide O_2 to the innermost embryos up to the time of hatching, but only because the egg mass spread and flattened and individual egg r_i increased as the embryos developed.

Pinder and Friet (1994) modified the global model of Seymour and Roberts (1991) for cylindrical egg masses of *Ambystoma maculatum,* which are up to 6 cm in diameter, have no channels for convection, and do not change shape during development (although r_i of individual capsules increases). Both the model and direct microelectrode measurements demonstrated that diffusion alone is inadequate for O_2 delivery to late embryonic stages. These egg masses (as well as those of some other amphibians) have symbiotic algae in them, so that there is local CO_2 consumption and O_2 production. Photosynthesis thus increases P_{O_2} to hyperoxia during the day, and at night respiration of both larvae and algae decreases P_{O_2} to anoxia in the center of the egg mass. Therefore, applying a steady-state

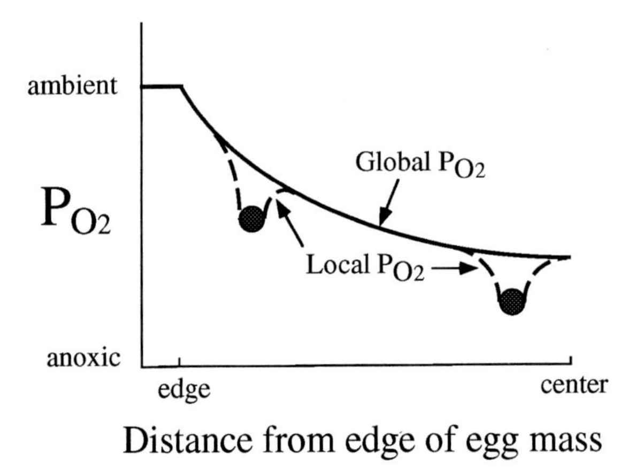

Figure 18.4 Global and local P_{O_2} profiles in an amphibian egg mass. "Global" P_{O_2} was calculated for the egg mass as a whole, assuming a homogeneous $\dot{M}_{O_2}$. "Local" P_{O_2} was calculated for the immediate vicinity of individual eggs and superimposed on the global P_{O_2} profile. (After Seymour, 1994.)

diffusion model is inappropriate, and these egg masses should be modeled as oxygen capacitors that are discharged and recharged daily. The relationship between embryo $\dot{M}_{O_2}$ and local P_{O_2} must be determined to realistically model these egg masses.

The start of internal convection: Parallel diffusion and convection. The preceding models deal with barriers external to the embryo. Only a few attempts have been made to model – or measure – gas exchange within the embryos of fish or amphibians. Gas exchange must be by direct diffusion to tissues in early stages before circulation starts. Direct diffusion has been modeled only as a simplified exercise assuming a spherical body form (or other simple geometric shape) and homogeneous M_{O2}, showing that the upper limit for direct diffusion is a 1 mm sphere of tissue (Krogh, 1941). In eggs, however, active tissue is wrapped around the outside of the metabolically inert yolk; therefore, diffusion distances are relatively small. Tissue $\dot{M}_{O_2}$ probably remains very heterogeneously distributed, with highly active tissues like the epithelium and the red muscle layer (in fish larvae) in close proximity to the external medium (El-Fiky & Wieser, 1988; Wells & Pinder, 1996a). Late embryos presumably resemble newly hatched larvae in which gas exchange is largely cutaneous (Burggren & Pinder, 1991; Rombough & Ure, 1991; Wells & Pinder, 1996b), changing over to predominantly branchial in fish as lamellae develop on gill filaments, and to a combination of cutaneous, branchial, and pulmonary exchange in amphibians (Burggren & Pinder, 1991).

Even after circulation starts, direct diffusion may be an important conductance parallel to circulatory oxygen transport, as in a model developed by Piiper, Gatz, and Crawford (1976) for gas exchange in a lungless salamander, *Desmognathus fuscus.* Although direct diffusion accounted for only 3–5% of $\dot{M}_{O_2}$ in a 6 g *Desmognathus,* direct diffusion is much more important in smaller larvae, which have much shorter diffusion distances (Mellish, Smith, & Pinder, in press). *Ambystoma mexicanum* embryos treated with CO to prevent circulatory oxygen transport, cardiac-lethal mutants (in which the heart never starts to beat), and embryos from which the heart anlagen had been removed earlier in development had the same critical P_{O_2} and overall G_{O_2} as normal embryos (Mellish, Smith, & Pinder, in press), suggesting that circulatory O_2 transport is a minor conductance in these 5–10 mg animals. Similarly, *Xenopus laevis* larvae grow and behave normally when raised in 2% CO (Burggren & Territo, 1995), and CO-treated zebra fish embryos (80–106 hours of development) have the same $\dot{M}_{O_2}$ and ventricular pressure as normal embryos (B. Pelster & W. Burggren, unpublished data). CO decreased G_{O_2} by 50% in brown trout larvae (50–140 mg), suggesting that direct diffusion and circulatory transport contribute equally to O_2 transport (A. Pinder, R. Sethi, & P. Wells, unpublished data).

Bird and reptile eggs and embryos

As in fish and amphibians, bird and reptile eggs have a large diffusive resistance to gas transfer imposed by the eggshell and associated shell membranes overlying the gas exchange surface (see also Chapters 14 and 15). The eggs are much larger than those of fish and amphibians, however, and so for most of development the embryos have a specialized gas exchange organ: first, the area vasculosa of the yolk sac and then the chorioallantois. For obvious reasons the chicken egg has been the exemplar of amniote egg development, and fortunately it is representative of birds in general (Rahn & Paganelli, 1990). Because oxygen consumption increases rapidly as yolk is converted to metabolizing tissue, whereas the gas exchange capacity of the egg remains relatively constant, oxygen consumption becomes limited by the gas exchange system, and $\dot{M}_{O_2}$ plateaus in late stages of development (references in Metcalfe & Stock, 1993), just before internal pipping and the start of lung use (around day 19 in chickens). The metabolic rate of avian embryos is proportional to body mass$^{0.71}$ but is only about 25% of that of an adult bird scaled to the same size (Hoyt & Rahn, 1980).

Gas exchange in late-stage bird eggs. Perhaps the closest adult vertebrate analog to a late-stage amniote egg is a lungless salamander, which has a high diffusion resistance of the gas exchanger and whose circulation is arranged with the gas exchanger parallel to tissues (Piiper & Scheid, 1984). Gases diffuse in gas phase through an outer barrier composed of pores in the eggshell, the outer shell membrane, and the inner shell membrane and then through an inner, fluid-filled barrier composed of a thin layer of the inner shell membrane, the living cells of the chorioallantoic epithelium and blood vessel endothelium, and finally into the blood (Figure 18.5). The resistance of each of these barriers in series has been estimated (Metcalfe & Stock, 1993). In the outer barrier, the eggshell has the lowest diffusive conductance. Adding the outer shell membrane decreases conductance by about 6% (Kayar et al., 1981), and the thin inner shell membrane (not including the fluid-filled layer) adds negligible resistance (Wangensteen & Weibel, 1982). The conductance of an individual pore, regardless of species or egg size, is the same for all eggs. Shell conductance is matched to metabolic rate by varying the density of pores so that the partial pressures under the shell just before internal pipping are $P_{O_2} = 104 \pm 1.3$ and $P_{CO_2} = 41 \pm 1.2$ torr (mean $\pm$ standard error in 25 species of birds; Rahn & Paganelli, 1990). The conductance of the gas-filled outer barrier is somewhat lower for CO_2 than for O_2 because of its higher molecular weight. The value of *RQ* is 0.71 (Ancel & Visschedijk, 1993). Exceptions to these "standard" conductances and ΔP occur owing to high altitude or requirements for water balance (Rahn & Paganelli, 1990), but the values are surprisingly uniform among most birds.

About two-thirds of the total resistance to oxygen movement from the environment to blood in bird eggs is in the fluid-filled inner barrier (Bissonnette & Metcalfe, 1978; Piiper et al., 1980). Resistance of the inner barrier to O_2 movement is divided between diffusion

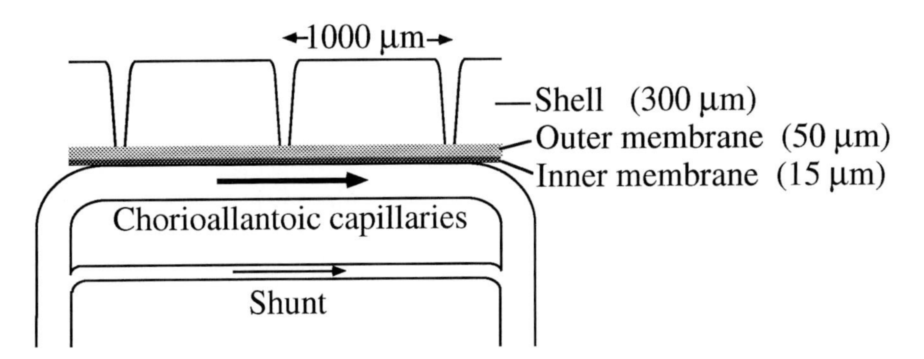

Figure 18.5 A general model of gas exchange in bird eggs, showing the three serial resistances of the diffusion barrier external to the chorioallantois: the eggshell and the outer and inner shell membranes.

through tissues and the reaction of O_2 with Hb. Wangensteen and Weibel (1982) calculated from morphometry that the major resistance was the reaction with Hb. The tissue barrier is only 0.4 μm thick but is limited to the surface area of the shell. In contrast, Bissonnette and Metcalfe (1978) calculated that both diffusion and reaction were significant resistances and that diffusion through tissue was the greater barrier during most of the development of bird embryos.

O_2 and CO_2 have similar diffusion coefficients in gas; thus, ΔP_{CO_2} is similar to ΔP_{O_2} across the shell and shell membranes. CO_2 has a much higher diffusion coefficient in aqueous solution, however, and so ΔP_{CO_2} across the inner barrier is much smaller than ΔP_{O_2}. Because of the slightly higher resistance to CO_2 diffusion in the gas phase and the much lower resistance to diffusion in aqueous solution compared with O_2, the shell and shell membranes represent about 86% of the total resistance to CO_2 excretion, and diffusion through the inner barrier represents only about 14%; total resistance to CO_2 transfer is only about half that for O_2 (Piiper et al., 1980).

Changes in shell and membrane conductance during development. The gradients across the eggshell and the conductances for gas exchange both increase during development because $\dot{M}$ increases much more than G. Most models assume a constant conductance throughout development, and this gives a fairly good approximation to the actual changes in partial pressure differences across the shell (e.g., Wangensteen & Rahn, 1970/71; Wangensteen, Wilson, & Rahn, 1970/71; Ancel & Visschedijk, 1993). ΔP_{O_2} and ΔP_{CO_2} across the eggshell and membranes, calculated from $\dot{M}_{O_2}$ and shell conductance, increase from close to 0 in a newly fertilized egg to around 40 torr just before internal pipping (Figure 18.6). A constant-conductance model is least accurate in the earliest stages of development, predicting much lower ΔP_{O_2} than measured, because the total conductance from atmo-

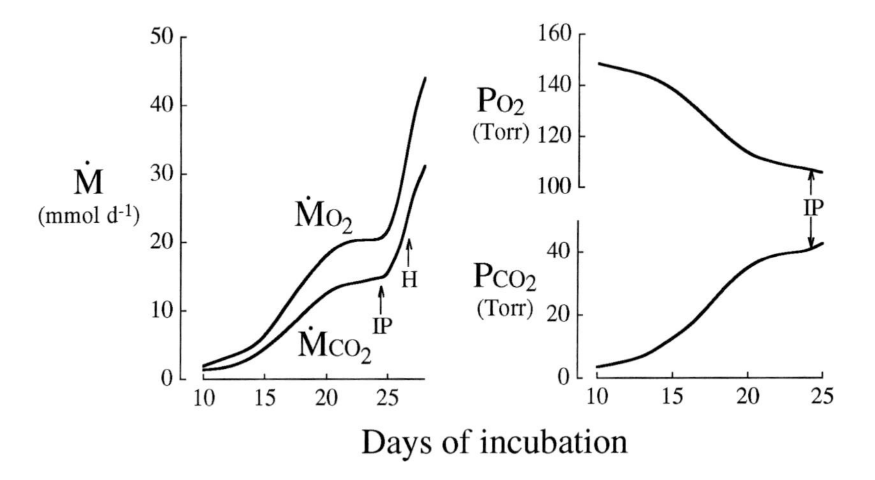

Figure 18.6 Changes in measured $\dot{M}$ and calculated P for the air space of a guinea fowl egg from 10 days of incubation to hatch, assuming a constant shell conductance (note that incubation is slightly longer than in chickens; data from Ancel & Visschedijk, 1993). (IP = internal pipping, H = hatch.)

sphere to blood increases during development (Bissonnette & Metcalfe, 1978; Kayar et al., 1981; Baumann & Meuer, 1992; Metcalfe & Stock, 1993). The shell conductance is indeed constant, set by the number and dimensions of pores when the egg is laid. Conductance of the shell membranes, however, increases 10-fold between days 6 and 10 of incubation as the water that initially fills the interstices between fibers in the outer and inner shell membranes is replaced with gas. Water evaporates, and the balance between capillary tension of fibers in the membranes and the colloid osmotic pressure of the albumen changes to stop replacement of water from the albumen (Seymour & Piiper, 1988). Thereafter, the shell membrane conductance is constant; the conductance of the outer barrier remains constant for the last half of incubation.

Before the chorioallantois becomes functional, the embryo uses the area vasculosa of the yolk sac for gas exchange (from about day 3 to day 6 in chickens; Baumann & Meuer, 1992). Conductance increases as the area vasculosa increases in area, the thickness of a layer of albumen between the area vasculosa and the shell decreases (Baumann & Meuer, 1992), and the heterogeneity of blood flow decreases (Meuer & Bertram, 1993; see below). Once the chorioallantois forms, its conductance increases over most of the rest of development (Bissonnette & Metcalfe, 1978). Its area increases to cover the entire surface of the egg by day 12. Conductance of the chorioallantois doubles between days 10 and 18 through a combination of increased blood oxygen carrying capacity and increased capillary volume in the chorioallantois (Bissonnette & Metcalfe, 1978).

Effects of inhomogeneities. The efficiency of any gas exchanger is decreased by inhomogeneities. One potential form of heterogeneity results from the structure of the eggshell: gases can diffuse only through pores; the 300 μm thick shell itself is impermeable. There are about 10,000 pores in a chicken egg, spaced about 1000 μm apart (Wangensteen, Wilson, & Rahn, 1970/71; Figure 18.5). An important function of the shell membranes is to provide a low-resistance diffusion pathway between pores under the shell. There is little difference in P_{O_2} immediately under a pore and that at a point halfway between pores (Visschedijk, Girard, & Ar, 1988). In addition, the P_{O_2} in the air cell should be similar to the P_{O_2} under the rest of the shell, as assumed in the models. K_{O_2} parallel to the egg surface has been measured to be 100 times the coefficient perpendicular to the surface (Ar & Girard, 1989), with a negligible drop in P_{O_2} between pores. Nonetheless, there is some $D/\dot{Q}$ heterogeneity, resulting in P_{O_2} differences between the air cell and the whole egg of about 4 torr (Paganelli et al., 1988).

Chorioallantoic exchange. Gas exchange through the shell and egg membranes is relatively uniform, but there are large inhomogeneities in chorioallantoic perfusion. Metcalfe and Bissonnette (1981) calculated that, given an ideal homogeneous exchanger, blood P_{O_2} should increase from 20 to 82 torr passing through the chorioallantois. They found, however, that the P_{O_2} of blood leaving the chorioallantois was only 62 torr. The difference could be explained by a 15% functional shunt. Using a very different approach, Piiper et al. (1980) also calculated a 15% functional shunt from an analysis of ΔP_{O_2} and ΔP_{CO_2} in normoxia, hypoxia, and hyperoxia. The functional shunt could be caused by arteriovenous anastomoses (or very fast transit time capillaries), O_2 extraction by the chorioallantois from oxygenated blood, or arteriovenous back-diffusion (Piiper et al., 1980; Metcalfe & Bissonnette, 1981). Evidence for inhomogeneous blood flow and a proportion of capillaries with fast transit times has been supplied by Meuer and Bertram (1993), who measured red cell velocities and transit times in the yolk sac area vasculosa of 4-day-old embryos. Transit times ranged from 0.1 to 17 seconds, with a median of 2.5 seconds. About one-third of the red cells had transit times too short to saturate the Hb. Meuer (1992) proposed that newly formed vasculature is laid down randomly and is later optimized during development, decreasing flow heterogeneity. The oxygen saturation of vitelline venous blood increases from 89% to almost 100% between day 4 and day 6 (Meuer, Sieger, & Baumann, 1989), but changes in heterogeneity between days 4 and 6 have not been measured.

Reptile eggs. Much less is known about respiratory gas exchange in reptile eggs than in bird eggs. The structure of reptile eggs is much more variable than that of bird eggs, ranging from rigid gecko eggs, with conductance lower than that of bird eggs, to flexible or parchmentlike shells in many squamates, with up to 2 orders of magnitude higher water conductance (Deeming & Thompson, 1991). Most reptile eggs appear to have a higher conductance than bird eggs, but O_2 and CO_2 conductances cannot be accurately calculated from water conductance (which is most easily and commonly measured), implying that a variable part of the diffusion barrier is through liquid (Deeming & Thompson, 1991). Few blood gas measurements have been made in developing reptiles, so chorioallantoic conductance has not been assessed. Birchard and Reiber (1993) found that the $\dot{M}_{O_2}$ of snapping turtle eggs was 20% of that of similar-sized quail eggs but that the chorioallantoic vascular density was almost 50% of the quail density. If both the shell conductance and the chorioallantoic membrane conductances per unit $\dot{M}_{O_2}$ are higher in reptiles, blood P_{O_2} should be higher and P_{CO_2} lower than in birds. Indeed, the ΔP_{O_2} across green turtle eggshells near hatch was calculated to be only 5.2 torr (Prange & Ackerman, 1974). Reptile eggs are often buried in clutches, however; thus, the gas partial pressures to which reptile embryos are exposed also depend on the porosity of the soil, the number of eggs in the clutch, and their aggregate metabolic rate. Ackerman (1977) measured P_{O_2} and P_{CO_2} in nests of two sea turtle species and fitted a two-compartment radial diffusion model to the experimental data. The two compartments were the nest and the surrounding sand. The sand was the source for O_2 and sink for CO_2 and was modeled to have no outer bound (an infinite medium). The apparent D_{O_2} and D_{CO_2} through the gas spaces between sand grains in the outer compartment and between eggs in the inner compartment were estimated by fitting the model results to experimental measurements. Diffusion coefficients within the nest were about 20% of those in air (similar to the estimated 26% gas volume between eggs packed in a tetrahedral pattern), whereas in sand diffusion coefficients were only 8–12% of those in air, depending on the average grain size of the sand. As a result of the fixed and relatively high external resistance of the sand, the turtle embryos were exposed to gas partial pressures similar to those inside the shells of bird eggs, with a P_{O_2} around 100 torr and a P_{CO_2} around 40 torr at hatching.

Placental gas exchange

Diversity of placentae. The chorioallantois remains the gas exchange organ in mammalian placental gas exchange, but the interface with the external environment is mediated through the physiological regulatory mechanisms of the mother rather than through simple unresponsive eggshells. Placentae have evolved independently many times in all vertebrate classes except birds (Blackburn, 1992). Not all placentae exchange nutrients between mother and fetus – many embryos and fetuses use yolk as their exclusive nutrient source –

but all must exchange respiratory gases. The details of placental gas exchange come from a few mammalian species, including humans and their most common animal model, sheep. There are comprehensive recent reviews of vertebrate placentation, especially morphology: Blackburn (1992) gives a short overview of vertebrates, Wourms (1994) reviews fish, and several papers in volume 266 of the *Journal of Experimental Zoology* review various taxonomic groups (e.g., reptiles: Blackburn, 1993, and Stewart, 1993; elasmobranchs: Hamlett et al., 1993; and teleosts: Schindler & Hamlett, 1993). In general, nonmammalian placentae appear to have much larger diffusion distances and lower surface areas than mammals, but nonmammalian fetuses also have lower metabolic rates. The placenta is derived from the yolk sac in anamniotes; it may be derived from either the yolk sac or the chorioallantois in reptiles. Gas exchange in utero in fish and amphibians may be augmented or entirely served by external or internal gills, fins, and general body surface (Webb & Brett, 1972a, 1972b).

Even in "placental" (eutherian) mammals, there is a wide variety of placental morphology, with from three to seven layers of tissue and basement membranes separating maternal and fetal blood. The relative directions of maternal and fetal blood are difficult to determine in the complex structure of the placenta: concurrent, crosscurrent, and countercurrent flows have all been proposed in different species (Figure 18.7). There is probably a mix of flow patterns in any one placenta (Meschia, Battaglia, & Bruns, 1967).

Paired capillary concurrent model. The best-developed models of placental gas exchange are based on data from sheep, with analysis of humans where data are available (Hill, Power, & Longo, 1972, 1973; Longo, Hill, & Power, 1972); the models and comparisons to experimental data are reviewed by Longo (1987) and Carter (1989). The placenta was simplified to parallel maternal and fetal capillaries of equal length and diameter and uniform diffusion barrier, with concurrent flow (Figure 18.8). Blood was assumed to be uniform (no red cells) and flow nonpulsatile, and only diffusion normal to the two capillaries was considered. The rate of reaction of Hb with O_2 was modeled, and Hb equilibrium curves were approximated with the Hill equation. O_2 exchange rate (i.e., fetal $\dot{M}_{O_2}$) was treated as the dependent variable, calculated from initial and final fetal O_2 contents, although it was recognized that in fact gas exchange would be regulated in vivo to maintain constant fetal $\dot{M}_{O_2}$.

The central prediction of the oxygen uptake model is that equilibration should be almost complete (99.4%) between fetal and maternal blood, although equilibration time was just adequate and thus would be sensitive to decreases in transit time for the blood (Hill, Power, & Longo, 1972). A similar model for CO_2 exchange (Hill, Power, & Longo, 1973) predicted equilibration of CO_2 as well. Although CO_2 has a diffusion constant about 20 times higher than O_2, the interdependence of CO_2 reactions and Hb–O_2 reactions (Bohr or Haldane effects) results in an exchange rate of CO_2 not much higher than for O_2. The rate of exchange of CO_2 was essentially independent of placental diffusing capacity; the exchange rate was determined by reaction rates.

Longo, Hill, and Power (1972) analyzed the sensitivity of the model to changes in various parameters. When $\dot{M}_{O_2}$ was calculated as the dependent variable, the model was most sensitive to umbilical arterial P_{O_2} and fetal P_{50}, less sensitive to maternal $\dot{Q}\beta$ and maternal P_{50}, and least sensitive to maternal arterial P_{O_2} and placental diffusing capacity. When fetal $\dot{M}_{O_2}$ was more realistically assumed to be regulated to be constant for each developmental stage, the model was most sensitive to maternal P_{50}, maternal $\dot{Q}\beta$, and total maternal and fetal blood flows; less sensitive to the ratio of maternal/fetal blood flow or maternal P_{O_2};

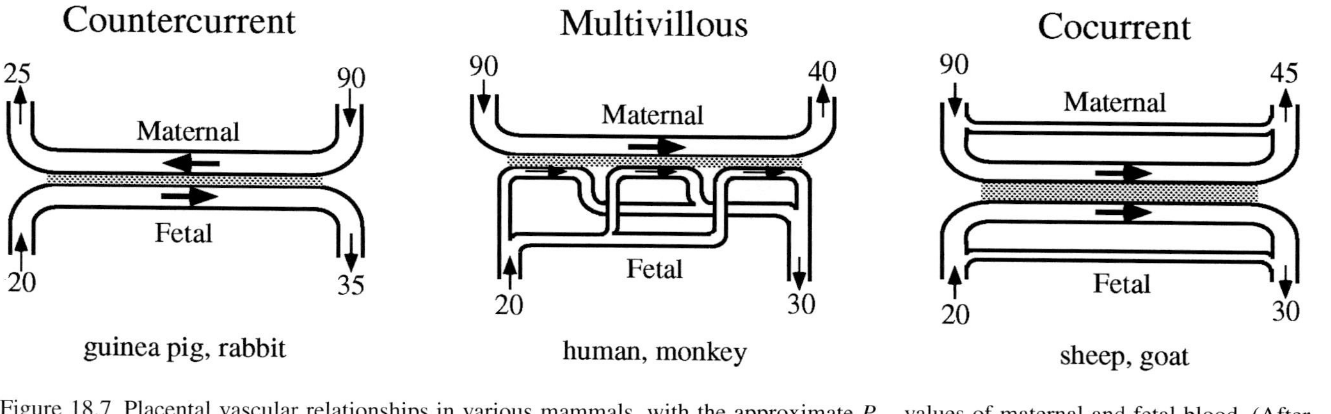

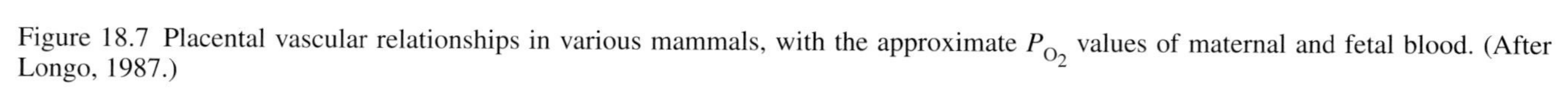

Figure 18.7 Placental vascular relationships in various mammals, with the approximate P_{O_2} values of maternal and fetal blood. (After Longo, 1987.)

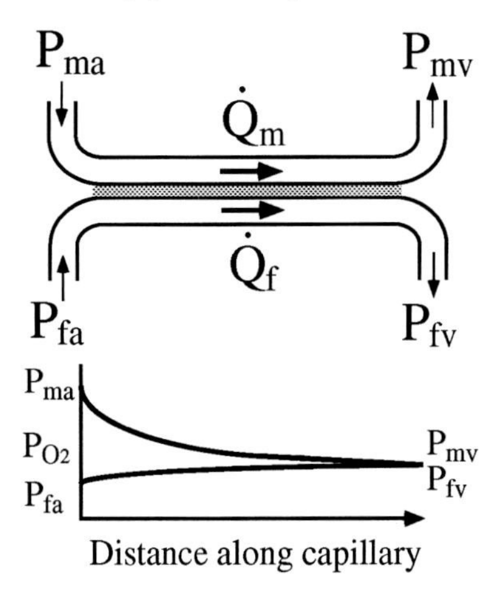

Figure 18.8 The concurrent flow model for gas equilibration across human and sheep placenta. The two blood flows have just adequate time to equilibrate ($P_{ma} = P$ in the uterine [maternal] artery, $P_{fa} = P$ in the umbilical [fetal] artery, $P_{mv} = P$ in the uterine vein, $P_{fv} = P$ in the umbilical vein). (After Hill, Power, & Longo, 1972.)

and little affected by fetal $\dot{Q}\beta$, fetal P_{50}, or placental diffusing capacity. There are strongly nonlinear responses in the model, presumably related to the nonlinear Hb–O_2 equilibrium curve. For example, maternal arterial P_{O_2} could be raised or lowered to about 70 torr without much effect, but below 70 torr, maintaining fetal $\dot{M}_{O_2}$ quickly became impossible (Longo, Hill, & Power, 1972). Interestingly, raising fetal P_{50} above maternal P_{50} had little effect on O_2 transfer; decreasing fetal P_{50} was more detrimental. O_2 transfer was proportional to blood flow, as expected in a perfusion-limited system.

Testing the concurrent model. Placental models were developed because the appropriate experimental measurements in vivo are extremely difficult. This problem still exists, and thus the prediction of gas equilibration has been tested only indirectly. One of the main drawbacks to modeling placental gas exchange is that the spatial relationship between maternal and fetal blood flow has a major influence on gas transfer (Longo, Hill, & Power, 1972; Faber, Thornburg, & Binder, 1992), but there are few data available on the actual vascular arrangements in vivo. The placenta is not as highly ordered a structure as a lung or gill, and the spatial relationships between maternal and fetal blood flows probably vary from one area to another. Other possible relationships between maternal and fetal blood flow, such as a perfused pool or crosscurrent exchanger, were not modeled to test whether concurrent flow gave the best fit to experimental data. Longo (1987) compared the predictions of the concurrent flow model with experimental data. In general, experimental manipulations of maternal O_2 delivery, P_{O_2}, and mutant Hbs with altered P_{50} are consistent with the model predictions, although interpretations of experiments are complicated by regulatory responses in vivo.

The conclusion that placental gas exchange has little diffusion limitation is not accepted by all workers; Meschia (1994) and Wilkening and Meschia (1992) believe that the number of simplifications necessary in the model make it unrealistic and that the experimental data are more consistent with a partly diffusion-limited placenta. The equilibration time available even in the completely homogeneous model is just adequate for complete equilibration of blood; thus, even moderate inhomogeneity of flow or diffusion barrier would make at least some capillaries diffusion limited.

The most obvious discrepancies between placental model predictions and experimental measurements are P_{O_2} and P_{CO_2} differences measured between the uterine vein and umbilical vein (10–15 torr and 4 torr, respectively, in sheep; ΔPs are similar in humans). These differences were speculated to be the result of a 26% functional shunt, perhaps due to $\dot{Q}_m/\dot{Q}_f$ heterogeneity (Figure 18.9A) or $D/\dot{Q}\beta$ heterogeneity (Figure 18.9B) rather than diffusion limitation (Hill, Power, & Longo, 1972). Placental oxygen consumption is unlikely to contribute significantly to vein–vein differences (Longo, 1987). Unfortunately, in contrast to $\dot{V}/Q$ inequality, $\dot{Q}_m/\dot{Q}_f$ heterogeneity cannot be assessed with the multiple inert gas equilibration technique (Wagner, Saltzman, & West, 1974) because the technique relies on using gases with different blood/gas partition coefficients. Blood/water (and obviously maternal blood/fetal blood) partition coefficients are always close to 1. It might be possible to experimentally manipulate maternal blood/fetal blood partition coefficients by replacing one with a perfluorocarbon blood substitute (Lowe, 1990) and thus use a version of multiple inert gas equilibration (Wagner, Saltzman, & West, 1974) to assess $\dot{Q}_m/\dot{Q}_f$ heterogeneity.

Blood flow heterogeneity was measured with microspheres by Power, Dale, and Nelson (1981), counted in very small (30 mg) samples. Maternal and fetal blood flows were normally distributed with a fairly high standard deviation of 44%, which was predicted to result in a vein–vein P_{O_2} difference of about half of that measured. Heterogeneity between samples decreased quickly with increasing sample size, indicating that heterogeneity was on a small spatial scale. Although Power, Dale, and Nelson (1981) believed that the sample size they measured was small enough to reflect the scale of placental exchange units, each sample still contained many thousands of capillaries. If there is large flow variation from one capillary to the next as measured by Meuer and Bertram (1993) in chick yolk sacs, even more of the vein–vein P_{O_2} difference could be explained by heterogeneity. Another proba-

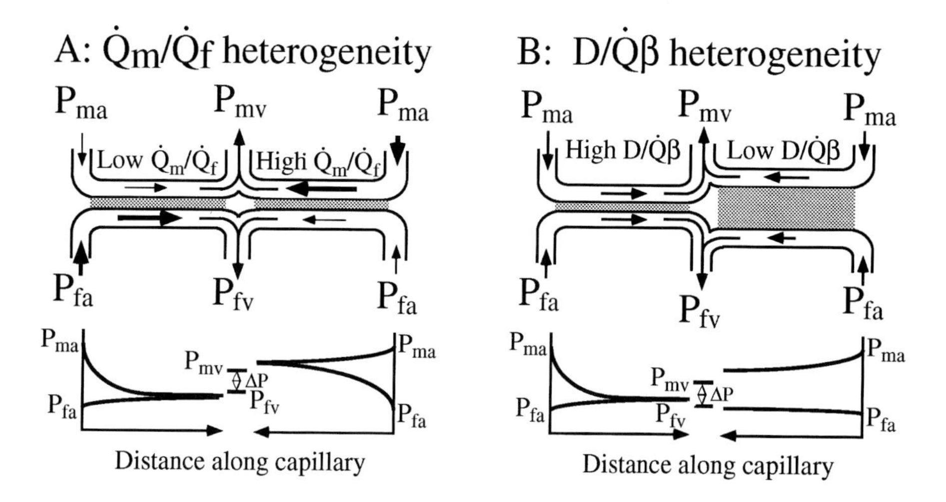

Figure 18.9 Two possible mechanisms accounting for the measured ΔP between blood leaving the placenta in the uterine vein (P_{mv}) and the umbilical vein (P_{fv}). (*A*) In $\dot{Q}_m/\dot{Q}_f$ heterogeneity, all blood flows may equilibrate, but the mixed P_{mv} is biased toward the relatively high P_{O_2} of high $\dot{Q}_m/\dot{Q}_f$ pathways, and P_{fv} is biased toward the relatively low P_{O_2} of low $\dot{Q}_m/\dot{Q}_f$ pathways. (*B*) In $D/\dot{Q}\beta$ heterogeneity, equilibrated blood from pathways with high $D/\dot{Q}\beta$ is mixed with nonequilibrated blood from pathways with low $D/\dot{Q}\beta$.

ble source of vein–vein P_{O_2} difference, $D/\dot{Q}\beta$ inequality, is unfortunately extremely difficult to assess (Yamaguchi et al., 1993), and no attempt has been made to measure it in placental gas exchange.

Alternative models. Other models of placental exchange have been developed. General models of placental exchange of inert molecules (which are neither produced nor consumed within the placenta nor bound to other molecules) demonstrate the theoretical efficiencies of different relationships of maternal and fetal blood flows (e.g., Meschia, Battaglia, & Bruns, 1967; Faber, Thornburg, & Binder, 1992), but because they assume a constant blood capacitance, they are not directly applicable to O_2 or CO_2, which exhibit strongly nonlinear capacitance curves. An ideal countercurrent exchanger is the most efficient, but it is also most sensitive to flow heterogeneities (Faber, Thornburg, & Binder, 1992; Powell, 1993). Diffusion limitation could not be distinguished from heterogeneity as a cause of uterine vein–umbilical vein differences (Faber, Thornburg, & Binder, 1992).

Costa, Costantino, and Fumero (1992) modeled the placenta as a "ventilated pool," with a constant maternal P_{O_2} external to flat-walled fetal capillaries. Separate terms for fetal capillaries and sinusoids were included (each fetal capillary has an average of four expanded regions, or sinusoids), but the model suffers from assuming "whirling" flow, or vortices, in sinusoids, resulting in zero resistance to diffusion in blood. Vortices seem unlikely, however, because blood flow is not turbulent at the low Reynolds number of flow in small vessels, estimated by the authors to be around 1. In any case, much of the resistance to gas movement in blood is due to reaction rates rather than diffusion through the blood. It is also not clear how maternal "pool" P_{O_2} was determined. This model suggests that oxygen uptake is partly diffusion limited.

There have been few attempts to extend the concurrent model of Hill, Power, and Longo (1972) to include different vascular arrangements, $\dot{Q}_m/\dot{Q}_f$ heterogeneity, $D/\dot{Q}\beta$ heterogeneity, and other factors, presumably because experimental data are lacking. Experimental measurements are currently limited to flows, gas partial pressures, and blood properties in the major vessels leading to and from the placenta. Because several different mechanisms could account for experimentally measurable phenomena, progress awaits development of new techniques to distinguish between models.

Summary and a perspective

The most successful gas exchange models, in the sense that they best explain experimental observations, are those in which the basic assumptions are least severely violated. A basic assumption of most models is homogeneity. External diffusion barriers such as egg capsules, the gelatinous matrix of amphibian egg masses, eggshells, and the gas spaces in reptile egg clutches and surrounding sand are relatively homogeneous, and thus the models yield realistic predictions. Perfusion of gas exchange surfaces, however, is not homogeneous, resulting in $D/\dot{Q}\beta$ heterogeneity and, in the placenta, $\dot{Q}_m/\dot{Q}_f$ heterogeneity. A major challenge in gas exchange modeling is to predict measurable effects of various forms of heterogeneity so that they can be experimentally distinguished and their contributions to "functional shunt" evaluated. Chorioallantoic gas exchange in bird eggs – in which there is no $\dot{V}/Q$ heterogeneity, red cell movements can be directly observed (e.g., Meuer & Bertram, 1993), and saturation of Hb may be measurable by microspectrophotometric techniques (e.g., Swain & Pittman, 1989) – is an attractive experimental system in which to test for effects of $D/\dot{Q}\beta$ heterogeneity. Basic questions also remain about oxygen uptake and transport in fish and amphibian larvae, such as how the relative importance of direct diffusion to circulatory transport is related to larval size, metabolic rate, and environmental P_{O_2}. The question of how heterogeneous $\dot{M}_{O_2}$ affects the importance of direct diffusion must await new ways to visualize metabolic rates of tissues in vivo. Placental gas exchange is likely to remain the least tractable system because of the complexity of the structure and its awkward position. Still unclear are the morphological relationships between fetal and maternal blood flows and the extent to which measured ΔP between umbilical venous and uterine venous blood is the result of diffusion limitation, shunt, or heterogeneity.

19

Principles of abnormal cardiac development

ADRIANA C. GITTENBERGER–DE GROOT AND
ROBERT E. POELMANN

Introduction

Standing on the shoulders of the many articles and reviews on cardiac development based upon descriptive approaches, and using tools developed with the aid of molecular biology (Pexieder, 1995), one can see new horizons. However, much to the chagrin of the increasing number of scientists equipped with molecular tools and interested in morphogenesis, the primary focus of research remains the refinement of descriptive events. Achieving significant new breakthroughs will necessarily require a move from the prevailing descriptive thinking to analysis identifying direct cause–effect relationships in order to expose the mechanisms underlying abnormal development.

The present chapter will focus on several crucial steps in normal cardiogenesis that have been identified as major developmental "bottlenecks." The impact of such a bottleneck is that distinct morphogenetic disturbances may at first sight appear to produce similar heart malformations. Critical analysis, however, shows that unique pathogenetic mechanisms produce a spectrum of unique results. Understanding of this principle is essential for both the basic scientist and the clinician. The following examples will provide insight into the complexities, the pitfalls, and – if well considered – the opportunities for improving the resolution of our perception.

The paradox of the double-outlet right ventricle

"Double-outlet right ventricle" (DORV) is a congenital cardiac anomaly that can be described using morphologic terms (Anderson, 1987). In this malformation one complete great artery (100% of the aorta or the pulmonary trunk) and more than 50% of the other great artery (pulmonary trunk or aorta) arise from the right ventricle. This nomenclature is sufficient if we can properly define a right ventricle. Hearts with this malformation can be classified into two basic subtypes: one with normally related great arteries with the pulmonary orifice 100% above the right ventricle, and the other with transposed arteries featuring the aorta 100% above this ventricle (Bartelings & Gittenberger–de Groot, 1991). Although this purely descriptive nomenclature is useful for clinicians and surgeons who have to repair or palliate the functioning heart, this nomenclature approach is of no use to scientists studying morphogenetic mechanisms. To define the pathogenesis of a specific structural anomaly, it is first necessary to describe normal heart morphogenesis and then determine the sequence of events that can lead to derailment. Defining the timing of these "events" is further confounded by the fact that we can study morphologic processes in animals and humans only when the events are not lethal to the embryo. In the human, only fetal and neonatal stages can be studied. Over the past several decades, early embryo harvest from animal has broadened the available investigative developmental window. In this setting we describe unique experimental paradigms that lead to DORV in two different model species.

Retinoic acid and the chick embryo

All-*trans*-retinoic acid given to chick embryos at Hamburger–Hamilton (HH) stage 15 results in a spectrum of heart malformations varying from an isolated, rightward-positioned aorta, through a small slitlike "subaortic" ventricular septal defect (VSD), to a large subaortic VSD (Figures 19.1a, 19.2a, 19.2b). If in the latter case the aorta is at least 50% above the right ventricle, this malformation fits the definition for a DORV (Broekhuizen et al., 1992, 1995; Bouman et al., 1995). Concomitant aortic arch anomalies show high variability. In hearts with severe DORV, careful morphologic evaluation reveals inlet abnormalities as well. This is consistent with the lack of expansion of the right atrioventricular (AV) canal, which results in a straddling tricuspid valve (Bouman et al., 1995). Myocardial contraction is impaired in treated embryos as early as HH stage 24 (Broekhuizen et al., 1995).

It is challenging to theorize which change or changes in the normal developmental cascade produce this specific spectrum of malformations. Current data from molecular biology and cell-tracing techniques can assist in this task. First, at HH stage 15 the embryo has an extensive set of retinoic acid (RA) receptors, expressed on the myocardial cells and on the neural crest cells migrating from the pharyngeal arches toward the heart. Our current hypothesis is that the first "effect" of RA is on the atrial and ventricular myocardium of the looping heart tube. This incursion then results in disturbed looping that is not sufficiently tight for normal development. The morphological spectrum from rightward-positioned aorta to the extreme DORV/double-inlet left ventricle can thus be explained. Because all cases contain an outlet septum derived from neural crest cells, it is assumed that neural crest cells do reach the outflow tract and account for the mesenchymal aortopulmonary cells. However, recent studies show the neuronal parasympathetic ganglion cells are disturbed after RA manipulation (M. L. A. Broekhuizen, R. E. Poelmann, & A. C. Gittenberger–de Groot, unpublished results). This implies that the alteration in cardiac function at later stages may have both myocardial and innervation-derived components. Important questions remain, including:

1. If neural crest cells do reach the heart as in normal development, what mechanism produces the spectrum of aortic arch malformations?
2. What impact does altered cardiac function have on aortic arch selection?

Retinoic acid and the mouse embryo

In this model, RA is applied to the mother by either nutrition or injection into the peritoneal cavity (Pexieder, Pfizenmaier Rousseil, & Prados-Frutos, 1992; Hierck et al., 1996). This teratogenic event produces a set of malformations that at first resembles the spectrum described in RA treatment of chick embryos. The time of application is comparable: embryo day 7.5 in the mouse and HH stage 15 in the chick. There are, however, several

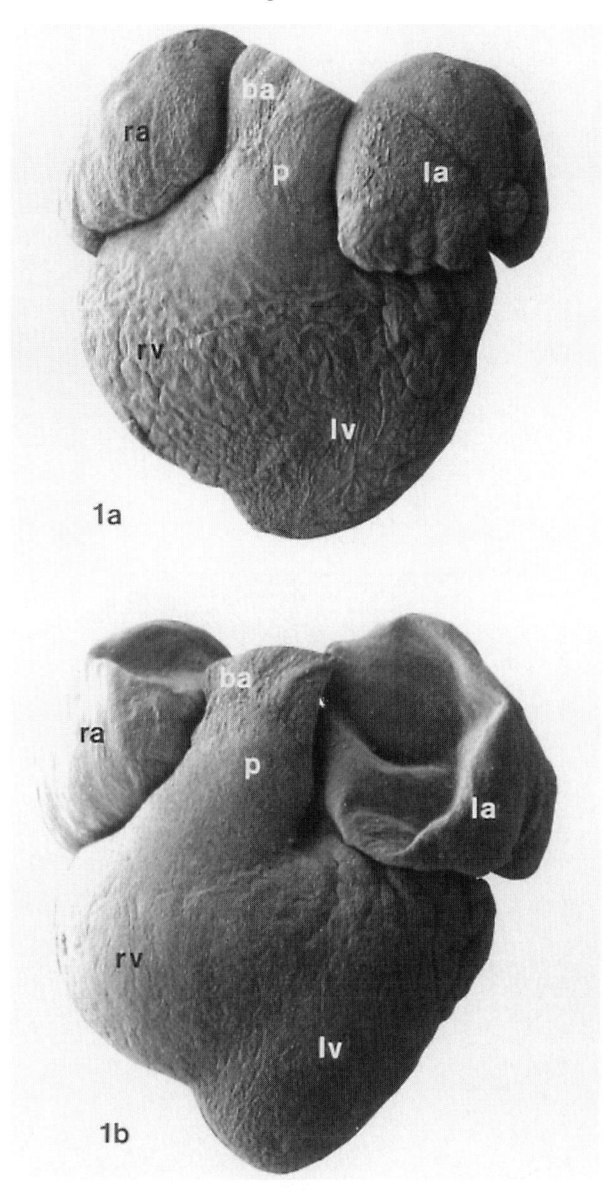

Figure 19.1 (*a*) Scanning electronmicrograph of the ventral aspect of a normal HH stage 29 embryonic chicken heart (×35). The arterial pole is nicely wedged between the right and the left atrium. (*b*) Scanning electronmicrograph of the ventral aspect of an HH stage 29 embryonic chicken heart after venous clipping at HH stage 17 (×40). The arterial pole is not well wedged in and is relatively dextroposed (rv = right ventricle, lv = left ventricle, ra = right atrium, la = left atrium, p = pulmonary trunk, ba = right and left branchiocephalic arteries.)

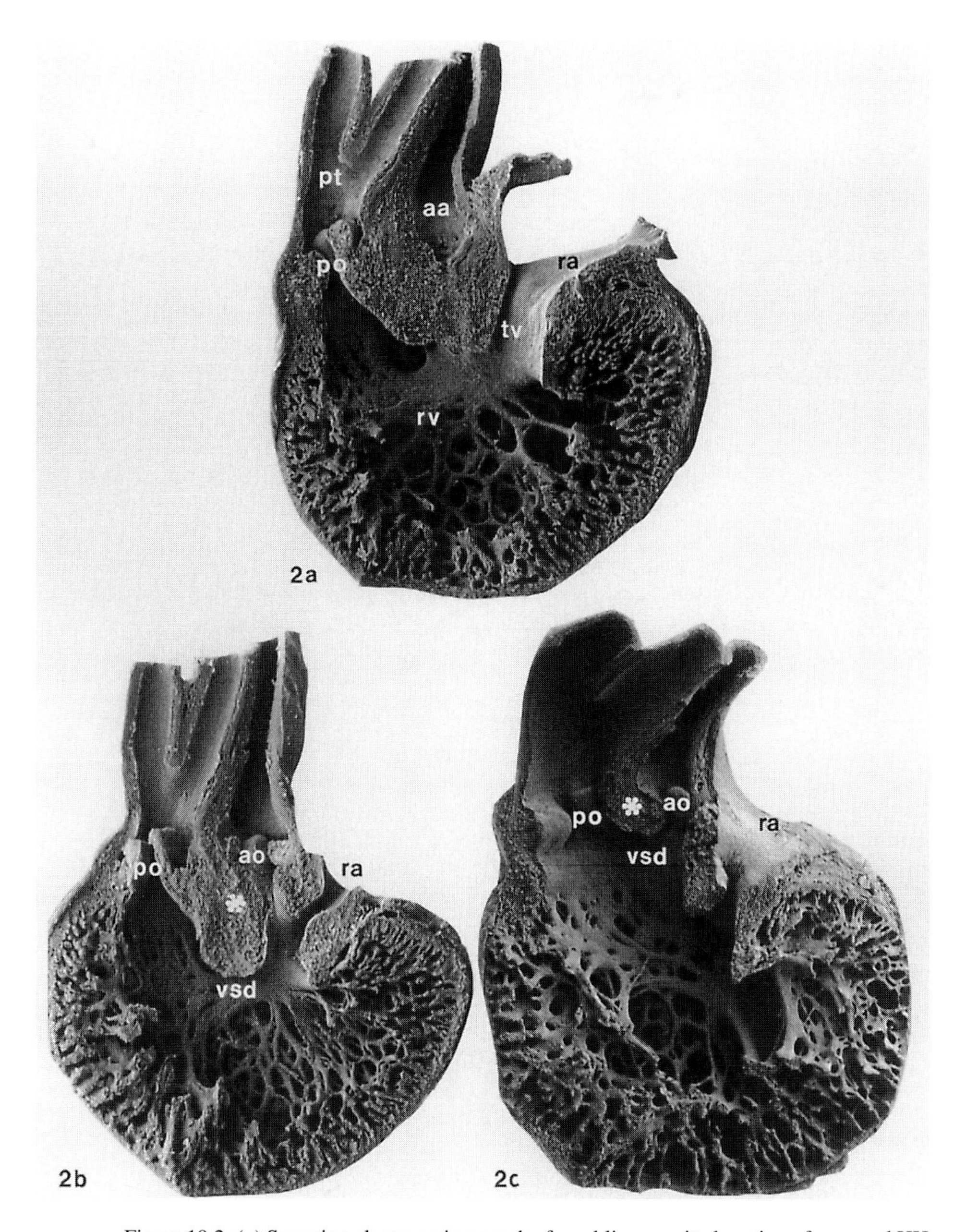

Figure 19.2 (*a*) Scanning electronmicrograph of an oblique sagittal section of a normal HH stage 33 chicken heart (×50). View toward the anterior aspect of the right ventricle (rv). The subpulmonary infundibulum connects by way of the pulmonary orifice (po) to the pulmonary trunk (pt) and arteries. The muscular tissue of the infundibulum acts as an outflow tract septum. The part of the ascending aorta (aa) connecting to the left ventricle (lv) has been cut off (tv = tricuspid orifice, ra = right atrium). (*b*) Identical section plane as in (*a*) (×50). This case (HH stage 34) was treated at HH stage 15 with retinoic acid. The resulting malformation is a DORV in which both great arteries and the tricuspid orifice connect to the right ventricle. The ventricular septal defect (vsd) is positioned relatively low. The semilunar valves of the aortic (ao) and the pulmonary orifices (po) are at about equal levels. (*c*) Identical section plane as in (*a*) and (*b*) in a venous clip model (HH stage 34) that was clipped at HH stage 17 (×50). The resultant DORV has a high subarterial ventricular septal defect (vsd). Histologic evaluation shows that the tissue separating the arterial orifices consists of endocardial cushion tissue and condensed mesenchyme (asterisk).

morphologic distinctions. First of all, the type of DORV does not contain a subaortic VSD as described for the chick embryo but instead contains transposed great arteries (TGA) and a subpulmonary VSD. In TGA the aorta arises 100% from the right ventricle, and the pulmonary trunk straddles above, in this case, a subpulmonary VSD. Again, the 50% rule decides whether we are dealing with a DORV setting or solely with a subpulmonary VSD. In the latter case, the pulmonary trunk arises for more than 50% from the left ventricle. This implies that the malformation in the mouse goes from complete TGA without a VSD through subpulmonary VSD to complete DORV. This spectrum is again understandable as deriving from degrees of tightness in looping. It has, however, an additional component in that the outflow tract rotation has been abnormal, leading to TGA. In addition, there is marked hypoplasia of the developing right ventricle, consistent with a direct effect of RA on the developing myocardium (Yasui et al., 1995). Biochemical studies of RA-treated rat hearts show a decrease in contractile proteins matched by an increase in structural proteins (Pelouch, 1995), supporting the direct myocardial route of interference. Preliminary investigations of myocardial function in mouse embryos following RA treatment reveal reduced right ventricular contribution to cardiac output (B. Keller, unpublished data). Aortic arch malformations again are present in this malformation sequence. Again, the idea that the neural crest cell population is altered in this malformation cannot be refuted. However, even experimental protocols that directly target neural crest cell populations are not expected to provide simple answers.

Neural crest ablation and the chick embryo

Kirby and colleagues (1993) have shown unequivocally that neural crest cells from the midotic to somite 3 regions migrate along pharyngeal arches 3, 4, and 6 to the heart. These neural crest cells have at least three proven fates following terminal differentiation: arterial smooth muscle cells, parasympathetic ganglia cells, and a mesenchymal subpopulation involved in aortopulmonary septation (Noden, Poelmann, & Gittenberger–de Groot, 1995). Interestingly, the neural crest cells have a short "time-out" during their passage through the pharyngeal arch area (Kirby, 1993). The cardiac malformation produced by neural crest excision but not described in the RA models is a persistent truncus arteriosus. This anatomic lesion is thought to be due to a failure of cardiac neural crest cells to migrate and form the aortopulmonary septum. However, recent neural crest ablation studies have shown that there are still ganglionic cells and condensed mesenchyme in the outflow tract ridges of embryos with persistent truncus arteriosus (Gittenberger–de Groot, Bartelings, & Poelmann, 1995b). A compensation mechanism after cardiac neural crest ablation has recently been described by Couley et al. (1996). The presence of neural crest–derived cells in the outflow tract cannot be explained by replacement of the excised population by cells from the adjacent nodose placode, because these ganglionic cells are present when the nodose placode is included in the ablation.

It is important to note that the pathogenesis of truncus arteriosus, which by definition contains a large VSD, is due to abnormal aortopulmonary septation rather than alterations in cardiac looping, as proposed for the previous experimental DORV models. However, a second subset of malformations after neural crest ablation falls within the category of nontransposed DORV, consistent with the RA chick embryo spectrum of defects. Therefore, Bockman et al. (1987) proposed that DORV results from a combination of altered looping and retarded septation. Of interest, in addition to the identification of DORV in some embryos following neural crest ablation, inflow malformations and alterations in myocardial function have also been described (Leatherbury, Connuck, & Kirby, 1993). The alteration in myocardial function may occur as early as HH stage 16 (L. Leatherbury, unpublished data), which is well before neural crest cells reach the outflow tract of the heart. This suggests a direct impact on the developing myocardium. It would then seem plausible that the alterations in myocardial contraction in the RA chick model and the neural crest ablation model could share a common mechanism. There are, however, marked differences in the time of insult (i.e., RA at HH stage 15 and ablation at HH stage 10), which make a common mechanism less likely. Comparable to the ablation model, the later application of RA to the chick results in a spectrum of aortic arch malformations.

Venous clip in the chick embryo

Investigators have long recognized the relationship between hemodynamic factors and myocardial growth (Clark, 1990). In an attempt to determine the specific morphologic consequences of a hemodynamic alteration on cardiac development, we devised an experimental paradigm that solely interferes with blood flow and avoids direct mechanical interference with neural crest cells or myocardial cells. One tenet of a "hemodynamic-molding" paradigm is that as soon as the heart begins to actively contract, hemodynamic forces related to wall stress and strain might influence the simultaneous morphologic processes of looping and septation. In our experimental model, we targeted a point in cardiac morphogenesis when the embryonic chick heart had completed the first stage of looping. Previous investigators have also explored "hemodynamic alteration models" (Rychter & Lemez, 1965). First, we determined the venous and arterial flow patterns of the developing yolk sac and their individual contributions to the intracardiac laminar currents (Hogers et al., 1995). Then we selected the right posterior cardinal vein for occlusion by clipping. So-called vein clipping resulted in an immediate opening up of capillary networks and rerouting of blood to the left side. Evaluation of the blood flow currents using vital dyes confirmed the alterations in venous return patterns after the vessel clipping. Treated, sham, and control embryos were then reincubated to allow the completion of normal septation. In the experimental cases, survival rate was 70% and the most frequent structural anomaly was within the spectrum comprising a small subaortic VSD to full-blown DORV (Figures 19.1b, 19.2c). Again, as was obvious in the chick RA model, the DORV included nontransposed great arteries. Interestingly, this hemodynamic intervention resulted in a type of VSD that crests high in the outflow tract. In the most severe cases, outflow tract septation at the semilunar valve level was accomplished only by the mesenchymal tissue of the aortic and pulmonary valve leaflets. In younger embryonic stages, the condensed mesenchyme of the aortopulmonary septum, derived from neural crest tissue (Noden, Poelmann, & Gittenberger–de Groot, 1995), seems to be abnormally positioned. Thus, the structural anomaly produced by venous clipping approaches the spectrum that includes truncus arteriosus. This is highly intriguing, as neural crest cell contribution can be only indirectly hampered by the events secondary to rerouted blood. In addition to the outlet anomalies, a set of aortic arch abnormalities correlate in severity to the degree of outflow tract abnormality. Additional experiments are in progress to determine the relationship between specific alterations in venous return and cardiac phenotype (B. Hogers, unpublished data). Preliminary results indicate again that the resultant cardiac phenotype is within the DORV spectrum. Also of interest is the fact that cardiac function appears to compensate for the alteration in venous flow patterns (M. L. A. Broekhuizen, unpublished data).

Fortunately, the repertoire of final cardiac phenotypes is limited despite the wide range of insults to the structural and functional maturation of the embryonic cardiovascular system. This refers, of course, to a macroscopic or microscopic pathophysiological definition of cardiac phenotype. Further refinement of the spectrum and, subsequently, the etiology of these malformations will require the application of new techniques to these old questions. One striking example of novel insights into clinical diseases has occurred in the category of cardiomyopathies. These abnormalities of cardiac muscle have been subdivided into three morphological categories: hypertrophic (HCM), dilated, and restrictive (Roberts, Marian, & Bachinski, 1995). Further, these cardiomyopathies can be sporadic or inherited. Molecular, genetic, and biological techniques have shown that the familial types can be inherited through abnormalities in human chromosomes 1, 7, 11, 14, or 15. Candidate affected genes include the genes for various contractile proteins such as β-myosin (chromosome 14), troponin T (chromosome 1), and tropomyosin (chromosome 15). Current research in this area is addressing the mechanisms responsible for variable penetrance and phenotypic variation within an affected family. Not surprisingly, this phenotypic variability is expressed at molecular, cellular, tissue, and organ levels, detected by altered functional parameters. In fact, with regard to the degree of myofiber disarray and precise degree of systolic and diastolic dysfunction, no two hearts are alike. Comparison of these genetically determined cardiomyopathies with acquired muscle diseases such as those associated with diabetes may provide insights into this phenotypic variability.

When comparing the pathogenesis of altered cardiac structure and/or function, it is essential to determine accurately the primary and secondary effects in order to define accurately the mechanisms responsible for the final phenotype. The two experimental paradigms that illustrate this point are the knockout and promoter-driven transgenic models (see also Chapters 1 and 7). These molecular techniques are responsible for the development of new models at unprecedented rates. Three recent issues of *Current Biology* (Brandon, Idzerda, & McKnight 1995a, 1995b, 1995c) listed 327 independently derived mutants with 263 fundamentally different modifications of gene products. This inventory can be accessed through the Internet and will be regularly updated. The lists include abnormal cardiac phenotypes associated with a wide variety of gene knockouts (e.g., integrin, fibronectin, *Hox* genes). A more sagacious interpretation of these lists suggests that there are probably limits to the detection of subtle alterations in cardiac phenotype by scientists less familiar with cardiac embryology, and embryo lethality may have been due to severe cardiovascular abnormalities that escaped detection.

It is also remarkable that targeted genetic alterations in many gene products known to have cardiac expression at certain stages of development have no associated discernible phenotype. This finding has kindled interest in the investigation of biological redundancy. The elegant studies of Chambon and colleagues (Mendelsohn et al., 1994a) show that knockouts of the individual RA receptors (RARs) do not lead to a phenotype. However, crossbreed knockouts, such as RARα (–/–) and RARβ (–/–), have marked cardiovascular anomalies. Another example of this kind is seen in the *Pax* gene knockout studies where there is at first sight an inconsistency between the immunohistochemical and in situ hybridization detection sites and the resulting phenotypes (Chalepakis & Gruss, 1995). This may be explained, in part, with a redundancy paradigm. However, unique events can be identified from confusing sets of similar phenotypes, as with the *Pax* 6 coding for eye development, a phenomenon highly conserved in various animal taxa.

A second important transgenic paradigm relates to coupling a gene with a promoter for which the organ specificity, time of onset, and duration of expression are known. These experiments, in which, for example, a gene can be expressed in an "untimely" and abnormal site, may more readily lead to abnormal cardiac phenotypes. Such experiments have been carried out with the *Hox* gene. *Hox A7* under α-actin promoter control (Kessel, Balling, & Gruss, 1990) and *Hox B8* under RARβ promoter control (Charité et al., 1994) both result in a spectrum of abnormalities that relate to disturbed patterning in the head and neck region, as well as limb abnormalities. The impacts of these altered patterning signals on cardiac morphogenesis are being studied.

It is also important to recognize where these "investigator-designed" paradigms may obscure the links between mechanism and cardiac phenotype. When the initial *Hox A7* model was published, no specific cardiac abnormalities were noted. However, subsequent study of the microscopic sections revealed myocardial thickening and tricuspid and mitral valve stenosis and even atresia (A. C. Gittenberger–de Groot, unpublished data). Early embryonic lethality and perhaps the lack of the experienced eye of a cardiac embryologist were the likely reasons why this important finding was initially overlooked. The *Hox B8* model also shows a very severe cardiac phenotype with a persistent truncus arteriosus. The accompanying VSD may be related to a derangement of the neural crest patterning (A. C. Gittenberger–de Groot, unpublished data). The accompanying myocardial and tricuspid and mitral valve abnormalities are not so easily explained. Nonetheless, the abnormal cardiac phenotypes indicate that a *Hox*-gene-coded mesodermal-patterning mechanism may be initiated by *Nkx*-controlled myocardial progenitors during development of the cardiogenic plates. Kern, Argao, and Potter (1995) present a recent review of the status of homeobox genes in heart development.

Future directions

The many descriptions of cardiac embryology reflect various schools of thought, each with anatomic and experimental paradigms to support its views on cardiovascular development. Several recent reviews of cardiac development incorporating molecular biological data were published in the proceedings of the Takao meetings on heart development (Markwald, 1995; Gittenberger–de Groot, Bartelings, & Poelmann, 1995a). The same volume contains an extensive gene product inventory at the cellular level (Pexieder, 1995). It is clear that our understanding of the morphogenetic alterations produced by either natural or transgenic approaches is highly influenced by our current knowledge and paradigms of cardiogenesis. We must recognize that, historically, cardiac morphogenesis has not been a well-defined field. It may be disappointing for some but challenging to others, that the underlying mechanisms for cardiac looping, septation, valve formation, and conduction system development are still insufficiently understood to provide mechanistic links from the gene level to the level of the functioning heart. One recent additional example may underscore the continuing new opportunities for insights into cardiovascular development. A recent genetic target is the protein neuregulin, and several knockout models of its receptors (Meyer & Birchmeier, 1995; Gassmann et al., 1995; Lee et al., 1995) show abnormalities of the cranial and cardiac neural crest derived structures. The fascinating cardiac phenotype is an absolute lack of myocardial trabeculation. This fundamental morphologic failure probably kills the embryo. Obvious questions include the following: Is there any link between the two affected cell types that early in embryonic development? Do these findings indicate that there is a specific gene for myocardial trabeculae? The answers to these questions might throw a completely new light on the interpretation of cardiac malfunctioning described in the neural crest ablation model (Leatherbury, Connuck, & Kirby, 1993) and RA-induced model (Broekhuizen et al., 1995).

What specific, etiologic conclusions can be drawn from the above examples to aid the molecular biologist who produces an altered cardiac phenotype in a transgenic animal and to aid the clinician who wants to sort out the causes for congenital heart malformations? Obviously, we have to work with paradigms that can acknowledge and incorporate a multilevel approach and analysis, as was elegantly proposed by Pexieder (1995). At least four such levels are relevant to the investigation of heart development: the molecular level, the cell level, the tissue level, and the organ level, each reflected in functional parameters. Scientific investigation at each level requires specific technical approaches, and the successful solution to these complex morphologic processes will likely require expertise in each of the above levels. Thus, it is unlikely that a single scientific center or group will be able to supply all the necessary expertise, which emphasizes the absolutely critical importance of extensive collaboration between the various disciplines as we enter one of the most exciting eras of cardiovascular investigation.

20

In utero and postnatal interventions for congenital cardiovascular malformations

V. MOHAN REDDY AND FRANK L. HANLEY

Introduction

Advances in the management of congenital heart lesions have been dramatic in the brief 50-year history of congenital heart surgery (Figure 20.1; Schumacker, 1992). These advances were driven largely by the development of technology in the early years and by a sophisticated understanding of physiology and developmental biology more recently. In the early years of management of congenital heart lesions, mortality was the main determinant in choosing a particular surgical strategy. However, with time mortality has been markedly reduced. As a result, other criteria for determining the success of a particular surgical approach have been developed, for example, postoperative complications, convalescence, and hospital stay. With further refinements, long-term functional results have become the criteria for judging the success of a particular approach. At the present time optimal preservation of the structure and function of all organs is becoming the primary goal of management. Along with the development of appropriate surgical techniques for repair of congenital heart defects, the concept of appropriate timing of repair has come into clear focus over the past several decades. Intervention early in life appears to prevent or minimize secondary morphologic and functional consequences in all organs (Castaneda et al., 1993).

Extrapolating from the benefits achieved with "early" repair (neonatal and early infancy) of congenital heart lesions, and combining the information gained from this experience with observations made by developmental biologists and physiologists, it appears that certain congenital heart lesions can be most effectively repaired in utero. In this chapter we will focus on the physiologic basis of neonatal intervention for congenital heart lesions and on the rationale and the potential benefits of intrauterine intervention for specific congenital heart lesions. Before delving into interventional details, a brief review of the unique features of cardiac development and fetal circulation will provide insight into the factors affecting the development of congenital heart lesions. The reader is also directed to Chapter 16 for additional details on mammalian circulation.

Fetal cardiovascular physiology

Circulation begins in the third week of human embryonic life, and at this stage, the heart is a straight tube. With further development of the cardiovascular system, separation of right and left heart structures begins to take place. Following the completion of cardiac morphogenesis, the fetal circulation has a number of unique features (Figure 20.2; Rudolph & Heymann, 1967). Two important unique aspects of the fetal circulation are (1) the placenta,

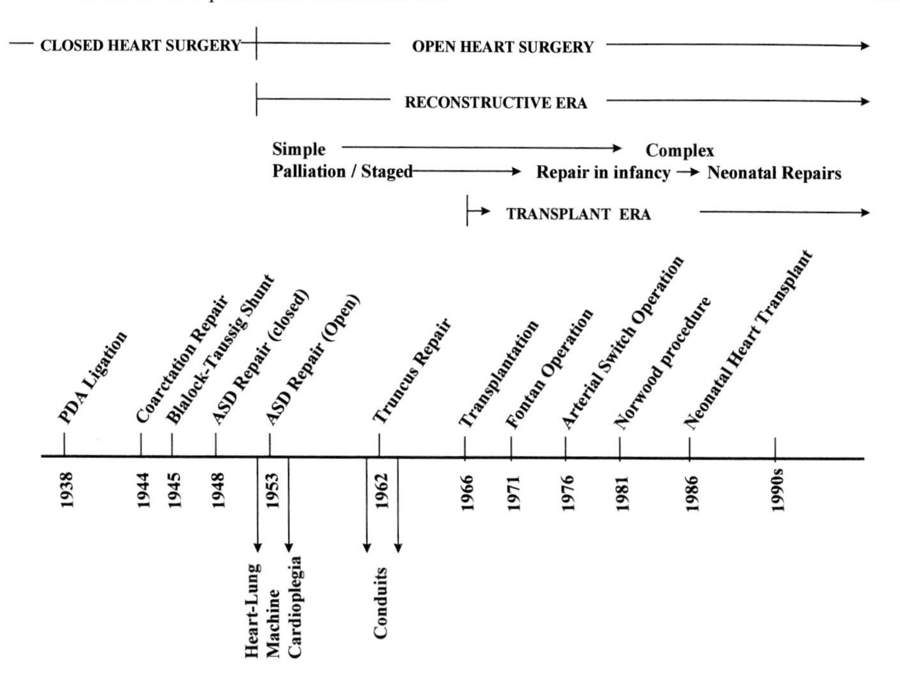

Figure 20.1 Time line chart illustrating the major developments in cardiac surgery for congenital cardiovascular malformations.

which functions as the organ of gas exchange while the lungs are nonfunctional as gas exchange units, and (2) three vascular shunts: the ductus venosus (at the venous level), foramen ovale (at the cardiac level), and ductus arteriosus (at the arterial level), which allow the right and left ventricles to function in parallel rather than in series. In utero the right ventricle pumps approximately 65% of the cardiac output and the left ventricle pumps about 35% (Heymann, Creasy, & Rudolph, 1973). Under these circumstances there is some degree of obligatory mixing of oxygenated and deoxygenated blood prior to perfusion of the systemic circulation. In the inferior vena cava the oxygenated blood from the umbilical veins crosses the ductus venosus and mixes with the deoxygenated blood from the systemic veins of the lower half of the body. This mixing is incomplete and the oxygenated blood preferentially streams across the foramen ovale and is delivered to the developing brain and heart. The less oxygenated blood from the superior vena cava is predominantly pumped by the right ventricle across the ductus arteriosus into the descending aorta (and to the placenta via the umbilical arteries).

The fetal vascular shunts therefore play a critical role in the developing heart by determining intracardiac blood flow. Alteration in intracardiac blood flow can profoundly alter cardiac morphogenesis (Icardo, 1989). According to the flow-related theory of cardiac development (Rose & Clark, 1992), normal flow and hemodynamics are necessary for normal cardiac morphogenesis (see Chapter 19). In the absence of normal flow and dynamics, the major cardiac structures, such as valves, chambers, and great vessels, will develop secondary abnormalities even though genetic programming is normal. This mechanism appears to play a significant role in the development of a group of congenital heart lesions (Clark, 1995).

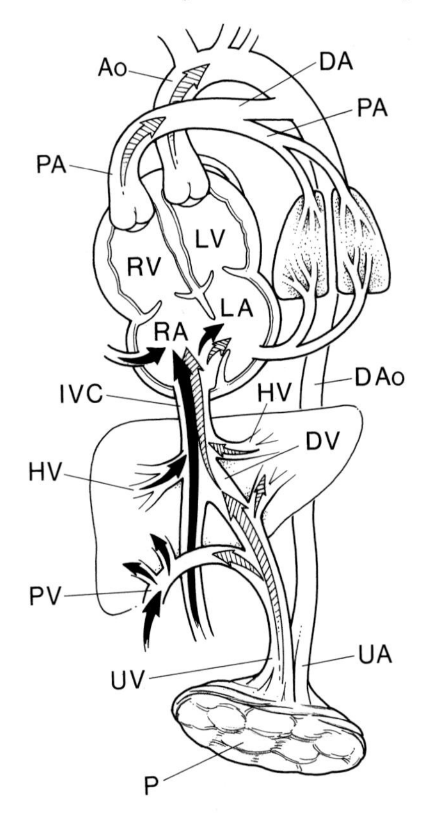

Figure 20.2 Fetal circulation (Ao = ascending aorta, DA = ductus arteriosus, DAo = descending aorta, DV = ductus venosus, HV = hepatic vein, IVC = inferior vena cava, LA = left atrium, LV = left ventricle, P = placenta, PA = main pulmonary artery, PV = portal vein, RA = right atrium, RV = right ventricle, UA = umbilical artery, UV = umbilical vein). Filled arrows show systemic venous return. Crosshatched arrows show oxygenated umbilical venous return.

A number of experimental and clinical observations suggest that alterations in intracardiac blood flow patterns influence cardiac morphogenesis (Lev et al., 1963; Jaffe, 1965; Sweeny, 1981; Allan, Crawford, & Tynan, 1986; Allan, 1988; Icardo, 1989; Rose & Clark, 1992; Hogers et al., 1995) For example, reduction of left atrial volume may produce varying degrees of left heart hypoplasia (Sweeny, 1981). A similar mechanism due to a restricted foramen ovale has been observed in some human fetuses that develop hypoplastic left heart syndrome (Lev et al., 1963). Also, the progression of severe pulmonary stenosis to pulmonary atresia has been observed in the human fetuses with subsequent underdevelopment of the right ventricle (Allan, Crawford, & Tynan, 1986). Refinements in fetal echocardiography will further define the pathophysiology of fetal heart disease and will help to identify the epigenetic factors that influence cardiovascular development, allowing fetal intervention aimed at these factors.

Transitional circulation

Dramatic changes in circulation take place at birth in humans, resulting in the transition from the "parallel" fetal circulation to the "in-series" circulation of the newborn (Klopfenstein & Rudolph, 1978; Heymann, 1989; see also Chapter 16, this volume). The function of gas exchange is transferred from the placenta to the neonatal lungs. This occurs with the rapid increase of pulmonary blood flow and closure of the ductus venosus, foramen ovale, and ductus arteriosus. The function of the placenta as the high-flow and low-resistance gas exchange organ is taken over by the lungs. With the first breath of the newborn, the pulmonary vascular resistance drops significantly and the pulmonary blood flow increases 8- to 10-fold. Pulmonary artery pressure falls rapidly, and by 24 hours after birth, it is approximately half that of the systemic arterial pressure (Moss, Emmanouilides, & Duffie, 1963). The pulmonary artery pressure and pulmonary vascular resistance gradually fall thereafter (Figure 20.3) and reach adult levels by about 2–6 weeks (Heymann, 1989).

These changes in pulmonary circulation at birth and during the first few weeks of life have tremendous pathophysiologic implications in the management of certain congenital heart defects (Rudolph, 1970). These implications can be highlighted by the following representative lesions: (1) transposition of great arteries (TGA), (2) left-to-right shunts, and (3) ductus arteriosus–dependent lesions with two functional ventricles.

Transposition of great arteries with intact ventricular septum

In normal hearts, the transition from fetal to neonatal circulation increases the left ventricular volume load by 25% and decreases right ventricular volume load by 30%. In addition, the left ventricle is presented with the sustained pressure load of the systemic circulation while the right ventricular pressure load decreases (Castaneda et al., 1993). This results in an increase in left ventricular mass and a relative decrease in right ventricular mass (Figure 20.4). In contrast, in TGA with intact ventricular septum, the postnatal left ventricle ejects into the low-resistance, low-pressure pulmonary vascular bed. As a result, the left ventricular muscle mass does not increase normally. Consequently, within weeks the left ventricle loses its capacity to maintain an adequate cardiac output against systemic afterload. This

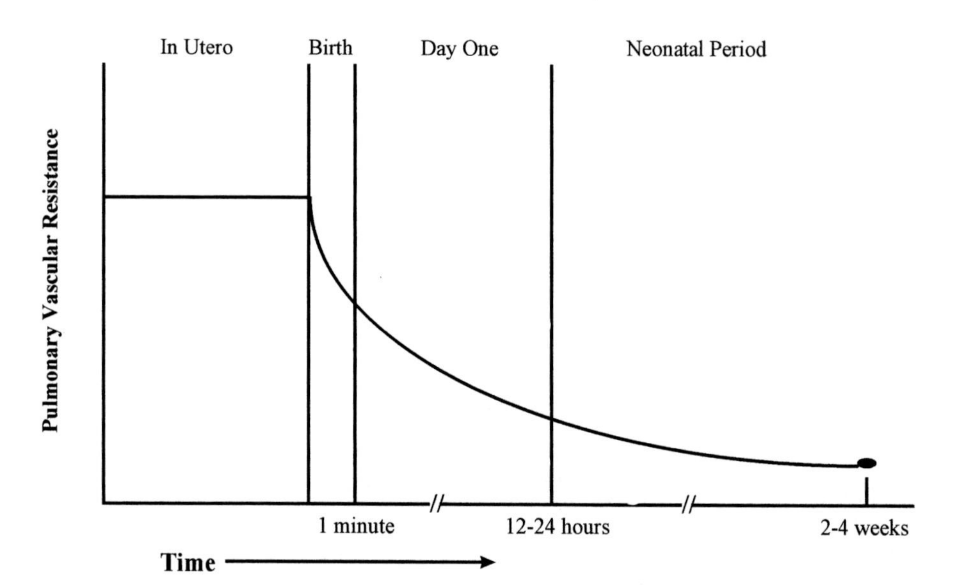

Figure 20.3 Perinatal changes in pulmonary vascular resistance.

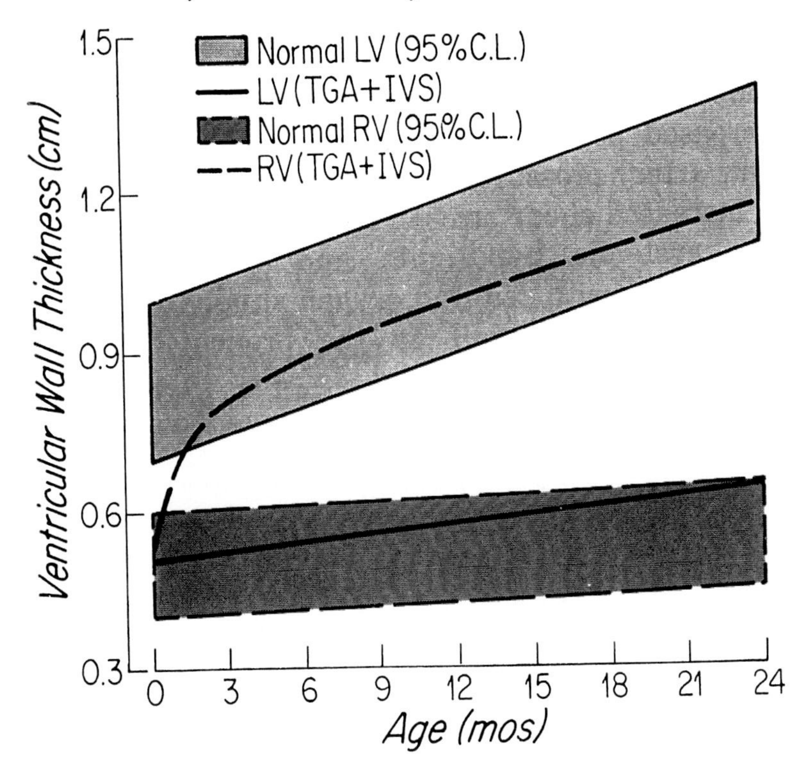

Figure 20.4 Left ventricular (LV) wall thickness changes in hearts with normally related and transposed great arteries. In transposition of great arteries with intact ventricular septum (TGA-IVS), the LV mass is normal at birth. The rapidly decreasing pulmonary vascular resistance results in a drop in peak LV pressure and, hence, decreased development of LV muscle mass (*solid line*). The upper bar shows the increase in LV free-wall thickness in normally related great arteries and the similar increase (*dashed line*) in right ventricular mass in TGA-IVS. LV muscle mass in TGA-IVS follows the pattern of right ventricular muscle mass in normally related great arteries. (From Castaneda et al., 1993. Reproduced with permission.)

reduction in left ventricular performance influences the type and timing of surgical intervention for this malformation, as discussed later in this chapter. On the other hand, the right ventricle, in assuming the role of systemic ventricle, increases its muscle mass.

Left-to-right shunt lesions

In these lesions, the volume of blood shunting left to right increases as the pulmonary vascular resistance falls. The symptomatology, morbidity, and mortality are higher in the first few weeks of life with lesions in which the left-to-right shunting occurs throughout the cardiac cycle, that is, at the arterial level (e.g., truncus arteriosus, large aortopulmonary window, large patent ductus arteriosus). On the other hand, in lesions with defects at the ventricular level, only a systolic shunt occurs. The morbidity in the first few weeks of life is less in these patients (e.g., ventricular septal defect), and the increase in cardiac work and pulmonary vascular congestion can be managed pharmacologically. Studies in lambs indicate that increased pulmonary blood flow causes pulmonary endothelial dysfunction as early as 4 weeks after birth (Reddy et al., 1995b), which may explain the increased incidence of pulmonary hypertensive crises seen in the postoperative period following repair of large left-to-right shunt lesions (Hanley et al., 1993a). Therefore, in lesions like truncus arteriosus, in which morbidity due to increasing pulmonary endothelial dysfunction is high, and where there is no chance of spontaneous correction, elective early neonatal repair may be advantageous (Hanley et al., 1993a). On the other hand, for lesions such as ventricular septal defect (VSD), with lower early morbidity, and where there is a significant incidence of spontaneous closure (Hoffman & Rudolph, 1965), routine neonatal repair is not warranted.

Ductus arteriosus–dependent lesions with two functional ventricles

In these lesions which cause severe obstruction of either the pulmonary or the systemic arterial circulation (e.g., critical aortic stenosis, critical pulmonary stenosis, interrupted aortic arch), postnatal survival is dependent on patency of the ductus arteriosus (Olley, Coceani, & Bodach, 1976). In this setting prostaglandin (PGE1) infusion is used to keep the ductus arteriosus open. This delays the complete transition of fetal to neonatal circulation and helps maintain adequate pulmonary and systemic perfusion, providing a period for stabilization and preparation for corrective intervention.

Myocardial development

In animal models some degree of hyperplastic growth of the myocardium persists in the early neonatal period (see Chapter 3, this volume; Anversa, Olivetti, & Loud, 1980). It is speculated that myocyte mitotic activity and the capacity for hyperplasia persist for 3–6 months after birth in humans (Brown, 1986). Even after myocyte mitotic activity ceases, hyperplasia of nonmyocyte components of the heart nevertheless continues. During physiologic hypertrophy, coronary capillary units continue to proliferate, maintaining constant intercapillary distance (Figure 20.5; Anversa et al., 1978; see also Chapter 4, this volume).

The potential of neonatal myocytes to undergo hyperplasia may be beneficial for myocardial remodeling after neonatal intervention. For example, it may be necessary for successful left ventricular remodeling in patients with TGA undergoing early neonatal anatomic repair, in which the left ventricle must assume the systemic workload.

Postnatal intervention

The timing of surgery is a critical component in outcomes of surgery for congenital heart defects. Neonatal or early-infancy repair of is now standard practice for many but not all congenital heart lesions (Table 20.1; Castaneda et al., 1993; Kirklin & Barratt-Boyes, 1993).

Neonatal intervention

A risk–benefit analysis aimed at determining the efficacy of neonatal repair must be performed in the management of each individual with congenital heart disease. The two major

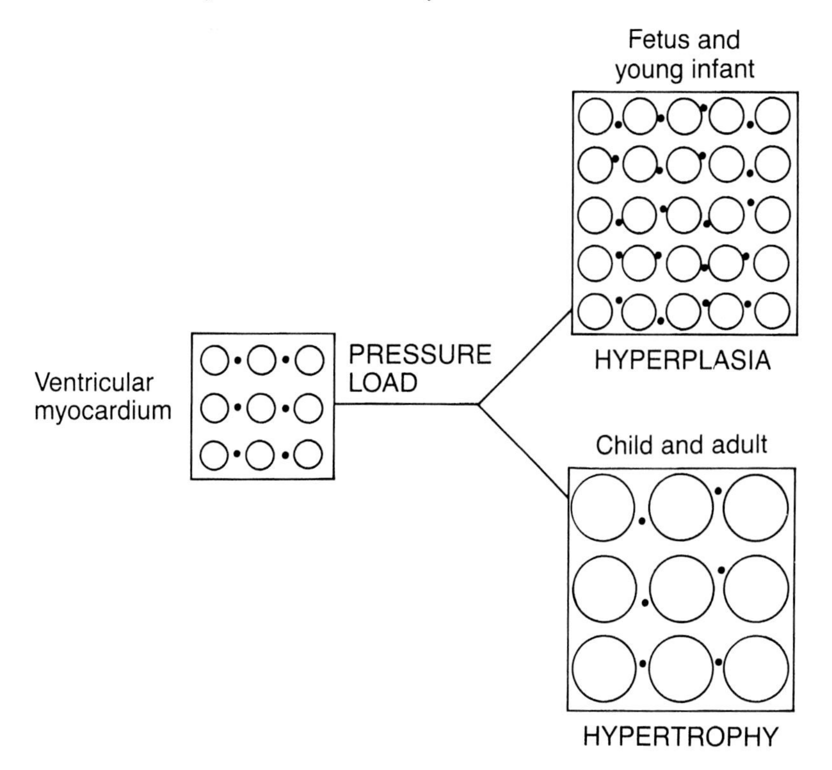

Figure 20.5 Schematic representation of myocardial hyperplasia and hypertrophy on exposure to pressure overload: In the fetus, the neonate, and the young infant, the increase in myocardial mass is due to myocyte hyperplasia and moderate hypertrophy and is accompanied by increased coronary angiogenesis. In the older infant, the child, and the adult, the increase in myocardial mass is due only to myocyte hypertrophy without accompanying coronary angiogenesis. Circles represent myocytes and dots represent coronary arteries. (From Castaneda et al., 1993. Reproduced with permission.)

reasons to defer intervention in neonates with congenital heart disease are (1) increased operative risk of early repair relative to later repair (this risk has been markedly reduced in recent years, although it has not been completely eliminated) and (2) the possibility of spontaneous closure of the defect, which occurs in some lesions, such as VSD, patent ductus arteriosus (PDA), and atrial septal defect (ASD). Thus, it follows that all congenital cardiac malformations will fall into one of the following categories as detailed in Table 20.1:

1. If the natural history of a lesion is associated with high mortality or morbidity in the neonatal period, then neonatal intervention is warranted, regardless of whether a slightly higher operative risk exists and regardless of whether there is potential for spontaneous correction over time. Many congenital cardiac malformations fall into this category, including TGA, truncus arteriosus, total anomalous pulmonary venous drainage, interrupted aortic arch, pulmonary atresia with or without VSD, critical aortic and pulmonary stenosis, large PDAs, aortopulmonary window, the majority of single-ventricle hearts, and large VSDs in severe congestive heart failure (Hoffman & Rudolph, 1965; Castaneda et al., 1993; Kirklin & Barratt-Boyes, 1993).

Table 20.1. *Surgical options for common congenital cardiovascular malformations*

Lesion	Operation	Typical timing
Patent ductus arteriosus (large)	PDA ligation	Neonatal
Coarctation of aorta	Resection and reanastomosis; SFA	Neonatal
Atrial septal defect	Closure: suture or patch	Infancy or preschool
Ventricular septal defect	Patch closure	Late infancy and early childhood
Tetralogy of Fallot	VSD closure and RVOT reconstruction	Childhood vs. neonatal, or infancy
Complete common atrioventricular canal defect	Single- or two-patch repair	Early infancy
Truncus arteriosus	VSD closure and RVOT reconstruction	Neonatal
Total anomalous pulmonary venous return	Anastomosis of CPV to LA	Neonatal
Transposition of great arteries	Arterial switch operation	Neonatal
Double-outlet right ventricle	Variable: depends on morphology	Neonatal or infancy
Hypoplastic left heart syndrome	Norwood stage I palliation	Neonatal
Single ventricle	Shunt or band ± atrial septectomy ± DKS or BVF resection	Neonatal

Note: BVF = bulboventricular foramen, CPV = common pulmonary vein, DKS = Damus–Kaye–Stansel, LA = left atrium, PDA = patent ductus arteriosus, RVOT = right ventricular outflow tract, SFA = subclavian flap angioplasty, VSD = ventricular septal defect.

2. If the natural history of a lesion in the neonatal period is more favorable and there is a possibility of spontaneous correction, then neonatal or early-infancy repair is not indicated. Such lesions include most VSDs, ASDs, and small PDAs (Hoffman & Rudolph, 1965; Kirklin & Barratt-Boyes, 1993).

3. If the natural history is more favorable but spontaneous correction is not possible, a judgment must be made that weighs the risks of early repair against the morbidity of delaying repair. Lesions in this category include tetralogy of Fallot, some forms of double-outlet right ventricle, partial atrioventricular canal defect, sinus venosus ASD, and single ventricle with well-balanced pulmonary blood flow (Castaneda et al., 1993; Kirklin & Barratt-Boyes, 1993).

Further discussion of some important representative lesions will offer a more meaningful insight into the evolution of and current trends in the timing of surgical management of congenital heart lesions.

Ventricular septal defect

The mortality and morbidity in the first few weeks and months of life are low for isolated VSDs. Also, spontaneous closure of VSDs has been well documented (Hoffman & Rudolph, 1965). These facts have been the main reasons for deferring surgical closure of VSDs until late infancy or early childhood. Traditionally, surgical closure of VSDs is advocated in the neonatal period or early infancy only if medical therapy fails to control the symptoms of congestive heart failure or, more rarely, if there is secondary progressive elevation of pulmonary vascular resistance. In all other cases, surgery is deferred. Most muscular and perimembranous VSDs have high rates of spontaneous closure in the first year of life and rarely require early surgery. In contrast, large muscular and perimembranous VSDs and subarterial VSDs rarely close spontaneously and are associated with up to 10% mortality if unrepaired during the first year of life (Hoffman & Rudolph, 1965). In the current era, mortality and morbidity of surgical closure of VSDs is less than 2% in centers experienced in neonatal and infant cardiac surgery. Given these facts, it seems appropriate to surgically close symptomatic large VSDs in early infancy and moderate VSDs in the second year of life. Increasing pulmonary vascular resistance or intractable congestive heart failure with any size VSD is an indication for prompt surgical closure.

Transposition of great arteries

Untreated, 50% of patients with TGA die in the first month of life and most (90%) die in the first year of life (Castaneda et al., 1993; Kirklin & Barratt-Boyes, 1993). Even with palliation to improve mixing of the oxygenated and deoxygenated blood by surgical removal or balloon tearing of the atrial septum, the mortality is high, prompting earlier repairs. The Senning and Mustard atrial baffle procedures were widely performed in early infancy during the 1960s and 1970s. Although these repairs achieved physiologic correction by redirecting the flow of the oxygenated and deoxygenated blood to the appropriate vascular bed, the repairs were not anatomic in that the right ventricle remained the systemic pumping chamber in these patients. More recently, the neonatal arterial switch operation has gained wide acceptance as the procedure of choice largely because the long-term follow-up of atrial repairs has shown a high incidence of atrial rhythm disturbances and frequent failure of the right ventricle as the systemic ventricle (Wernovsky et al., 1988; Chang et al., 1992; Castaneda et al., 1993; Kirklin & Barratt-Boyes, 1993).

Tetralogy of Fallot

Traditional management of tetralogy of Fallot (TOF) in patients with symptoms involved a first-stage palliation with a systemic to pulmonary artery shunt to increase pulmonary blood flow, followed at a later age by complete repair (Kirklin & Barratt-Boyes, 1993). This approach was developed at a time when complete repair in infancy carried a prohibitive risk. Primary repair of TOF in the neonatal or early-infancy period has been advocated only recently (DiDonato et al., 1991; Hennein et al., 1995; Reddy et al., 1995a). Unlike patients with VSD, all patients with TOF will eventually require surgical correction. This necessity for intervention therefore tends to favor a strategy of primary neonatal repair (Reddy et al., 1995a). On the other hand, similar to the case with VSD, the probability of survival of patients with unrepaired TOF in the infancy period is often acceptable with palliation or medical treatment, making neonatal repair an option but not mandatory. The reluctance to repair all patients in the neonatal or early-infancy period stems from concerns about the potential for increased surgical risk associated with primary neonatal repair in a setting where the natural history of the lesion does not demand an immediate repair. At present, there is no consensus on the best approach.

Other complex lesions

The examples just given illustrate the advantages and disadvantages of neonatal primary repair for several representative lesions. Because highly successful strategies exist for these lesions, there is no physiologic rationale for even earlier intervention, that is, in utero repair. There are, however, a number of complex lesions – pulmonary atresia with intact ventricular septum (PA-IVS), critical aortic stenosis, absent pulmonary valve syndrome, and others – for which neonatal intervention has not yielded a satisfactory long-term outcome (Castaneda et al., 1993; Kirklin & Barratt-Boyes, 1993; Hanley, 1993). Each of these lesions exhibits a primary defect that alters fetal blood flow patterns and results in more severe, secondary morphologic changes as fetal life progresses. These lesions require careful consideration for possible in utero intervention.

In utero interventions

Fetal heart lesions can be arbitrarily divided into two categories (Table 20.2): (1) lesions resulting in fetal compromise or secondary morphologic abnormalities and (2) lesions without fetal compromise. The parallel nature of the fetal circulation, as described earlier in the chapter, is such that the great majority of congenital heart defects do not compromise fetal hemodynamics. In certain lesions, however, morphologic progression of a defect will occur over time during fetal development. Although this morphologic progression does not compromise the fetus per se, it may have important negative consequences following the transition to postnatal circulation. Therefore, fetal intervention is less likely to be necessary because of fetal instability but more likely to be necessary to reduce the subsequent postnatal compromise, that is, to alter the natural history of the disease.

In considering fetal intervention for congenital heart defects, three essential criteria should be fulfilled. These are (1) the technical ability to diagnose the defects accurately in utero, (2) the potential benefit of correcting the lesion in utero compared with neonatal correction (otherwise there would be little reason for fetal intervention), and (3) the technical ability to safely perform the surgical procedure in utero without maternal risk. Fetal echocardiography can accurately diagnose many congenital heart defects between 18 and 24 weeks of gestation and probably as early as 12 weeks (Dolkart & Reimers, 1991; Brook & Silverman, 1993). The potential benefit of correcting some heart defects in utero can be

Table 20.2. *Fetal heart defects*

Type	Examples
Fetal compromise	Complete heart block, some forms of pulmonary atresia, tricuspid insufficiency
No fetal compromise	
Ongoing secondary morphologic abnormalities	Pulmonary atresia with intact ventricular septum, restrictive foramen, critical aortic stenosis, absent pulmonary valve
No significant secondary abnormalities	Ventricular septal defect, truncus arteriosus, transposition of great arteries, complete common atrioventricular canal defect

convincingly argued based on their in utero pathogenesis and relatively poor outcome following neonatal intervention (Hanley, 1994; Bull et al., 1994). According to the flow-related theory of cardiac development (Rose & Clark, 1992), a relatively simple primary defect that occurs during primary morphogenesis may lead, in the developing heart, to altered pressure and flow patterns that then gradually induce secondary hypoplasia or maldevelopment of major cardiac structures such as heart chambers or great vessels. For example, in PA-IVS many of the important morphologic abnormalities, including hypoplasia of the right ventricle and tricuspid valve, ventricular hypertrophy, myocardial sinusoids, and coronary abnormalities, develop secondarily to the primary lesion of right ventricular outflow tract atresia. Relief of outflow tract obstruction, especially in the neonatal period (when there is still potential for myocyte hyperplasia), allows for the growth of the hypoplastic right ventricle. The earlier this is achieved, the greater the benefit. If this concept is extrapolated to fetal life, it is logical to assume that following relief of right ventricular outflow tract obstruction in the fetus, the secondary morphologic consequences could be minimized, prevented, or even allowed to regress. Similar arguments based on the flow-related theory of cardiac development can be used for other lesions as well, as outlined in Figure 20.6.

Having established that fetal diagnosis is possible and that fetal intervention is potentially beneficial, it is necessary to focus on the last criterion: the ability to safely perform the necessary cardiac surgical procedure in utero. From Figure 20.6 it is clear that the surgical procedures that are anticipated to be performed in the fetus are technically simple, such as pulmonary valvotomy, aortic valvotomy, and atrial septectomy. Based on clinical experience with corrective surgery in premature and low-birth-weight babies frequently weighing less than 2000 grams, it is obvious that the limitations to fetal cardiac surgery would not likely be technical surgical issues. This was confirmed in the early attempts at experimental cardiac surgery in the sheep fetus.

Although technical issues were easily mastered, it became clear that important physiologic roadblocks existed based on the responses of the fetus to the stress of intervention and on the response of the placenta to extracorporeal circulation (Hanley, 1993). These initial studies in experimental fetal cardiac surgery demonstrated very dramatically that the present status of fetal intervention for cardiac defects is reminiscent of the overall field of cardiac surgery in the late 1940s and early 1950s. It was obvious at that time that relatively simple and easily achieved procedures like closure of an ASD or a VSD would have great clinical benefit. However, the major obstacle was the lack of a safe and effective method of gaining intracardiac access. Subsequent development of clinically applicable cardiopulmonary bypass and myocardial protection techniques allowed a reliable way to achieve intracardiac access and has resulted in rapid progress in cardiac surgery (Figure 20.1). The rapid accumulation of knowledge regarding the physiologic responses of the organism to extracorporeal circulation allowed development of techniques that minimized these pathophysiologic responses. As a result, the use of extracorporeal circulation soon became the technique of choice, quickly outperforming other ingenious but much more limited techniques for gaining intracardiac access.

At the present time, lack of a sufficient understanding of the fetal responses to stress, intervention, and extracorporeal circulation limits clinical application of fetal cardiac intervention. Early models of fetal extracorporeal circulation suggest that progressive metabolic acidosis causing fetal death during fetal bypass might be due to depressed or redirected cardiac output. Subsequent studies identified that several factors are probably responsible for this (Hanely, 1993). Mid- to late-gestation fetuses are capable of mounting an immense

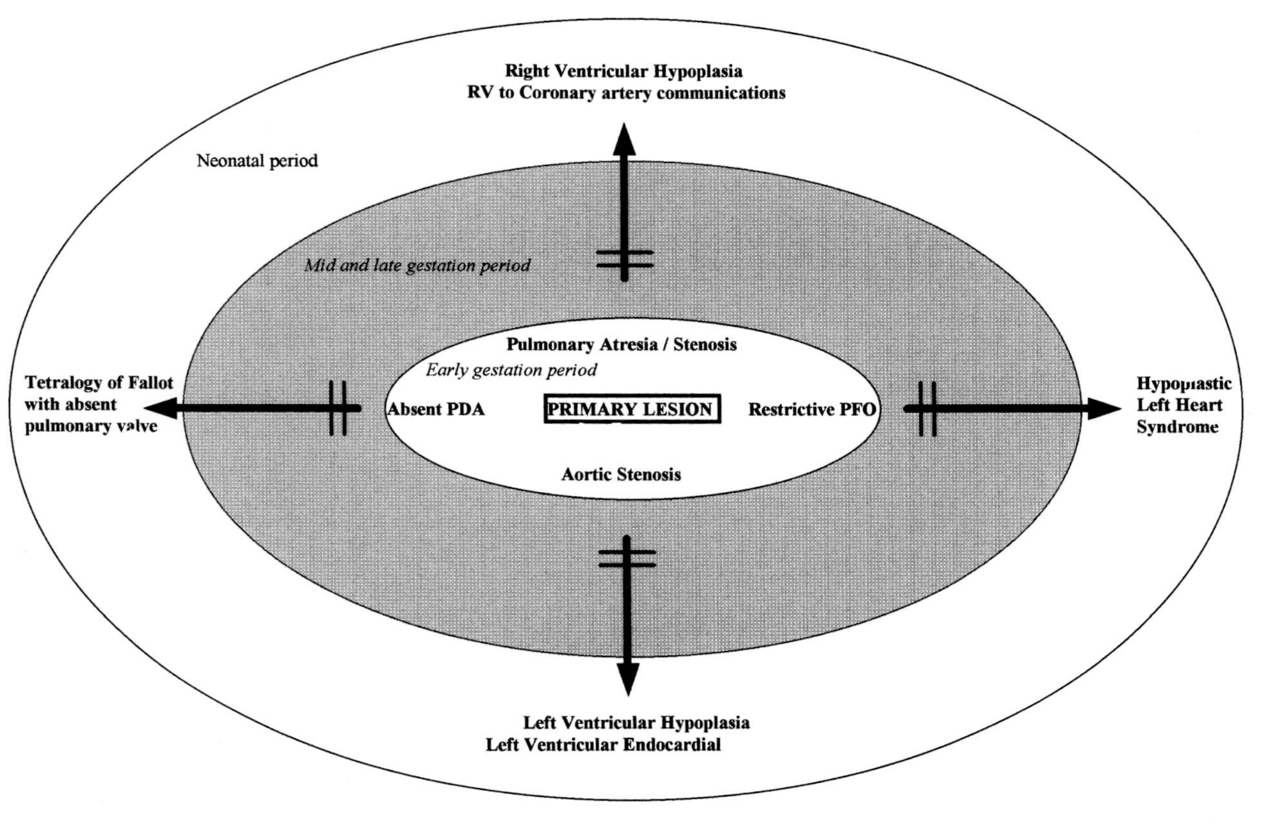

Figure 20.6 Lesions that may potentially benefit from fetal intervention. The inner ring indicates the early primary lesion. The middle ring indicates the mid- and late-gestation periods during which serious secondary morphologic changes occur. The outer ring indicates the fully developed lesion at birth. Intervention (indicated by) in the late second trimester or early third trimester may alter the secondary morphologic consequences.

stress response. Catecholamine levels increase by over 50-fold during fetal exposure and during fetal bypass. This results in significant elevation of the total fetal vascular resistance, presenting the fetal heart with an increased afterload that is not tolerated by the immature contractile apparatus of the fetal myocardium. Additionally, inhalational anesthetic agents (e.g., halothane) given to the mother not only are ineffective in blunting the fetal stress response but also cause fetal and maternal myocardial depression (Hanley, 1993). Total spinal anesthesia, which does not have the aforementioned detrimental effects, has been shown to improve fetal hemodynamics and placental gas exchange function before, during, and after fetal bypass (Fenton et al., 1994). Although total spinal anesthesia is very effective as a research tool, this technique of anesthesia is not practical in the human fetus. Further studies in a more appropriate primate animal model using clinically applicable anesthetics such as high-dose narcotics are needed.

It was also evident from early studies of fetal bypass that placental vascular resistance rises and placental gas exchange function deteriorates during and after fetal bypass, resulting in early fetal death from hypercarbia, acidosis, and, finally, ventricular fibrillation. Studies addressing the mechanism of this placental dysfunction have suggested that vasoactive products of the arachidonic acid cascade have an important role in this dysfunction. An improvement in placental function has been observed during and after bypass when using indomethacin and methylprednisolone to block the synthesis of prostaglandins (Sabik et al., 1994). In spite of these recent advances, at the present time the ideal method of handling the placenta during fetal bypass remains uncertain.

The future of in utero cardiac intervention

Recent experimental protocols using the accumulating knowledge of the fetus and placenta under stress have resulted in survival of 90% of fetuses to full term following fetal bypass without any deleterious effects (Reddy et al., in press). Given these advances, it seems likely that fetal cardiac surgery may take its place as a useful form of therapy for certain forms of congenital heart defects in the near future.

Another alternative intervention for fetal cardiovascular malformations is catheter-directed therapy. Although this has been attempted in human fetuses with very limited success (Maxwell, Allan, & Tynan, 1991), the major technical difficulty with this approach is gaining safe access to the fetal cardiovascular system. Improvements in this area are being investigated and hold promise for the future.

In addition, other novel molecular approaches may emerge from the better understanding of the genetic causes and mechanisms of normal and abnormal cardiac development. For example, a number of single-gene defects have been implicated in the development of congenital heart defects (Bristow, 1995), which may in future be favorably influenced by gene therapy. Knockout gene models have begun to provide more insight into cardiac development. For example, the connexin gene *Cx43* is implicated in the development of pulmonary stenosis (Reaume et al., 1995) and has also been linked to the development of visceroatrial heterotaxia. Manipulation of such genes may indeed provide a breakthrough in the management of congenital heart defects in the future.

21

Applying the science of cardiovascular development to congenital cardiovascular malformations

EDWARD B. CLARK

Introduction

Every adult once had the heart of a child and before that the heart of a fetus and an embryo. This fact delineates the continuum from primary cardiovascular development through maturity and senescence. Although studies of cardiovascular development date back to Aristotle, advances over the last two decades are now defining the processes that link pathogenesis to the prevention of human disease (Neill & Clark, 1995; Clark, Markwald, & Takao, 1995).

In this century, the enterprise of children's cardiovascular health has broadened from the medical management of acute rheumatic fever to the surgical repair of congenital cardiovascular malformations (Moller et al., 1993). In spite of surgical advances, cardiovascular malformations, with a prevalence of 8 per 1000 live births, are the leading cause of infant mortality from congenital defects (Clark, 1994). As other causes of infant morbidity and mortality have yielded to therapy, teams of clinical and basic-science investigators are searching for ways to reduce the prevalence of cardiac defects.

Now, we are in an era reexamining cardiac embryology and defining the etiology and pathogenesis of cardiac disease (Clark, 1986). Advances in molecular and cellular biology have provided tools for the diagnosis of cardiac defects such as hypertrophic cardiomyopathy, Marfan syndrome, and familial arrhythmias. These gains will eventually lead to the prevention of heart disease (Deitz & Pyeritz, 1995). Thus, our management of congenital defects and adult-onset cardiovascular diseases will benefit from this expanded knowledge base regarding cardiovascular development.

The origin of a considerable proportion of cardiovascular disease is in the developmental processes that form the heart and the vascular bed. In an elegant series of epidemiologic studies, Barker and colleagues demonstrated that factors affecting the fetus and infant dramatically influence adult-onset diseases such as coronary atherosclerosis and stroke (Barker, 1992). Each advancing stage of mechanistic understanding leads to new discoveries that link specific gene abnormalities and environmental influences to a range of cardiovascular diseases.

The clinician's interest in development further expands the enterprise of children's cardiovascular health. The noninvasive ultrasound assessment of the human developmental process now begins during the early fetal stage (Wladimiroff et al., 1992). Fetal echocardiography can now routinely diagnose most structural cardiovascular malformations at 20 weeks of gestation. Congenital cardiovascular malformations, cardiomyopathies and associated arrhythmias, coronary artery disease and the homeostasis of blood pressure, and the etiology of stroke likely have their basis in disordered developmental cascades. These developmental cascades are now accessible with the tools of the molecular geneticist and developmental biologist. Within the next decade, clinicians will intervene to treat, modify, or prevent diseases, including those of the fetal patient.

Cardiac development and the genetic etiology of congenital cardiovascular malformations

The etiology of cardiac defects has been obscure and the topic of much conjecture and supposition. Some defects arise by chance and are due in part to the complex temporal spatial coordination of heart morphogenesis. For example, perimembranous ventricular septal defects occur when the three coincident septation processes – conotruncal, muscular, and atrioventricular – fail to converge at the proper place and time. Such stochastic events likely account for a proportion of defects for which an etiology will never be defined. However, recent advances in molecular biology indicate a genetic basis for a wide range of abnormalities.

Although initially thought to be multifactorial in etiology, the majority of congenital cardiovascular malformations are likely single-gene or closely related gene defects (Burn, 1995; Payne et al., 1995). Conotruncal defects with or without the accompanying signs or symptoms of DiGeorge syndrome (hypoparathyroidism, depressed immunity, branchial arch defects) correlate with deletions on the Chromosome 22q11.2 locus (Burn et al., 1995; Driscoll, Goldmuntz, & Emanuel, 1995). The search, however, for the conotruncal gene continues. Identifying the specific gene responsible for disordered conotruncal development would offer the possibility of identifying a range of mutations that correlate with the diverse phenotype. Recently, an open reading frame of the *ADU/VDU* gene cloned from a balanced translocation was identified in a mother and daughter with DiGeorge phenotype (Budarf et al., 1995). This development may lead to the identification of additional point mutations similar to those found in other dominant negative mutations.

The etiology of conotruncal defects, like that of other congenital cardiac malformations, is heterogeneous (MIM, 1995). A locus on Chromosome 10 is also associated with a group of conotruncal defects. Most syndromes that have an autosomal dominant pattern of transmission have yet to be specifically cloned. Holt-Oram syndrome (digitalization of the thumb and atrial septal defect) lies on Chromosome 12, atrioventricular canal defects in Down patients resides on a locus on the long arm of Chromosome 21, and atrioventricular septation defects in patients without Down syndrome have been identified on Chromosome 8. Alligiel syndrome of hepatic dysfunction and vascular abnormalities of the pulmonary bed is associated with a mutation in the gene that encodes the ES/130 protein integral to the differentiation of vascular endothelium and endocardial cushion mesenchyme (Markwald et al., 1995).

Left heart defects ranging from bicuspid aortic valve to hypoplastic left heart syndrome also likely have a single-gene etiology. A recent phenotypic study defined an increased association of bicuspid aortic valve in families of children with hypoplastic left heart syndrome (Sill et al., 1993).

Cardiovascular development and surgical therapy

For 50 years, the surgical treatment of congenital heart disease has focused on closing holes, opening valves, and creating new connections. As pioneering patients approach their middle years of life, it is clear that their prognosis is dependent upon the health of their myocardium. Heart failure and arrhythmias are the limiting factors in many patients 30 and 40 years after repair of tetralogy of Fallot, ventricular septal defect, and other malformations (Morris & Menashe, 1991; Moller & Anderson, 1992; Murphy et al., 1993).

With advances in the molecular basis of cell replication and the control of cell proliferation, new opportunities are emerging for treatments focused on the myocardium and vascular bed (Nabel, 1995). The emerging understanding of development will also affect events in the operating room. In addition to the patch and suture, the surgeon's armamentarium will include growth factors and computerized models of the heart. These models will gauge how ventricles must be reshaped to improve the match of ventricle to vascular bed. The surgeon will resect tissue and implant in the myocardium beads carrying growth factors to stimulate remodeling, augment vascular supply, and repair the conduction system (see Chapter 20).

Advances will not only be limited to the myocardium. Prosthetic heart valves are mechanical wonders but pose a problem. Currently, a patient requiring an artificial valve has a choice between a mechanical valve, with risks of thrombosis, embolism, and mechanical trauma to blood cells, and valve tissue from pig, cow, or human, with risk of early calcification. Using tissue-engineering techniques, new valves will be grown and covered with a patient's own endothelium. These biologically engineered valves will reduce the risks of thrombosis, embolism, fibrosis, and calcification (Nemecek, 1995).

Cardiovascular development and opportunities for prevention of congenital cardiovascular malformations

Prevention is the most important outcome of improving our understanding of cardiovascular development. As documented in this book, cardiovascular development is a redundant and highly conserved process. The evolutionary changes in development are well preserved from class to class and species to species (see Chapter 9). For some abnormalities, prevention will involve the "up-regulation" of dormant systems using peptides delivered at critical periods of development. For example, proliferation of the myocardium may be regulated by growth factors targeted at the interventricular septum, thus minimizing the phenotype of hypertrophic cardiomyopathy.

Other malformations may be preventable through application of sound public health principles. One opportunity may involve dietary supplementation with folate. For 20 years, physicians have known that nearly 60% of neural tube defects can be prevented by supplementing folic acid in the diet of women during early pregnancy (Canadian Task Force, 1994). Similar data are emerging about the prevention of some forms of congenital cardiac defects, specifically, conotruncal abnormalities. In the California Birth Defects Surveillance System, infants of women who received 0.4 mg of folate in the periconceptional period had a 40% reduction in the prevalence of conotruncal heart defect (Shaw et al., 1995). Since conotruncal defects account for the vast majority of costs and mortality from cardiac defects, the reduction in neural tube, as well as cardiac, defects will have a personal and societal benefit (Waitzman, Romano, & Scheffler, 1994). A crude cost–benefit analysis suggests that for a yearly investment of $1.5 million, medical care and educational expense savings would exceed $120 million a year in California alone.

Patients with congenital cardiovascular malformations, the leading cause of infant mortality from congenital defects, are poised to benefit from the advances in our understanding of the biology of cardiac development. Although molecular diagnostic strategies will not eradicate congenital cardiovascular malformations entirely, the application of epidemiology, developmental biology, and clinical intervention will establish etiologic diagnoses, improve surgical outcome, and reduce the prevalence of congenital cardiovascular malformations.

Epilogue: Future directions in developmental cardiovascular sciences

BRADLEY B. KELLER AND WARREN W. BURGGREN

The preceding 21 chapters have provided a broad overview to improve the reader's grasp of the molecular, cellular, and integrative mechanisms determining cardiovascular development across a diverse range of species, with a focus on the impact of environment and disease on heart development. The question now remains, Where do we go from here? In all the chapters, the authors have attempted to outline the limits of current understanding and to highlight areas of future investigation. Several common themes deserve reinforcement, perhaps coming together in full force only after a complete review of the text.

Species diversity provides a unique basic science paradigm for developing and testing novel hypotheses regarding the mechanisms that determine cardiovascular morphogenesis. Nature provides a wide range of elegant biological methods to solve specific structure–function problems, one obvious example being the ability to shunt blood flow from the developing skin to lungs on transition to air-breathing. Invertebrate and nonmammalian cardiovascular systems are every bit as complex and adaptable to alterations in environmental forces as the popular "mature-mammalian" model for the cardiovascular system.

A multilevel analysis of cardiovascular development must be pursued for individual models. Coincident structural and functional maturation occurs at the subcellular, cellular, tissue, and organ levels during heart development in all species. It is not sufficient to limit developmental investigations to a single analysis plane, and we must define developmental mechanisms from the initial expression of DNA coding through the topological distribution of macromolecules into the mature 3-dimensional structure. In many instances, this multilevel analysis will demand both high technology and high cost, and each individual investigator will possess only a subset of the needed tools and skills. This demand for integrated technology and research will continue to drive the current research trend of collaborative research.

Collaborative research teams that participate in multi-investigator, multidisciplinary research will continue to grow in number. The coordination of skilled investigators will maximize the expertise applied to answer specific questions and will limit the duplication of expensive technology and highly specialized personnel. Despite rumors to the contrary, the future for extramural funding of collaborative research teams remains bright! However, the successful funding days of the solo investigator are under increasing challenge. Cardiovascular development, by nature a multidisciplinary field, is particularly well suited to collaborative research ventures incorporating anatomists, biologists, bioengineers, embryologists, geneticists, teratologists, physician-scientists, and more. The extent to which these ventures thrive will likely be a marker of the forward progress of the general field of research into the pathogenesis of cardiovascular diseases.

Maternal–fetal interactions and comparable nonmammalian paradigms remain a vastly underexplored area related to cardiovascular development. The developing cardiovascular system is exquisitely sensitive to alterations in both loading conditions and environment, and these factors have been shown to have both subtle and substantial impacts on the final cardiac phenotype. New technologies and paradigms will be required for the investigation of maternal–fetal pairs, and the statistical analysis will be yet another challenge. However, as medical therapy moves toward fetal interventions, there is a critical need for quantitative, hypothesis-driven research on these relationships.

New models of development and disease will be necessary to provide solutions to some of the most difficult problems under investigation. Novel animals are now under genetic design to target specific genes and pathways in the search for relevant regulatory mechanisms. These novel animals will provide the ability to test hypotheses that were previously unsuitable for quantitative investigation in isolation from other, confounding variables. However, the ability to target and disrupt specific genes to create an abnormal "phenotype" will not provide sufficient proof of the causes of specific cardiovascular diseases until similar genetic markers are identified in vivo. In fact, a wide range of primary molecular insults may result in a relatively small number of final phenotypes owing to the limited developmental repertoire of each individual species. Thus, although new molecular markers become available almost daily, the proof of molecular mechanisms is likely to require years of careful investigation.

The functional and structural maturation of embryonic cardiovascular systems in a broad range of species is now under investigation. For many species, these studies remain "embryonic," with most of the exciting and important work yet to be done. The experimental paradigms need to move from technical investigations to integrated paradigms that link structure, function, and morphogenesis. Dramatic and inspiring questions regarding the regulation of the 3-dimensional transformation of the simple, muscle-wrapped tube into the complex three-, four-, or five-chambered heart await investigation. Given this exciting environment, the demand for bright and motivated people to join collaborative research teams has never been higher.

As with all intensely active fields of science, we must recognize that the knowledge base and paradigms currently in use in cardiovascular development are by definition limited and subject to constant change. Each area of investigation gains new talents and new perspectives and shifts focus in predictable and unpredictable ways. Thus, although there is no simple answer to "Where do we go from here?" the most important first step is to be intrigued, stimulated, and inspired by the question. Good luck!

References

Abman, S. H., Chatfield, B. A., Hall, S. L., et al. (1990). Role of endothelium derived relaxing factor during transition of pulmonary circulation at birth. *American Journal of Physiology* 259:H1921–H1927. (Ch. 6)

Abman, S. H., Chatfield, B. A., Rodman, D. M., et al. (1991). Maturation changes in endothelium derived relaxing factor activity of ovine pulmonary arteries *in vitro. American Journal of Physiology* 260:L280–L285. (Ch. 6)

Abrass, C. K., Spicer, D., & Raugi, G. J. (1994). Insulin induces changes in extracellular matrix glycoproteins synthesized by rat mesangial cells in culture. *Kidney International* 46:613–620. (Ch. 5)

Aciniegas, E., Pulido, M., & Pereyra, B. (1988). Growth response of endothelial and smooth muscle cells to collagen matrix throughout development. *Atherosclerosis* 73:71–80. (Ch. 6)

Ackerman, R. A. (1977). The respiratory gas exchange of sea turtle nests. *Respiration Physiology* 31:19–38. (Ch. 18)

Adair, T. H., Guyton, A. C., Montani, J.-P., Lindsay, H. L., & Stanek, K. A. (1987). Whole body structural vascular adaptation to prolonged hypoxia in chick embryos. *American Journal of Physiology* 252:H1228–H1234. (Ch. 17)

Adams, J., & Watt, F. M. (1993). Regulation of development and differentiation by the extracellular matrix. *Development* 117:1183–1198. (Ch. 5)

al-Adhami, M. A., & Kunz, Y. W. (1976). Haemopoietic centres in the developing anglefish, *Pterophyllum scalare* (Cuvier & Valenciennes). *Wilhelm Roux's Archives* 179:393–401. (Ch. 12)

al-Adhami, M. A., & Kunz, Y. W. (1977). Ontogenesis of haematopoietic sites in *Brachydanio rerio* (Hamilton-Buchanan) (Teleostei). *Development, Growth, and Differentiation* 19:171–179. (Ch. 12)

Adolph, E. F. (1979). Development of dependence on oxygen in embryo salamanders. *American Journal of Physiology* 236:R282–R291. (Ch. 17)

Ager, A. (1990). Dynamic interactions between lymphocytes and vascular endothelial cells. In *The Endothelium: An Introduction to Current Research,* ed. J. B. Warren, pp. 229–252. New York: Wiley-Liss. (Ch. 6)

Agnisola, C., & Tota, B. (1994). Structure and function of the fish cardiac ventricle: flexibility and limitations. *Cardioscience* 5:145–153. (Ch. 4)

Aiba, S., & Creazzo, T. L. (1993). Comparison of the number of dihydropyridine receptors with the number of functional L-type calcium channels in embryonic heart. *Circulation Research* 72:396–402. (Ch. 2)

Akhurst, R. J., Lehnert, S. A., Faissner, A., & Duffie, E. (1990). TGF beta in murine morphogenetic processes: the early embryo and cardiogenesis. *Development* 108:645–656. (Ch. 10)

Akiyama, R., Ono, H., Höchel, J., Pearson, J. T., and Tazawa, H. (in press). Noninvasive determination of instantaneous heart rate by means of acoustocardiogram in developing avian embryos. *Medical and Biological Engineering and Computing.* (Ch. 15)

Akiyama, S. K., Nagata, K., & Yamada, K. M. (1990). Cell surface receptors for extracellular matrix components. *Biochimica et Biophysica Acta* 1031:91–110. (Ch. 5)

Albelda, S. M. (1991). Endothelial and epithelial cell adhesion molecules. *American Journal of Respiratory Cell and Molecular Biology* 4:195–203. (Ch. 6)

Albelda, S. M., & Buck, C. A. (1990). Integrins and other cell adhesion molecules. *FASEB Journal* 4:2868–2880. (Chs. 5, 6)

Alderdice, D. F. (1988). Osmotic and ionic regulation in teleost eggs and larvae. In *Fish Physiology.* Vol. 11, *The Physiology of Developing Fish,* ed. W. S. Hoar & D. J. Randall, pp. 163–251. San Diego & New York: Academic Press. (Ch. 17)

Alderdice, D. F., & Forrester, C. R. (1971). Effects of salinity, temperature, and dissolved oxygen on early development of the Pacific cod (*Gadus macrocephalus*). *Journal Fisheries Research Board of Canada* 28:883–902. (Ch. 17)

Alderdice, D. F., & Velsen, F. P. J. (1971). Some effects of salinity and temperature on early development of Pacific herring (*Clupea pallasi*). *Journal Fisheries Research Board of Canada* 28:1545–1562. (Ch. 17)

Alderdice, D. F., Wickett, W. P., & Brett, J. R. (1958). Some effects of temporary exposure to low dissolved oxygen levels on Pacific salmon eggs. *Journal Fisheries Research Board of Canada* 15:229–249. (Ch. 17)

Alexander, C. M., & Werb, Z. (1989). Proteinases and extracellular matrix remodeling. *Current Opinions in Cell Biology* 1(5):974–982. (Ch. 5)

Allan, L. D. (1988). Development of congenital heart lesions in mid to late gestation. *International Journal of Cardiology* 19:36–43. (Ch. 20)

Allan, L. D., Crawford, D. C., & Tynan, M. J. (1986). Pulmonary atresia in prenatal life. *Journal of the American College of Cardiology* 8:1131–1135. (Ch. 20)

Almatar, S. M. (1984). Effects of acute changes in temperature and salinity on the oxygen uptake of larvae of herring (*Clupea harengus*) and plaice (*Pleuronectes platessa*). *Marine Biology* 80:117–124. (Ch. 17)

Amento, E. P., Ehsani, N., Palmer, H., & Libby, P. (1991). Cytokines and growth factors positively and negatively regulate interstitial collagen gene expression in human vascular smooth muscle cells. *Arteriosclerosis and Thrombosis* 11:1223–1230. (Ch. 5)

Ancel, A., & Visschedijk, A. H. J. (1993). Respiratory exchanges in the incubated egg of the domestic guinea fowl. *Respiration Physiology* 91:31–42. (Ch. 18)

Anderson, D. F., Bissonnette, J. M., Faber, J. J., & Thornburg, K. L. (1981). *American Journal of Physiology* 241:H60–H66. (Ch. 16)

Anderson, D. T. (1959). The development of the polychaete *Scoloplos armiger. Quarterly Journal of Microscopical Science* 100:89–166. (Ch. 11)

Anderson, D. T. (1966). The comparative embryology of the Polychaeta. *Acta Zoologica* 47:1–42. (Ch. 11)

Anderson, D. T. (1973). *Embryology and Phylogeny in Annelids and Arthropods.* Oxford: Pergamon Press. (Ch. 11)

Anderson, P. A. W., Glick, K. I., Manring, A., & Crenshaw, C., Jr. (1984). Developmental changes in cardiac contractility in fetal and postnatal sheep: *In vitro* and *in vivo. American Journal of Physiology* 247:H371–H379. (Ch. 16)

Anderson, P. A. W., Greig, A., Mark, T. M., Malouf, N. N., Oakeley, A. E., Ungerleider, R. M., Allen, P. D., & Kay, B. K. (1995). Molecular basis of human cardiac troponin T isoforms expressed in the developing, adult, and failing heart. *Circulation Research* 76:681–686. (Ch. 3)

Anderson, P. A. W., Malouf, N. N., Oakeley, A. E., Pagani, E. D., & Allen, P. D. (1991). Troponin T isoform expression in humans: A comparison among normal and failing adult heart, fetal heart, and adult and fetal skeletal muscle. *Circulation Research* 69:1226–1233. (Ch. 3)

Anderson, P. A. W., & Oakeley, A. E. (1989). Immunological identification of five troponin T isoforms reveals an elaborate maturational troponin T profile in rabbit myocardium. *Circulation Research* 65:1087–1093. (Ch. 3)

Anderson, R. H. (1987). Terminology. In *Paediatric Cardiology,* ed. R. H. Anderson, F. J. Macartney, E. A. Shinebourne, & M. Tynan, pp. 65–82. New York & Edinburgh: Churchill Livingstone. (Ch. 19)

Andreadis, A., Gallego, M. E., & Nadal-Ginard, B. (1987). Generation of protein isoform diversity by alternative splicing: Mechanistic and biological implications. *Annual Reviews of Cell Biology* 3:207–242. (Ch. 3)

Andrikopoulos, K., Liu, X., Keene, D. R., Jaenisch, R., & Ramirez, F. (1995). Targeted mutation in the *col5a2* gene reveals a regulatory role for type V collagen during matrix assembly. *Nature Genetics* 9:31–36. (Ch. 5)

Anteunis, A., Leclerc, M., Vial, M., Brillouet, C., Luquot, G., Robinaux, R., & Binaghi, R. A. (1985). Immunocompetent cells in the starfish *Asterias rubens:* An ultrastructural study. *Cell Biology International Reports* 9:663–670. (Ch. 11)

Anversa, P., Loud, A. V., Giacomelli, F., et al. (1978). Absolute morphometric study of myocardial hypertrophy in experimental hypertension. *Circulation Research* 38:597–603. (Ch. 20)

Anversa, P., Olivetti, G., & Loud, A. V. (1980). Morphometric study of early postnatal development in the left and right ventricular myocardium of the rat. *Circulation Research* 46:495. (Ch. 20)

Ar, A., & Girard, H. (1989). Anisotropic gas diffusion in the shell membranes of the hen's egg. *Journal of Experimental Zoology* 251:20–26. (Ch. 18)

Ar, A., & Mover, H. (1994). Oxygen tensions in developing embryos: System inefficiency or system requirement? *Israel Journal of Zoology* 40:307–326. (Ch. 18)

Arai, M., Otsu, K., MacLennan, D. H., & Periasamy, M. (1992). Regulation of sarcoplasmic reticulum gene expression during cardiac and skeletal muscle development. *American Journal of Physiology* 262:C614–C620. (Ch. 2)

Arceci, R. J., King, A. A. J., Simon, M. C., Orkin, S. H., & Wilson, D. B. (1993). Mouse GATA-4: A retinoic acid-inducible GATA-binding transcription factor expressed in endodermally derived tissues and heart. *Molecular and Cell Biology* 13:2235–2246. (Ch. 3)

Armstrong, P. B. (1931). Functional reactions in the embryonic heart accompany the ingrowth and development of the vagus innervation. *Journal of Experimental Zoology* 58:43–67. (Ch. 12)

Armstrong, P. B., & Armstrong, M. T. (1984). A role for fibronectin in cell sorting. *Journal of Cell Science* 69:179–197. (Ch. 5)

Armstrong, P. B., & Armstrong, M. T. (1990). An instructive role for the interstitial matrix in tissue patterning: Tissue segregation and intercellular invasion. *Journal of Cell Biology* 110:1439–1449. (Ch. 5)

Armstrong, P. B., & Child, J. S. (1965). Stages in the normal development of *Fundulus heteroclitus. Biological Bulletin (Woods Hole)* 128:143–168. (Ch. 12)

Arnold, J. M., & Carlson, B. A. (1986). Living *Nautilus* embryos: Preliminary observations. *Science* 233:73–76. (Ch. 11)

Artman, M. (1992). Sarcolemmal Na^+–Ca^{2+} exchange activity and exchanger immunoreactivity in developing rabbit hearts. *American Journal of Physiology* 63:H1506–H1513. (Ch. 2)

Artman, M., Graham, T. P., Jr., & Boucek, R. J., Jr. (1985). Effects of postnatal maturation on myocardial contractile responses to calcium antagonists and changes in contraction frequency. *Journal of Cardiovascular Pharmacology* 7:850–855. (Ch. 2)

Artman, M., Ichikawa, H., Avkiran, M., & Coetzee, W. A. (1995). Na^+–Ca^{2+} exchange current density in cardiac myocytes from rabbits and guinea pigs during postnatal development. *American Journal of Physiology* 68:H1714–H1722. (Ch. 2)

Assali, N. S., Holm, L. W., & Sehgal, N. (1962). Regional blood flow and vascular resistance of the fetus in utero. *American Journal of Obstetrics and Gynecology* 83:809–817. (Ch. 16)

Atkinson, J. W. (1971). Organogenesis in normal and lobeless embryos of the marine prosobranch gastropod *llyanassa obsoleta. Journal of Morphology* 133:339–352. (Ch. 11)

Avila, C., Arigue, A., Tamse, C. T., & Kuzirian, A. M. (1994). *Hermissenda crassicornis* larvae metamorphose in the laboratory in response to artificial and natural inducers. *Biological Bulletin (Woods Hole)* 187:252–253. (Ch. 11)

Azpiazu, N., & Frasch, M. (1993). *Tinman* and *bagpipe:* Two homeobox genes that determine cell fates in the dorsal mesoderm of Drosophila. *Genes and Development* 7:1325–1340. (Ch. 1)

Bachmann, S., Pohla, H., & Goldschmid, A. (1980). Phagocytes in the axial complex of the sea urchin, *Sphaerechinus granularis* (Lam.): Fine structure and X-ray microanalysis. *Cell and Tissue Research,* 213:109–120. (Ch. 11)

Baer, K. E., von. (1828). Ueber Entwicklungsgeschichte der Thiere. *Beobachtung und Reflexion* Konisberg. (Ch. 9)

Baghdassarian, D., Toru-Delbauffe, D., Gavaret, J. M., & Pierre, M. (1993). Effects of transforming growth-beta 1 on the extracellular matrix and cytoskeleton of cultured astrocytes. *GLIA* 7:193–202. (Ch. 5)

Baird, B. H., & Thistle, D. (1986). Uptake of bacterial extracellular polymer by a deposit-feeding holothurian (*Isostichopus badionotus*). *Marine Biology* 92:183–187. (Ch. 11)

Bajelan, M., & Leclerc, M. (1990). Properties of sea star phagocytes in axial organ cell subpopulations. *Cell Biology International Reports* 14:1129–1132. (Ch. 11)

Balachov, N. V., & Soltitskij, V. V. (1991). On the development of cardiac muscle reactivity in Black Sea mullets at early stages of ontogenesis (in Russian; English abstract). In *Mullet Culture in the Azov and Black Sea Basins,* ed. N. G. Kulikova, pp. 51–59. Moscow: VNIRO. (Ch. 12)

Balashov, N. V., Soltitskii, V. V., & Makukhina, L. I. (1991). The effect of noradrenaline and acetylcholine on cardiac muscle of the golden mullet *Liza auratus* at early stages of ontogenesis (in Russian: English abstract). *Zhurnal Evolyutsionnoi Biokhimii i Fiziologii* 27(2):255–257. (Ch. 12)

Baldwin, H. S., & Buck, C. A. (1994). Integrins and other cell adhesion molecules in cardiac development. *Trends in Cardiovascular Medicine* 4:178–187. (Chs. 5, 10)

Baldwin, H. S., Lloyd, T. R., & Solursh, M. (1994). Hyaluronate degradation affects ventricular function of the early postlooped embryonic rat heart *in situ. Circulation Research* 74:244–252. (Ch. 7)

Baldwin, H. S., & Solursh, M. (1989). Degradation of hyaluronic acid does not prevent looping of the mammalian heart *in situ. Developmental Biology* 136:555–559. (Ch. 10)

Baldwin, H. S., Suzuki, H., & Solursh, M. (1990). *In situ* hybridization localizes fibronectin RNA to the precardiac mesoderm in the rat. *Anatomical Record* 226:10A. (Ch. 10)

Ball, K. L., Johnson, M. D., & Solaro, R. J. (1994). Isoform specific interactions of troponin I and troponin C determine pH sensitivity of myofibrillar Ca^{2+} activation. *Biochemistry* 33:8464–8471. (Ch. 3)

Ballard, W. W. (1964). *Comparative Anatomy and Embryology.* New York: Ronald Press. (Ch. 12)

Ballard, W. W. (1973a). Normal embryonic stages for salmonid fishes, based on *Salmo gairdneri* (Richardson) and *Salvelinus fontinalis* (Mitchil). *Journal of Experimental Zoology* 184:7–26. (Ch. 12)

Ballard, W. W. (1973b). A new fate map for *Salmo gairdneri. Journal of Experimental Zoology* 184:49–73. (Ch. 12)

Ballard, W. W. (1982). Morphogenetic movements and fate map of *Catostomus commersoni. Journal of Experimental Zoology* 219:301–321. (Ch. 12)

Ballard, W. W. (1986). Morphogenetic movements and a provisional fate map of development in the holostean fish *Amia calva. Journal of Experimental Zoology* 238:355–372. (Ch. 12)

Ballard, W. W., & Ginsburg, A. S. (1980). Morphogenetic movements in Acipenserid embryos. *Journal of Experimental Zoology* 213:69–103. (Ch. 12)

Ballard, W. W., & Needham, R. G. (1964). Normal embryonic stages of *Polyodon spathula. Journal of Morphology* 114:465–478. (Ch. 12)

Balon, E. K. (1975). Reproductive guilds of fishes: A proposal and definition. *Journal of the Fisheries Research Board of Canada* 32:821–864. (Chs. 12, 17)

Balon, E. K. (1980). Early ontogeny of lake charr – *Salvelinus (Cristivomer) namaycush.* In *Charrs: Salmonid Fishes of the Genus* Salvelinus, ed. E. K. Balon, pp. 485–562. The Hague: Dr. W. Junk. (Ch. 12)

Balon, E. K. (1981). Additions and amendments to the classification of reproductive styles in fishes. *Environmental Biology of Fishes* 6:377–389. (Ch. 12)

Balser, E. J., & Ruppert, E. E. (1990). Structure, ultrastructure, and function of the preoral heart–kidney in *Saccoglossus kowalevskii* (Hemichordata, Enteropneusta), including new data on the stomochord. *Acta Zoologica* 71:235–249. (Ch. 11)

Balter, M. (1995). Zebrafish yields a bumper crop of genes. *Science* 269:480. (Ch. 12)

Banai, S., Jaklitsch, M. T., Shou, M., Lazarous, D. F., Scheinowitz, M., Biro, S., Epstein, S. E., & Unger, E. F. (1994). Angiogenic-induced enhancement of collateral blood flow to ischemic myocardium by vascular endothelial growth factor in dogs. *Circulation* 89:2183–2189. (Ch. 4)

Barany, M. (1967). ATPase activity of myosin correlated with speed of muscle shortening. *Journal of General Physiology* 50:197–218. (Ch. 3)

Barbee, R. W., Perry, B. D., Re, R. N., Murgo, J. P., & Field L. J. (1994). Hemodynamics in transgenic mice with overexpression of atrial natriuretic factor. *Circulation Research* 74:740–746. (Ch. 7)

Bargmann, W., & von Heln, G. (1968). Ueber des Axialorgan ("Mysterious Gland") von *Asterias rubens* L. *Zeitschrift für Zellforschung und Mikvoskopische Anatomie* 88:262–277. (Ch. 11)

Barkero D. J. P. (1992). *Fetal and Infant Origins of Adult Disease.* London: British Medical Journal. (Ch. 21)

Barnes, A. E., & Jensen, W. N. (1959). Blood volume and red cell concentration in the normal chick embryo. *American Journal of Physiology* 197:403–405. (Chs. 7, 15)

Barry, A. (1942). The intrinsic pulsation rates of fragments of embryonic chick heart. *Journal of Experimental Zoology* 91:119–130. (Ch. 10)

Barry, A. (1948). The functional significance of the cardiac jelly in the tubular heart of the chick embryo. *Anatomical Record* 102:289–298. (Chs. 7, 15)

Barry, W. H., & Bridge, J. H. B. (1993). Intracellular calcium homeostasis in cardiac myocytes. *Circulation* 87:1806–1815. (Ch. 2)

Bartelings, M. M., & Gittenberger–de Groot, A. C. (1991). Morphogenetic considerations on congenital malformations of the outflow tract. Part 2, Complete transposition of the great arteries and double outlet right ventricle. *International Journal of Cardiology* 33:5–26. (Ch. 19)

Barth, E., Stammler, G., Speiser, B., & Schaper, J. (1992). Ultrastructural quantitation of mitochondria and myofilaments in cardiac muscle from 10 different animal species including man. *Journal of Molecular and Cellular Cardiology* 24:669–681. (Ch. 2)

Bartlett, G. R. (1980). Phosphate compounds in vertebrate red blood cells. *American Zoologist* 20:103–114. (Ch. 17)

Barton, P. J. R., Cohen, A., Robert, B., Fiszman, M. Y., Bonhomme, F., Guenet, J. L., Leader, D. P., & Buckingham, M. E. (1985). The myosin alkali light chains of mouse ventricular and slow skeletal muscle are indistinguishable and are encoded by the same gene. *Journal of Biological Chemistry* 260:8578–8584. (Ch. 3)

Barton, P. J. R., Robert, B., Cohen, A., Garner, I., Sassoon, D., Weydert, A., & Buckingham M. E. (1988). Structure and sequence of the myosin alkali light chain gene expressed in adult cardiac atria and fetal striated muscle. *Journal of Biological Chemistry* 263:12669–12676. (Ch. 3)

Bashey, R. I., Martinez-Hernandez, A., & Jiminez, S. A. (1992). Isolation, characterization, and localization of cardiac collagen type VI: Association with other matrix components. *Circulation Research* 70:1006–1017. (Ch. 5)

Bassani, J. W. M., Bassani, R. A., & Bers, D. M. (1994). Relaxation in rabbit and rat cardiac cells: Species-dependent differences in cellular mechanisms. *Journal of Physiology* 476:279–293. (Ch. 2)

Bate, M. (1993). The mesoderm and its derivatives. In *The Development of* Drosophila melanogaster, ed. M. Bate and A. Martinez Arias, vol. 1, pp. 1029–1039. Plainview, N.Y.: Cold Spring Harbor Laboratory Press. (Ch. 1)

Bauer, C., & Jelkmann, W. (1977). Carbon dioxide governs the oxygen-affinity of crocodile blood. *Nature* 269:825–827. (Ch. 14)

Baumann, R. (1984). Regulation of oxygen affinity of embryonic blood during hypoxic incubation. In *Respiration and Metabolism of Embryonic Vertebrates,* ed. R. S. Seymour, pp. 221–230. Dordrecht, Boston, & London: Dr. W. Junk. (Ch. 17)

Baumann, R., & Meuer, H.-J. (1992). Blood oxygen transport in the early avian embryo. *Physiological Reviews* 72:941–965. (Chs. 14, 17, 18)

Beijnink, F. B., Walker, C. W., & Voogt, P. A. (1984). An ultrastructural study of relationships between the ovarian haemal system, follicle cells, and primary oocytes in the sea star, *Asterias rubens*: Implications for oocyte nutrition. *Cell and Tissue Research* 238:339–347. (Ch. 11)

Benetti, D. D., & Martinez, L. (1993). Respiratory distress in dolphin, *Coryphaena hippurus,* larvae. (Abstract.) In *From Discovery to Commercialization,* ed. M. Carrillo, L. Dahle, J. Morales, P. Sorgeloos, N. Svennevig, & J. Wyban, p. 312. Special Publication of the European Aquaculture Society, no. 19. Oostende, Belgium. (Ch. 12)

Bennet, H. S. (1936). The development of the blood supply to the heart in the embryo pig. *American Journal of Anatomy* 60:27–53. (Ch. 4)

Benson, D. W., Jr., Hughes, S. F., Hu, N., & Clark, E. B. (1989). Effect of heart rate increase on dorsal aortic flow before and after volume loading in the stage 24 chick embryo. *Pediatric Research* 26:245–249. (Ch. 7)

Berezovsky, V. A., Goida, E. A., Mukalov, I. O., & Sushdo, B. S. (1979). Experimental study of oxygen distribution in *Misgurnis fossilis* eggs. *Fiziolohichny Zhurnal* (Kiev) 25:379–389. *Canadian Translations of Fisheries and Aquatic Sciences* 5209 (1986). (Chs. 17, 18)

Berger, R., Albelda, S. M., Berd, D., et al. (1993). Expression of platelet-endothelial cell adhesion molecules (PECAM-1) during melanoma-induced angiogenesis. *Journal of Cutaneous Pathology* 20:399–406. (Ch. 6)

Bernard, C. (1975). Establishment of ionic permeabilities of the myocardial membrane during embryonic development of the rat. In *Development and Physiological Correlates of Cardiac Muscle,* ed. M. Lieberman & T. Sano, pp. 169–184. New York: Raven Press. (Ch. 2)

Berne, R. M., & Rubio, R. (1979). Coronary circulation. In *Handbook of Physiology,* sec. 2, *The Cardiovascular System, vol. 1, The Heart,* ed. R. M. Berne, N. Sperelakis, & S. R. Geiger, pp. 873–952. Bethesda, Md.: American Physiological Society. (Chs. 16, 25)

Bers, D. M., Bridge, J. H., & MacLeod, K. T. (1987). The mechanism of ryanodine action in rabbit ventricular muscle evaluated with Ca^{2+}-selective microelectrodes and rapid cooling contractures. *Canadian Journal of Physiology and Pharmacology* 65:610–618. (Ch. 2)

Bers, D. M., Lederer, W. J., & Berlin, J. R. (1990). Intracellular Ca transients in rate cardiac myocytes: Role of Na–Ca exchange in excitation–contraction coupling. *American Journal of Physiology* 258:C944–C954. (Ch. 2)

Bertossi, M., Roncali, L., Nico, B., et al. (1993). Perivascular astrocytes and endothelium in the development of the blood–brain barrier in the optic tectum of the chick embryo. *Anatomy Embryology* (West Germany) 188:21–29. (Ch. 6)

Beukelmann, D. J., & Wier, W. G. (1988). Mechanism of release of calcium from sarcoplasmic reticulum of guinea pig cells. *Journal of Physiology* 405:233–255. (Ch. 2)

Bevan, J. A., & Laher, I. (1991). Pressure and flow-dependent vascular tone. *FASEB Journal* 5:2267–2271. (Ch. 6)

Bevilacqua, M. P., Stengalin, S., Gimbrone, M. A., et al. (1989). Endothelial leukocyte adhesion molecule 1: An inducible receptor for neutrophils related to complement regulatory proteins and lectins. *Science* 243:1160–1165. (Ch. 6)

Bhavsar, P. K., Dhoot, G. K., Cumming, D. V., Butler-Browne, G. S., Yacoub, M. H., & Barton, P. J. (1991). Developmental expression of troponin I isoforms in fetal human heart. *FEBS Letters* 292:5–8. (Ch. 3)

Binyon, J. (1972). *Physiology of Echinoderms.* Oxford: Pergamon Press. (Ch. 11)

Birchard, G. F., & Reiber, C. L. (1993). A comparison of avian and reptilian chorioallantoic vascular density. *Journal of Experimental Biology* 178:245–249. (Chs. 14, 18)

Bird, D. J., Lutz, P. L., & Potter, I. C. (1976). Oxygen dissociation curves of the blood of larval and adult lampreys (*Lampetra fluviatilis*). *Journal of Experimental Biology* 65:449–458. (Ch. 12)

Bissonnette, J. M., & Metcalfe, J. (1978). Gas exchange of the fertile-hen's egg: Components of resistance. *Respiration Physiology* 34:209–218. (Chs. 15, 18)

Black, I. B., Bohn, M. C., Jonakait, G. M., & Kessler, J. A. (1981). Transmitter phenotypic expression in the embryo. *Ciba Foundation Symposium* 83:177–193. (Ch. 8)

Blackburn, D. G. (1992). Convergent evolution of viviparity, matrotrophy, and specializations for fetal nutrition in reptiles and other vertebrates. *American Zoologist* 32:313–321. (Chs. 14, 18)

Blackburn, D. G. (1993). Chorioallantoic placentation in squamate reptiles: Structure, function, development, and evolution. *Journal of Experimental Zoology* 266:414–430. (Ch. 18)

Blaxter, J. H. S. (1988). Pattern and variety in development. In *Fish Physiology,* vol. 11A, *The Physiology of Developing Fish,* ed. W. S. Hoar & D. J. Randall, pp. 1–58. San Diego: Academic Press. (Chs. 12, 18)

Bockman, D. E., Redmond, M. E., Waldo, K., Davis, H., & Kirby, M. L. (1987). Effect of neural crest ablation on development of the heart and arch arteries in the chick. *American Journal of Anatomy* 180:332–341. (Ch. 19)

Bodmer, R. (1993). The gene *tinman* is required for specification of the heart and visceral muscles in *Drosophila. Development* 118:719–729. (Ch. 1)

Bodmer, R. (1995). Heart development in *Drosophila* and its relationship to vertebrates. *Trends in Cardiovascular Medicine* 5:21–28. (Ch. 1)

Bodmer, R., Jan, L. Y., & Jan, Y. N. (1990). A new homeobox-containing gene, *msh-2,* is transiently expressed early during mesoderm formation of *Drosophila. Development* 110:661–669. (Ch. 1)

Bodo, M., Becchetti, E., Giammarioli, M., Baroni, T., Bellucci, C., Pezzetti, F., Calvitti, M., & Carinci, P. (1994). Interleukin-1 and interleukin-6 differentially regulate the accumulation of newly synthesized extracellular matrix components and the cytokine release by developing chick embryo skin fibroblasts. *International Journal of Developmental Biology* 38:535–542. (Ch. 5)

Boerth, S. R., Zimmer, D. B., & Artman, M. (1994). Steady-state mRNA levels of the sarcolemmal Na^+–Ca^{2+} exchanger peak near birth in developing rabbit and rat hearts. *Circulation Research* 74:354–359. (Ch. 2)

Bogers, A. J. J. C., Gittenberger–de Groot, A. C., Poelmann, R. E., Peault, B. M., & Huysmans, H. A. (1989). Development of the origin of the coronary arteries: A matter of ingrowth or outgrowth? *Anatomy and Embryology* (West Germany) 180:437–441. (Ch. 4)

Boheler, K. R., Carrier, L., de la Bastie, D., Allen, P. D., Komajda, M., Mercadier, J. J., and Schwartz, K. (1991). Skeletal actin mRNA increases in the human heart during ontogenic development and is the major isoform of control and failing adult hearts. *Journal of Clinical Investigation* 88:323–330. (Ch. 3)

Boheler, K. R., Chassagne, C., Martin, X., Wisnewsky, C., & Schwartz, K. (1992). Cardiac expressions of α- and β-myosin heavy chains and sarcomeric α-actins are regulated through transcriptional mechanisms. *Journal of Biological Chemistry* 267:12979–12985. (Ch. 3)

Boletzky, S. V. (1987a). Embryonic phase. In *Cephalopod Life Cycles,* vol. 2, *Comparative Reviews,* ed. P. R. Boyle, pp. 5–31. New York: Academic Press. (Ch. 11)

Boletzky, S. V. (1987b). Ontogenetic and phylogenetic aspects of the cephalopod circulatory system. *Experientia* 43:478–483. (Ch. 11)

Bond, A. N. (1960). An analysis of the response of salamander gills to changes in the oxygen concentration of the medium. *Developmental Biology* 2:1–20. (Ch. 17)

Boolootian, R. A., & Campbell, J. L. (1964). A primitive heart in the echinoid *Strongylocentrotus purpuratus. Science* 145: 173–175. (Ch. 11)

Boolootian, R. A., & Campbell, J. L. (1966). The axial gland complex. *Nature* 212:946–947. (Ch. 11)

Bordzilovskaya, N. P., Dettlaff, T. A., Huhon, S. T., & Malacinski, G. M. (1989). Developmental-stage series of axolotl (*Ambystoma mexicanum*) embryos. In *Developmental Biology of Axolotl,* ed. J. B. Armstrong & G. M. Malacinski, pp. 201–219. Oxford: Oxford University Press. (Ch. 17)

Borg, T. K. (1982). Development of the connective tissue network in the neonatal hamster heart. *Anatomical Record* 165:435–444. (Ch. 5)

Borg, T. K., & Burgess, M. L. (1993). Holding it all together. Organization and function(s) of the extracellular matrix in the heart. *Heart Failure* 8:230–238. (Ch. 5)

Borg, T. K., & Caulfield, J. B. (1979). Collagen in the heart. *Texas Reports on Biology and Medicine* 39:321–333. (Ch. 5)

Borg, T. K., Gay, R., & Johnson, L. D. (1982). Changes in the distribution of fibronectin and collagen during development of the neonatal heart. *Collagen Related Research* 2:2111–2118. (Chs. 5, 7)

Borg, T. K., Johnson, L. D., & Lill, P. H. (1984). Specific attachment of collagen to cardiac myocytes: *In vivo* and *in vitro. Developmental Biology* 97:417–423. (Ch. 5)

Borg, T. K., Rasso, D. S., & Terracio, L. (1990). Potential role of the extracellular matrix in the postseptation development of the heart. *Annals of the New York Academy of Sciences* 588:87–92. (Ch. 5)

Borg, T. K., & Teracio, L. (1990). Interaction of the extracellular matrix with cardiac myocytes during development and disease. In *Issues in Biomedicine,* ed. T. Robinson, pp. 113–129. Basel: Karger Publishers. (Chs. 5, 7)

Borg, T. K., Xuehui, M., Hilenski, L., Vinson, N., & Terracio, L. (1990). The role of the extracellular matrix on myofibrillogenesis *in vitro.* In *Developmental Cardiology: Morphogenesis and Function,* ed. E. B. Clark and A. Takao, pp. 175–190. Mount Kisco, N.Y.: Futura. (Ch. 5)

Bork, P. (1992). The modular architecture of vertebrate collagens. *FEBS Letters* 307:49–54. (Ch. 5)

Born, G. V. R., Dawes, G. S., Mott, J. C., & Widdicombe (1954). Changes in the heart and lung at birth. *Cold Spring Harbor Symposia on Quantitative Biology* 19:102–108. (Ch. 16)

Boucek, R. J., Jr., Shelton, M., Artman, M., Mushlin, P. S., Starnes, V. A., & Olson, R. D. (1984). Comparative effects of verapamil, nifedipine, and diltiazem on contractile function in the isolated immature and adult rabbit heart. *Pediatric Research* 18:948–952. (Ch. 2)

Bouman, H. G. A., Broekhuizen, M. L. A., Baasten, A. M. J., Gittenberger–de Groot, A. C., & Wenink, A. C. G. (1995). Spectrum of looping disturbances in stage 34 chicken hearts after retinoic acid treatment. *Anatomical Record* 243(1):101–108. (Ch. 19)

Bourne, G. B., Redmond, J. R., & Jorgensen, D. D. (1990). Dynamics of the molluscan circulatory system: Open versus closed. *Physiological Zoology* 63:140–166. (Ch. 11)

Bourne, G. H. (1980). *Hearts and Heart-like Organs.* Vol. 1. New York: Academic Press. (Ch. 9)

Bouvagnet, P., Neveu, S., Montoya, M., & Leger, J. J. (1987). Development changes in the human cardiac isomyosin distribution: An immunohistochemical study using monoclonal antibodies. *Circulation Research* 61:329–336. (Ch. 3)

Bowers, P., Tinney, J. P., & Keller, B. B. (1996). Nitroprusside selectively reduces preload in the stage 21 chick embryo. *Cardiovascular Research* 31:E132–138. (Chs. 6, 7)

Bradford, D. F. (1990). Incubation time and rate of embryonic development in amphibians: The influence of ovum size, temperature, and reproductive mode. *Physiological Zoology* 63:1157–1180. (Ch. 17)
Bradford, D. F., & Seymour, R. S. (1988). Influence of environmental P_{O_2} on embryonic oxygen consumption, rate of development, and hatching in the frog *Pseudophryne bibroni. Physiological Zoology* 61:475–482. (Ch. 17)
Brandon, E. P., Idzerda, R. L., & McKnight, G. S. (1995a). Targeting the mouse genome: A compendium of knockouts. *Current Biology* 5(6):625–634. (Ch. 19)
Brandon, E. P., Idzerda, R. L., & McKnight, G. S. (1995b). Targeting the mouse genome: A compendium of knockouts. *Current Biology* 5(7):758–765. (Ch. 19)
Brandon, E. P., Idzerda, R. L., & McKnight, G. S. (1995c). Targeting the mouse genome: A compendium of knockouts. *Current Biology* 5(8):873–881. (Ch. 19)
Brannan, C. I., Perkins, A. S., Vogel, K. S., Ratner, N., Nordlund, M. L., Reid, S. W., Buchberg, A. M., Jenkins, N. A., Parada, L. F., & Copeland, N. G. (1994). Targeted disruption of the neurofibromatosis type-1 gene leads to developmental abnormalities in heart and various neural crest–derived tissues. *Genes & Development* 8:1019–1029. (Ch. 1)
Braunwald, E., Ross, J., Jr., & Sonnenblick, E. H. (1976). *Mechanisms of Contraction of the Normal and Failing Heart.* 2nd ed. Boston: Little, Brown. (Ch. 16)
Breier, G., Albrecht, U., Sterrer, S., et al. (1992). Expression of vascular endothelial growth factor during embryonic angiogenesis and endothelial cell differentiation. *Development* 114:521–532. (Ch. 6)
Brettell, L. M., & McGowan, S. E. (1994). Basic fibroblast growth factor decreases elastin production by neonatal rat lung fibroblasts. *American Journal of Respiratory Cell Molecular Biology* 10:306–315. (Ch. 5)
Bride, M. (1975). Establissement de l'innervation dans le coeur du têtard de Xénope et ses répercussions sur le fonctionnement de l'organe. *Comptes Rendus Séances, Société de Biologie* (Paris) 169:1265–1271. (Ch. 13)
Bridge, J. H. B., Smolley, J. R., & Spitzer, K. W. (1990). The relationship between charge movements associated with I_{Ca} and $I_{Na\text{-}Ca}$ in cardiac myocytes. *Science* 248:376–378. (Ch. 2)
Brilla, C. G., Maisch, B., & Weber, K. T. (1993). Renin-angiotensin system and myocardial collagen matrix remodeling in hypertensive heart disease: *In vivo* and *in vitro* studies on collagen matrix regulation. *Clinical Investigation* 71:S35–S41. (Ch. 5)
Brillantes, A. B., Bezprozvannaya, S., & Marks, A. R. (1994). Developmental and tissue-specific regulation of rabbit skeletal and cardiac muscle calcium channels involved in excitation–contraction coupling. *Circulation Research* 75:503–510. (Ch. 2)
Brillouet, C., Leclerc, M., Binagi, R. A., & Luquet, G. (1984). Specific immune response in the sea star *Asterias rubens;* production of "antibody-like" factors. *Cellular Immunology* 84:138–144. (Ch. 11)
Bristow, J. (1995). The search for genetic mechanisms of congenital heart disease. *Cellular and Molecular Biology Research* 41:307–319. (Ch. 20)
Brittain, T., & Wells, R. M. G. (1991). An investigation of the cooperative functioning of the haemoglobin of the crocodile, *Crocodile porosus. Comparative Biochemistry and Physiology* 98B:641–646. (Ch. 14)
Broekhuizen, M. L. A., Bouman, H. G. A., Mast, F., Mulder, P. G. H., Gittenberger–de Groot, A. C., & Wladimiroff, J. W. (1995). Hemodynamic changes in HH stage 34 chick embryos after treatment with all-*trans*-retinoic acid. *Pediatric Research* 38(3):342–348. (Ch. 19)
Broekhuizen, M. L. A., Mast, F., Struijk, P. C., van der Bie, W., Mulder, P. G. H., Gittenberger–de Groot, A. C., & Wladimiroff, J. W. (1993). Hemodynamic parameters of stage 20 to stage 35 chick embryo. *Pediatric Research* 34:44–46. (Ch. 7)
Broekhuizen, M. L. A., Wladimiroff, J. W., Tibboel, D., Poelmann, R. E., Wenink, A. C. G., & Gittenberger–de Groot, A. C. (1992). Induction of cardiac anomalies with all-*trans*-retinoic acid in the chick embryo. *Cardiology in the Young* 2:311–317. (Ch. 19)
Broertjes, J. J. S., De Waard, P., & Voogt, P. A. (1984). On the presence of vitellogenic substances in the starfish, *Asterias rubens* (L.). *Journal of the Marine Biological Association of the United Kingdom* 64:261–269. (Ch. 11)
Broertjes, J. J. S., Posthuma, G., Beijnink, F. B., & Voogt, P. A. (1980a). The admission of nutrients from the digestive system into the haemal channels in the sea-star *Asterias rubens* (L.). *Journal of the Marine Biological Association of the United Kingdom* 60:883–890. (Ch. 11)

Broertjes, J. J. S., Posthuma, G., Den Breejen, P., & Voogt, P. A. (1980b). Evidence for an alternative transport route for the use of vitellogenesis in the sea-star *Asterias rubens* (L.). *Journal of the Marine Biological Association of the United Kingdom* 60:157–162. (Ch. 11)

Brook, M. M., & Silverman, N. H. (1993). Echocardiographic techniques for assessing normal and abnormal fetal cardiac anatomy. *Western Journal of Medicine* 159:286–300. (Ch. 20)

Brooke, L. T., & Colby, P. J. (1980). Development and survival of embryos of lake herring at different constant oxygen concentrations and temperatures. *Progressive Fish-Culturist* 42:3–9. (Ch. 17)

Brown, A. L. (1986). Morphologic factors in cardiac hypertrophy. In *Cardiac Hypertrophy,* ed. N. R. Alpert, p. 11. New York: Academic Press. (Ch. 20)

Bruyere, H. J., Folts, J. D., & Gilbert, E. F. (1984). Hemodynamic mechanisms in the pathogenesis of cardiovascular malformations in the chick embryo: Cardiac function changes following epinephrine stimulation in chick embryos. In *Congenital Heart Disease: Cause and Process,* ed. J. J. Nora and A. Takao, pp. 279–292. Mount Kisco, N.Y.: Futura. (Ch. 8)

Budarf, M. L., Collins, J., Gong, W., Roe, B., Wand, Z., Bailey, L. C., Sellinger, B., Michaud, D., Driscoll, D. A., & Emanuel, B. S. (1995). Cloning a balanced translocation associated with DiGeorge syndrome and identification of a disrupted candidate gene. *Nature Genetics* 10:269–277. (Ch. 21)

Bull, C., Kostelka, M., Sorensen, K., et al. (1994). Outcome measures for the neonatal management of pulmonary atresia with intact ventricular septum. *Journal of Thoracic and Cardiovascular Surgery* 107:359–366. (Ch. 20)

Burgess, M. L., Carver, W. E., Terracio, L., Wilson, S. P., Wilson, M. A., & Borg, T. K. (1994). Integrin-mediated collagen gel contraction by cardiac fibroblasts: Effects of angiotensin II. *Circulation Research* 74:291–298. (Ch. 5)

Burgess, W. H., & Maciag, T. (1989). The heparin-binding (fibroblast) growth factor family of proteins. *Annual Review of Biochemistry* 58:575–606. (Ch. 5)

Burggren, W. W. (1976). The persistence of a patent ductus arteriosus in an adult specimen of the tortoise *Testudo graeca. Copeia,* pp. 405–407. (Ch. 14)

Burggren, W. W. (1977). The pulmonary circulation of the chelonian reptile: Morphology, haemodynamics, and pharmacology. *Journal of Comparative Physiology* 116:303–324. (Ch. 14)

Burggren, W. W. (1984). Transition of respiratory processes during amphibian metamorphosis: From egg to adult. In *Respiration and Metabolism of Embryonic Vertebrates,* ed. R. S. Seymour, pp. 31–53. Dordrecht, Boston, & London: Dr. W. Junk. (Ch. 17)

Burggren, W. W. (1985a). Gas exchange, metabolism, and "ventilation" in gelatinous frog egg masses. *Physiological Zoology* 58:503–514. (Chs. 17, 18)

Burggren, W. W. (1985b). Hemodynamics and regulation of central cardiovascular shunts in reptiles. In *Cardiovascular Shunts,* ed. K. Johansen & W. W. Burggren, pp. 121–142. Copenhagen: Munksgaard. (Ch. 14)

Burggren, W. W. (1989). Lung structure and function: Amphibians. In *Comparative Pulmonary Physiology: Current Concepts,* vol. 39, *Lung Biology in Health and Disease,* ed. S. C. Wood, pp. 153–192. New York: Marcel Dekker. (Ch. 13)

Burggren, W. W. (1995). Central cardiovascular function in amphibians: Qualitative influences of phylogeny, ontogeny, and seasonality. In *Mechanisms of Systemic Regulation,* vol. 1, *Respiration and Circulation,* ed. N. Heisler, pp. 175–197. Berlin: Springer-Verlag. (Ch. 13)

Burggren, W. W., Bicudo, J. E., Glass, M. L., & Abe, A. S. (1992). Development of blood pressure and cardiac reflexes in the frog *Pseudis paradoxus. American Journal of Physiology* 263:R602–R608. (Chs. 13, 17)

Burggren, W. W., & Doyle, M. (1986a). The action of acetylcholine upon heart rate changes markedly with development in the bullfrog. *Journal of Experimental Zoology* 240:137–140. (Ch. 13)

Burggren, W. W., & Doyle, M. (1986b). Ontogeny of regulation of gill and lung ventilation in the bullfrog, *Rana catesbeiana. Respiration Physiology* 66:279–291. (Chs. 8, 13, 17)

Burggren, W. W., Farrell, A. P., & Lillywhite, H. (1997). Vertebrate cardiovascular systems. In *Handbook of Comparative Physiology,* ed. W. Danzler. Washington: American Physiological Society. (Ch. 9)

Burggren, W. W., & Fritsche, R. (1995). Cardiovascular measurements in animals in the milligram body mass range. *Brazilian Journal of Medical and Biological Research* 28:1291–1305. (Chs. 12, 13, 15, 17)

Burggren, W. W., & Infantino, R. L., Jr. (1994). The respiratory transition from water to air breathing during amphibian metamorphosis. *American Zoologist* 34:238–246. (Ch. 13)
Burggren, W. W., Infantino, R. L., & Townsend, D. S. (1990). Developmental changes in cardiac and metabolic physiology of the direct-developing tropical frog *Eleutherodactylus coqui. Journal of Experimental Biology* 152:129–147. (Ch. 17)
Burggren, W. W., & Johansen, K. (1986). Circulation and respiration in lungfishes (Dipnoi). *Journal of Morphology Supplement* 1:217–236. (Ch. 17)
Burggren, W. W., & Just, J. J. (1992). Developmental changes in amphibian physiological systems. In *Environmental Physiology of the Amphibia,* ed. M. E. Feder & W. W. Burggren, pp. 467–530. Chicago: University of Chicago Press. (Chs. 7, 13, 17)
Burggren, W. W., & Mwalukoma, A. (1983). Respiration during chronic hypoxia and hyperoxia in larval and adult bullfrogs (*Rana catesbeiana*). I, Morphological responses of lungs, skin, and gills. *Journal of Experimental Biology* 105:191–203. (Chs. 13, 17)
Burggren, W. W., & Pinder, A. W. (1991). Ontogeny of cardiovascular and respiratory physiology in lower vertebrates. *Annual Reviews of Physiology* 53:107–135. (Chs. 12, 17, 18)
Burggren, W. W., Tazawa, H., & Thompson, D. (1994). Genetic and maternal environmental influences on embryonic physiology: Intraspecific variability in avian embryonic heart rates. *Israel Journal of Zoology* 40:351–362. (Ch. 15)
Burggren, W. W., & Territo, P. (1995). Early development of blood oxygen transport. In *Hypoxia and Brain,* ed. J. Houston & J. Coates, pp. 45–56. Burlington, Vt.: Queen City Printers. (Chs. 12, 13, 17, 18)
Burggren, W. W., & Warburton, S. (1994). Patterns of form and function in developing hearts: Contributions from non-mammalian vertebrates. *Cardioscience* 5:183–191. (Ch. 17)
Burleson, M. L., Smatresk, N. J., & Milsom, W. K. (1992). Afferent inputs associated with cardioventilatory control. In *Fish Physiology,* vol. 12B, *The Cardiovascular System,* ed. W. S. Hoar, D. J. Randall, & A. P. Farrell, pp. 390–426. San Diego: Academic Press. (Ch. 12)
Burn, J. (1995). Heart malformation: The human model. In *Developmental Mechanisms of Heart Disease,* ed. E. B. Clark, R. Markwald, & A. Takao, pp. 489–504. Armonk: Futura. (Ch. 21)
Burn, J. H., Wilson, D. I., Cross, I., Atif, U., Scambler, P., Takao, A., & Goodship, J. (1995). The clinical significance of 22q11.2 deletion. In *Developmental Mechanisms of Heart Disease,* ed. E. B. Clark, R. Markwald, & A. Takao, pp. 559–568. Armonk: Futura. (Ch. 21)
Bury, H. (1896). The metamorphosis of echinoderms. *Quarterly Journal of Microscopical Science* 38:45–131. (Ch. 11)
Byrne M. (1988). Evidence for endocytotic incorporation of nutrients from the haemal sinus by the oocytes of the brittlestar *Ophiolepis paucispina.* In *Echinoderm Biology,* ed. R. D. Burke, P. V. Mladenov, & R. L. Parsley, pp. 557–563. Rotterdam: Balkama. (Ch. 11)
Byrne, M. (1994). Ophiuroidea. In *Microscopic Anatomy of Invertebrates,* vol. 14, *Echinodermata,* ed. F. W. Harrison & F. S. Chia, pp. 247–343. New York: Alan-Liss. (Ch. 11)
Cain, J. R., Abbott, U. K., & Rogallo, V. L. (1967). Heart rate of the developing chick embryo. *Proceedings for the Society for Experimental Biology and Medicine* 126:507–510. (Chs. 15, 17)
Calabrese, R. L., & Arbas, E. A. (1989). Central and peripheral oscillators generating heartbeat in the leech *Hirudo medicinalis.* In *Neuronal and Cellular Oscillators,* ed. J. W. Jacklet, pp. 237–267. New York: Marcel Dekker. (Ch. 11)
Campbell, K. A., Hu, N., Clark, E. B., & Keller, B. B. (1992). Analysis of dynamic atrial dimension and function during early cardiac development in the chick embryo. *Pediatric Research* 32:333–337. (Ch. 7)
Canadian Task Force on the Periodic Health Examination (1994). 1, Periodic health examination. 2, 1994 update. 3, Primary and secondary prevention of neural tube defects. *Canadian Medical Association Journal* 151:159–160. (Ch. 21)
Cannel, M. B. (1991). Contribution of sodium–calcium exchange to calcium regulation in cardiac muscle. *Annals of the New York Academy of Sciences* 639:428–443. (Ch. 2)
Carey, D. J. (1991). Control of growth and differentiation of vascular cells by extracellular matrix proteins. *Annual Reviews of Physiology* 53:161–177. (Ch. 4)
Carey, D. J., Evans, D. M., Stahl, R. C., Asundi, V. K., Conner, K. J., Garbes, P., Smith-Cizmeci, G. (1992). Molecular cloning and characterization of *N*-syndecan, a novel transmembrane heparan sulfate proteoglycan. *Journal of Cell Biology* 117:191–201. (Ch. 5)
Carlson, A. J., & Meek, W. J. (1908). On the mechanism of the embryonic heart rhythm in *Limulus. American Journal of Physiology* 21:1–10. (Ch. 11)

Carlson, A. R., & Siefert, R. E. (1974). Effects of reduced oxygen on the embryos and larvae of lake trout (*Salvelinus namaycush*) and largemouth bass (*Micropterus salmoides*). *Journal of the Fisheries Research Board of Canada* 31:1393–1396. (Ch. 17)

Carlson, B. M. (1988). *Patten's Foundations of Embryology.* New York, & St. Louis: McGraw-Hill. (Ch. 17)

Caroni, P., Zurini, M., Clark, A., & Carafoli, E. (1983). Further characterization and reconstitution of the purified Ca^{2+}-pumping ATPase of heart sarcolemma. *Journal of Biological Chemistry* 58:7305–7310. (Ch. 2)

Carrier, L., Boheler, K. R., Chassagne, C., de la Bastie, D., Wisnewsky, C., Lakatta, E. G., & Schwartz, K. (1992). Expression of the sarcomeric actin isogenes in the rat heart with development and senescence. *Circulation Research* 70:999–1005. (Ch. 3)

Carroll, S. B. (1995). Homeotic genes and the evolution of arthropods and chordates. *Nature* 376:479–485. (Ch. 9)

Carter, A. M. (1989). Factors affecting gas transfer across the placenta and the oxygen supply to the fetus. *Journal of Developmental Physiology* 12:305–322. (Ch. 18)

Carver, W., Molano, I., Reaves, T. A., Borg, T. K., & Terracio, L. (1995). Role of the α_1 integrin chain in collagen gel contraction *in vitro* by rat heart fibroblasts. *Journal of Cellular Physiology* 165:425–437. (Ch. 5)

Carver, W., Nagpal, M., Nachtigal, M., Borg, T. K., & Terracio, L. (1991). Collagen expression in mechanically stimulated cardiac fibroblasts. *Circulation Research* 69:113–119. (Ch. 5)

Carver, W., Price, R. L., Raso, D., Terracio, L., & Borg, T. K. (1994). Distribution of β_1 integrin in the developing rat heart. *Journal of Histochemistry and Cytochemistry* 42:167–175. (Ch. 5)

Carver, W., & Terracio, L. (1993). Integrin-mediated cell–matrix interactions in heart development and disease. *Heart Failure* 8:255–263. (Ch. 5)

Carver, W., Terracio, L., & Borg, T. K. (1993). Collagen expression and accumulation in the neonatal rat heart. *Anatomical Record* 236:511–520. (Ch. 5)

Casillas, C. B., Tinney, J. P., Clark, E. B., & Keller, B. B. (1993). Heart rate reserve during cardiovascular morphogenesis. *Pediatric Research* 33:245A. (Chs. 7, 17)

Casini, A., Pinzani, M., Milani, S., Grappone, C., Galli, G., Jezequel, A. M., Schuppan, D., Rotella, C.M., & Surrenti, C. (1993). Regulation of extracellular matrix synthesis by transforming growth factor beta 1 in human fat-storing cells. *Gastroenterology* 105:245–253. (Ch. 5)

Castaneda, A. R., Jonas, R. A., Mayer, J. E., Jr., & Hanley, F. L. (1993). *Cardiac Surgery in the Neonate and Infant.* New York: Saunders. (Ch. 20)

Caulfield, J. B., & Borg, T. K. (1979). The collagen network of the heart. *Laboratory Investigation* 40:354–371. (Ch. 5)

Chalepakis, G., & Gruss, P. (1995). Identification of DNA recognition sequences for the Pax3 paired domain. *Gene* 162:267–270. (Ch. 19)

Chang, A. C., Wernovsky, G., Wessel, D. L., et al. (1992). Late right ventricular failure after Mustard/Senning repair: Surgical implications and outcome. *Circulation* 86(suppl. 2):140–148. (Ch. 20)

Chan-Thomas, P. S., Thompson, R. P., Rober, B., Yacoub, M. H., & Barton, P. J. R. (1993). Expression of homeobox genes *Msx-1* (*Hox-7*) and *Msx-2* (*Hox-8*) during cardiac development in the chick. *Developmental Dynamics* 197:203–216. (Ch. 1)

Charité, J. DeGraaff, W., Shen, S., & Deschamps, J. (1994). Ectopic expression of the HoxB-8 causes duplication of the ZPA in the forelimb and homeotic transformation of axial structures. *Cell* 78:589–601. (Ch. 19)

Charlemagne, D., Mayoux, E., Poyard, M., Oliviero, P., & Geering, K. (1987). Identification of two isoforms of the catalytic subunit of Na K–ATPase in myocytes from adult rat heart. *Journal of Biological Chemistry* 262:8941–8943. (Ch. 2)

Charron, J., Malynn, B. A., Fisher, P., Stewart, V., Jeannotte, L., Goff, S. P., Robertson, E. J., & Alt, F. W. (1992). Embryonic lethality in mice homozygous for a targeted disruption of the *N-myc* gene. *Genes & Development* 6:2248–2257. (Ch. 1)

Cheah, K. S., Lau, E. T., & Au, P. K. (1991). Expression of the mouse alpha 1 (II) collagen gene is not restricted to cartilage during development. *Development* 111:945–953. (Ch. 5)

Cheanvechai, V., Hughes, S. F., & Benson, D. W., Jr. (1992). Relationship between cardiac cycle length and ventricular relaxation rate in the chick embryo. *Pediatric Research* 31:480–482. (Chs. 2, 7)

Chen, F., Mottino, G., Klitzner, T. S., Phillipson, K. D., & Frank, J. S. (1995). Distribution of Na^+/Ca^{2+} exchange protein in developing rabbit myocytes. *American Journal of Physiology* 268:C1126–C1132. (Ch. 2)

Cherry, R., Nielsen, H., Reed, E., et al. (1992). Vascular (humoral) rejection in human cardiac allograft biopsies: Relation to circulating anti-HLA antibodies. *Journal of Heart Transplantation* 11:24–30. (Ch. 6)

Chien, K. R., Zhu, H., Knowlton, K. U., Miller-Hance, W., van-Bilsen, M., O'Brien, T. X., & Evans, S. M. (1993). Transcriptional regulation during cardiac growth and development. *Annual Reviews of Physiology* 55:77–95. (Ch. 3)

Chin, T. K., Friedman, W. F., & Klitzner, T. S. (1990). Developmental changes in cardiac myocyte calcium regulation. *Circulation Research* 67:574–579. (Ch. 2)

Chiquet-Ehrismann, R., Hagios, C., & Matsumoto, K. (1994). The tenascin gene family. *Perspectives on Developmental Neurobiology* 2:3–7. (Ch. 5)

Chisaka, O., & Capecchi, M. R. (1991). Regionally restricted developmental defects resulting from targeted disruption of the mouse homeobox gene *hox-1.5*. *Nature* 350:473–479. (Ch. 1)

Chizzonite, R. A., & Zak, R. (1984). Regulation of myosin isoenzyme composition in fetal and neonatal rat ventricle by endogenous thyroid hormones. *Journal of Biological Chemistry* 259:12628–12632. (Ch. 3)

Chua, C. C., Chua, B. H., Zhoa, Z. Y., Krebs, C., Diglio, C., & Perrin, E. (1991). Effect of growth factors on collagen metabolism in cultured heart fibroblasts. *Connective Tissue Research* 26:271–281. (Ch. 5)

Chung, A. E., Dong, L. J., Wu, C., & Durkin, M. E. (1993). Biological functions of entactin. *Kidney International* 43:13–19. (Ch. 5)

Cirotto, C., & Arangi, I. (1989) Chick embryo survival under acute carbon monoxide challenges. *Comparative Biochemistry and Physiology* 94A(1):117–123. (Chs. 13, 17)

Clark, E. B. (1984). Functional aspects of cardiac development. In *Growth of the Heart in Health and Disease,* ed. R. Zak, pp. 81–103. New York: Raven Press. (Ch. 17)

Clark, E. B. (1986). Cardiac embryology–Its relevance to congenital heart disease. *American Journal of Diseases of Children* 140:41–44. (Ch. 21)

Clark, E. B. (1990). Hemodynamic control of the embryonic circulation. In *Developmental Cardiology: Morphogenesis and Function,* ed. E. B. Clark & A. Takao, pp. 291–304. Mount Kisco, N.Y.: Futura. (Chs. 7, 19)

Clark, E. B. (1994). Epidemiology of congenital cardiovascular malformations. In *Heart Disease in Infants, Children, and Adolescents,* 5th ed., ed. G. C. Emmanouilides, T. A. Riemenschneider, H. H. Allen, & H. P. Gutgesell, pp. 60–69. Baltimore: Williams & Wilkins. (Ch. 21)

Clark, E. B., & Hu, N. (1982). Developmental hemodynamic changes in the chick embryo stages 18 to 27. *Circulation Research* 51:810–815. (Chs. 7, 17)

Clark, E. B., Hu, N., & Dooley, J. B. (1985). The effect of isoproterenol on cardiovascular function in the stage 24 chick embryo. *Teratology* 31:41–47. (Chs. 7, 8, 15)

Clark, E. B., Hu, N., Dummett, J. L., Vandekieft, G. K., Olson, C., & Tomanek, R. (1986). Ventricular function and morphology in chick embryo from stages 18 to 29. *American Journal of Physiology* 250 (Heart and Circulatory Physiology 19):H407–H413. (Chs. 7, 15)

Clark, E. B., Hu, N., Frommelt, P., Vandekieft, G. K., Dummett, J. L., & Tomanek, R. J. (1989). Effect of increased ventricular pressure on heart growth in chick embryo. *American Journal of Physiology* 257:H55–H61. (Ch. 7)

Clark, E. B., Hu, N., Turner, D. R., Litter, J. E., & Hansen, J. (1991). Effect of chronic verapamil treatment on ventricular function and growth in the chick embryo. *American Journal of Physiology* 261 (Heart and Circulatory Physiology 30):H166–H171. (Ch. 7)

Clark, E. B., Markwald, R. R., & Takao, A. (eds.) (1995). *Developmental Mechanisms of Heart Disease.* Armonk, N.Y.: Futura. (Ch. 21)

Clark, E. B., & Van Mierop, L. H. S. (1989). Development of the cardiovascular system. In *Moss' Heart Disease in Infants, Children, and Adolescents,* 4th ed., ed. F. H. Adams, G. C. Emmanouilides, & T. A. Riemenschneider, pp. 2–15. Baltimore: Williams & Wilkins. (Ch. 7)

Clark, E. R. (1918). Studies on the growth of blood vessels in the tail of the frog larva by observation and experiment on the living animal. *American Journal of Anatomy* 23:37–88. (Ch. 6)

Clark, R. B. (1956). The blood vascular system of *Nephtys. Quarterly Journal of Microscopical Science* 97:235–249. (Ch. 11)

Cleeman, L., & Morad, M. (1991). Role of Ca^{2+} channel in cardiac excitation–contraction coupling in the rat: Evidence from Ca^{2+} transients and contraction. *Journal of Physiology* 432:283–312. (Ch. 2)

Cleveland, L., Little, E. E., Ingersoll, C. G., Wiedmeyer, R. H., & Hunn, J. B. (1991). Sensitivity of brook trout to low pH, low calcium, and elevated aluminum concentrations during laboratory pulse exposures. *Aquatic Toxicology* 19:303–318. (Ch. 17)

Coffin, J. D., & Poole, T. J. (1991). Endothelial cell origin and migration in embryonic heart and cranial blood vessel development. *Anatomical Record* 232:383–395. (Chs. 4, 6)

Cole, H. A., & Perry, S. V. (1975). The phosphorylation of troponin I from cardiac muscle. *Journal of Biochemistry* 149:525–533. (Ch. 3)

Colmorgen, M., & Paul, R. J. (1995). Imaging of physiological functions in transparent animals (*Agonus cataphractus, Daphnia magna, Pholcus phalangioides*) by video microscopy and digital image processing. *Comparative Biochemistry and Physiology* 111A:583–595. (Ch. 17)

Comeau, S. G., & Hicks, J. W. (1994). Regulation of central vascular blood flow in the turtle. *American Journal of Physiology* 267:R569–R578. (Ch. 14)

Concas, V., Laurent, S., Brisac, A. M., Perret, C., & Safer, M. (1989). Endothelin has potent direct inotropic and chronotropic effects in cultured heart cells. *Journal of Hypertension* 7(suppl.):S96–S97. (Ch. 8)

Conrad, G. W., Stephens, A. P., Schwarting, S. S., & Conrad, A. H. (1994). Effect of clinostat rotation on larval heart development in *llyanassa obsoleta. Bulletin of the Mount Desert Island Biological Laboratory* 33:17–18. (Ch. 11)

Consigli, S. A., & Joseph-Silverstein, J. (1991). Immunolocalization of basic fibroblast growth factor during chicken cardiac development. *Journal of Cell Physiology* 146:379–385. (Ch. 4)

Cooke, I. M. (1988). Studies on the crustacean cardiac ganglion. *Comparative Biochemistry and Physiology* 91C:205–218. (Ch. 11)

Cooper, T. A., & Ordahl, C. P. (1985). A single cardiac troponin T gene generates embryonic and adult isoforms via developmentally regulated alternate splicing. *Journal of Biological Chemistry* 260:11140–11148. (Ch. 3)

Copenhaver, W. M. (1926). Experiments on the development of the heart of *Amblystoma punctatum. Journal of Experimental Zoology* 43:321–371. (Ch. 10)

Costa, A., Costantino, L. L., & Fumero, R. (1992). Oxygen exchange mechanisms in the human placenta: Mathematical modelling and simulation. *Journal of Biomedical Engineering* 14:385–389. (Ch. 18)

Couchman, J. R., Austria, R., Woods, A., & Hughes, R. C. (1988). Adhesion defective BHK cell mutant has cell surface heparan sulfate proteoglycan of altered properties. *Journal of Cellular Physiology* 136:226–236. (Ch. 5)

Couley, G., Grazin-Botton, A., Coltey, P. I., & Le Douarin, N. M. (1996). The regeneration of the cephalic neural crest, a problem revisited: The regeneration cells originate from the contralateral or from the anterior and posterior neural fold. *Development* 122:3393–3407. (Ch. 19)

Crawford, D. C., Chobanian, A. V., & Brecher, P. (1994). Angiotensin II induces fibronectin expression associated with cardiac fibrosis in the rat. *Circulation Research* 74:727–739. (Ch. 5)

Crossin, K. L., & Hoffman, S. (1991). Expression of adhesion molecules during the formation and differentiation of the avian endocardial cushion tissue. *Developmental Biology* 145:277–286. (Chs. 5, 6)

Crossman, D. C., & Tuddenham, E. D. G. (1990). Procoagulant functions of endothelium. In *The Endothelium: An Introduction to Current Research,* ed. J. B. Warren, pp. 119–128. New York: Wiley-Liss. (Ch. 6)

Crumb, W. J., Jr., Pigott, J. D., & Clarkson, C. W. (1995). Comparison of I_{to} in young and adult human atrial myocytes: Evidence for developmental changes. *American Journal of Physiology* 68:H1335–H1342. (Ch. 2)

Crutchley, D. J. (1987). Hemostatic potential of the pulmonary endothelium. In *Pulmonary Endothelium in Health and Disease,* ed. U. S. Ryan, pp. 237–273. New York: Dekker. (Ch. 6)

Cuénot, L. (1948). Anatomie, éthologie et systématique des Echinodermes. In *Traité de Zoologie,* vol. 11, ed. P. P. Grassé, pp. 3–275. Paris: Masson. (Ch. 11)

Cummins, P., & Perry, S. V. (1973). The subunits and biological activity of polymorphic forms of tropomyosin. *Biochemical Journal* 133:765–777. (Ch. 3)

Cuneo, B., Hughes, S., & Benson, D. W., Jr. (1991). Heart rate perturbation in the chick embryo: A comparison of two methods. *American Journal of Physiology* 260:H1864–H1869. (Ch. 7)

Cunningham, J. E. R., & Balon, E. K. (1985). Early ontogeny of *Adinia xenica* (Pisces, Cyprinodontiformes): 1, The development of embryos in hiding. *Environmental Biology of Fishes* 14:115–166. (Ch. 12)
Cunningham, J. E. R., & Balon, E. K. (1986a). Early ontogeny of *Adinia xenica* (Pisces, Cyprinodontiformes): 2, Implications of embryonic resting interval for larval development. *Environmental Biology of Fishes* 15:15–45. (Ch. 12)
Cunningham, J. E. R., & Balon, E. K. (1986b). Early ontogeny of *Adinia xenica* (Pisces, Cyprinodontiformes): 3, Comparison and evolutionary significance of some patterns in epigenesis of egg-scattering, hiding, and bearing cyprinodontiforms. *Environmental Biology of Fishes* 15:91–105. (Chs. 12, 17)
Damen, P., & Dictus, W. J. (1994). Cell lineage of the protobranch of *Patella vulgata* (Gastropoda, Mollusca). *Developmental Biology* 162:364–383. (Ch. 11)
Damsky, C. H., & Werb, Z. (1992). Signal transduction by integrin receptors for extracellular matrix: Cooperative processing of extracellular information. *Current Opinions in Cell Biology* 4:772–781. (Chs. 5, 6)
Danilo, P., Jr., Reder, R. F., Binah, O., & Legato, M. J. (1984). Fetal canine cardiac Purkinje fibers: Electrophysiology and ultrastructure. *American Journal of Physiology* 246:H250–H260. (Ch. 2)
Danos, M. C., and Yost, H. J. (1995). Linkage of cardiac left–right asymmetry and dorsal–anterior development in *Xenopus. Development* 121:1467–1474. (Ch. 13)
Davie, P. S., & Farrell, A. P. (1991). The coronary and luminal circulations of the myocardium of fishes. *Canadian Journal of Zoology* 69:1993–2001. (Ch. 9)
Davies, D. F., & Tripathi, S. C. (1993). Mechanical stress mechanisms and the cell: An endothelial paradigm. *Circulation Research* 72:239–245. (Ch. 13)
Davies, P. F., Robotewsky, A., Griem, M. L., et al. (1992). Hemodynamics forces and vascular cell communication in arteries. *Archives of Pathology and Laboratory Medicine* 116:1301–1306. (Ch. 6)
Davis, A. C., Wims, M., Spotts, G. D., Hann, S. R., & Bradley, A. (1993). A null *c-myc* mutation causes lethality before 10.5 days of gestation in homozygotes and reduced fertility in heterozygous female mice. *Genes & Development* 7:671–682. (Ch. 1)
Davis, E. C. (1993). Endothelial cell connecting filaments anchor endothelial cells to the subjacent elastic lamina in the developing aortic intima of the mouse. *Cell and Tissue Research* 272:211–219. (Ch. 6)
Dawes, G. D. (1968). *Foetal and Neonatal Physiology.* Chicago: Year Book Medical Publishers. (Ch. 16)
Daykin, P. N. (1965). Application of mass transfer theory to the problem of respiration of fish eggs. *Journal of the Fisheries Research Board of Canada* 22:159–171. (Ch. 18)
Dbaly, J., Ošťádal, B., & Rychter, Z. (1968). Development of the coronary arteries in rat embryos. *Acta Anatomica* 71:209–222. (Ch. 4)
De Beer, G. (1958). *Embryos and Ancestors.* 3rd ed. Oxford: Oxford University Press. (Ch. 9)
Deeming, D. C., & Thompson, M. B. (1991). Gas exchange across reptilian eggshells. In *Egg Incubation: Its Effects on Embryonic Development in Birds and Reptiles,* ed. D. C. Deeming & M. W. J. Ferguson, pp. 277–284. Cambridge: Cambridge University Press. (Ch. 18)
DeHaan, R. L. (1965). Morphogenesis of the vertebrate heart. In *Organogenesis,* ed. R. L. DeHaan & H. Ursprung, pp. 377–419. New York: Holt, Rinehart & Winston. (Ch. 10)
DeHaan, R. L., Fujii, S., & Satin, J. (1990). Cell interactions in cardiac development. *Development, Growth, and Differentiation* 32:233–241. (Ch. 10)
Deitz, H. C., & Pyeritz, R. E. (1995). Mutations in the human gene for fibrillin-1 (*FBN1*) in the Marfan syndrome and related disorders. *Human Molecular Genetics* 4:1799–1809. (Ch. 21)
Dejours, P. (1981). *Principles of Comparative Respiratory Physiology.* 2nd ed. Amsterdam: Elsevier/North-Holland Biomedical Press. (Chs. 17, 18)
Delmotte, F., Brillouet, C., Leclerc, M., Luquet, T. G., & Kader, J. (1986). Purification of an "antibody-like" protein from the seastar *Asterias rubens. European Journal of Immunology* 16:1325–1330. (Ch. 11)
DeLonay, A. J., Little, E. E., Woodward, D. F., Brumbaugh, W. G., Farag, A. M., & Rabeni, C. F. (1993). Sensitivity of early-life-stage golden trout to low pH and elevated aluminum. *Environmental Toxicology and Chemistry* 12:1223–1232. (Ch. 17)

Demian, E. S., & Yousif, F. (1973). Embryonic development and organogenesis in the snail *Marisa cornuarietis* (Mesogastropoda: Ampullariidae). III, Development of the renal and circulatory systems. *Malacologia* 12:175–194. (Ch. 11)

Denekamp, J. (1993). Review article: Angiogenesis, neovascular proliferation, and vascular pathophysiology as targets for cancer therapy. *British Journal of Radiology* 66:181–196. (Ch. 6)

DeRuiter, M. C., Poelmann, R. E., Mentink, M. M., Vaniperen, L., & Gittenberger–de Groot, A. C. (1993). Early formation of the vascular system in quail embryos. *Anatomical Record* 235:261–274. (Ch. 4)

DeSilva, C. (1974). Development of the respiratory system in herring and plaice larvae. In *The Early Life History of Fish,* ed. J. S. Baxter, pp. 465–485. Berlin: Springer-Verlag. (Ch. 12)

Diaz, A., Munoz, E., Johnston, R., Korn, J. H., & Jiminez, S. A. (1993). Regulation of human lung fibroblast alpha 1(I) procollagen gene expression by tumor necrosis factor alpha, interleukin-1 beta, and prostaglandin E2. *Journal of Biological Chemistry* 268:10364–10371. (Ch. 5)

DiDonato, R. M., Jonas, R. A., Lang, P., Rome, J. J., Mayer, J. E., & Castaneda, A. R. (1991). Neonatal repair of tetralogy of Fallot. *Journal of Thoracic Cardiovascular Surgery* 101:126–137. (Ch. 20)

Dilly, P. N., Welsch, U., & Rehkämper, G. (1986). Fine structure of heart, pericardium, and glomerular vessel in *Cephalodiscus gracilis* M'Intosh, 1882 (Pterobranchia, Hemichordata). *Acta Zoologica* 67:173–179. (Ch. 11)

Docherty, R., Forrester, J. V., Lackie, J. M., & Gregory, D. W. (1989). Glycosaminoglycans facilitate the movement of fibroblasts through three-dimensional collagen matrices. *Journal of Cell Science* 92:263–270. (Ch. 5)

Dolan, L. M., & Dobrozsi, D. J. (1987). Atrial natriuretic polypeptide in the fetal rat: Ontogeny and characterization. *Pediatric Research* 22:115–117. (Ch. 8)

Dolkart, L. A., & Reimers, F. T. (1991). Transvaginal fetal echocardiography in early pregnancy: Normative data. *American Journal of Obstetrics and Gynecology* 165:688–694. (Ch. 20)

Dowell, R. T., & McManus, R. E. (1978). Pressure-induced enlargements in neonatal and adult rats: Left ventricular functional characteristic and evidence of cardiac muscle cell proliferation in the neonate. *Circulation Research* 42:303–310. (Ch. 9)

Drake, C. J., Davis, L. A., & Little, C. D. (1992). Antibodies to beta 1–integrins cause alterations of aortic vasculogenesis, *in vivo. Developmental Dynamics* 193:83–91. (Ch. 6)

Drake, C. J., Davis, L. A., Walters, L., & Little, C. D. (1990). Avian vasculogenesis and the distribution of collagens I, IV, laminin, and fibronectin in the heart primordia. *Journal of Experimental Zoology* 255:309–322. (Chs. 6, 10)

Driscoll, D. A., Goldmuntz, E., & Emanuel, B. S. (1995). Detection of *22q11.2* deletions in patients with conotruncal cardiac malformations, DiGeorge, velocardiofacial, and conotruncal anomaly face syndromes. In *Developmental Mechanisms of Heart Disease,* ed. E. B. Clark, R. Markwald, & A. Takao, pp. 569–576. Armonk: Futura. (Ch. 21)

Duellman, W. E., & Trueb, L. (1986). *Biology of Amphibians.* New York: McGraw-Hill. (Ch. 18)

Dumont, D. J., Auerbach, A. Breitman, M. L., Gradwohl, G., Fong, G. H., Puri, M. C., & Gertsenstein, M. (1994). Dominant-negative and targeted null mutations in the endothelial receptor tyrosine kinase, *tek,* reveal a critical role in vasculogenesis of the embryo. *Genes & Development* 8:1897–1909. (Ch. 1)

Dunnigan, A., Hu, N., Benson, D. W., Jr., & Clark, E. B. (1987). Effect of heart rate increase on dorsal aortic flow in the stage 24 chick embryo. *Pediatric Research* 22:442–444. (Ch. 7)

Dusseau, J. W., & Hutchins, P. M. (1988). Hypoxia-induced angiogenesis in chick chorioallantoic membranes: A role for adenosine. *Respiration Physiology* 71:33–44. (Ch. 17)

Dzau, V. J., & Gibbons, G. H. (1993). Vascular remodeling: Mechanisms and implications. *Journal of Cardiovascular Pharmacology* 21:S1–S5. (Ch. 6)

Earley, J. J. (1991). Simple harmonic motion of tropomyosin: Proposed mechanism for length-dependent regulation of muscle active tension. *American Journal of Physiology* 261:C1184–C1195. (Ch. 3)

Easton, H. S., Bellairs, R., & Lash, J. W. (1990). Is chemotaxis a factor in the migration of precardiac mesoderm in the chick? *Anatomy and Embryology* 181:461–468. (Ch. 10)

Eckelbarger, K. J., & Young, C. M. (1992). Ovarian ultrastructure and vitellogenesis in ten species of shallow-water and bathyal sea cucumbers (Echinodermata: Holothuroidea). *Journal of the Marine Biological Association of the United Kingdom* 72:759–781. (Ch. 11)

Edmondson, D. G., Lyons, G. E., Martin, J. F., & Olson, E. N. (1994). *Mef2* gene expression marks the cardiac and skeletal muscle lineages during mouse embryogenesis. *Development* 120:1251–1263. (Ch. 1)

Egginton, S., & Johnston, I. A. (1984). Effects of acclimation temperature on routine metabolism muscle mitochondrial volume density and capillary supply in the elver (*Anguilla anguilla* L.). *Journal of Thermal Biology* 9:165–170. (Ch. 17)

Eghbali, M., Tomek, R., Sukhatme, V. P., Woods, C., & Bhambi, B (1991). Differential effects of transforming growth-beta 1 and phorbol myristate acetate on cardiac fibroblasts: Regulation of collagen mRNAs and expression of early transcription factors. *Circulation Research* 69:483–490. (Ch. 5)

Eghbali-Webb, M. (1995). *Molecular Biology of the Collagen Matrix in the Heart.* Austin: R. G. Landes. (Ch. 5)

Eichmann, A., Marcelle, C., Breant, C., et al. (1993). Two molecules related to the VEGF receptor are expressed in early endothelial cells during avian embryonic development. *Mechanisms of Development* 42:33–48. (Ch. 6)

El-Fiky, N., & Wieser, W. (1988). Life styles and patterns of development of gills and muscles in larval cyprinids (Cyprinidae; Teleostei). *Journal of Fish Biology* 33:135–145. (Chs. 17, 18)

Elias, J. A., Freundlich, B., Adams, S., & Rosenbloom, J. (1990). Regulation of human lung fibroblast collagen production by recombinant interleukin-1, tumor necrosis factor, and interferon-gamma. *Annals. of the New York Academy of Sciences* 580:233–244. (Ch. 5)

Elisabetta, D. (1993). Endothelial cell adhesive receptors. *Journal of Cardiovascular Pharmacology* 21:S18–S21. (Ch. 6)

El-Saleh, S. C., Warber, K. D., & Potter, J. D. (1986). The role of tropomyosin-troponin in the regulation of skeletal muscle contraction. *Journal of Muscle Research and Cell Motility* 7:387–404. (Ch. 3)

Engelmann, G. L. (1993). Coordinate gene expression during neonatal rat heart development: A possible role for the myocyte in extracellular matrix biogenesis and capillary angiogenesis. *Cardiovascular Research* 27:1598–1605. (Ch. 5)

Engelmann, G. L., Dionne, C. A., & Jaye, M. C. (1993). Acidic fibroblast growth factor and heart development: Role in myocyte proliferation and capillary angiogenesis. *Circulation Research* 72:7–19. (Ch. 4)

Erickson, H. P. (1994). Evolution of the tenascin family implications for function of the C-terminal fibrinogen-like domain. *Perspectives on Developmental Neurobiology* 2:9–19. (Ch. 5)

Escande, D., Loisance, D., Planche, C., & Coraboeuf, E. (1985). Age-related changes of action potential plateau shape in isolated human atrial fibers. *American Journal of Physiology* 249:H843–H850. (Ch. 2)

Evans, H. M. (1909). On the development of the aortae, cardinal and umbilical veins, and other blood vessels of vertebrate embryos from capillaries. *Anatomical Record* 3:498–518. (Ch. 6)

Ewart, A. K., Jin, W., Atkinson, D., Morris, C. A., & Keating, M. T. (1994). Supravalvular aortic stenosis associated with a deletion disrupting the elastin gene. *Journal of Clinical Investigation* 93:1071–1077. (Ch. 5)

Ewert, M. A. (1985). Embryology of turtles. In *Biology of the Reptilia,* vol. 14A, ed. C. Gans, F. Billett, & P. F. A. Maderson, pp. 75–268. New York: John Wiley & Sons. (Ch. 14)

Faber, J. J. (1968). Mechanical function of the septating embryonic heart. *American Journal of Physiology* 214:475–481. (Ch. 15)

Faber, J. J., Green, T. J., & Thornburg, K. L. (1974). Embryonic stroke volume and cardiac output in the chick. *Developmental Biology* 41:14–21. (Chs. 7, 15, 16)

Faber, J. J., Thornburg, K. L., & Binder, N. D. (1992). Physiology of placental transfer in mammals. *American Zoologist* 32:343–354. (Ch. 18)

Fabiato, A. (1983). Calcium-induced release of calcium from the cardiac sarcoplasmic reticulum. *American Journal of Physiology* 245:C1–C88. (Ch. 2)

Fabiato, A. (1989). Appraisal of the physiological relevance of two hypotheses for the mechanism of calcium release from the mammalian cardiac sarcoplasmic reticulum: Calcium-induced release versus charge-coupled release. *Molecular and Cellular Biochemistry* 89:135–140. (Ch. 2)

Fabiato, A., & Fabiato, F. (1978). Calcium-induced release of calcium from the sarcoplasmic reticulum of skinned cells from adult human, dog, cat, rabbit, rat, and frog hearts and from fetal and newborn rat ventricles. *Annals of the New York Academy of Sciences* 307:491–522. (Ch. 2)

Fajardo, L. F. (1989). Special report: The complexity of endothelial cells. *American Journal of Clinical Pathology* 92:241–250. (Ch. 6)

Farrell, A. P. (1987). Coronary flow in a perfused rainbow trout heart. *Journal of Experimental Biology* 129:107–123. (Ch. 9)

Farrell, A. P. (1990a). Circulation of body fluids. In *Comparative Animal Physiology,* ed. C. L. Prosser, pp. 509–558. New York: John Wiley & Sons. (Ch. 9)

Farrell, A. P. (1990b). Comparative biology of the pulmonary circulation. In *The Pulmonary Circulation: Normal and Abnormal,* ed. A. P. Fishman, pp. 17–30. Philadelphia: University of Pennsylvania Press. (Ch. 9)

Farrell, A. P. (1991). From hagfish to tuna: A perspective on cardiac function in fish. *Physiological Zoology* 64:1137–1164. (Ch. 9)

Farrell, A. P., Gamperl, A. K., & Francis, E. T. B. (in press). The heart. In *Biology of the Reptilia,* vol. 20, ed. C. Gans and A. S. Gaunt. Chicago: University of Chicago Press. (Ch. 9)

Farrell, A. P., Hammons, A. M., Graham, M. S., & Tibbits, G. F. (1988). Cardiac growth in rainbow trout, *Salmo gairdneri. Canadian Journal of Zoology* 66:2368–2373. (Ch. 9)

Farrell, A. P., & Jones, D. R. (1992). The heart. *In Fish Physiology, vol. 12A, The Cardiovascular System,* ed. W. S. Hoar, D. J. Randall, & A. P. Farrell, pp. 1–88. San Diego: Academic Press. (Chs. 9, 12)

Feder, M. E. (1983). Responses to acute aquatic hypoxia in larvae of the frog *Rana berlandieri. Journal of Experimental Biology* 104:79–95. (Ch. 17)

Feder, M. E., & Burggren, W. W. (1985). Cutaneous gas exchange in vertebrates: Design, patterns, control, and implications. *Biological Reviews* 60:1–45. (Chs. 13, 18)

Fenton, K. M., Heineman, M. K., Hickey, P. R., et al. (1994). Inhibition of fetal stress response improves cardiac output and gas exchange after fetal cardiac bypass. *Journal of Thoracic and Cardiovascular Surgery* 107:1416–1423. (Ch. 20)

Ferguson, J. C. (1984). Translocative functions of the enigmatic organs of starfish – The axial organ, hemal vessels, Tiedenmann's bodies, and rectal caeca: An autoradiographic study. *Biological Bulletin* 166:140–155. (Ch. 11)

Ferguson, J. C. (1985). Hemal transport of ingested nutrients by the ophiuroid, *Ophioderma brevispinum.* In *Echinodermata,* ed. B. F. Keegan & B. D. S. O'Connor, pp. 623–626. Rotterdam: Balkama. (Ch. 11)

Fernández, J., Téllez, V., & Olea, N. (1992). Hirudinea. In *Microscopic Anatomy of Invertebrates,* vol. 7, *Annelida,* ed. F. W. Harrison & S. L. Gardiner, pp. 323–394. New York: Wiley-Liss. (Ch. 11)

Filep, J. G. (1992). Endothelin peptides: Biological actions and pathophysiological significance in the lung. *Life Sciences* 52:119–133. (Ch. 6)

Fineman, J. R., Reddy, M. V., & Wong, J. (1994). Early impairment of endothelium-dependent vasoactive responses in lambs with increased pulmonary blood flow and pulmonary hypertension. *Pediatric Research* 34:34A. (Ch. 6)

Fineman, J. R., Wong, J., Morin, F. C., et al. (1994). Chronic nitric oxide inhibition in utero produces persistent pulmonary hypertension in newborn lambs. *Journal of Clinical Investigation* 93:2675–2683. (Ch. 6)

Fisher, D. J., Heymann, M. A., & Rudolph, A. M. (1980). Myocardial oxygen and carbohydrate consumption in fetal lambs in utero and in adult sheep. *American Journal of Physiology* 238:H399–H405. (Ch. 2)

Fisher, D. J., Heymann, M. A., & Rudolph, A. M. (1981). Myocardial consumption of oxygen and carbohydrates in newborn sheep. *Pediatric Research* 15:843–846. (Ch. 2)

Fisher, D. J., Heymann, M. A., & Rudolph, A. M. (1982a). Fetal myocardial oxygen and carbohydrate consumption during acutely induced hypoxemia. *American Journal of Physiology* 242:H657–H661. (Ch. 16)

Fisher, D. J., Heymann, M. A., & Rudolph, A. M. (1982b). Regional myocardial blood flow and oxygen delivery in fetal, newborn, and adult sheep. *American Journal of Physiology* 243:H729–H731. (Ch. 16)

Fisher, D. J., Tate, C. A., & Phillips, S. (1992). Developmental regulation of the sarcoplasmic reticulum calcium pump in the rabbit heart. *Pediatric Research* 31:474–479. (Ch. 2)

Fisher, K. C. (1942). The effect of temperature on the critical oxygen pressure for heart beat frequency in embryos of Atlantic salmon and speckled trout. *Canadian Journal of Research,* sec. D, 20:1–12. (Ch. 12)

Fisher, S. A., & Periasamy, M. (1994). Collagen synthesis inhibitors disrupt embryonic cardiocyte myofibrillogenesis and alter the expression of cardiac specific genes. *Journal of Molecular Cellular Cardiology* 26:721–731. (Ch. 10)

Fishman, M. C., & Stainier, D. Y. R. (1994). Cardiovascular development: Prospects for a genetic approach. *Circulation Research* 74:757–763. (Chs. 1, 12)

Floege, J., Eng, E., Young, B. A., Alpers, C. E., Barrett, T. B., Bowen-Pope, D. F., & Johnson, R. J. (1993). Infusion of platelet-derived growth factor or basic fibroblast growth factor induces selective glomerular mesangial cell proliferation and matrix accumulation in rats. *Journal of Clinical Investigation* 92:2952–2962. (Ch. 5)

Florez, F. (1972). Influence of oxygen concentration on growth and survival of larvae and juveniles of the ide, *Idus idus* (L.). *Report of the Institute of Freshwater Research, Drottningholm* 52:65–73. (Ch. 17)

Folkman, J., & Klagsbrun, M. (1987). Angiogenic factors. *Science* 235:442–447. (Ch. 6)

Fong, G.-H., Rossant, J., Gertssenstein, M., & Breltman, M. L. (1995). Role of the *Flt-1* receptor tyrosine kinase in regulating the assembly of vascular endothelium. *Nature* 376:66–70. (Ch. 1)

Fox, H. (1984). *Amphibian Morphogenesis.* Clifton, N.J.: Humana Press. (Ch. 13)

Fozzard, H., & Sheu, S. S. (1980). Intracellular potassium and sodium activities of chick ventricular muscle during embryonic development. *Journal of Physiology* 306:579–586. (Ch. 2)

Frampton, G., Chan, D., Khouri, V., et al. (1993). Pathogenetic aspects of autoantibodies to endothelial cells in systemic vasculitis. *Advances in Experimental Medicine and Biology* 336:5–44. (Ch. 6)

Frangos, J. A., McIntire, L. V., Eskin, S. G., et al. (1985). Flow effects on prostacyclin production by cultured human endothelial cells. *Science* 227:1477–1479. (Ch. 6)

Frank, J. S., Mottino, G., Reid, D., Molday, R. S., & Phillipson, K. D. (1992). Distribution of the Na^{+}-Ca^{2+} exchange protein in mammalian cardiac myocytes: An immunofluorescence and immunocolloidal gold-labeling study. *Journal of Cell Biology* 117:337–345. (Ch. 2)

Franklin, C. E., & Davie, P. S. (1992). Sexual maturity can double heart mass and cardiac power output in male rainbow trout. *Journal of Experimental Biology* 171:139–148. (Ch. 9)

Frasch, M. (1995). Induction of visceral and cardiac mesoderm by ectodermal *Dpp* in the early *Drosophila* embryo. *Nature* 374:464–467. (Chs. 1, 10)

Freeman, B. M. (1964). The emergence of the homeothermic-metabolic response in the fowl (*Gallus domesticus*). *Comparative Biochemistry and Physiology* 13:413–422. (Ch. 17)

Freeman, B. M. (1970). Thermoregulatory mechanisms of the neonate fowl. *Comparative Biochemistry and Physiology* 33:219–230. (Ch. 17)

Friedman, W. F. (1972). The intrinsic physiologic properties of the developing heart. *Progress in Cardiovascular Disease* 15:87–95. (Ch. 16)

Frieswick, G. M., Danielson, T., & Shideman, F. E. (1979). Adrenergic inotropic responsiveness of embryonic chick and rat hearts. *Developmental Neuroscience* 2:276–285. (Ch. 8)

Fritsche, R., & Burggren, W. (1996). Development of cardiovascular responses to hypoxia in larvae of the frog. *Xenopus laevis. American Journal of Physiology* 271 (Regulatory, Integrative, and Comparative Physiology, 40):R912–R917. (Chs. 13, 17)

Fukui, M., Nakamura, T., Ebihara, I., Shirato, I., Tomino, Y., & Koide, H. (1992). ECM gene expression and its modulation by insulin in diabetic rats. *Diabetes* 41:1520–1527. (Ch. 5)

Gallagher, J. T., Lyon, M., & Steward, W. P. (1986). Structure and function of heparan sulfate proteoglycans. *Biochemical Journal* 236:313–325. (Ch. 5)

Galloway, R., Potter, I. C., Macey, D. J., & Hilliard, R. W. (1987). Oxygen consumption and responses to hypoxia of ammocoetes of the Southern Hemisphere lamprey *Geotria australis. Fish Physiology and Biochemistry* 4:63–72. (Ch. 12)

Ganim, J., Luo, W., Ponniah, S., Grupp, I., Kim, H. W., Ferguson, D. G., Kadambi, V., Neumann, J. C., Doetschman, T., & Kranias, E. G. (1992). Mouse phospholamban gene expression during development *in vivo* and *in vitro. Circulation Research* 71:1021–1030. (Ch. 2)

Garcia-Martinez, V., & Schoenwolf, G. C. (1993). Primitive-streak origin of the cardiovascular system in avian embryos. *Developmental Biology* 159:706–719. (Ch. 10)

Garcia-Pelaez, I., & Arteaga, M. (1993). Experimental study of the development of the truncus arteriosus of the chick embryo heart. I, Time of appearance. *Anatomical Record* 237:378–384. (Ch. 10)

Garside, E. T. (1959). Some effects of oxygen in relation to temperature on the development of lake trout embryos. *Canadian Journal of Zoology* 37:689–698. (Ch. 17)

Garside, E. T. (1966). Effects of oxygen in relation to temperature on the development of embryos of brook trout and rainbow trout. *Journal of the Fisheries Research Board of Canada* 23:1121–1134. (Ch. 17)

Garstang, W. (1992). The theory of recapitulation: A critical restatement of the biogenic law. *Journal of the Linnean Society of London, Zoology* 35:81–101. (Ch. 9)

Gassmann, M., Casagranda, F., Orioli, D., Simon, H., Lai, C., Klein, R., & Lemke, G. (1995). Aberrant neural and cardiac development in mice lacking the ErbB4 neuregulin receptor. *Nature* 378:390–394. (Ch. 19)

Gemmill, J. F. (1914). The development and certain points in the adult structure of the starfish *Asterias rubens* L. *Philosophical Transactions of the Royal Society of London,* ser. B, 205:213–294. (Ch. 11)

Gemmill, J. F. (1915). The larva of the starfish *Porania pulvillus* (O.F.M). *Quarterly Journal of Microscopical Science* 61:27–50. (Ch. 11)

Gemmill, J. F. (1919). Rhythmic pulsation in the madreporic vesicle of young ophiuroids. *Quarterly Journal of Microscopical Science* 63:537–540. (Ch. 11)

George, E. L., Georges-Labouesse, E. N., Patel-King, R. S., Rayburn, H., & Hynes, R. O. (1993). Defects in mesoderm, neural tube, and vascular development in mouse embryos lacking fibronectin. *Development* 119:1079–1091. (Chs. 1, 5)

Gerdes, A. M., & Capasso, J. M. (1995). Structural remodelling and mechanical dysfunction of cardiac myocytes in heart failure. *Journal of Molecular and Cellular Cardiology* 27:849–856. (Ch. 9)

Geesin, J. C., Hendricks, L. J., Gordon, J. S., & Berg, R. A. (1991). Modulation of collagen synthesis by growth factors: The role of ascorbate-stimulated lipid peroxidation. *Archives of Biochemistry and Biophysics* 289:6–11. (Ch. 5)

Giese, A. C., Pearse, J. S., & Pearse, V. B. (eds.) (1991). *Reproduction of Marine Invertebrates.* Vol. 6, *Echinoderms and Lophophorates.* Pacific Grove, Calif.: Boxwood Press. (Ch. 11)

Gilbert, J. C., & Glantz, B. A. (1989). Determinants of left ventricular filling and of the diastolic pressure–volume relation. *Circulation Research* 64:827–852. (Ch. 7)

Gilbert, R. D. (1982). Effects of afterload and baroreceptors on cardiac function in fetal sheep. *Journal of Developmental Physiology* 4:299–309. (Ch. 16)

Girard, H. (1973a). Arterial pressure in the chick embryo. *American Journal of Physiology* 224:454–460. (Chs. 7, 15, 17)

Girard, H. (1973b). Adrenergic sensitivity of circulation in the chick embryo. *American Journal of Physiology* 224:461–469. (Chs. 8, 15)

Girard, P. R., & Nerem, R. M. (1993). Endothelial cell signalling and cytoskeletal changes in response to shear stress. *Frontiers of Medical and Biological Engineering* 5:31–36. (Ch. 6)

Gittenberger–de Groot, A. C., Bartelings, M. M., Oddens, J. R., Kirby, M. L., & Poelmann, R. E. (1995). Coronary artery development and neural crest. In *Developmental Mechanisms of Heart Disease,* ed. E. B. Clark, R. R. Markwald, & A. Takao, pp. 291–294. Armonk, N.Y.: Futura. (Ch. 4)

Gittenberger–de Groot, A. C., Bartelings, M. M., & Poelmann, R. E. (1995a). Cardiac morphogenesis. In *Developmental Mechanisms of Heart Disease,* ed. E. B. Clark, R. R. Markwald & A. Takao, pp. 157–168. Mount Kisco, N.Y.: Futura. (Ch. 19)

Gittenberger–de Groot, A. C., Bartelings, M. M., & Poelmann, R. E. (1995b). Normal and abnormal morphogenesis of the outflow tract. In *Developmental Mechanisms of Heart Disease,* ed. E. B. Clark, R. R. Markwald, & A. Takao, pp. 249–253. Mount Kisco, N.Y.: Futura. (Ch. 19)

Goldsmith, J. E., & Butler, H. V. (1937). The development of the cardiac coronary circulation system. *American Journal of Anatomy* 60:185–201. (Ch. 4)

Goldspink, D. F., Lewis, S. E. M., & Merry, B. J. (1986). Effects of aging and long term dietary intervention on protein turnover and growth of ventricular muscle in the rat heart. *Cardiovascular Research* 20:672–678. (Ch. 9)

Goldstein, M. A., & Traeger, L. (1985). Ultrastructural changes in postnatal development of the cardiac myocyte. In *The Developing Heart,* ed. M. L. Legato, pp. 1–20. Boston: Martinus Nijhoff. (Ch. 2)

Goldstein, R. H., Poliks, C. F., Pilch, P. F., Smith, B. D., & Fine, A. (1989). Stimulation of collagen formation by insulin and insulin-like growth factor I in cultures of human lung fibroblasts. *Endocrinology* 124:964–970. (Ch. 5)

Gonzalez-Sanchez, A., & Bader, D. M. (1990). *In vitro* analysis of cardiac progenitor cell differentiation. *Developmental Biology* 139:197–209. (Ch. 10)

Goodbody, I. (1974). The physiology of ascidians. *Advances in Marine Biology* 12:1–149. (Ch. 11)

Goodrich, E. S. (1930). *Studies on the Structure and Development of Vertebrates.* Chicago: University of Chicago Press. (Ch. 14)

Gorza, L., Ausoni, S., Merciai, N., Hastings, K. E., & Schiaffino, S. (1993). Regional differences in troponin I isoform switching during rat heart development. *Developmental Biology* 156:253–264. (Ch. 3)

Gotoh, T. (1983). Quantitative studies on the ultrastructural differentiation and growth of mammalian cardiac muscle cells: The atria and ventricles of the cat. *Acta Anatomica* 115:168–177. (Ch. 2)

Grant, R. T. (1926). Development of the cardiac coronary vessels in the rabbit. *Heart* 13:261–271. (Ch. 4)

Grant, R. T., & Regnier, M. (1926). The comparative anatomy of the cardiac coronary vessels. *Heart* 14:285–317. (Ch. 9)

Greig, A., Hirschberg, Y., Anderson, P. A. W., Hainsworth, C., Malouf, N. N., Oakeley, A. E., & Kay, B. K. (1994). Molecular basis of cardiac troponin T isoform heterogeneity in rabbit heart. *Circulation Research* 74:41–47. (Ch. 3)

Greil, A. (1903). Vergleichende Anatomie und Entwicklungsgeschichte des Herzens und des Truncus Arteriosus der Wrbeltiere. *Morphologisches Jahrbuch* 31:7–21. (Ch. 14)

Grepin, C., Dagnino, L., Robitaille, L., Haberstroh, L., Antakly, T., & Nemer, M. (1994). A hormone-encoding gene identifies a pathway for cardiac but not skeletal muscle gene transcription. *Molecular and Cellular Biology* 14:3115–3129. (Ch. 3)

Grigg, G. C., & Harlow, P. (1981). A fetal-maternal shift in the oxygen equilibrium of hemoglobin from the viviparous lizard, *Sphenomorphus quoyii* (Reptilia, Scincidae). *Journal of Comparative Physiology* 142:495–499. (Ch. 14)

Grigg, G. C., Wells, R. M. G., & Beard, L. A. (1993). Allosteric control of oxygen binding by haemoglobin during embryonic development in the crocodile *Crocodylus porosus:* The role of red cell organic phosphates and carbon dioxide. *Journal of Experimental Biology* 175:15–32. (Ch. 14)

Grimmer, J. D., & Holland, N. D. (1979). Haemal and coelomic circulatory systems in the arms and pinnules of *Florometra serratissima* (Echinodermata: Crinoidea). *Zoomorphology* 94:93–109. (Ch. 11)

Grodzinski, Z. (1954). Initiation of contractions in the heart of the sea-trout, *Salmo trutta* L. *Bulletin de l'Academie Polonaise des Sciences Classe* 2:127–130. (Ch. 12)

Guidry, C., & Grinnell, F. (1985). Studies of the mechanisms of hydrated collagen gel reorganization by human skin fibroblasts. *Journal of Cell Science* 79:67–81. (Ch. 5)

Guimond, R. W., & Hutchison, V. H. (1976). Gas exchange of the giant salamanders of North America. In *Respiration of Amphibious Vertebrates,* ed. G. M. Hughes, pp. 313–338. London & New York: Academic Press. (Ch. 17)

Gulidov, M. V. (1974). The effect of different oxygen conditions during incubation on the survival and some of the developmental characteristics of the "Verkhovka" (*Leucaspius delineatus*) in the embryonic period. *Journal of Ichthyology* 14:393–397. (Ch. 17)

Gullberg, D., Tingstrom, A., Thuresson, A., Olsson, L., Terracio, L., Borg, T. K., & Rubin, K. (1990). β_1 integrin-mediated collagen gel contraction is stimulated by PDGF. *Experimental Cell Research* 186:264–272. (Ch. 5)

Guo, X., Wattanapermpool, J., Palmiter, K. A., Murphy, A. M., & Solaro, R. J. (1994). Mutagenesis of cardiac troponin I: Role of the unique NH2-terminal peptide in myofilament activation. *Journal of Biological Chemistry* 269:15210–15216. (Ch. 3)

Haeckel, E. (1866). *Generelle Morphologie der Organismen.* Berlin. (Ch. 9)

Hailstones, D., Barton, P., Chan-Thomas, P., Sasse, S., Sutherland, C., Hardeman, E., & Gunning, P. (1992). Differential regulation of the atrial isoforms of the myosin light chains during striated muscle development. *Journal of Biological Chemistry* 267:23295–23300. (Ch. 3)

Hall, A. K., & Landis, S. C. (1991). Principal neurons and small intensely fluorescent (SIF) cells in the rat superior cervical ganglion have distinct developmental histories. *Journal of Neuroscience* 11:472–484. (Ch. 8)

Hall, D. E., Reichardt, L. F., Crowley, E., Holley, B., Moezzi, H., Sonnenberg, A., & Damsky, C. H. (1990). The α_1/β_1 and α_6/β_1 integrin heterodimers mediate cell attachment to distinct sites on laminin. *Journal of Cell Biology* 110:2175–2184. (Ch. 5)
Halpern, M. H., & May, M. M. (1958). Phylogenetic study of the extracardiac arteries to the heart. *American Journal of Anatomy* 102:469–480. (Ch. 9)
Hamburger, V., & Hamilton, H. L. (1951). A series of normal stages in the development of the chick embryo. *Journal of Morphology* 88:49–92. (Chs. 7, 8, 10, 15)
Hamdorf, K. (1961). Die Beeinflussung der Embryonal- und Larvalentwicklung der Regenbogenforelle (*Salmo irideus* Gibb.) durch die Umweltfaktoren O_2-Partialdruck und Temperatur. *Zeitschrift für Vergleichende Physiologie* 44:523–549. (Ch. 17)
Hamlett, W. C., Eulitt, A. M., Jarrell, R. L., & Kelly, M. A. (1993). Uterogestation and placentation in elasmobranchs. *Journal of Experimental Zoology* 266:347–367. (Ch. 18)
Hamor, T., & Garside, E. T. (1976). Developmental rates of embryos of Atlantic salmon, *Salmo salar* L., in response to various levels of temperature, dissolved oxygen, and water exchange. *Canadian Journal of Zoology* 54:1912–1917. (Ch. 17)
Hamor, T., & Garside, E. T. (1977). Size relations and yolk utilization in embryonated ova and alevins of Atlantic salmon *Salmo salar* L. in various combinations of temperature and dissolved oxygen. *Canadian Journal of Zoology* 55:1892–1898. (Ch. 17)
Han, Y., Dennis, J. E., Cohen-Gould, L., Bader, D. M., & Fischman, D. A. (1992). Expression of sarcomeric myosin in the presumptive myocardium of chicken embryos occurs within six hours of myocardial commitment. *Developmental Dynamics* 193:257–265. (Ch. 10)
Hanley, F. L. (1993). Fetal responses to extracorporeal circulatory support. *Cardiology in the Young* 3:263–272. (Ch. 20)
Hanley, F. L. (1994). Fetal cardiac surgery. *Advances in Cardiac Surgery* 5:47–74. (Ch. 20)
Hanley, F. L., Heinemen, M. K., Jonas, R. A., et al. (1993a). Repair of truncus arteriosus in the neonate. *Journal of Thoracic and Cardiovascular Surgery* 105:1047–1056. (Ch. 20)
Hanley, F. L., Sade, R. M., Blackstone, E. H., et al. (1993b). Outcomes in neonatal pulmonary atresia with intact ventricular septum: A multiinstitutional study. *Journal of Thoracic and Cardiovascular Surgery* 105:406–423. (Ch. 20)
Hanson, J. (1949). The histology of the blood system in Oligochaeta and Polychaeta. *Biological Reviews* 24:127–173. (Ch. 11)
Haque, M. A., Pearson, J. T., Hou, P.-C. L., & Tazawa, H. (1996). Effects of pre-incubation egg storage on embryonic functions and growth. *Respiration Physiology* 103:89–98. (Ch. 15)
Haque, M. A., Watanabe, W., Ono, H., Sakamoto, Y., & Tazawa, H. (1994). Comparisons between invasive and noninvasive determinations of embryonic heart rate in chickens. *Comparative Biochemistry and Physiology* 108A:221–227. (Chs. 7, 15)
Har-el, R., & Anzer, M. L. (1993). Extracellular matrix 3: Evolution of the extracellular matrix in invertebrates. *FASEB Journal* 7:1115–1123. (Ch. 6)
Harh, J. Y., Paul, M. H., Gallen, W. J., Friedberg, D. Z., & Kaplan, S. (1973). Experimental production of hypoplastic left heart syndrome in the chick embryo. *American Journal of Cardiology* 31:51–56. (Ch. 7)
Harris, D. E., & Warshaw, D. M. (1993). Smooth and skeletal muscle actin are mechanically indistinguishable in the *in vitro* motility assay. *Circulation Research* 72:219–224. (Ch. 3)
Hartvig, M., & Weber, R. E. (1984). Blood adaptations for maternal–fetal oxygen transfer in the viviparous teleost, *Zoarces viviparus* L. In *Respiration and Metabolism of Embryonic Vertebrates,* ed. R. S. Seymour, pp. 17–30. Dordrecht, Boston, & London: Dr. W. Junk. (Ch. 17)
Hartzell, H. C. (1984). Phosphorylation of C-protein in intact amphibian cardiac muscle. *Journal of General Physiology* 83:563–588. (Ch. 3)
Hartzell, H. C. (1985). Effects of phosphorylated and unphosphorylated C-protein on cardiac actomyosin ATPase. *Journal of Molecular Biology* 186:185–195. (Ch. 3)
Hashimoto, E., Ogita, T., Nakaoka, T., Matsuoka, R., Takao, A., & Kira, Y. (1994). Rapid induction of vascular endothelial growth factor expression by transient ischemia in rat heart. *American Journal of Physiology* 267:H1948–H1954. (Ch. 4)
Hashimoto, Y., Narita, T., & Tazawa, H. (1991). Cardiogenic ballistograms of chicken eggs: Comparison of measurements. *Medical and Biological Engineering and Computing* 29:393–397. (Ch. 15)

Hastings, D., & Burggren, W. W. (1995). Developmental changes in oxygen consumption in larvae of the South African clawed frog *Xenopus laevis. Journal of Experimental Biology* 198:2465–2475. (Chs. 13, 17)

Hatem, S. N., Sweeten, T., Vetter, V., & Morad, M. (1995). Evidence for presence of Ca^{2+} channel-gated Ca^{2+} stores in neonatal human atrial myocytes. *American Journal of Physiology* 268:H1195–H1201. (Ch. 2)

Hay, D. A., & Low, F. N. (1970). The fusion of dorsal and ventral endocardial cushions in the embryonic chick heart: A study in fine structure. *American Journal of Anatomy* 133:1–24. (Ch. 7)

Hayes, F. R., Wilmot, I. R., & Livingstone, D. A. (1951). The oxygen consumption of the salmon egg in relation to development and activity. *Journal of Experimental Zoology* 116:377–395. (Ch. 17)

Heath, A. G. (1980). Cardiac responses of larval and adult tiger salamanders to submergence and emergence. *Comparative Biochemistry and Physiology* 65A:439–444. (Ch. 17)

Heintzberger, C. F. M. (1978). Development of the vessels of the heart in the chicken and rat. *Acta Morphologia Neerlando-Scandinavica* 16:140–141. (Ch. 4)

Heintzberger, C. F. M. (1983). Development of myocardial vascularization in the rat. *Acta Morphologia Neerlando-Scandinavica* 21:267–284. (Ch. 4)

Heip, J., Moens, L., Joniau, M., & Kondo, M. (1978). Ontogenetical studies on extracellular hemoglobins of *Artemia salina. Developmental Biology* 64:73–81. (Ch. 11)

Heisler, N., & Glass, M. L. (1985). Mechanisms and regulation of central vascular shunts in reptiles. In *Cardiovascular Shunts,* ed. K. Johansen and W. W. Burggren, pp. 334–347. Copenhagen: Munksgaard. (Ch. 14)

Hemmingsen, E. A., Douglas, E. L., Johansen, K., & Millard, R. W. (1972). Aortic blood flow and cardiac output in the hemoglobin-free fish *Chaenocephalus aceratus. Comparative Biochemistry and Physiology* 43A:1045–1051. (Ch. 9)

Henderson, S. A., Spencer, M., Sen, A., Kumar, C., Siddiqui, M. A., & Chien, K. R. (1989). Structure, organization, and expression of the rat cardiac myosin light chain-2 gene: Identification of a 250-base pair fragment which confers cardiac-specific expression. *Journal of Biological Chemistry* 264:18142–18148. (Ch. 3)

Hennein, H. A., Mosca, R. S., Urcelay, G., et al. (1995). Intermediate results after complete repair of tetralogy of Fallot in neonates. *Journal of Thoracic and Cardiovascular Surgery* 109:332–344. (Ch. 20)

Herbert, C. V., & Jackson, D. C. (1985). Temperature effects on the responses to prolonged submergence in the turtle *Chrysems picta bellii.* II, Metabolic rate, blood acid-base and ionic changes, and cardiovascular function in aerated and anoxic water. *Physiological Zoology* 58:670–681. (Ch. 9)

Herreid, C. F., II, DeFesi, C. R., & LaRussa, V. F. (1977). Vascular follicle system of the sea cucumber, *Stichopus moebii. Journal of Morphology* 154:19–38. (Ch. 11)

Herreid, C. F., II, LaRussa, V. F., & DeFesi, C. R. (1976). Blood vascular system of the sea cucumber, *Stichopus moebii. Journal of Morphology* 150:423–451. (Ch. 11)

Herzberg, O. (1986). A model for the Ca^{2+}-induced conformational transition of troponin C. *Journal of Biological Chemistry* 261:2638–2644. (Ch. 3)

Hewett, T. E., Grupp, I. L., Grupp, G., & Robbins, J. (1994). α-skeletal actin is associated with increased contractility in the mouse heart. *Circulation Research* 74:740–746. (Ch. 3)

Heymann, M. A. (1989). Fetal and neonatal circulations. In *Moss' Heart Disease in Infants, Children, and Adolescents,* 4th ed., ed. F. H. Adams, G. C. Emmanouilides, & T. A. Riemenschneider, pp. 23–35. Baltimore: Williams & Wilkins. (Ch. 20)

Heymann, M. A., Creasy, R. K., & Rudolph, A. M. (1973). Quantitation of blood flow patterns in utero. In *Foetal and Neonatal Physiology,* ed. K. S. Comline, K. W. Cross, G. S. Dawes, & P. W. Nathanielsz, pp. 129–135. Cambridge, U.K.: Cambridge University Press. (Chs. 16, 20)

Hicks, J. W., & Malvin, G. M. (1992). Mechanism of intracardiac shunting in the turtle *Pseudemys scripta. American Journal of Physiology* 262:R986–R992. (Ch. 14)

Hierck, B. P., Gittenberger–de Groot, A. C., Van Iperen, L., Brouwer, A., & Poelmann, R. E. (1996). Differential expression of α_6 and other subunits of laminin binding integrins during development of the murine heart. *Developmental Dynamics* 207:89–104. (Ch. 19)

Higgins, D., & Pappano, A. J. (1981). Developmental changes in the sensitivity of the chick embryo ventricle to beta-adrenergic agonist during adrenergic innervation. *Circulation Research* 48:245–253. (Ch. 8)

Hilenski, L. L., Terracio, L., & Borg, T. K. (1991). Myofibrillar and cytoskeletal assembly in neonatal rat cardiac myocytes cultured on laminin and collagen. *Cell and Tissue Research* 264:577–587. (Ch. 5)

Hilenski, L. L., Terracio, L., Sawyer, R., & Borg, T. K. (1989). Effects of extracellular matrix on cytoskeletal and myofibrillar organization *in vitro*. *Scanning Microscopy* 3:535–548. (Ch. 5)

Hilenski, L. L., Xuehui, M., Vinson, N., Terracio, L., & Borg, T. K. (1992). The role of β_1 integrin in spreading and myofibrillogenesis in neonatal at cardiomyocytes *in vitro*. *Cell Motility and Cytoskeleton* 21:87–100. (Ch. 5)

Hill, D. J., Logan, A., McGarry, M., & De Sousa, D. (1992). Control of protein and matrix-molecule synthesis in isolated ovine fetal growth-plate chondrocytes by the interactions of basic fibroblast growth factor, insulin-like growth factors-I and -II, insulin, and transforming growth factor-beta 1. *Journal of Endocrinology* 133:363–373. (Ch. 5)

Hill, E. P., Power, G. G., & Longo, L. D. (1972). A mathematical model of placental O_2 transfer with consideration of hemoglobin reaction rates. *American Journal of Physiology* 222:721–729. (Ch. 18)

Hill, E. P., Power, G. G., & Longo, L. D. (1973). A mathematical model of carbon dioxide transfer in the placenta and its interaction with oxygen. *American Journal of Physiology* 224:283–299. (Ch. 18)

Hirakow, R. (1983). Development of the cardiac blood vessels in staged human embryos. *Acta Anatomica* 115:220–230. (Ch. 4)

Hirakow, R., & Gotoh, T. (1980). Quantitative studies on the ultrastructural differentiation and growth of mammalian cardiac muscle cells. II, The atria and ventricles of the guinea pig. *Acta Anatomica* 108:230–237. (Ch. 2)

Hiruma, T., & Hirakow, R. (1989). Epicardial formation in embryonic chick heart: Computer-aided reconstruction, scanning, and transmission electron microscopic studies. *American Journal of Anatomy* 184:129–138. (Ch. 4)

Hisaoka, K. K., & Firtlit, C. F. (1960). Further studies on the embryonic development of the zebrafish, *Brachydanio rerio* (Hamilton-Buchanan). *Journal of Morphology* 107:205–225. (Ch. 12)

Hoar, W. S., Randall, D. J., & Farrell, A. P. (eds.) (1992). *Fish Physiology*, vol. 12A, *The Cardiovascular System*. New York: Academic Press. (Ch. 7)

Hochstetter, F. (1906). Beiträge zur Anatomie und Entwicklungsgeschichte des Blutgefäßsystemes der Krocodile. In *Reise in Ostafrika in den Jahren, 1903–1905: Anatomie und Entwicklungsgeschichte*, vol. 4, ed. A. Voeltzkow, pp. 141–206. Stuttgart: Wissenschaft. (Ch. 14)

Hoerter, J., Mazet, F., & Vassort, G. (1981). Perinatal growth of the rabbit cardiac cell: Possible implications for the mechanism of relaxation. *Journal of Molecular and Cellular Cardiology* 13:725–740. (Ch. 2)

Hoffman, J. I. E., & Rudolph, A. M. (1965). The natural history of ventricular septal defects in infancy. *American Journal of Cardiology* 16:634–640. (Ch. 20)

Hoffman, L. E., Jr, & van Mierop, L. H. S. (1971). Effect of epinephrine on heart rate and arterial blood pressure of the developing chick embryo. *Pediatric Research* 5:472–477. (Ch. 8)

Hofmann, P. A., Greaser, M. L., & Moss, R. L. (1991). C-protein limits shortening velocity of rabbit skeletal muscle fibres at low levels of Ca^{2+} activation. *Journal of Physiology* (London) 439:701–715. (Ch. 3)

Hofmann, P. A., Hartzell, H. C., & Moss, R. L. (1991). Alterations in the Ca^{2+} sensitive tension due to partial extraction of C-protein from rat skinned cardiac myocytes and rabbit skeletal muscle fibers. *Journal of General Physiology* 97:1141–1163. (Ch. 3)

Hofmann, P. A., Metzger, J. M., Greaser, M. L., & Moss, R. L. (1990). Effects of partial extraction of light chain 2 on the Ca^{2+} sensitivities of isometric tension, stiffness, and velocity of shortening in skinned skeletal muscle fibers. *Journal of General Physiology* 95:477–498. (Ch. 3)

Hogers, B., DeRuiter, M. C., Baastern, A. M. J., Gittenberger–de Groot, A. C., & Poelman, R. E. (1995). Intracardiac blood flow patterns are related to the yolk sac circulation of the chick embryo. *Circulation Research* 76:871–877. (Chs. 7, 19, 20)

Hoh, J. F. Y., McGrath, P. A., & Hale, P. T. (1978). Electrophoretic analysis of multiple forms of rat cardiac myosin: Effects of hypophysectomy and thyroxine replacement. *Journal of Molecular and Cellular Cardiology* 10:1053–1076. (Ch. 3)

Holeton, G. F. (1971). Respiratory and circulatory responses of rainbow trout larvae to carbon monoxide and to hypoxia. *Journal of Experimental Biology* 55:683–694. (Chs. 12, 17)

Holland, R. A. B. (1994). Special oxygen carrying properties of embryonic blood. *Israel Journal of Zoology* 40:401–416. (Ch. 18)

Holliday, F. G. T., Blaxter, J. H. S., & Lasker, R. (1964). Oxygen uptake of developing eggs and larvae of the herring (*Clupea harengus*). *Journal of the Marine Biology Association of the United Kingdom* 44:711–723. (Ch. 17)

Holroyde, M. J., Howe, E., & Solaro, R. J. (1979). Modification of calcium requirements for activation of cardiac myofibrillar ATPase by cAMP dependent phosphorylation. *Biochimica et Biophysica Acta* 586:63–69. (Ch. 3)

Honda, M., Goto, Y., Kuzou, H., Ishikawa, S., Morioka, S., Yamori, Y., & Moriyama, K. (1993). Biochemical remodeling of collagen in the heart of spontaneously hypertensive rats–Prominent increase in type V collagen. *Japanese Circulation Journal* 57:434–441. (Ch. 5)

Hood, L. C., & Rosenquist, T. H. (1992). Coronary artery development in the chick: Origin and development of smooth muscle cells, and the effect of neural crest ablation. *Anatomical Record* 234:291–300. (Ch. 4)

Hou, P.-C. L. (1992). Development of hemodynamic regulation in larvae of the African clawed toad, *Xenopus laevis.* Ph.D. diss., University of Massachusetts, Amherst. (Chs. 12, 17)

Hou, P.-C. L., & Burggren, W. W. (1995a). Blood pressure and heart rate during larval development in the anuran amphibian *Xenopus laevis. American Journal of Physiology* 269:R1120–R1125. (Chs. 7, 13, 17)

Hou, P.-C. L., & Burggren, W. W. (1995b). Cardiac output and peripheral resistance during larval development in the anuran amphibian *Xenopus laevis. American Journal of Physiology* 269:R1126–R1132. (Chs. 7, 12, 13)

Houde, E. D. (1987). Fish early life dynamics and recruitment variability. *American Fisheries Society Symposium* 2:17–29. (Ch. 17)

Houlihan, D. F., Agnisola, C., Hamilton, N. M., & Genoino, I. T. (1987). Oxygen consumption of the isolated heart of *Octopus:* Effects of power output and hypoxia. *Journal of Experimental Biology* 131:175–187. (Ch. 9)

Howe, R. S., Burggren, W. W., & Warburton, S. J. (1995). Fixed patterns of bradycardia during late embryonic development in domestic fowl with C locus mutations. *American Journal of Physiology* 268:H56–H60. (Ch. 15)

Hoyle, C., Brown, N. A., & Wolpert, L. (1992). Development of left/right handedness in the chick heart. *Development* 115:1071–1078. (Ch. 10)

Hoyt, D. F. (1987). A new model of avian embryonic metabolism. *Journal of Experimental Zoology* 1(suppl):127–138. (Ch. 15)

Hoyt, D. F., & Rahn, H. (1980). Respiration of avian embryos: A comparative analysis. *Respiration Physiology* 39:255–264. (Ch. 18)

Hoyt, R. W., Eldridge, M., & Wood, S. C. (1984). Noninvasive pulsed Doppler determination of cardiac output in an unanesthetized neotenic salamander, *Ambystoma tigrinum. Journal of Experimental Zoology* 230:491–493. (Ch. 13)

Hu, N., & Clark, E. B. (1989). Hemodynamics of the stage 12 to 29 chick embryo. *Circulation Research* 65:1665–1670. (Chs. 4, 7, 15)

Hu, N., Connuck, D. M., Keller, B. B., & Clark, E. B. (1991). Diastolic filling characteristics in the stage 12 to 27 chick embryo ventricle. *Pediatric Research* 29(4):334–337. (Ch. 7)

Hu, N., Hansen, A. L., & Clark, E. B. (1996). Distribution of blood flow between embryo and vitelline bed in the stage 18, 21, and 24 chick embryo. *Cardiovascular Research* 31:E127–E131. (Ch. 7)

Hu, N., Hansen, A. L., Clark, E. B., & Keller, B. B. (1994). Effect of atrial natriuretic peptide on diastolic filling in the stage 21 chick embryo. *Pediatric Research* 37:465–468. (Chs. 7, 8)

Hu, N., & Keller, B. B. (1995). Relationship of simultaneous atrial and ventricular pressures in the stage 16 to 27 chick embryo. *American Journal of Physiology* 269:H1359–1362. (Ch. 7)

Hu, N., Keller, B. B., & Clark, E. B. (1989). Relationship of cycle length, diastolic filling, and ventricular performance in the stage 21 chick. *Pediatric Research* 25:IIA–25A. (Ch. 7)

Hu, N., Keller, B. B., & Clark, E. B. (1991). The effect of bradycardia on systolic and diastolic time intervals in the stage 16 to 27 chick embryo. *Pediatric Research* 29(4):19A. (Ch. 7)

Hu, N., Keller, B. B., & Clark, E. B. (1992). Effect of chronic verapamil treatment on ventricular relaxation rate in stage 21 to 29 chick embryo. *Circulation* 86(4):I–843. (Ch. 7)

Hu, N., Keller, B. B., Taber, L. A., & Clark, E. B. (1990). Diastolic wall modulus and ventricular stiffness in the stage 16 to 27 chick embryo. *Circulation* 82(4):III–605. (Ch. 7)

Hu, N., Ngo, T. D., & Clark, E. B. (1996). Distribution of blood flow between embryo and vitelline bed in the stage 18, 21, and 24 chick embryo. *Cardiovascular Research* 31:E127–E131. (Ch. 7)

Hu, N., Taber, L. A., Keller, B. B., & Clark, E. B. (1992). Maintenance of ventricular relaxation rate (t) in the growth accelerated embryo heart in stage 21 to 29 chick. *Pediatric Research* 31:19A. (Ch. 7)

Hu, N., Tinney, J. P., & Keller, B. B. (1994). Atrial pressure-volume loop characteristics in the stage 24 chick embryo. *Pediatric Research* 37:27A. (Ch. 7)

Hudlicka, O. (1988). Capillary growth: Role of mechanical factors. *News in Physiological Science* 3:117–120. (Ch. 4)

Hudlicka, O., & Brown, M. (1993). Physical forces and angiogenesis. In *Mechanoreception by the Vascular Wall,* ed. G. M. Rubanyi, pp. 197–241. Mount Kisco, N.Y.: Futura. (Ch. 4)

Hudlicka, O., Brown, M., & Egginton, S. (1992). Angiogenesis in skeletal and cardiac muscle. *Physiological Reviews* 72:369–417. (Ch. 17)

Hudlicka, O., & Tyler, K. R. (1986). *Angiogenesis: The Growth of the Vascular System.* London: Academic Press. (Ch. 17)

Hughes, A. F. W. (1949). The heart output of the chick embryo. *Journal of the Royal Microscopy Society* 69:145–152. (Ch. 15)

Hughes, G. M. (1984). General anatomy of the gills. In *Fish Physiology,* vol. 10A, *Gills,* ed. W. S. Hoar & D. J. Randall, pp. 1–72. San Diego: Academic Press. (Ch. 12)

Hughes, G. M., & al-Kadhomiy, N. K. (1988). Changes on scaling of respiratory systems during the development of fishes. *Journal of Marine Biology Association of the United Kingdom* 68:489–498. (Ch. 12)

Hughes, G. M., & Morgan, M. (1973). The structure of fish gills in relation to their respiratory function. *Biological Review* 48:419–475. (Ch. 12)

Humphreys, J. E., & Cummins, P. (1984). Regulatory proteins of the myocardium: Atrial and ventricular tropomyosin and troponin I in the developing and adult bovine and human heart. *Journal of Molecular and Cellular Cardiology* 16:643–657. (Ch. 3)

Humphries, M. J. (1990). The molecular basis and specificity of integrin–ligand interactions. *Journal of Cell Science* 97:585–592. (Ch. 5)

Hunkeler, N. M., Kullman, J., & Murphy, A. M. (1991). Troponin I isoform expression in human heart. *Circulation Research* 69:1409–1414. (Ch. 3)

Hurle, J. M., Kitten, G. T., Sakai, L. Y., Volpin, D., & Solursh, M. (1994). Elastic extracellular matrix of the embryonic chick heart: An immunohistological study using laser confocal microscopy. *Developmental Dynamics* 200:321–332. (Ch. 5)

Hutchins, G. M., Kessler-Hanna, A., & Moore, G. (1988). Development of the coronary arteries in the embryonic human heart. *Circulation* 77:1250–1257. (Ch. 4)

Hutchison, H., & Dupré, R. K. (1992). Thermoregulation. In *Environmental Physiology of the Amphibians,* ed. M. E. Feder & W. W. Burggren, pp. 206–249. Chicago: University of Chicago Press. (Ch. 17)

Huynh, T. V., Chen, F., Wetzel, G. T., Friedman, W. F., & Klitzner, T. S. (1992). Developmental changes in membrane Ca^{2+} and K^{+} currents in fetal, neonatal, and adult rabbit ventricular myocytes. *Circulation Research* 70:508–515. (Ch. 2)

Hyman, L. H. (1955). *The Invertebrates.* Vol 4, *Echinodermata.* New York: McGraw-Hill. (Ch. 11)

Hyman, L. H. (1959). *The Invertebrates.* Vol. 5, *Smaller Coelomate Groups.* New York: McGraw-Hill. (Ch. 11)

Hynes, R. O. (1992). Integrins: Versatility, modulation, and signaling in cell adhesion. *Cell* 69:11–25. (Ch. 5)

Icardo, J. M. (1984). The growing heart: An anatomical perspective. In *Growth of the Heart in Health and Disease,* ed. R. Zak, pp. 41–80. New York: Raven Press. (Ch. 10)

Icardo, J. M. (1988). Heart anatomy and developmental biology. *Experientia* 44:910–919. (Ch. 10)

Icardo, J. M. (1989). Endocardial cell arrangement: Role of hemodynamics. *Anatomical Record* 225:150–155. (Chs. 6, 20)

Icardo, J. M. (1996). Developmental biology of the early vertebrate heart. *Journal of Experimental Zoology* 275:144–161. (Ch. 10)

Icardo, J. M., Arrechedera, H., & Colvee, E. (1995). Atrioventricular endocardial cushions in the pathogenesis of common atrioventricular canal: Morphological study in the iv/iv mouse. In *Developmental Mechanisms of Heart Disease,* ed. E. B. Clark, R. R. Markwald, & A. Takao, pp. 529–544. Armont, N.Y.: Futura. (Ch. 10)

Icardo, J. M., Fernandez-Teran, M. A., & Ojeda, J. L. (1990). Early cardiac structure and developmental biology. In *Handbook of Human Growth and Developmental Biology,* ed. E. Meisami & P. Timiras, vol. 3, pt. B, pp. 3–24. Boca Raton, Fla.: CRC Press. (Ch. 10)

Icardo, J. M., Hurle, J. M., & Ojeda, J. L. (1982). Endocardial cell polarity during the looping of the heart in the chick embryo. *Developmental Biology* 90:203–209. (Ch. 10)

Icardo, J. M., & Manasek, F. J. (1983). Fibronectin distribution during early chick embryo heart development. *Developmental Biology* 95:19–30. (Ch. 10)

Icardo, J. M., & Manasek, F. J. (1991). Cardiogenesis: Developmental mechanisms and embryology. In *The Heart and Cardiovascular System,* 2nd ed., ed. H. A. Fozzard, E. Haber, R. B. Jennings, A. M. Katz, & H. E. Morgan, vol. 2, pp. 1563–1586. New York: Raven Press. (Ch. 10)

Icardo, J. M., Nakamura, A., Fernandez-Teran, M. A., & Manasek, F. J. (1992). Effects of injecting fibronectin and antifibronectin antibodies on cushion mesenchyme formation in the chick: An *in vivo* study. *Anatomy and Embryology* 185:239–247. (Ch. 5)

Icardo, J. M., & Ojeda, J. L. (1984). Effects of colchicine on the formation and looping of the tubular heart of the embryonic chick. *Acta Anatomica* 119:1–9. (Ch. 10)

Ignotz, R. A., Heino, J., & Massague, J. (1989). Regulation of cell adhesion receptors by transforming growth factor-b. *Journal of Biological Chemistry* 264:389–392. (Ch. 5)

Immergluck, K., Lawrence, P. A., & Bienz, M. (1990). Induction across germ layers in *Drosophila* mediated by a genetic cascade. *Cell* 62:261–268. (Ch. 1)

Inagaki, Y., Truter, S., Tanaka, S., Di Liberto, M., & Ramirez, F. (1995). Overlapping pathways mediate the opposing actions of tumor necrosis factor-alpha and transforming growth factor-beta on alpha 2(I) collagen gene transcription. *Journal of Biological Chemistry* 270:3353–3358. (Ch. 5)

Ingermann, R. L. (1992). Maternal–fetal oxygen transfer in lower vertebrates. *American Zoologist* 32:322–330. (Chs. 14, 18)

Ingermann, R. L., Berner, N. J., & Ragsdale, F. R. (1991). Effect of pregnancy and temperature on red cell oxygen-affinity in the viviparous snake *Thamnophis elegans. Journal of Experimental Biology* 156:399–406. (Ch. 14)

Ingermann, R. L., & Terwilliger, R. C. (1984). Facilitation of maternal–fetal oxygen transfer in fishes: Anatomical and molecular specializations. In *Respiration and Metabolism of Embryonic Vertebrates,* ed. R. S. Seymour, pp. 1–15. Dordrecht, Boston, & London: Dr. W. Junk. (Ch. 17)

Ingermann, R. L., Terwilliger, R. C., & Roberts, M. S. (1984). Foetal and adult blood oxygen affinities of the viviparous seaperch, *Embiotoca lateralis. Journal of Experimental Biology* 108:453–458. (Ch. 17)

Ingersoll, C. G., Gulley, D. D., Mount, D. R., Mueller, M. E., Fernandez, J. D., Hockett, J. R., & Bergman, H. L. (1990a). Aluminum and acid toxicity to two strains of brook trout (*Salvelinus fontinalis*). *Canadian Journal of the Fisheries Aquatic Society* 47:1641–1648. (Ch. 17)

Ingersoll, C. G., Mount, D. R., Gulley, D. D., La Point, T. W., & Bergman, H. L. (1990b). Effects of pH, aluminum, and calcium on survival and growth of eggs and fry of brook trout (*Salvelinus fontinalis*). *Canadian Journal of the Fisheries Aquatic Society* 47:1580–1592. (Ch. 17)

Ip, H. S., Wilson, D. B., Heikinheimo, M., Tang, Z., Ting, C.-N., Simon, M. C., Leiden, J. M., & Parmacek, M. S. (1994). The GATA-4 transcription factor transactivates the cardiac muscle-specific troponin C promoter–enhancer in nonmuscle cells. *Molecular Cellular Biology* 14:7517–7526. (Ch. 3)

Iruela-Arispe, M. L., & Sage, E. H. (1991). Expression of type VIII collagen during morphogenesis of the chicken and mouse heart. *Developmental Biology* 144:107–118. (Ch. 5)

Irwin, C. R., Schor, S. L., & Ferguson, M. W. (1994). Effects of cytokines on gingival fibroblasts *in vitro* are modulated by the extracellular matrix. *Journal of Periodontal Research* 29:309–317. (Ch. 5)

Isaacks, R. E., Harkness, D. R., & Witham, P. R. (1978). Relationship between the major phosphorylated metabolic intermediates and oxygen-affinity of whole blood in the loggerhead (*Caretta caretta*) and the green sea turtle (*Chelonia mydas*) during development. *Developmental Biology* 62:344–353. (Ch. 14)

Ishida, J. (1985). Hatching enzyme: Past, present, and future. *Zoological Science* 2:1–10. (Ch. 17)
Itasaki, N., Nakamura, H., Sumida, H., & Yasuda, M. (1991). Actin bundles on the right side of the caudal part of the heart tube play a role in dextro-looping in the embryonic chick heart. *Anatomy and Embryology* 183:29–39. (Ch. 10)
Iuchi, I. (1973). Chemical and physiological properties of the larval and the adult hemoglobins in rainbow trout, *Salmo gairdneri irideus. Comparative Biochemistry and Physiology* 44B:1087–1101. (Ch. 12)
Izumo, S., Nadal-Ginard, B., & Mahdavi, V. (1986). All members of the MHC multigene family respond to thyroid hormone in a highly tissue-specific manner. *Science* 231:597–600. (Ch. 3)
Jackman, H. L., Morris, P. W., Rabito, S. F., et al. (1993). Inactivation of endothelin-1 by an enzyme of the vascular endothelial cells. *Hypertension* 21:925–928. (Ch. 6)
Jackman, L. B., Armanini, M., Phillips, H. S., et al. (1993). Developmental expression of binding sites and messenger ribonucleic acid for vascular endothelial growth factor suggests a role for this protein in vasculogenesis and angiogenesis. *Endocrinology* 133:848–859. (Ch. 6)
Jackson, R. L., Busch, S. J., & Cardin, A.D. (1991). Glycosaminoglycans: Molecular properties, protein interactions, and role in physiological processes. *Physiological Reviews* 71:481–522. (Ch. 5)
Jacobson, A. G., & Sater, A. K. (1988). Features of embryonic induction. *Development* 104:341–359. (Ch. 10)
Jaffe, O. C. (1965). Hemodynamic factors in the development of the chick embryo heart. *Anatomical Record* 151:69–75. (Ch. 20)
Jaffredo, T., Molina, R. M., Al Moustafa, A.-E., Gautier, R., Cosset, F.-L., Verdier, G., & Dieterlen-Lievre, F. (1993). Patterns of integration and expression of retroviral, non-replicative vectors in avian embryos: Embryo developmental stage and virus subgroup envelope modulate tissue-tropism. *Cell Adhesion and Communication* 1:119–132. (Ch. 10)
Jahroudi, A., & Lynch, D. C. (1994). Endothelial cell specific regulation of von Willebrand factor gene expression. *Molecular and Cell Biology* 14:999–1008. (Ch. 6)
James, P., Inui, M., Tada, M., Chiesi, M., & Carafoli, E. (1989). Nature and site of phospholamban regulation of the Ca^{2+} pump of sarcoplasmic reticulum. *Nature* 342:90–92. (Ch. 2)
Janzer, R. C., & Raff, M. C. (1987). Astrocytes induce blood–brain properties in endothelial cells. *Nature* 325:253–257. (Ch. 6)
Jeck, C. D., & Boyden, P. A. (1992). Age-related appearance of outward currents may contribute to developmental differences in ventricular repolarization. *Circulation Research* 71:1390–1403. (Ch. 2)
Jeck, C. D., & Rosen, M. R. (1990). Use-dependent effects of lidocaine in neonatal and adult ventricular myocardium. *Journal of Pharmacology and Experimental Therapeutics* 255:738–743. (Ch. 2)
Jessell, T. M., & Melton, D. A. (1992). Diffusible factors in vertebrate embryonic induction. *Cell* 68:257–270. (Ch. 10)
Jin, J. P., Huang, Q. Q., Yet, H. I., & Lin, J. J. (1992). Complete nucleotide sequence and structural organization of rat cardiac troponin T gene: A single gene generates embryonic and adult isoforms via developmentally regulated alternative splicing. *Journal of Molecular Biology* 227:1269–1276. (Ch. 3)
Johansen, K., & Burggren, W. W. (1980). Cardiovascular function in lower vertebrates. In *Hearts and Heart-like Organs,* ed. G. Bourne, pp. 61–117. New York: Academic Press. (Ch. 9)
Johansen, K., & Burggren, W. (1985). *Cardiovascular Shunts: Phylogenetic, Ontogenetic, and Clinical Aspects.* Copenhagen: Munksgaard. (Ch. 9)
Johnston, I. A. (1982). Capillarisation, oxygen diffusion distances, and mitochondrial content of carp muscles following acclimation to summer and winter temperatures. *Cell and Tissue Research* 222:325–337. (Ch. 17)
Jonakait, G. M., Rosenthal, M., & Morrell, J. I. (1989). Regulation of tyrosine hydroxylase mRNA in catecholaminergic cells of embryonic rat: Analysis by in situ hybridization. *Journal of Histochemistry and Cytochemistry* 37:1–5. (Ch. 8)
Jones, D. R., & Milsom, W. K. (1982). Peripheral receptors affecting breathing and cardiovascular function in non-mammalian vertebrates. *Journal of Experimental Biology* 100:59–91. (Chs. 9, 17)
Jones, D. R., & Shelton, G. (1993). The physiology of the alligator heart: Left aortic flow patterns and right-to-left shunts. *Journal of Experimental Biology* 176:247–269. (Ch. 14)

Jones, J. C. (1971). On the heart of the orange tunicate, *Ecteinascidia turbinata* Herdman. *Biological Bulletin* 141:130–145. (Ch. 11)

Joseph-Silverstein, J., Consigli, S. A., Lyser, K. M., & VerPault, C. (1989). Basic fibroblast growth factor in the chick embryo: Immunolocalization to striated muscle cells and their precursors. *Journal of Cell Biology* 108:2459–2466. (Ch. 4)

Juliano, R. L., & Haskill, S. (1993). Signal transduction from the extracellular matrix. *Journal of Cell Biology* 120:577–585. (Ch. 5)

Justus, J. T. (1978). The cardiac mutant: An overview. *American Zoologist* 18:321–326. (Ch. 17)

Kam, Y. (1993). Physiological effects of hypoxia on metabolism and growth of turtle embryos. *Respiration Physiology* 92:127–138. (Chs. 14, 17)

Kamino, K., Hirota, A., & Fujii, S. (1981). Localization of pacemaking activity in early embryo heart monitored using voltage-sensitive dye. *Nature* 260:595–597. (Chs. 7, 17)

Karttunen, P., & Tirri, R. (1986). Isolation and characterisation of single myocytes from the perch, *Perca fluviatilis. Comparative Biochemistry and Physiology* A84:181–188. (Ch. 9)

Kasahara, H., Itoh, M., Sugiyama, T., Kido, N., Hayashi, H., Saito, H., Tsukita, S., & Kato, N. (1994). Autoimmune myocarditis induced in mice by cardiac C-protein: Cloning of complementary DNA encoding murine cardiac C-protein and partial characterization of the antigenic peptides. *Journal of Clinical Investigation* 94:1026–1036. (Ch. 3)

Kass, D. A., Beyar, R., Lankford, E., Heard, M., Maughan, W. L., & Sagawa, K. (1989). Influence of contractile state of curvilinearity of in situ end-systolic pressure–volume relations. *Circulation* 79:167–178. (Ch. 7)

Katchman, S. D., Hsu-Wong, S., Ledo, I., Wu, M., & Uitto, J. (1994). Transforming growth factor-beta up-regulates human elastin promoter activity in transgenic mice. *Biochemical Biophysical Research Communications* 203:485–490. (Ch. 5)

Kato, H., Suzuki, H., Tajima, S., Ogato, Y., Tominaga, T., Sato, A., & Satura, T. (1991). Angiotensin II stimulates collagen synthesis in cultured vascular smooth muscle cells. *Journal of Hypertension* 9:17–22. (Ch. 5)

Kaufman, M. H. (1992). *The Atlas of Mouse Development.* New York: Harcourt, Brace. (Ch. 7)

Kaufman, T. M., Horton, J. W., White, D. J., & Mahony, L. (1990). Age-related changes in myocardial relaxation and sarcoplasmic reticulum function. *American Journal of Physiology* 259 (Heart Circ. Physiol. 28):H309–H316. (Chs. 2, 7)

Kay, E. P., Gu, X., Ninomiya, Y., & Smith, R. E. (1993). Corneal endothelial modulation: A factor released by leukocytes induces basic fibroblast growth factor that modulates cell shape and collagen. *Investigative Ophthalmology and Visual Science* 34:663–672. (Ch. 5)

Kayar, S. R., Snyder, G. K., Birchard, G. F., & Black, C. P. (1981). Oxygen permeability of the shell and membranes of chicken eggs during development. *Respiration Physiology* 46:209–221. (Ch. 18)

Kazazoglou, T., Schmid, A., Renaud, J., & Lazdunski, M. (1983). Ontogenic appearance of Ca^{2+} channels as binding sites for nitrendipine during development of nervous skeletal and cardiac muscle systems in the rat. *FEBS Letters* 164:75–79. (Ch. 2)

Keen, J. E., Farrell, A. P., Tibbits, G. F., & Brill, R. W. (1992). Cardiac physiology in tunas. II, Effect of ryanodine, calcium, and adrenaline on force–frequency relationships in atrial strips from skipjack tuna, *Katsuwonus pelamis. Canadian Journal of Zoology* 70:1211–1217. (Ch. 9)

Keen, J. E., Vianzon, D.-M., Farrell, A. P., & Tibbits, G. F. (1994). Effect of acute temperature change and temperature acclimation on excitation–contraction coupling in trout myocardium. *Journal of Comparative Physiology B, Metabolic and Transport Functions* 164:438–443. (Ch. 9)

Keller, B. B. (1995). Functional maturation and coupling of the embryonic cardiovascular system. In *Developmental Mechanisms of Heart Disease,* ed. E. B. Clark, R. R. Markwald, & A. Takao, pp. 367–386. Mount Kisco, N.Y.: Futura. (Chs. 7, 15, 17)

Keller, B. B., & Clark, E. B. (1993). Cardiovascular structural and functional maturation. *Current Opinions in Cardiology* 8:98–107. (Ch. 7)

Keller B, B., Heller, F. A., and Tinney, J. P. (1993). Effect of isoproterenol induced vasodilation on ventricular pressure–volume relations in the stage 21 chick embryo. *Pediatric Research* 33:248A. (Chs. 7, 8)

Keller, B. B., Hu, N., & Clark, E. B. (1990). Correlation of ventricular area, perimeter, and conotruncal diameter with ventricular mass and function in the stage 12 to 24 chick embryo. *Circulation Research* 66:109–114. (Ch. 7)

Keller, B. B., Hu, N., & Clark, E. B. (1991). Cardiac mechanics in the developing heart during normal and accelerated ventricular growth. *Pediatric Research* 29:20A. (Ch. 7)
Keller, B. B., Hu, N., Serrino, P. J., & Clark, E. B. (1991a). Ventricular pressure–area loop characteristics in the stage 16 to 24 chick embryo. *Circulation Research* 68(1):226–231. (Chs. 7, 15)
Keller, B. B., Hu, N., & Tinney, J. P. (1994). Embryonic ventricular diastolic and systolic pressure–volume relation. *Cardiology in the Young* 4:19–27. (Chs. 7, 15, 16)
Keller, B. B., Tinney, J. P., & Hu, N. (1992). Arterial elastance in the stage 16 to 21 chick embryo. *Circulation* 86(4):I–205A. (Ch. 7)
Keller, B. B., Tinney, J. P., Hu, N., & Clark, E. B. (1991b). Ventricular end-diastolic and end-systolic relations in the embryonic chick heart. *Circulation* 84(4):II–188. (Ch. 7)
Keller, J. M., & Keller, B. B. (1995). Effect of altered temperature on heart rate in the stage 21 chick embryo. Project presented at Park Road Elementary School Science Fair, Pittsfords, N.Y. (Ch. 7)
Kempski, M. H., Kibler, N., Blackburn, J. L., Dzakowic, J., Hu, N., & Clark, E. B. (1993). Hemodynamic regulation in the chick embryo. *Journal of Biomechanical Engineering* 24:119–122. (Ch. 7)
Kern, M. J., Argao, E. A., & Potter, S. S. (1995). Homeobox genes and heart development. *Trends in Cardiovascular Medicine* 5:47–54. (Chs. 1, 9, 19)
Kessel, M., Balling, R., & Gruss, P. (1990). Variations in cervical vertebrae after expression of a *Hox* 1.1 transgene in mice. *Cell* 61:301–308. (Ch. 19)
Kiceniuk, J. W., & Jones, D. R. (1977). The oxygen transport system in trout (*Salmo gairdneri*) during exercise. *Journal of Experimental Biology* 69:247–260. (Ch. 12)
Kieval, R. S., Bloch, R. J., Lindenmayer, G. E., Ambesi, A., & Lederer, W. J. (1992). Immunofluorescence localization of the Na–Ca exchanger in heart cells. *American Journal of Physiology* 63:C545–C550. (Ch. 2)
Kilborn, J. M., & Fedida, D. (1990). A study of the developmental changes in outward currents of rat ventricular myocytes. *Journal of Physiology* 430:37–60. (Ch. 2)
Kimmel, P. (1990). Ontogeny of cardiovascular control mechanisms in the bullfrog, *Rana catesbeiana.* Ph.D. diss., University of Massachusetts, Amherst. (Ch. 13)
Kimmel, P. (1992). Adrenergic receptors and the regulation of vascular resistance in bullfrog tadpoles (*Rana catesbeiana*). *Journal of Comparative Physiology B, Metabolic and Transport Functions* 162:455–462. (Chs. 13, 17)
Kind, C. (1975). The development of the circulating blood volume of the chick embryo. *Anatomy and Embryology* (West Germany) 147:127–132. (Chs. 7, 15)
King, L. A. (1994). Adult and fetal hemoglobins in the oviparous swell shark, *Cephaloscyllium ventriusum. Comparative Biochemistry and Physiology B* 109:237–243. (Ch. 12)
Kinne, O., & Kinne, E. M. (1962). Rates of development in embryos of a cyprinodont fish exposed to different temperature–salinity–oxygen combinations. *Canadian Journal of Zoology* 40:231–253. (Ch. 17)
Kirby, M. L. (1988a). Nodose placode contributes autonomic neurons to the heart in the absence of cardiac neural crest. *Journal of Neuroscience* 8:1089–1095. (Ch. 7)
Kirby, M. L. (1988b). Role of extracardiac factors in heart development. *Experientia* 44:944–951. (Ch. 10)
Kirby, M. L. (1993). Cellular and molecular contributions of the cardiac neural crest to cardiovascular development. *Trends in Cardiovascular Medicine* 3:18–23. (Chs. 10, 19)
Kirby, M. L., & Waldo, K. L. (1990). Role of neural crest in congenital heart disease. *Circulation* 82:332–340. (Ch. 1)
Kirklin, J. W., & Barratt-Boyes, B. G. (1993). *Cardiac Surgery.* 2nd ed. New York: Churchill Livingstone. (Ch. 20)
Kirkpatrick, S. E., Covell, J. W., & Friedman, W. F. (1973). A new technique for the continuous assessment of fetal and neonatal cardiac performance. *American Journal of Obstetrics and Gynecology* 116:963–967. (Ch. 16)
Kitoh, K., & Oguri, M. (1985). Differentiation of the compact layer in the heart ventricle of rainbow trout. *Bulletin of the Japanese Society for Scientific Fisheries* 51:539–542. (Ch. 12)
Kitten, G. T., Markwald, R. R., & Bolender, D. L. (1987). Distribution of basement membrane antigens in cryopreserved early embryonic hearts. *Anatomical Record* 217:379–390. (Ch. 5)
Kjellen, L., & Lindahl, U. (1991). Proteoglycans: Structures and interactions. *Annual Review of Biochemistry* 60:443–475. (Ch. 5)

Klagsbrun, M., & D'Amore, P. A. (1991). Regulators of angiogenesis. *Annual Reviews of Physiology* 53:217–239. (Chs. 4, 6)

Klinkhardt, M. B., Straganov, A. A., & Pavlov, D. A. (1987). Motoricity of Atlantic salmon embryos (*Salmo salar* L.) at different temperatures. *Aquaculture* 64(3):219–236. (Ch. 12)

Klitzner, T. S., & Friedman, W. F. (1988). Excitation–contraction coupling in developing mammalian myocardium: Evidence from voltage clamp studies. *Pediatric Research* 23:428–432. (Ch. 2)

Klitzner, T. S., & Friedman, W. F. (1989). A diminished role for the sarcoplasmic reticulum in newborn myocardial contraction: Effects of ryanodine. *Pediatric Research* 26:98–101. (Ch. 2)

Klopfenstein, H. S., & Rudolph, A. M. (1978). Postnatal changes in the circulation and responses to volume overloading in sheep. *Circulation Research* 42:839–845. (Ch. 20)

Kluft, C. (1990). Endothelium as a source of tissue-type plasminogen activator (t-PA) for fibrinolysis. In *The Endothelium: An Introduction to Current Research,* ed. J. B. Warren, pp. 129–139. New York: Wiley-Liss. (Ch. 6)

Kohmoto, O., Levi, A. J., & Bridge, J. H. B. (1994). Relation between reverse sodium–calcium exchange and sarcoplasmic reticulum calcium release in guinea pig ventricular cells. *Circulation Research* 74:550–554. (Ch. 2)

Kollmann, J. (1882). Das Uberwintern von Frosch- und Tritonlarven und die Umwandlung des mexicanischen axolotl. *Verhandlungen der Naturforschenden Gesellschaft in Basel* 7:387. (Ch. 9)

Komoro, T., Akiyama, R., Pearson, J., Burggren, W. W., & Tazawa, H. (1995). Development of cardiogenic rhythm in early chick embryos within the shell. *Physiological Zoology* 68:71. (Ch. 15)

Komuro, I., & Izumo, S. (1993). *Csx:* A murine homeobox-containing gene specifically expressed in the developing heart. *Proceedings of the National Academy of Sciences USA* 90:8145–8149. (Ch. 1)

Komuro, I., & Yazaki, Y. (1994). Intracellular signaling pathways in cardiac myocytes induced by mechanical stress. *Trends in Cardiovascular Medicine* 4:117–121. (Ch. 5)

Korhonen, J., Dolvi, A., Partanen, J., & Alitalo, K. (1994). The mouse ties receptor tyrosine kinase genes: Expression during embryonic angiogenesis. *Oncogene* 9:395–403. (Ch. 6)

Korsgaard, B., & Weber, R. E. (1989). Maternal–fetal trophic and respiratory relationships in viviparous ectothermic vertebrates. *Advances in Comparative and Environmental Physiology* 5:209–233. (Ch. 17)

Kovacs, J., Carone, F. A., Liu, Z. Z., Nakamura, S., Kumar, A., & Kanwar, Y. S. (1994). Differential growth factor–induced modulation of proteoglycans synthesized by normal human renal versus cyst-derived cells. *Journal of the American Society of Nephrology* 5:47–54. (Ch. 5)

Krijgsman, B. J. (1956). Contractile and pacemaker mechanisms of the heart of tunicates. *Biological Reviews* 31:288–312. (Ch. 11)

Krogh, A. (1941). *The Comparative Physiology of Respiratory Mechanisms.* Philadelphia: University of Pennsylvania Press. (Ch. 18)

Krug, E. L., Mjaavedt, C. H., & Markwald, R. R. (1987). Extracellular matrix from embryonic myocardium elicits an early morphogenetic event in cardiac endothelial differentiation. *Developmental Biology* 120:348–355. (Ch. 5)

Kumar, C. C., Cribbs, L., Delaney, P., Chien, K. R., & Siddiqui, M. A. Q. (1986). Heart myosin light chain 2 gene: Nucleotide sequence of full length cDNA and expression in normal and hypertensive rat. *Journal of Biological Chemistry* 261:2866–2872. (Ch. 3)

Kurabayashi, M., Tsuchimochi, H., Komuro, I., Takaku, F., & Yazaki, Y. (1988). Molecular cloning and characterization of human cardiac α- and β-form myosin heavy chain complementary DNA clones: Regulation of expression during development and pressure overload in human atrium. *Journal of Clinical Investigation* 82:524–531. (Ch. 3)

Kurihara, Y., Kurihara, H., Suzuki, H., Kodama, T., Maemura, K., Nagai, R., Oda, H., Kuwaki, T., Cao, W. H., Kamada, N., Jishage, K., Ouchi, Y., Azuma, S., Toyoda, Y., Ishikawa, T., Kumada, M., & Yazaki, Y. (1994). Elevated blood pressure and craniofacial abnormalities in mice deficient in endothelin-1. *Nature* 368:703–710. (Ch. 8)

Kuroda, O., Matsunaga, C., Whittow, G. C., & Tazawa, H. (1990). Comparative metabolic responses to prolonged cooling in precocial duck (*Anas domestica*) and altricial pigeon (*Columba domestica*) embryos. *Comparative Biochemistry and Physiology* 95A:407–410. (Ch. 17)

Kurosawa, S., Kurosawa, H., & Becker, A. E. (1986). The coronary arterioles in newborns, infants, and children: A morphometric study of normal hearts and hearts with aortic atresia and complete transposition. *International Journal of Cardiology* 10:43–56. (Ch. 4)

Kwee-Isaac, L., Baldwin, H. S., Steward, C. W., Buck, C. E., Buck, C. A., & Labow, M. A. (1995). Defective development of the embryonic and extraembryonic circulatory systems in vascular cell adhesion molecule (VCAM)-1 deficient mice. *Development* 121:489–503. (Chs. 1, 5)

Laale, H. W. (1984). Fish embryo culture: Cardiac monolayers and contractile activity in embryo explants from the zebrafish, *Brachydanio rerio. Canadian Journal of Zoology* 62:878–885. (Ch. 12)

Labastie, M. C., Poole, T. J., Peault, B. M., & LeDouarin, N. M. (1986). MB1, a quail leukocyte-endothelium antigen: Partial characterization of the cell surface and secreted forms in cultured endothelial cells. *Proceedings of the National Academy of Sciences USA* 83:9016–9020. (Ch. 6)

Lamontagne, D., Pohl, U., & Busse, R. (1992). Mechanical deformation of vessel wall and shear stress determine basal release of endothelium-derived relaxing factor in the intact rabbit coronary vascular bed. *Circulation Research* 70:123–130. (Ch. 6)

Langer, G. A. (1992). Calcium and the heart: Exchange at the tissue cell and organelle levels. *FASEB Journal* 6:893–902. (Ch. 2)

Langille, B. L. (1993). Remodelling of developing and mature arteries: Endothelium, smooth muscle, and matrix. *Journal of Cardiovascular Pharmacology* 21:S11–S17. (Ch. 6)

Larsen, T. H. (1990). Atrial natriuretic factor in the heart of the human embryo. *Acta Anatomica* 138:132–136. (Ch. 8)

Larson, M. K., & Bayne, C. J. (1994). Evolution of immunity – Potential immunocompetence of the echinoid axial organ. *Journal of Experimental Zoology* 270:474–485. (Ch. 11)

Lau, V. K., & Sagawa, K. (1979). Model analysis of the contribution of atrial contraction to ventricular filling. *Annals of Biomedical Engineering* 7:167–201. (Ch. 7)

Lau, V. K., Sagawa, K., & Suga, H. (1979). Instantaneous pressure–volume relationship of the right atrium during isovolumic contraction. *American Journal of Physiology* 236 (Heart Circulation Physiology 5):H672–H679. (Ch. 7)

Laughlin, K. F., Lundy, H., & Tait, J. A. (1976). Chick embryo heart rate during the last week of incubation: Population studies. *British Poultry Science* 17:293–301. (Ch. 15)

Laurence, G. C. (1975). Laboratory growth and metabolism of the winter flounder *Pseudopleuronectes americanus* from hatching through metamorphosis at three temperatures. *Marine Biology* 32:223–229. (Ch. 17)

Lawrence, J. (1987). *A Functional Biology of Echinoderms.* Baltimore: Johns Hopkins University Press. (Ch. 11)

Layton, W. M., Jr. (1978). Heart malformations in mice homozygous for a gene causing situs inversus. In *Morphogenesis and Malformation of the Cardiovascular System,* ed. G. C. Rosenquist and D. Bergsma, pp. 277–294. New York: Alan R. Liss. (Ch. 10)

Leatherbury, L., Connuck, D. M., & Kirby, M. L. (1993). Neural crest ablation versus sham surgical effects in a chick embryo model of defective cardiovascular development. *Pediatric Research* 33:628–631. (Ch. 19)

Leblanc, N., & Hume, J. R. (1990). Sodium current-induced release of calcium from cardiac sarcoplasmic reticulum. *Science* 248:372–376. (Ch. 2)

Leclerc, M., Arneodo, V. J., Legac, E., Bajelan, M., & Vaugier, G. L. (1993). Identification of T-like and B-like lymphocyte subsets in sea star *Asterias rubens* by monoclonal antibodies to human leukocytes. *Thymus* 21:133–139. (Ch. 11)

Leclerc, M., Brillouet, G., Luquet, C., & Binaghi, R. A. (1986). The immune system of invertebrates: The sea star *Asterias rubens* (Echinodermata) as a model of study. *Bulletin, Institut Pasteur* (Paris) 84:311–330. (Ch. 11)

Leclerc, M., Maitre, F., & Contrepois, L. (1994). Identification of a macrophage-like subset in the axial organ sea star *Asterias rubens. Cell Biology International* 18:835–837. (Ch. 11)

Lee, K. F., Simon, H., Chen, H., Bates, B., Hung, M. C., & Hauser, C. (1995). Requirement for neuregulin receptor erbB2 in neural and cardiac development. *Nature* 378:394–398. (Ch. 19)

Lee, K. J., Hickey, R., Zhu, H., & Chien, K. R. (1994). Positive regulatory elements (HF-1a and HF-1b) and a novel negative regulatory element (HF-3) mediate ventricular muscle specific expression of myosin light-chain 2-luciferase fusion genes in transgenic mice. *Molecular and Cellular Biology* 14:1220–1229. (Ch. 3)

Lee, K. J., Ross, R. S., Rockman, H. A., Harris, A. N., O'Brien, T. X., van Bilsen, M., Shubeita, H. E., Kandolf, R., Brem, G., Price, J., Evans, S. M., Zhu, H., Franz, W. M., & Chien, K. R. (1992). Myosin light chain-2 luciferase transgenic mice reveal distinct regulatory programs for

cardiac and skeletal muscle-specific expression of a single contractile protein gene. *Journal of Biological Chemistry* 267:15875–15885. (Ch. 3)
Lee, R. K. K., Stainier, D. Y. R., Weinstein, B. M., & Fishman, M. C. (1994). Cardiovascular development in the zebrafish. II, Endocardial progenitors are sequestered within the heart field. *Development* 120:3361–3366. (Chs. 1, 2, 10, 12)
Leenstra, S., Troost, D., Das, P. K., et al. (1993). Endothelial cell marker PAL-E reactivity in brain tumor, developing brain, and brain disease. *Cancer* 72:3061–3067. (Ch. 6)
Legato, M. J. (1979). Cellular mechanisms of normal growth in the mammalian heart. II, A quantitative and qualitative comparison between the right and left ventricular myocytes in the dog from birth to five months of age. *Circulation Research* 44:263–279. (Ch. 2)
Lehrer, S. S. (1994). The regulatory switch of the muscle thin filament: Ca^{2+} or myosin heads? *Journal of Muscle Research and Cell Motility* 15:232–236. (Ch. 3)
Lemez, L. (1972). Thrombocytes of chick embryos from the 2nd day of incubation till the 1st postembryonic day. *Acta Universitatis Carolinae, Medica Monographia,* 53–54:365–371. (Ch. 15)
Lev, M., Arcilla, R., Rimaldi, H. J., et al. (1963). Premature narrowing or closure of foramen ovale. *Am Heart J* 65:638–647. (Ch. 20)
Levavasseur, F., Mayer, U., Guillouzo, A., & Clement B. (1994). Influence of nidogen complexed or not with laminin on attachment, spreading, and albumin and laminin B2 mRNA levels of rat hepatocytes. *Journal of Cellular Physiology* 161:257–266. (Ch. 5)
Levesque, M. J., Liepsch, F. D., Moravec, S., et al. (1986). Correlation of endothelial cell shape and wall stress in a stenosed dog aorta. *Arteriosclerosis* 6:220–229. (Ch. 6)
Levesque, M. J., & Nerem, R. M. (1985). The elongation and orientation of cultured endothelial cells in response to shear stress. *Arteriosclerosis* 9:439–445. (Ch. 6)
Levesque, P. C., Leblanc, N., & Hume, J. R. (1994). Release of calcium from guinea pig cardiac sarcoplasmic reticulum induced by sodium calcium exchange. *Cardiovascular Research* 28:370–378. (Ch. 2)
Levi, A. J., Spitzer, K. W., Kohomoto, O., & Bridge, J. H. B. (1994). Depolarization-induced Ca entry via Na–Ca exchange triggers SR release in guinea pig cardiac myocytes. *American Journal of Physiology* 266:H1422–H1433. (Ch. 2)
Levin, M., Johnson, R. L., Stern, C. D., Kuehn, M., & Tabin, C. (1995). A molecular pathway determining left–right asymmetry in chick embryogenesis. *Cell* 82:803–814. (Ch. 10)
Levinsohn, E. M., Packard, D. S., Jr., West, E. M., & Hootnick, D. R. (1984). Arterial anatomy of chicken embryo and hatchling. *American Journal of Anatomy* 169:377–405. (Ch. 15)
Lewis, F. T. (1904). The question of sinusoids. *Anatomischer Anzeiger: Zentralblatt für die Gesamte Wissenschaftliche Anatomie* 25:261–279. (Ch. 4)
Lewis, S. E. M., Kelly, F. J., & Goldspink, D. F. (1984). Pre- and post-natal growth and protein turnover in smooth muscle, heart, and slow- and fast-twitch skeletal muscles of the rat. *Biochemical Journal* 217:517–526. (Ch. 9)
Liebersbach, B. F., & Sanderson, R. D. (1994). Expression of syndecan-1 inhibits cell invasion into type I collagen. *Journal of Biological Chemistry* 269:20013–20019. (Ch. 5)
Liem, K. F. (1981). Larvae of air-breathing fishes as countercurrent flow devices in hypoxic environments. *Science* 211:1177–1179. (Chs. 12, 17)
Lilly, B., Calewsky, S., Firulli, A. B., Schulz, R. A., & Olson, E. (1994). *D-MEF2:* A MADS box transcription factor expressed in differentiating mesoderm and muscle cell lineages during *Drosophila* embryogenesis. *Proceedings of the National Academy of Sciences USA* 91:5662–5666. (Ch. 1)
Lilly, B., Zhao, B., Ranganayakulu, G., Paterson, B. M., Schulz, R. A., & Olson, E. (1995). Requirement of MADS domain transcription factor *D-MEF2* for muscle formation in *Drosophila. Science* 267:688–693. (Ch. 1)
Lin, C. Q., & Bissell, M. J. (1993). Multi-faceted regulation of cell differentiation by extracellular matrix. *FASEB Journal* 7:737–743. (Ch. 6)
Lin, S., Gaiano, N., Culp, P., Burns, J. C., Friedmann, T., Yee, J. K., & Hopkins, N. (1994). Integration and germ-line transmission of a pseudotyped retroviral vector in zebrafish. *Science* 265:666–669. (Ch. 1)
Linask, K. K. (1992). N-cadherin localization in early heart development and polar expression of Na^+, K^+-ATPase and integrin during pericardial coelom formation and epithelialization of the differentiating myocardium. *Developmental Biology* 151:213–224. (Ch. 10)

Linask, K. K., & Lash, J. W. (1988). A role for fibronectin in the migration of avian precardiac cells. I, Dose-dependent effects of fibronectin antibody. *Developmental Biology* 129:315–323. (Chs. 5, 10)

Linask, K. K., & Lash, J. W. (1993). Early heart development: Dynamics of endocardial cell sorting suggests a common origin with cardiomyocytes. *Developmental Dynamics* 195:62–69. (Ch. 10)

Lindemann, J. P., Jones, L. R., Hathaway, D. R., Henry, B. G., & Watanabe, A. M. (1983). Beta-adrenergic stimulation of phospholamban phosphorylation and Ca^{2+}-ATPase activity in guinea pig ventricles. *Journal of Biological Chemistry* 258:464–471. (Ch. 2)

Linsenmayer, T. F., Gibney, E., Igoe, F., Gordon, M. K., Fitch, J. M., Fessler, L. I., & Birk, D. E. (1993). Type V collagen: Molecular structure and fibrillar organization of the chicken alpha 1(V) NH2-terminal domain, a putative regulator of corneal fibrillogenesis. *Journal of Cell Biology* 121:1181–1189. (Ch. 5)

Lints, T. J., Parsons, L. M., Hartley, L., Lyons, I., & Harvey, R. P. (1993). *Nkx-2.5:* A novel murine homeobox gene expressed in early heart progenitor cells and their myogenic descendants. *Development* 119:419–431. (Ch. 1)

Lipke, D. W., McCarthy, K. J., Elton, T. S., Arcot, S. S., Oparil, S., & Couchman, J. R. (1993). Coarctation induces alterations in basement membranes in the cardiovascular system. *Hypertension* 22:743–753. (Ch. 5)

Lipshultz, S., Shanfeld, J., & Chacko, S. (1981). Emergence of beta-adrenergic sensitivity in the developing chick heart. *Proceedings of the National Academy of Sciences USA* 78:288–292. (Ch. 8)

Little, C. D., Piquet, D. M., Davis, L. A., Walters, L., & Drake, C. J. (1989). Distribution of laminin, collagen type IV, collagen type I, and fibronectin in chicken cardiac jelly/basement membrane. *Anatomical Record* 224:417–425. (Ch. 5)

Liu, B., Harvey, C. S., & McGowan, S. E. (1993). Retinoic acid increases elastin in neonatal rat lung fibroblast cultures. *American Journal of Physiology* 265:L430–L437. (Ch. 5)

Loeber, C. P., & Runyan, R. B. (1990). A comparison of fibronectin, laminin, and galactosyltransferase adhesion mechanisms during embryonic cardiac mesenchymal cell migration *in vitro.* *Developmental Biology* 140:401–412. (Ch. 5)

Loeffler, C. A. (1971). Water exchange in the pike egg. *Journal of Experimental Biology* 55:797–811. (Ch. 17)

Logan, M., & Mohun, T. (1993). Induction of cardiac muscle differentiation in isolated animal pole explants of *Xenopus laevis* embryos. *Development* 118:865–875. (Ch. 10)

Lompre, A. M., Nadal-Ginard, B., & Mahdavi, V. (1984). Expression of the cardiac ventricular α- and β-myosin heavy chain genes *is* developmentally and hormonally regulated. *Journal of Biological Chemistry* 259:6437–6446. (Ch. 3)

Longo, L. D. (1987). Respiratory gas exchange in the placenta. In *Handbook of Physiology,* sec. 3, *The Respiratory System,* vol. 4, *Gas Exchange,* ed. L. E. Farhi & S. M. Tenney, pp. 351–401. Bethesda, Md.: American Physiological Society. (Ch. 18)

Longo, L. D., Hill, E. P., & Power, G. G. (1972). Theoretical analysis of factors affecting placental O_2 transfer. *American Journal of Physiology* 222:730–739. (Ch. 18)

Lowe, K. C. (1990). Perfluorochemical oxygen carriers and ischaemic tissues. In *Oxygen Transport to Tissue,* ed. J. Piiper, T. K. Goldstick, & M. Meyer, vol. 12. New York: Plenum Press. (Ch. 18)

Lowe, L. A., Supp, D. M., Sampath, K., Yokoyama, T., Wright, C. V. E., Potter, S. S., Overbeek, P., & Kuehn, M. R. (1996). Conserved left–right asymmetry of nodal expression and alterations in murine situs inversus. *Nature* 381:158–161. (Ch. 10)

Lowey, S., Waller, G. S., & Trybus, K. M. (1993). Skeletal muscle myosin light chains are essential for physiological speeds of shortening. *Nature* 365:454–456. (Ch. 3)

Lucchesi, P. A., & Sweadner, K. J. (1991). Postnatal changes in Na K-ATPase isoform expression in rat cardiac ventricle: Conservation of biphasic ouabain activity. *Journal of Biological Chemistry* 266:9327–9331. (Ch. 2)

Luo, W., Grupp, I. L., Harrer, J., Ponniah, S., Grupp, G., Duffy, J. J., Doetschman, T., & Kranias, E. G. (1994). Target ablation of the phospholamban gene is associated with markedly enhanced myocardial contractility and loss of β-agonist stimulation. *Circulation Research* 75:401–409. (Ch. 2)

Luscinskas, F., & Lawler, J. (1994). Integrins as dynamic regulation of vascular function. *FASEB Journal* 8:929–938. (Ch. 4)

Lyons, G. E. (1994). In situ analysis of the cardiac muscle gene program during embryogenesis. *Trends in Cardiovascular Medicine* 4:70–77. (Chs. 3, 10)

Lyons, G. E., Schiaffino, S., Sassoon, D., Barton, P., & Buckingham, M. (1990). Developmental regulation of myosin gene expression in mouse cardiac muscle. *Journal of Cell Biology* 111:2427–2436. (Ch. 3)

Lyons, I., Parsons, L., Hartley, L., Li, R., Andrews, J. E., Tobb, L., & Harvey, R. P. (1995). Myogenic and morphogenetic defects in the heart tubes of murine embryos lacking the homeobox gene *Nkx2.5. Genes and Development* 9:1654–1666. (Ch. 1)

Lytton, J., & MacLennan, D. H. (1992). Sarcoplasmic reticulum. In *The Heart and Cardiovascular System: Scientific Foundations,* 2nd ed., ed. H. A. Fozzard, E. Haber, M. B. Jennings, A. M. Katz, & H. E. Morgan, pp. 1203–1222. New York: Raven Press. (Ch. 2)

Macey, D. J., & Potter, I. C. (1982). The effect of temperature on the oxygen dissociation curves of whole blood of larval and adult lampreys (*Geotria australis*). *Journal of Experimental Biology* 97:253–261. (Ch. 12)

Mackie, E. J. (1994). Tenascin in connective tissue development and pathogenesis. *Perspectives on Developmental Neurobiology* 2:125–132. (Ch. 5)

MacKinnon, M. R., & Heatwole, H. (1981). Comparative cardiac anatomy of the reptilia. IV, The coronary arterial circulation. *Journal of Morphology* 170:1–27. (Ch. 9)

Madri, J. A., & Williams, S. K. (1983). Capillary endothelial cell cultures: Phenotypic modulation by matrix components. *Journal of Cell Biology* 97:153–165. (Ch. 6)

Mahdavi, V., Chambers, A. P., & Nadal-Ginard, B. (1984). Cardiac a- and b-myosin heavy chain genes are organized in tandem. *Proceedings of the National Academy of Sciences USA* 81:2626–2630. (Ch. 3)

Mahdavi, V., Izumo, S., & Nadal-Ginard, B. (1987). Developmental and hormonal regulation of sarcomeric myosin heavy chain gene family. *Circulation Research* 60:804–814. (Ch. 3)

Maheshwari, R. K., Kedar, V. P., Coon, H. C., & Bhartiya, D. (1991). Regulation of laminin expression by interferon. *Journal of Interferon Research* 11:75–80. (Ch. 5)

Mahon, E. F., & Hoar, W. S. (1956). The early development of the chum salmon, *Oncorhynchus keta. Journal of Morphology* 98(1):1–47. (Ch. 12)

Mahony, L. (1988). Maturation of calcium transport in cardiac sarcoplasmic reticulum. *Pediatric Research* 24:639–643. (Ch. 2)

Mahony, L., & Jones, L. R. (1986). Developmental changes in cardiac sarcoplasmic reticulum in sheep. *Journal of Biological Chemistry* 261:15257–15265. (Ch. 2)

Majors, A., & Ehrhart, L. A. (1993). Basic fibroblast growth factor in the extracellular matrix suppresses collagen synthesis and type III procollagen mRNA levels in arterial smooth muscle cell cultures. *Arteriosclerosis Thrombosis* 13(5):680–686. (Ch. 5)

Maltsev, V. A., Wobus, A. M., Rohwedel, J., Bader, D. M., & Hescheler, J. (1994). Cardiomyocytes differentiated *in vitro* from embryonic stem cells developmentally express cardiac-specific genes and ionic currents. *Circulation Research* 75:233–244. (Ch. 10)

Malvin, G. M. (1985). Cardiovascular shunting during amphibian metamorphosis. In *Cardiovascular Shunts: Ontogenetic, Phylogenetic, and Clinical Aspects,* ed. K. Johansen & W. W. Burggren, pp. 163–178. Alfred Benzon Symposium 21. Copenhagen: Munksgaard. (Chs. 9, 13)

Malvin, G. M. (1988). Microvascular regulation of cutaneous gas exchange in amphibians. *American Zoologist* 28:999–1007. (Ch. 12)

Malvin, G. M., & Heisler, N. (1988). Blood flow patterns in the salamander, *Ambystoma tigrinum,* before, during, and after metamorphosis. *Journal of Experimental Biology* 137:53–74. (Ch. 17)

Manasek, F. J. (1969). The appearance of granules in the Golgi complex of embryonic cardiac myocyte. *Journal of Cell Biology* 43:605–609. (Ch. 8)

Manasek, F. J. (1970). Histogenesis of the embryonic myocardium. *American Journal of Cardiology* 25:149–168. (Ch. 7)

Manasek, F. J., Icardo, J. M., Nakamura, A., & Sweeney, L. J. (1986). Cardiogenesis: Developmental mechanisms and embryology. In *The Heart and Cardiovascular System,* ed. H. A. Fozzard, E. Haber, R. B. Jennings, A. M. Katz, & H. E. Morgan, vol. 2, pp. 965–986. New York: Raven Press. (Ch. 10)

Manasek, F. J., Isobe, Y., Shimada, Y., & Hopkins, W. (1984a). The embryonic myocardial cytoskeleton, interstitial pressure, and the control of morphogenesis. In *Congenital Heart Disease: Causes and Processes,* ed. J. J. Nora & A. Takao, pp. 359–376. New York: Futura. (Ch. 10)

Manasek, F. J., Kulikowski, R. R., Nakamura, A., Nguyenphuc, Q., & Lacktis, J. W. (1984b). Early heart development: A new model of cardiac morphogenesis. In *Growth of the Heart in Health and Disease,* ed. R. Zak, pp. 105–130. New York: Raven Press. (Ch. 10)

Mandarim-de-Lacerda, C. A. (1990). Development of the coronary arteries in staged human embryos (the Paris Embryological Collection revisited). *Anais Academia Brasileira de Ciencias* 62:79–84. (Ch. 4)

Mann, K. H. (1961). *Leeches (Hirudinea): Their Structure, Physiology, Ecology, and Embryology.* New York: Pergamon Press. (Ch. 11)

Manning, A., & McLachlan, J. C. (1990). Looping of chick embryo hearts *in vitro. Journal of Anatomy* 168:257–263. (Ch. 10)

Manwell, C. (1958). Ontogeny of hemoglobin in the skate *Raja binoculata. Science* 128:419–420. (Ch. 14)

March, H. W. (1961). Persistence of a functioning bulbus cordis homologue in the turtle heart. *American Journal of Physiology* 201:1109–1112. (Ch. 14)

Margossian, S. S., White, H. D., Caulfield, J. B., Norton, P., Taylor, S., & Slayter, H. S. (1992). Light chain 2 profile and activity of human ventricular myosin during dilated cardiomyopathy: Identification of a causal agent for impaired myocardial function. *Circulation* 85:1720–1733. (Ch. 3)

Markiewicz, F. (1960). Development of blood vessels in branchial arches of the pike (*Esox lucius* L.). *Acta Biologica Cracoviensia, Series Zoologia* 3:4–12. (Ch. 12)

Markwald, R. R. (1995). Overview: Formation and early morphogenesis of the primary heart tube. In *Developmental Mechanisms of Heart Disease,* ed. E. B. Clark, R. R. Markwald, & A. Takao, pp. 149–155. Armonk, N.Y.: Futura. (Chs. 4, 19)

Markwald, R. R., Fitzharris, T. P., & Manasek, F. J. (1977). Structural development of endocardial cushions. *American Journal of Anatomy* 148:85–120. (Ch. 5)

Markwald, R. R., Krug, E. L., Runyan, R. B., & Kitten, G. T. (1985). Proteins in cardiac jelly which induce mesenchyme formation. In *Cardiac Morphogenesis,* ed. V. Ferrans, G. Rosenquist, & C. Weinstein, pp. 60–69. New York: Alan R. Liss. (Ch. 5)

Markwald, R. R., Mjaavedt, C. H., & Krug, E. L. (1990). Induction of endocardial cushion tissue formation by adheron-like molecular complexes derived from the myocardial basement membrane. In *Developmental Cardiology: Morphogenesis and Function,* ed. E. B. Clark and A. Takao, pp. 191–204. New York: Futura. (Ch. 5)

Markwald, R. R., Rezaee, M., Nakajima, Y., Wunsch, A., Isokawa, K., Litke, L., & Krug, E. (1995). Molecular basis for the segmental pattern of cardiac cushion mesenchyme formation: Role of ES/130 in the embryonic chick heart. In *Developmental Mechanisms of Heart Disease,* ed. E. B. Clark, R. R. Markwald, & A. Takao, pp. 185–196. Armonk, N.Y.: Futura. (Ch. 21)

Markwald, R. R., Runyan, R. B., Kitten, G. T., Funderburg, F. M., Bernanke, D. H., & Brauer, P. R. (1984). Use of collagen gel cultures to study heart development: Proteoglycan and glycoprotein interactions during the formation of endocardial cushion tissue. In *The Role of the Extracellular Matrix in Development,* ed. Robert L. Telstad, pp. 323–350. New York: Alan R. Liss. (Ch. 5)

Marsh, J. D., & Allen, P. D. (1989). Developmental regulation of cardiac calcium channels and contractile sensitivity to $[Ca]_o$. *American Journal of Physiology* 256:H179–H185. (Ch. 2)

Martin, A. W. (1980). Some invertebrate myogenic hearts: The hearts of worms and molluscs. In *Hearts and Heart-like Organs,* ed. G. H. Bourne, pp. 1–39. London: Academic Press. (Ch. 11)

Martin, G. G., Hose, J. H., & Corzine, C. J. (1989). Morphological comparison of major arteries in the ridgeback prawn *Sicyonia ingentis. Journal of Morphology* 200:175–183. (Ch. 11)

Masuda, H., & Sperelakis, N. (1993). Inwardly rectifying potassium current in rat fetal and neonatal ventricular cardiomyocytes. *American Journal of Physiology* 265:H1107–H1111. (Ch. 2)

Matsushima, K., Bano, M., Kidwell, W. R., & Oppenheim, J. J. (1985). Interleukin 1 increases collagen type IV production by murine mammary epithelial cells. *Journal of Immunology* 134:904–909. (Ch. 5)

Mattfeldt, T., & Mall, G. (1987). Growth of capillaries and myocardial cells in the normal rat heart. *Journal of Molecular and Cellular Cardiology* 19:1237–1246. (Ch. 4)

Maughan, W. L., Sunagawa, K., Burkhoff, D., Graves, W. L., Jr., Hunter, W. C., & Sagawa, K. (1985a). Effect of heart rate on the canine end-systolic pressure–volume relationship. *Circulation* 72(3):586–589. (Ch. 7)

Maughan, W. L., Sunagawa, K., Burkhoff, D., & Sagawa, K. (1985b). Effect of arterial impedance changes on the end-systolic pressure–volume relation. *Circulation Research* 54:595–602. (Ch. 7)

Maxwell, D., Allan, L., & Tynan, M. J. (1991). Balloon dilatation of the aortic valve in the fetus: A report of two cases. *British Heart Journal* J65:256–258. (Ch. 20)

Mayer, Y., Czosnek, H., Zeelon, P. E., Yaffe, D., & Nudel, U. (1983). Expression of the genes coding for the skeletal muscle and cardiac actins in the heart. *Nucleic Acids Research* 12:1087–1100. (Ch. 3)

Maylie, J. G. (1982). Excitation–contraction coupling in neonatal and adult myocardium of cat. *American Journal of Physiology* 242:H834–H843. (Ch. 2)

Maynard, D. M. (1960). Circulation and heart function. In *The Physiology of Crustacea,* ed. H. P. Wolverkamp & T. H. Waterman. New York: Academic Press, pp. 161–72. (Ch. 12)

McAuliffe, J. J., Gao, L. Z., & Solaro, R. J. (1990). Changes in myofibrillar activation and troponin C Ca^{2+} binding associated with troponin T isoform switching in developing rabbit heart. *Circulation Research* 66:1204–1216. (Ch. 3)

McCain, E. R., & Cather, J. N. (1989). Regulative and mosaic development of *Ilyanassa obsoleta* embryos lacking the A and C quadrants. *Invertebrate Reproduction and Development* 15:185–192. (Ch. 11)

McCormick, D. A., & Zetter, B. R. (1992). Adhesive interactions in angiogenesis and metastasis. *Pharmacology and Therapeutics* 53:239–260. (Ch. 6)

McCutcheon, I. E., Metcalfe, J., Metzenberg, A. B., & Ettinger, T. (1982). Organ growth in hyperoxic and hypoxic chick embryos. *Respiration Physiology* 50:153–163. (Ch. 17)

McDonald, D. G., & McMahon, B. R. (1977). Respiratory development in arctic char *Salvelinus alpinus* under conditions of normoxia and chronic hypoxia. *Canadian Journal of Zoology* 55:1461–1467. (Chs. 12, 17)

McElman, J. F. (1983). Comparative embryonic ecomorphology and reproductive guild classification of walleye, *Stizostedion vitreum,* white sucker, *Catostomus commersoni. Copeia* 1983:246–250. (Ch. 12)

McElman, J. F., & Balon, E. K. (1979). Early ontogeny of walleye, *Stizostedion vitreum,* with steps of saltatory development. *Environmental Biology of Fishes* 4:309–348. (Ch. 12)

McElman, J. F., & Balon, E. K. (1980). Early ontogeny of white sucker, *Catostomus commersoni,* with steps of saltatory development. *Environmental Biology of Fishes* 5:191–224. (Chs. 12, 17)

McLaughlin, P. A. (1983). Internal anatomy. In *Internal Anatomy and Physiological Regulation,* ed. L. Mantel & D. Bliss, pp. 1–53. Biology of Crustacea, vol. 5. New York: Academic Press. (Ch. 11)

McMahon, B. R., Chu, K. H., & Mak, E. M. T. (1995). Development of the heart in the shrimp *Metapenaeus ensis.* In *Larvi 95,* pp. 472–474. European Aquaculture Society Special Publication, no. 24. (Ch. 11)

McMahon, B. R., Smith, P. S., & Wilkens, J. (1997). Invertebrate circulatory systems. In *Handbook of Comparative Physiology,* ed. W. Danzler, vol. II, section 13. Washington: American Physiological Society, Oxford University Press, pp. 931–1008. (Chs. 9, 11)

Mellish, J. A., Pinder, A. W., & Smith, S. C. (1994). You've got to have heart . . . or do you? *Axolotl Newsletter* 23:34–38. (Chs. 12, 13)

Mellish, J. A., Smith, S. C., & Pinder, A. W. (in press). Oxygen transport in heartless vertebrates. *Physiological Zoology.* (Chs. 13, 17, 18)

Mendelsohn, C., Larkin, S., Mark, M., LeMeur, M., Clifford, J., Zelent, A., & Chambon, P. (1994a). RARβ isoforms: Distinct transcriptional control by retinoic acid and specific spatial patterns of promoter activity during mouse embryonic development. *Mechanisms of Development* 45:227–241. (Ch. 19)

Mendelsohn, C., Lohnes, D., Decimo, D., Lufkin, T., LeMeur, M., Chambon, P., & Mark, M. (1994b). Function of the retinoic acid receptors (*RARs*) during development. II, Multiple abnormalities at various stages of organogenesis in *RAR* double mutants. *Development* 120:2749–2771. (Ch. 1)

Mendes, E. G. (1945). Contribuicao para a fisiologica dos sistemas respiratório e circulatoria de *Siphonops annulatus* (Amphibia–Gymnophiona). *Boletins da Faculdade de Filosofia, Ciéncias, e Letras, Universidade de São Paulo* 1945:25–64. (Ch. 13)

Meno, H., Jarmakani, J. M., & Phillipson, K. D. (1989). Developmental changes of sarcolemmal Na^+–H^+ exchange. *Journal of Molecular and Cellular Cardiology* 21:1179–1185. (Ch. 2)

Mercadier, J. J., Bouveret, P., Gorza, L., Schiaffino, S., Clark, W. A., Zak, R., Swynghedauw, B., & Schwartz, K. (1983). Myosin isoenzymes in normal and hypertrophied human ventricular myocardium. *Circulation Research* 53:52–62. (Ch. 3)

Mercadier, J. J., Lompre, A.-M., Wisnewsky, C., Samuel, J.-L., Bercovici, J., Swynghedauw, B., & Schwartz, K. (1981). Myosin isoenzymic changes in several models of rat cardiac hypertrophy. *Circulation Research* 49:525–532. (Ch. 3)

Meschia, G. F. (1994). Placental respiratory gas exchange and fetal oxygenation. In *Maternal–Fetal Medicine, Principles and Practice,* ed. R. K. Creasy & R. Resnik, pp. 288–297. Philadelphia: W. B. Saunders. (Ch. 18)

Meschia, G. F., Battaglia, F. C., & Bruns, P. D. (1967). Theoretical and experimental study of transplacental diffusion. *Journal of Applied Physiology* 22:1171–1178. (Chs. 16, 18)

Metcalfe, J., & Bissonnette, J. M. (1981). A comparison of chorioallantoic and placental respiration. In *Advances in Physiological Sciences,* ed. I. Hutas & L. A. Debreczeni, pp. 127–134. Budapest: Pergamon Press. (Ch. 18)

Metcalfe, J., Dhindsa, D. S., & Novy, M. J. (1972). General aspects of oxygen transport in maternal and fetal blood. In *Respiratory Gas Exchange and Blood Flow in the Placenta,* ed. L. D. Longo & H. Bartels, pp. 63–77. Bethesda, Md.: U.S. Department of Health, Education, and Welfare. (Ch. 14)

Metcalfe, J., & Stock, M. K. (1993). Oxygen exchange in the chorioallantoic membrane, avian homologue of the mammalian placenta. *Placenta* 14:605–613. (Ch. 18)

Meuer, H.-J. (1992). Erythrocyte velocity and total blood flow in the extraembryonic circulation of early chick embryos determined by digital video techniques. *Microvascular Research* 44:286–294. (Ch. 18)

Meuer, H.-J., & Bertram, C. (1993). Capillary transit times and kinetics of oxygenation in the primary respiratory organ of early chick embryo. *Microvascular Research* 45:302–313. (Ch. 18)

Meuer, H.-J., Sieger, U., & Baumann, R. (1989). Measurement of pH in blood vessels and interstitium of 4 and 6 day old chick embryos. *Journal of Developmental Physiology* 11:354–359. (Ch. 18)

Meyer, D., & Birchmeier, C. (1995). Multiple essential functions of neuregulin in development. *Nature* 378:386–390. (Ch. 19)

Mikawa, T., Borisov, A., Brown, A. M. C., & Fischman, D. A. (1992). Clonal analysis of cardiac morphogenesis in the chicken embryo using a replication-defective retrovirus. I, Formation of the ventricular myocardium. *Developmental Dynamics* 193:11–23. (Ch. 10)

Mikawa, T., & Fischman, D. (1992) Retroviral analysis of cardiac morphogenesis: Discontinuous formation of coronary vessels. *Proceedings of the National Academy of Sciences USA* 89:9504–9508. (Ch. 4)

Milev, P., Friedlander, D. R., Sakurai, T., Karthikeyen, L., Flad, M., Margolis, R. K., Grumet, M., & Margolis, R. U. (1994). Interactions of the chondroitin sulfate proteoglycan phosphacan, the extracellular domain of a receptor-type protein tyrosine phosphatase, with neurons, glia, and neural cell adhesion molecules. *Journal of Cell Biology* 127:1703–1715. (Ch. 5)

Miller, O. E. Vanni, M. A., Taber, L. A., & Keller, B. B. (in press). Passive stress-strain measurements in the stage 16 and stage 18 embryonic chick heart. *Journal of Biomechanical Engineering.* (Ch. 7)

MIM (Mendelian Inheritance in Man) (1995). Human genetic disorders. *Journal of NIH Research* 7:115–134. (Ch. 21)

Miner, J. H., Miller, J. B., & Wold, B. J. (1992) Skeletal muscle phenotypes initiated by ectopic MyoD in transgenic mouse heart. *Development* 114:853–860. (Ch. 10)

Mirsky, I. (1974) Review of various theories for the evaluation of left ventricular wall stresses. In *Cardiac Mechanics: Physiological, Clinical, and Mathematical Considerations,* ed. I. Mirsky, D. N. Ghista, & H. Sandler, pp. 381–409. New York: Wiley. (Ch. 7)

Mishra, A. P., & Singh, B. R. (1979). Oxygen uptake through water during early life of *Anabas testudineus. Hydrobiologia* 66:129–134. (Ch. 12)

Miyabara, S. (1990). Cardiovascular malformations of mouse trisomy 16: Pathogenic evaluation as an animal model for human trisomy 21. In *Developmental Cardiology: Morphogenesis and Function,* ed. E. B. Clark & A. Takao, pp. 219–234. Mount Kisco, N.Y.: Futura. (Ch. 5)

Mjaavedt, C. H., Krug, E. L., & Markwald, R. R. (1991). An antiserum (ES1) against a particulate form of extracellular matrix blocks the transition of cardiac endothelium into mesenchyme in culture. *Developmental Biology* 145:219–230. (Ch. 5)

Mjaavedt, C. H., Lepera, R. C., and Markwald, R. R. (1987). Myocardial specificity for initiating endothelial–mesenchymal cell transition in embryonic chick heart correlates with a particulate distribution of fibronectin. *Developmental Biology* 119:59–67. (Ch. 5)

Mjaavedt, C. H., & Markwald, R. R. (1989). Induction of an epithelial-mesenchymal transition by an *in vivo* adhero-like complex. *Developmental Biology* 136:118–128. (Ch. 5)

Moens, C. B., Stanton, B. R., Parada, L. F., & Rossant, J. (1993). Defects in heart and lung development in compound heterozygotes for two different targeted mutations at the *N-myc* locus. *Development* 119:485–499. (Ch. 1)

Molkentin, J. D., Kalvakolanu, D. V., & Markham, B. E. (1994). Transcription factor GATA-4 regulates cardiac muscle-specific expression of the a-myosin heavy-chain gene. *Molecular and Cellular Biology* 14:4947–4957. (Ch. 3)

Moller, J. H., Allen, H. D., Clark, E. B., Dajani, A. S., Golden, A., Hayman, L. L., Lauer, R. M., Marmer, E. L., McAnulty, J. H., Oparil, S., Strauss, A. W., Taubert, K. A., & Wagner, A. (1993). Report of the task force on children and youth. *Circulation* 88:2479–2486. (Ch. 21)

Moller, J. H., & Anderson, R. C. (1992). 1,000 consecutive children with a cardiac malformation with 26- to 37-year follow-up. *American Journal of Cardiology* 70:661–667. (Ch. 21)

Moncada, S., Palmer, R. M. J., & Higgs, E. A. (1991). Nitric oxide: Physiology, pathophysiology, and pharmacology. *Pharmacological Reviews* 43:109–142. (Ch. 6)

Morano, I., Hadicke, K., Grom, S., Koch, A., Schwinger, R. H., Bohm, M., Bartel, S., Erdmann, E., & Krause, E. G. (1994). Titin, myosin light chains and C-protein in the developing and failing human heart. *Journal of Molecular and Cellular Biology* 26:361–368. (Ch. 3)

Morgan, J. P., MacKinnon, R., Briggs, M., & Gwathmey, J. K. (1988). Calcium and cardiac relaxation. In *Diastolic Relaxation of the Heart,* ed. W. Grossman & B. H. Lorell, pp. 17–26. Boston: Martinus Nijhoff. (Ch. 7)

Morgan, M. (1974a). The development of gill arches and gill blood vessels of the rainbow trout, *Salmo gairdneri. Journal of Morphology* 142:351–364. (Ch. 12)

Morgan, M. (1974b). Development of secondary lamellae of the gills of the trout, *Salmo gairdneri* (Richardson). *Cell and Tissue Research* 151:509–523. (Ch. 12)

Morgan, T. H. (1894). The development of *Balanoglossus. Journal of Morphology* 9:1–86. (Ch. 11)

Morikawa, Y, & Rosen, M. R. (1995). Developmental changes in the effects of lidocaine on the electrophysiological properties of canine Purkinje fibers. *Circulation Research* 55:633–641. (Ch. 2)

Morris, C. D., & Menashe, V. D. (1991). 25-year mortality after surgical repair of congenital heart defect in childhood. *Journal of the American Medical Association* 266:3447–3452. (Ch. 21)

Morton, M. J., Pinson, C. W., & Thornburg, K. L. (1987). In utero ventilation with oxygen augments left ventricular stroke volume in lambs. *Journal of Physiology* 383:413–424. (Ch. 16)

Moss, A. J., Emmanouilides, G., & Duffie, E. R., Jr. (1963). Closure of the ductus arteriosus in the newborn infant. *Pediatrics* 32:25–30. (Ch. 20)

Moss, R. L. (1992). Ca^{2+} regulation of mechanical properties of striated muscle: Mechanistic studies using extraction and replacement of regulatory proteins. *Circulation Research* 70:865–884. (Ch. 3)

Mossman, H. W. (1948). Circulatory cycles in the vertebrates. *Biological Reviews* 23:237–255. (Ch. 12)

Mukai, H., Sugimoto, K., & Taneda, Y. (1978). Comparative studies on the circulatory system of the compound ascidians, *Botryllus, Botrylloides,* and *Symplegma. Journal of Morphology* 157:49–78. (Ch. 11)

Mullins, M. C., Hammerschmidt, M., Haffter, P., & Nüsslein-Volhard, C. (1994). Large-scale mutagenesis in the zebrafish: In search of genes controlling development in a vertebrate. *Current Biology* 4:189–202. (Ch. 1)

Murdoch, A. D., Liu, B., Schwarting, R., Tuan, R. S., & Iozzo, R. V. (1994). Widespread expression of perlecan proteoglycan in basement membranes and extracellular matrices of human tissues as detected by a novel monoclonal antibody against domain III and by in situ hybridization. *Journal of Histochemistry and Cytochemistry* 42:239–249. (Ch. 5)

Murphy, A. M., Jones, L., Sims, H. F., & Strauss A. W. (1991). Molecular cloning of rat cardiac troponin I and analysis of troponin I isoform expression in developing rat heart. *Biochemistry* 30:707–712. (Ch. 3)

Murphy, A. M., & Thompson, W. R. (1995). GATA-4 transcription factor regulates the cardiac troponin I gene *in vitro. Pediatric Research* 37:31A (abstract). (Ch. 3)

Murphy, J. G., Gersh, B. J., Mair, D. D., Fuster, V., McGoon, M. D., Ilstrup, D. M., McGoon, D. C., Kirklin, J. W., & Danielson, G. K. (1993). Long term outcome in patients undergoing surgical repair of tetralogy of Fallot. *New England Journal of Medicine* 329:593–599. (Ch. 21)

Muslin, A. J., & Williams, L. T. (1991). Well-defined growth factors promote cardiac development in axolotl mesodermal explants. *Development* 112:1095–1101. (Ch. 10)

Muthuchamy, M., Pajak, L., Howles, P., Doetschman, T., & Wieczorek, D. F. (1993). Developmental analysis of tropomyosin gene expression in embryonic stem cells and mouse embryos. *Molecular and Cellular Biology* 13:3311–3323. (Ch. 3)

Nabäuer, M., Callewaert, G., Cleeman, L., & Morad, M. (1989). Regulation of calcium release is gated by calcium current, not gating charge, in cardiac myocytes. *Science* 244:800–803. (Ch. 2)

Nabel, E. G. (1995). Gene therapy for cardiovascular disease. *Circulation* 91:541–548. (Ch. 21)

Nadal-Ginard, B., Smith, C. W. J., Patton, J. G., & Breitbart, R. E. (1991). Alternative splicing is an efficient mechanism for the generation of protein diversity: Contractile protein genes as a model system. *Advances in Enzyme Regulation* 31:261–286. (Ch. 3)

Nakagawa, M., Birkedal-Hansen, H., Terracio, L., Burgess, W., & Borg, T. K. (1990). Expression of collagenase in developing rat hearts *in vivo* and *in vitro. Journal of Cell Biology* 111:16A. (Ch. 5)

Nakagawa, M., Terracio, L., Carver, W., Birkedel-Hansen, H., & Borg, T. K. (1992). Expression of collagenase and IL-1a in developing rat hearts. *Developmental Dynamics* 195:87–99. (Ch. 5)

Nakamura, A., & Manasek, F. J. (1981). An experimental study of the relation of cardiac jelly to the shape of the early chick embryonic heart. *Journal of Embryology and Experimental Morphology* 65:235–256. (Chs. 7, 10)

Nakanishi, T., & Jarmakani, J. M. (1981). Effect of extracellular sodium on mechanical function in the newborn rabbit. *Developmental Pharmacology and Therapeutics* 2:188–200. (Ch. 2)

Nakanishi, T., & Jarmakani, J. M. (1984). Developmental changes in myocardial mechanical function and subcellular organelles. *American Journal of Physiology* 246:H615–H625. (Ch. 2)

Nakanishi, T., Okuda, H., Kamata, K., Abe, K., Sekiguchi, M., & Takao, A. (1987). Development of the myocardial contractile system in the fetal rabbit. *Pediatric Research* 22:201–207. (Ch. 2)

Nakazawa, M., Clark, E. B., Hu, N., & Wise, J. (1985) Effect of environmental hypothermia on vitelline artery blood pressure and vascular resistance in the stage 18, 21, and 24 chick embryo. *Pediatric Research* 19:651–654. (Chs. 7, 8)

Nakazawa, M., Kajio, F., Ikeda, K., & Takao, A. (1990a). Effect of atrial natriuretic peptide on hemodynamics of the stage 21 chick embryo. *Pediatric Research* 27:557–560. (Chs. 7, 8)

Nakazawa, M., Miyagawa, S., Morishima, M., Takao, A., Ohno, T., Ikeda, K., & Mori, K. (1990b). Hemodynamic effects of cardiovascular agents on the embryonic circulation: A comparative study in chick and rat embryos. In *Developmental Cardiology: Morphogenesis and Function,* ed. E. B. Clark & A. Takao, pp. 315–323. Mount Kisco, N.Y.: Futura. (Ch. 8)

Nakazawa, M., Miyagawa, S., Ohno, T., Miura, S., & Takao, A. (1988). Developmental hemodynamic changes in rat embryos at 11 to 15 days of gestation: Normal data of blood pressure and the effect of caffeine compared to data from chick embryo. *Pediatric Research* 23:200–205. (Chs. 7, 8)

Nakazawa, M., Miyagawa, S., & Takao, A. (1986). Cardiovascular effects of isoproterenol in the stage 21 chick embryo (in Japanese). *Archives of Japanese Pediatric Cardiology* 2:238–243. (Ch. 8)

Nakazawa, M., Miyagawa, S., Takao, A., Hu, N., & Clark, E. B. (1986). Hemodynamic effects of environmental hyperthermia in the stage 18, 21, and 24 chick embryo. *Pediatric Research* 20:1213–1215. (Chs. 7, 8)

Nakazawa, M., Morishima, M., Miyagawa-Tomita, S., & Kajio, F. (1992). Effect of atrial natriuretic peptide on cardiovascular function in the rat embryo (in Japanese). *Archives of Japanese Pediatric Cardiology* 7:659–662. (Ch. 8)

Nakazawa, M., Morishima, M., Tomita, H., Tomita, S. M., & Kajio, F. (1995). Hemodynamics and ventricular function in the day-12 rat embryo: Basic characteristics and the responses to cardiovascular drugs. *Pediatric Research* 37:117–123. (Ch. 8)

Narasimhamurti, N. (1932). The development and function of the heart and pericardium in *Echinodermata. Proceedings of the Royal Society of London,* ser. B, 109:471–487. (Ch. 11)

Nassar, R., Malouf, N. N., Kelly, M. B., Oakeley, A. E., & Anderson, P. A. (1991). Force–*p*Ca relation and troponin T isoforms of rabbit myocardium. *Circulation Research* 69:1470–1475. (Ch. 3)

Nassar, R., Reedy, M. C., & Anderson, P. A. W. (1987). Developmental changes in the ultrastructure and sarcomere shortening of the isolated rabbit ventricular myocyte. *Circulation Research* 61:465–483. (Ch. 2)

Navarkasattusas, S., Zhu, H., Garcia, A. V., Evans, S. M., & Chien, K. R. (1992). A ubiquitous factor (HF-1a) and a distinct muscle factor (HF-1b/MEF-2) form an E-box-independent pathway for cardiac muscle gene expression. *Molecular and Cellular Biology* 12:1469–1479. (Ch. 3)

Nechaev, I. V., Labas, Y. A., & Denisov, V. A. (1992). Influence of catecholaminergic compounds upon the embryonic motor system of the bony fish *Aequidens pulchen* (Cichlidae) (in Russian; English abstract). *Zhurnal Obshchei Biologii* 53(2):258–271. (Ch. 12)

Neill, C. A., & Clark, E. B. (1995). *The Developing Heart: A History of Pediatric Cardiology.* Dordrecht: Kluwer Academic. (Ch. 21)

Nelsen, O. E. (1953). *Comparative Embryology of the Vertebrates.* New York: McGraw-Hill. (Ch. 12)

Nemecek, S. (1995). Have a heart: Tissue-engineered valves may offer a transplant alternative. *Science* 265:46. (Ch. 21)

Nerem, R. M., Harrison, D. G., Taylor, W. R., et al. (1993). Hemodynamics and vascular endothelial biology. *Journal of Cardiovascular Pharmacology* 21:S6–S10. (Ch. 6)

Newman, P. J., Berndt, M. C., Gorsky, J., et al. (1990). PECAM-1 (CD31): Cloning and relation to adhesion molecules of the immunoglobulin gene superfamily. *Science* 247:1219–1222. (Ch. 6)

Nicoll, P. A. (1954). The anatomy and behaviour of the vascular systems in *Nereis virens* and *Nereis limbata. Biological Bulletin* 106:69–82. (Ch. 11)

Nicosia, R. F., Bonanno, E., & Smith, M. (1993). Fibronectin promotes the elongation of microvessels during angiogenesis *in vitro. Journal of Cell Physiology* 154:654–661. (Ch. 6)

Nieuwkoop, P. D., & Faber, J. (1967). *Normal Table of* Xenopus laevis. Amsterdam: North-Holland. (Ch. 13)

Niggli, E., & Lederer, W. J. (1990). Voltage-independent calcium release in heart muscle. *Science* 250:565–568. (Ch. 2)

Nilsson, S. (1983). *Autonomic Nerve Function in the Vertebrates.* Berlin: Springer-Verlag. (Ch. 9)

Nishibatake, M., Kirby, M. L., & van Mierop, L. H. S. (1987). Pathogenesis of persistent truncus arteriosus and dextroposed aorta in the chick embryo after neural crest ablation. *Circulation* 75:225–264. (Ch. 7)

Nishibatake, M., Nakazawa, M., Tomita, H., Ikeda, K., & Takao, A. (1990). Image analysis of cardiac contraction in the early stages of the chick embryo. In *Developmental Cardiology: Morphogenesis and Function,* ed. E. B. Clark & A. Takao, pp. 305–314. Mount Kisco, N.Y.: Futura. (Ch. 7)

Noden, D. M. (1989). Embryonic origins and assembly of blood vessels. *American Review of Respiratory Disease* 140:1097–1103. (Ch. 6)

Noden, D. M. (1991). Origins and patterning of avian outflow tract endocardium. *Development* 111:867–876. (Ch. 10)

Noden, D. M., Poelmann, R. E., & Gittenberger–de Groot, A. C. (1995). Cell origins and tissue boundaries during outflow tract development. *Trends in Cardiovascular Medicine* 5:69–75. (Ch. 19)

Noland, T. A., Jr., & Kuo, J. F. (1991). Protein kinase C phosphorylation of cardiac troponin I or troponin T inhibits Ca^{2+}-stimulated actomyosin MgATPase activity. *Journal of Biological Chemistry* 266:4974–4978. (Ch. 3)

Oberpriller, J. O., & Oberpriller, J. C. (1974). Response of the adult newt ventricle to injury. *Journal of Experimental Zoology* 187:249–260. (Foreword)

Oberpriller, J. O., Oberpriller, J. C., Matz, D. G., & Soonpaa, M. H. (1995). Stimulation of proliferative events in the adult amphibian cardiac myocyte. Annals of the New York Academy of Sciences 752:30–46. (Foreword)

O'Brien, R. N., Visaisouk, S., Raine, R., & Alderdice, D. F. (1978). Natural convection: A mechanism of transporting oxygen to incubating salmon eggs. *Journal of the Fisheries Research Board of Canada* 35:1316–1321. (Ch. 18)

O'Brien, T. X., Lee, K. J., & Chien, K. R. (1993). Positional specification of ventricular myosin light chain 2 expression in the primitive murine heart tube. *Proceedings of the National Academy of Sciences USA* 90:5157–5161. (Chs. 3, 10)

Ocklund, S., Tjonneland, A., Larsen, L. N., & Nylund A. (1982). Heart ultrastructure in *Branchionecta paludosa, Artemia salina, Branchipus schaefferi,* and *Streptocephalus* sp. (Crustacea, Anostraca). *Zoomorphology* 101:71–81. (Ch. 11)

O'Donoghue, C. H. (1917). A note on the ductus caroticus and ductus arteriosus and their distribution in the reptiles. *Journal of Anatomy* 51:137–149. (Ch. 14)

Oikawa, S., & Itazawa, Y. (1985). Gill and body surface areas of the carp in relation to body mass, with special reference to the metabolism–size relationship. *Journal of Experimental Biology* 117:1–14. (Ch. 12)

Okada, Y., Katsuda, S., Watanabe, H., & Kakanisha, I. (1993). Collagen synthesis of human arterial smooth muscle cells: Effects of platelet-derived growth factor, transforming growth factor-beta 1 and interleukin-1. *Acta Pathological Japonica* 43:160–167. (Ch. 5)

Okuda, H., Nakanishi, T., Nakazawa, M., & Takao, A. (1987). Effect of isoproterenol on myocardial mechanical function and cyclic AMP content in the fetal rabbit. *Journal of Molecular and Cellular Cardiology* 19:151–157. (Ch. 2)

Olley, P. M., Cocceani, F., & Bodach, F. (1976). E-type prostaglandins a new emergency therapy for certain cyanotic congenital heart malformations. *Circulation* 53:728–731. (Ch. 20)

Ono, H., Akiyama, R., Sakamoto, Y., Pearson, J. T., & Tazawa, H. (in press). Ballistocardiogram of avian eggs determined by an electromagnetic induction coil. *Medical and Biological Engineering and Computing.* (Ch. 15)

Ono, H., Hou, P.-C. L., & Tazawa, H. (1994). Responses of developing chicken embryos to variations of ambient temperature: Noninvasive study of heart rate. *Israel Journal of Zoology* 40:467–480. (Ch. 15)

Orlando, K., & Pinder, A. W. (1995). Larval cardiorespiratory ontogeny and allometry in *Xenopus laevis. Physiological Zoology* 68:63–75. (Chs. 12, 13, 17)

Orlowski, J., and Lingrel, J. B. (1988). Tissue-specific and developmental regulation of rat Na-K-ATPase catalytic a isoform and b subunit mRNAs. *Journal of Biological Chemistry* 263:10436–10442. (Ch. 2)

O'Rourke, M. F. (1982). Ventricular impedance in studies of arterial and cardiac function. *Physiological Reviews* 62:570–623. (Ch. 7)

O'Rourke, M. F., & Taylor, M. G. (1967). Input impedance of the systemic circulation. *Circulation Research* 10:365–380. (Ch. 7)

Orts-Llorca, F., & Jimenez-Collado, J. (1967). Determination of heart polarity (arteriovenous axis) in the chicken embryo. *Roux Archiv Entwicklung* 158:147–163. (Ch. 10)

Osaka, T., and Joyner, R. W. (1991). Developmental changes in calcium currents of rabbit ventricular cells. *Circulation Research* 68:788–796. (Ch. 2)

Oštádal, B., Rychter, Z., & Poupa, O. (1970). Comparative aspects of the development of the terminal vascular bed in the myocardium. *Physiologica Bohemoslovaca* 19:1–7. (Ch. 4)

Owen, R. (1834). On the structure of the heart in the perennibranchiate *Batrachia. Transactions of the Zoological Society of London* 1:213–220. (Ch. 13)

Paff, G. H., Boucek, R. J., & Klopfenstein, H. S. (1948). Experimental heart-block in the chick embryo. *Anatomical Record* 149:217–224. (Ch. 7)

Paganelli, C. V., Sotherland, P. R., Olszowka, A. J., & Rahn, H. (1988). Regional differences in diffusive conductance/perfusion ratio in the shell of the hen's egg. *Respiration Physiology* 71:45–56. (Ch. 18)

Page, C., Rose, M., Yacoub, M., et al. (1992). Antigenic heterogeneity of vascular endothelium. *American Journal of Pathology* 141:673–683. (Ch. 6)

Page, E. (1978). Quantitative ultrastructural analysis in cardiac membrane physiology. *American Journal of Physiology* 63:C147–C158. (Ch. 2)

Page, S. G., & Niedergerke, R. (1972). Structures of physiological interest in the frog heart ventricle. *Journal of Cell Science* 11:179–203. (Ch. 9)

Paine, M. D., & Balon, E. K. (1984). Early development of the northern logperch, *Percina caprodes semifasciata,* according to the theory of saltatory ontogeny. *Environmental Biology of Fishes* 11:173–190. (Ch. 12)

Panganiban, G. E., Reuter, R., Scott, M. P., & Hoffman, F. M. (1990). A *Drosophila* growth factor homologue, *decapentaplegic,* regulates homeotic gene expression within and across germ layers during midgut morphogenesis. *Development* 110:1041–1050. (Ch. 1)

Pappano, A. J. (1977). Ontogenic development of autonomic neuroeffector transmission and transmitter reactivity in embryonic and fetal hearts. *Pharmacological Reviews* 29:3–33. (Chs. 7, 8, 17)

Pardanaud, L., & Dieterlen-Lievre, F. (1993). Emergence of endothelial and hemopoietic cells in the avian embryo. *Anatomy and Embryology* 187:107–114. (Ch. 6)

Pardanaud, L., Yassine, F., & Dieterlen-Lievre, F. (1989). Relationship between vasculogenesis, angiogenesis, and hematopoiesis during avian ontogeny. *Development* 105:473–485. (Ch. 6)

Park, I. S., Michael, L. H., & Driscoll, D. J. (1982). Comparative response of the developing canine myocardium to inotropic agents. *American Journal of Physiology* 242:H13–H18. (Ch. 2)

Parlow, M. H., Bolender, D. L., Kokan-Moore, N. P., & Lough, J. (1991). Localization of bFGF-like proteins as punctate inclusions in the presentation myocardium of the chicken embryo. *Developmental Biology.* 146:139–147. (Ch. 4)

Parmacek, M. S., Ip, H. S., Jung, F., Shen, T., Martin, J. F., Vora, A. J., Olson, E. N., & Leiden, J. M. (1994). A novel myogenic regulatory circuit controls slow/cardiac troponin C gene transcription in skeletal muscle. *Molecular and Cellular Biology* 14:1870–1885. (Ch. 3)

Parmacek, M. S., Vora, A. J., Shen, T., Barr, E., Jung, F., & Leiden, J. M. (1992). Identification and characterization of a cardiac-specific transcriptional regulatory element in the slow/cardiac troponin C gene. *Molecular and Cellular Biology* 12:1967–1976. (Ch. 3)

Patten, B. M., & Kramer, T. C. (1933). The initiation of contraction in the embryonic chick heart. *American Journal of Anatomy* 53:349–375. (Ch. 7)

Paul, R. J., Bihlmayer, S., Colmorgen, M., & Zahler, S. (1994). The open circulatory system of spiders (*Eurypelma californicum, Pholcus phalangioides*): A survey of functional morphology and physiology. *Physiological Zoology* 67:1360–1382. (Ch. 9)

Payne, M. R., Johnson, M. C., Grant, J. W., & Strauss, A. W. (1995). Toward a molecular understanding of congenital heart disease. *Circulation* 91:494–504. (Ch. 21)

Pearson, C. A., Pearson, D., Shibahara, S., Hofsteenge, J., & Chiquet-Ehrismann, R. (1988). Tenascin: cDNA cloning and induction by TGF-β. *European Molecular Biology Organization Journal* 7:2977–2982. (Ch. 5)

Pearson, J. D. (1991). Endothelial cell biology. *Radiology* 179:9–14. (Ch. 6)

Pearson, J. T., Haque, M. A., Hou, P.-C. L., & Tazawa, H. (1996). Developmental patterns of O_2 consumption, heart rate, and O_2 pulse in unturned eggs. *Respiration Physiology* 103:53–67. (Ch. 15)

Pegg, W., and Michalak, M. (1987). Differentiation of sarcoplasmic reticulum during cardiac myogenesis. *American Journal of Physiology* 252:H22–H31. (Ch. 2)

Pegram, B. L., Trippodo, N. C., Natsume, T., Kardon, M. B., Frohlich, E. D., Cole, F. E., & MacPhee, A. A. (1986). Hemodynamic effects of atrial natriuretic hormone. *Federation Proceedings* 45:2382–2386. (Ch. 7)

Pelouch, V. (1995). Molecular aspects of regulation of cardiac contraction. *Physiological Research* 44:53–60. (Ch. 19)

Pelster, B. (1991). Ontogeny of the central circulatory system in the little skate, *Raja erinacea. Verhandlungen der Deutschen Zoologischen Gesellschaft* 84:419. (Ch. 17)

Pelster, B., & Bemis, W. E. (1991). Ontogeny of heart function in the little skate *Raja erinacea. Journal of Experimental Biology* 156:387–398. (Chs. 12, 17)

Pelster, B., & Bemis, W. E. (1992). Structure and function of the external gill filaments of embryonic skates (*Raja erinacea*). *Respiration Physiology* 89:1–13. (Chs. 12, 17)

Pelster, B., & Burggren, W. W. (1991). Central arterial hemodynamics in larval bullfrogs (*Rana catesbeiana*): Developmental and seasonal influences. *American Journal of Physiology* 260:R240–R246. (Ch. 13)

Pelster, B., & Burggren, W. W. (1996). Disrupting hemoglobin oxygen transport does not impact oxygen-dependent physiological processes in developing embryos of the zebrafish (*Danio rerio*). *Circulation Research* 79:358–362. (Chs. 12, 13, 17)

Pelster, B., Burggren, W. W., Petrou, S., & Wahlqvist, I. (1993). Developmental changes in the acetylcholine influence on heart muscle of *Rana catesbeiana: In situ* and *in vitro* effects. *Journal of Experimental Zoology* 267:1–8. (Chs. 8, 13)

Percy, L. R., & Potter, I. C. (1986). Description of the heart and associated blood vessels in larval lampreys. *Journal of Zoology* 208A:479–492. (Ch. 12)

Percy, L. R., & Potter, I. C. (1991). Aspects of the development and functional morphology of the pericardia, heart, and associated blood vessels of lampreys. *Journal of Zoology* 223:49–66. (Ch. 12)

Peters, K. G., De Vires, C., & Williams, L. T. (1993). Vascular endothelial growth factors receptor expression during embryogenesis and tissue repair suggests a role in endothelial differentiation and blood vessel growth. *Proceedings of the National Academy of Sciences USA* 90:8915–8919. (Ch. 6)

Peterson, R. H. (1975). Pectoral fin and opercular movements of Atlantic salmon (*Salmo salar*) alevins. *Journal of Fisheries Resources Board of Canada* 32:643–648. (Ch. 12)

Petery, L. B., Jr., & van Mierop, L. H. S. (1977). Evidence for the presence of adrenergic receptors in 3-day-old chick embryo. *American Journal of Physiology* 232:H250–H254. (Ch. 8)

Pettit, T. N., & Whittow, G. C. (1983). Water loss from pipped wedge-tailed shearwater eggs. *Condor* 85:107–109. (Ch. 15)

Pexieder, T. (1986). Standardized method for study of normal and abnormal cardiac development in chick, rat, mouse, dog, and human embryos. *Teratology* 33:91C–92C. (Ch. 7)

Pexieder, T. (1995). The conotruncus and its septation at the advent of the molecular biology era. In *Developmental Mechanisms of Heart Disease,* ed. E. B. Clark, R. R. Markwald, & A. Takao. Mount Kisco, N.Y.: Futura, pp. 227–247. (Ch. 19)

Pexieder, T., Blanc, O., Pelouch, V., Ostadalova, I., Milerova, M., & Ostadal, B. (1995). Late fetal development of retinoic acid induced transposition of great arteries: Morphology, physiology, and biochemistry. In *Developmental Mechanisms of Heart Disease,* ed. E. B. Clark, R. R. Markwald, & A. Takao, pp. 297–307. Mount Kisco, N.Y.: Futura. (Ch. 19)

Pexieder, T., Christen, Y., Vuillemin, M., & Patterson, D. F. (1984). Comparative morphometric analysis of cardiac organogenesis in chick, mouse, and dog embryos. In *Congenital Heart Disease: Causes and Processes,* ed. A. Takao & I. Nora. Mount Kisco, N.Y.: Futura, pp. 423–438. (Ch. 7)

Pexieder, T., & Janecek, P. (1984). Organogenesis of the human embryonic and early fetal heart as studied by microdissection and SEM. In *Congenital Heart Disease: Causes and Processes,* ed. A. Takao & I. Nora. Mount Kisco, N.Y.: Futura, pp. 401–442. (Ch. 7)

Pexieder, T., Pfizenmaier Rousseil, M., & Prados-Frutos, J. C. (1992). Prenatal pathogenesis of the transposition of great arteries. In *Transposition of the Great Arteries 25 Years after Rashkind Balloon Septostomy,* ed. Vogel, K. Bühlmeyer. Darmstadt: Steinkopff; New York: Springer, pp. 11–27. (Ch. 19)

Pickering, J. W. (1893). Observations on the physiology of the embryonic heart. *Journal of Physiology* 14:383–466. (Ch. 10)

Pierce, G. F., Tarpley, J. E., Yanagihara, D., Mustoe, T. A., Fox, G. M., and Thomason, A. (1992). Platelet-derived growth factor (BB homodimer), transforming growth factor-beta 1, and basic fibroblast growth factor in dermal wound healing: Neovessel and matrix formation and cessation of repair. *American Journal of Pathology* 140:1375–1388. (Ch. 5)

Piiper, J. (1961). Unequal distribution of pulmonary diffusing capacity and the alveolar-arterial PO_2 difference: Theory. *Journal of Applied Physiology* 16:493–498. (Ch. 18)

Piiper, J. (1993). Medium–blood gas exchange: Diffusion, distribution, and shunt. In *The Vertebrate Gas Transport Cascade: Adaptations to Environment and Mode of Life,* ed. J. E. P. W. Bicudo, pp. 106–120. Boca Raton, Fla.: CRC Press. (Ch. 18)

Piiper, J., Gatz, R. N., & Crawford, E. C. (1976). Gas transport characteristics in an exclusively skin-breathing salamander, *Desmognathus fuscus* (Plethodontidae). In *Respiration of Amphibious Vertebrates,* ed. G. M. Hughes, pp. 339–356. London: Academic Press. (Ch. 18)

Piiper, J., & Scheid, P. (1982). Models for a comparative functional analysis of gas exchange organs in vertebrates. *Journal of Applied Physiology: Respiratory, Environmental, and Exercise Physiology* 53:1321–1329. (Ch. 18)

Piiper, J., & Scheid, P. (1984). Respiratory gas transport system: Similarities between avian embryos and lungless salamanders. In *Respiration and Metabolism of Embryonic Vertebrates,* ed. R. S. Seymour, pp. 181–191. Dordrecht: Junk. (Ch. 18)

Piiper, J., & Scheid, P. (1989). Gas exchange: Theory, models, and experimental data. In *Comparative Pulmonary Physiology,* ed. S. C. Wood, pp. 369–416. New York: Marcel Dekker. (Ch. 18)

Piiper, J., Tazawa, H., Ar, A., & Rahn, H. (1980). Analysis of chorioallantoic gas exchange in the chick embryo. *Respiration Physiology* 39:273–284. (Ch. 18)

Pinder, A. W. (1985). Respiratory physiology of two amphibians, *Rana pipiens* and *Rana catesbeiana.* Ph.D. diss., University of Massachusetts, Amherst. (Ch. 13)

Pinder, A. W., & Burggren, W. (1983). Respiration during chronic hypoxia and hyperoxia in larval and adult bullfrogs (*Rana catesbeiana*). II, Changes in respiratory properties of whole blood. *Journal of Experimental Biology* 105:205–229. (Chs. 13, 17)

Pinder, A. W., Clemens, D., & Feder, M. E. (1991). Gas exchange in isolated perfused frog skin as a function of perfusion rate. *Respiration Physiology* 85:1–14. (Ch. 18)

Pinder, A. W., & Feder, M. E. (1990). Effect of boundary layers on cutaneous gas exchange. *Journal of Experimental Biology* 143:67–80. (Ch. 12)

Pinder, A. W., & Friet, S. C. (1994). Oxygen transport in egg masses of the amphibians *Rana sylvatica* and *Ambystoma maculatum:* Convection, diffusion, and oxygen production by algae. *Journal of Experimental Biology* 197:17–30. (Ch. 18)

Pinson, C. W., Morton, M. J., & Thornburg, K. L. (1987). An anatomic basis for fetal right ventricular dominance and arterial pressure sensitivity. *Journal of Developmental Physiology* 9:253–269. (Ch. 16)

Pinson, C. W., Morton, M. J., & Thornburg, K. L. (1991). Mild pressure loading alters right ventricular function in fetal sheep. *Circulation Research* 68:947–957. (Ch. 16)

Pittman, K., Skiftesvik, A. B., & Berg, L. (1990). Morphological and behavioral development of halibut, *Hippoglossus hippoglossus* (L.) larvae. *Journal of Fish Biology* 37:455–472. (Ch. 12)

Poelmann, R. E., et al. (1990). Endothelial cell migration in the developing blood vessel system of chicken/quail chimeras. *European Journal of Cell Biology* 31:48–49. (Ch. 6)

Poelmann, R. E., Gittenberger–de Groot, A. C., Mentink, M. M. T., Bökenkamp, R., & Hogers, B. (1993). Development of the cardiac vascular endothelium, studied with anti-endothelial antibodies in chicken–quail chimeras. *Circulation Research* 73:559–568. (Ch. 4)

Pohlman, A. G. (1909). The course of blood through the heart of the fetal mammal, with a note on the reptilian and amphibian circulations. *Anatomical Record* 3:75–109. (Ch. 16)

Poole, T. J., & Coffin, D. J. (1989). Vasculogenesis and angiogenesis: Two distinct morphogenetic mechanisms establish embryonic vascular patterns. *Journal of Experimental Zoology* 251:224–231. (Ch. 6)

Porter, G. A., & Bankston, P. W. (1987a). Maturation of myocardial capillaries in the fetal and neonatal rat: An ultrastructural study with a morphometric analysis of the vesicle populations. *American Journal of Anatomy* 178:116–125. (Ch. 4)

Porter, G. A., & Bankston, P. W. (1987b). Functional maturation of the capillary wall in the fetal and neonatal rat heart: Permeability characteristics of developing myocardial capillaries. *American Journal of Anatomy* 180:323–331. (Ch. 4)

Postelthwait, J. H., Johnson, S. L., Midson, C. N., Talbot, W. S., Gates, M., Ballinger, E. W., Africa, D., Andrews, R., Carl, T., Eisen, J. S., Horne, S., Kimmel, C. B., Hutchinson, M., Johnson, M., & Rodriguez, A. (1994). A genetic linkage map for zebrafish. *Science* 264:699–703. (Ch. 1)

Potts, J. D., & Runyan, R. B. (1989). Epithelial–mesenchymal cell transformation in the embryonic heart can be mediated, in part, by transforming growth factor β. *Developmental Biology* 134:392–401. (Ch. 5)

Potts, J. R., & Campbell, I. D. (1994). Fibronectin structure and assembly. *Current Opinions in Cell Biology* 6:648–655. (Ch. 5)

Powell, F. L. (1993). Comparing the effects of diffusion and heterogeneity on vertebrate gas exchange. In *The Vertebrate Gas Transport Cascade: Adaptations to Environment and Mode of Life,* ed. J. E. P. W. Bicudo, pp. 121–131. Boca Raton, Fla.: CRC Press. (Ch. 18)

Power, G. G., Dale, P. S., & Nelson, P. S. (1981). Distribution of maternal and fetal blood flow within cotyledons of the sheep placenta. *American Journal of Physiology* 241:H486–H496. (Ch. 18)

Prange, H. D., & Ackerman, R. A. (1974). Oxygen consumption and mechanisms of gas exchange of green turtle (*Chelonia mydas*) eggs and hatchlings. *Copeia,* pp. 759–763. (Ch. 18)

Prasad, M. S., & Prasad, P. (1984). Morphometrics of water–blood diffusion barrier at the secondary gill lamellae during early life of *Channa punctatus* (Bloch). *Archives of Biology* 95:91–100. (Ch. 12)

Price, J. W. (1934a). The embryology of the whitefish, *Coregonus clupeaformis,* pt. 1. *Ohio Journal of Science* 34:287–305. (Ch. 12)

Price, J. W. (1934b). The embryology of the whitefish, *Coregonus clupeaformis,* pt. 2. *Ohio Journal of Science* 34:399–414. (Ch. 12)
Price, K. M., Littler, W. A., & Cummins, P. (1980). Human atrial and ventricular myosin light-chain subunits in the adult and during development. *Biochemical Journal* 191:571–580. (Ch. 3)
Price, R. L., Nakagawa, M., Terracio, L., and Borg, T. K. (1992). Ultrastructural localization of laminin on in vivo embryonic, neonatal, and adult rat cardiac myocytes and in early rat embryos raised in whole embryo culture. *Journal of Histochemistry and Cytochemistry* 40:1373–1381. (Ch. 5)
Prober, J. S., & Cotran, R. S. (1991). Immunologic interactions of T lymphocytes with vascular endothelium. *Advances in Immunology* 50:261–303. (Ch. 6)
Protas, L. L., & Leontieva, G. R. (1992). Ontogeny of cholinergic and adrenergic mechanisms in the frog (*Rana temporaria*) heart. *American Journal of Physiology* 262:R150–R161. (Chs. 13, 17)
Putkey, J. A., Sweeney, H. L., & Campbell, S. T. (1989). Site-directed mutation of the trigger calcium-binding sites in cardiac troponin C. *Journal of Biological Chemistry* 264:12370–12378. (Ch. 3)
Putnam, J. L. (1975). Septation in the ventricle of the heart of *Siren intermedia. Copeia,* pp. 773–774. (Ch. 13)
Putnam, J. L., & Dunn, J. F. (1978). Septation in the ventricle of the heart of *Necturus maculosus. Herpetologica* 34:292–297. (Ch. 13)
Putnam, J. L., & Parkerson, J. B. (1985). Anatomy of the heart of the Amphibia. II, *Cryptobranchus alleganiensis. Herpetologica* 41:287–298. (Ch. 13)
Rabinovitch, M., Bothwell, T., Hayakawa, B. N., et al. (1986). Pulmonary artery endothelial abnormalities in patients with congenital heart defects and pulmonary hypertension. *Laboratory Investigation* 55:632–653. (Ch. 6)
Ragsdale, F. R., Imel, K. M., Nilsson, E. E., & Ingermann, R. L. (1993). Pregnancy-associated factors affecting organic phosphate levels and oxygen affinity of garter snake red cells. *General and Comparative Endocrinology* 91:181–188. (Ch. 14)
Ragsdale, F. R., & Ingermann, R. L. (1991). Influence of pregnancy on the oxygen affinity of red cells from the northern Pacific rattlesnake *Crotalus viridis oreganus. Journal of Experimental Biology* 159:501–505. (Ch. 14)
Ragsdale, F. R., & Ingermann, R. L. (1993). Biochemical bases for difference in oxygen affinity of maternal and fetal red blood cells of rattlesnakes. *American Journal of Physiology* 264:R481–R486. (Ch. 17)
Rahn, H., Ar, A., & Paganelli, C. V. (1979). How bird eggs breathe. *Scientific American* 240:46–55. (Ch. 15)
Rahn, H., Matalon, S., & Sotherland, P. R. S. (1985). Circulatory changes and oxygen delivery in the chick embryo prior to hatching. In *Cardiovascular Shunts,* ed. K. Johansen & W. Burggren, pp. 199–215. Copenhagen: Munksgaard. (Ch. 15)
Rahn, H., & Paganelli, C. V. (1990). Gas fluxes in avian eggs: Driving forces and the pathway for exchange. *Comparative Biochemistry and Physiology* 95A:1–15. (Ch. 18)
Rahn, H., Poturalski, S. A., & Paganelli, C. V. (1990). The acoustocardiogram: A noninvasive method for measuring heart rate of avian embryos *in ovo. Journal of Applied Physiology* 69:1546–1548. (Ch. 15)
Rakusan, D., & Poupa, O. (1963). Changes in the diffusion distance in the rat heart muscle during development. *Physiologica Bohemoslovaca* 12:220–226. (Ch. 4)
Rakusan, K., Jelinek, J., Korecky, B., Soukupova, M., & Poupa, O. (1964). Postnatal development of muscle fibers and capillaries in the rat heart. *Physiologica Bohemoslovaca* 14:32–37. (Ch. 4)
Rakusan, K., & Turek, Z. (1985). Protamine inhibits capillary formation in growing rat hearts. *Circulation Research* 57(3):393–398. (Ch. 4)
Randall, D. (1982). The control of respiration and circulation in fish during exercise. *Journal of Experimental Biology* 100:275–288. (Ch. 17)
Randall, D., Lin H., & Wright, P. A. (1991). Gill water flow and the chemistry of the boundary layer. *Physiological Zoology* 64:26–38. (Ch. 12)
Randall, D. J., & Davie, P. S. (1980). The hearts of urochordates and cephalochordates. In *Hearts and Heart-like Organs,* vol. 1, ed. G. H. Bourne, pp. 41–59. New York: Academic Press. (Chs. 9, 11)

Randall, D. J., Farrell, A. P., Haswell, M. S., & Burggren, W. W. (1981). *Evolution of Air Breathing in Vertebrates.* Cambridge: Cambridge University Press. (Ch. 9)
Ratajska, A., Torry, R. J., Kitten, G. T., Kolker, S. J., & Tomanek, R. J. (1995). Modulation of cell migration and vessel formation by vascular endothelial growth factor and basic fibroblast growth factor in cultured embryonic heart. *Developmental Dynamics* 203:399–407. (Ch. 4)
Raven, C. P. (1966). *Morphogenesis: The Analysis of Molluscan Development,* 2nd ed. New York: Pergamon Press. (Ch. 11)
Reaume, A. G., de Sousa, P. A., Kulkarni, S., Langille, B. L., Zhu, D., Davies, T. C., Juneja, S. C., Kidder, G. M., & Rossant, J. (1995). Cardiac malformation in neonatal mice lacking *connexin43. Science* 267:1831–1834. (Chs. 1, 20)
Reddy, V. M., Liddicoat, J. R., Klein, J. R., Wampler, R. K., & Hanley, F. L. (in press). Chronic survival to term after cardiac bypass from midline sternotomy using hemopump in fetal lambs. *Journal of Thoracic and Cardiovascular Surgery.* (Ch. 20)
Reddy, V. M., Liddicoat, J. R., McElhinney, D. B., et al. (1995a). Results of routine primary repair of tetralogy of fallot in neonates and infants under 3 months of age. *Annals of Thoracic Surgery* 60:S592–S596. (Ch. 20)
Reddy, V. M., Meyrick, B., Wong, J., et al. (1995b). *In utero* placement of aortopulmonary shunts: A model of postnatal pulmonary hypertension with increased pulmonary blood flow in lambs. *Circulation* 92:606–613. (Ch. 20)
Redmond, J. R., Jorgensen, D. D., & Bourne, G. B. (1972). Circulatory physiology of *Limulus. Progress in Clinical and Biological Research* 81:133–146. (Ch. 11)
Regenass, S., Resink, T. J., Kern, F., Buhler, F. R., & Hahn, A. W. (1994). Angiotensin-II-induced expression of laminin complex and laminin-A-chain-related transcripts in vascular smooth muscle cells. *Journal of Vascular Research* 31:163–172. (Ch. 5)
Reinach, F. C., Masaki, T., Shafiq, S., Obinata, T., & Fischman, D. A. (1982). Isoforms of C-protein in adult chicken skeletal muscle: Detection with monoclonal antibodies. *Journal of Cell Biology* 95:78–84. (Ch. 3)
Reller, M. D., Burson, M. A., Lohr, J. L., Morton, M. J., & Thornburg, K. L. (1995). Nitric oxide is an important determinant of coronary flow at rest and during hypoxemic stress in fetal lambs. *American Journal of Physiology* 269:H2074–H2081. (Ch. 16)
Reller, M. D., Morton, M. J., Giraud, G. D., Wu, D. E., & Thornburg, K. L. (1992a). Maximal myocardial blood flow is enhanced by chronic hypoxemia in late gestation fetal sheep. *American Journal of Physiology* 263:H1326–H1329. (Ch. 16)
Reller, M. D., Morton, M. J., Giraud, G. D., Wu, D. E., & Thornburg, K. L. (1992b). Severe right ventricular pressure loading in fetal sheep augments global myocardial blood flow to submaximal levels. *Circulation* 86:581–588. (Ch. 16)
Reller, M. D., Morton, M. J., Reid, D. L., & Thornburg, K. L. (1987). Fetal lamb ventricles respond differently to filling and arterial pressures and to in utero ventilation. *Pediatric Research* 22:621–626. (Ch. 16)
Renaud, J. F., Fosset, M., Kazazoglou, T., Lazdunski, M., & Schmid, A. (1989). Appearance and function of voltage-dependent Ca^{2+} channels during pre- and postnatal development of cardiac and skeletal muscles. *Annals of the New York Academy of Sciences USA* 560:418–425. (Ch. 2)
Renaud, J. F., Kazazoglou, T., Schmid, A., Romey, G., & Lazdunski, M. (1984). Differentiation of receptor sites of [^{3}H] nitrendipine in chick hearts and physiological relation to the slow Ca^{2+} channel and to excitation–contraction coupling. *European Journal of Biochemistry* 139:673–681. (Ch. 2)
Renter, D. J., Schilingemann, R. O., Westphal, J. R., et al. (1993). Angiogenesis in wound healing and tumor metastasis. *Behring Institut Mitteilungen* (West Germany) 92:258–272. (Ch. 6)
Resnik, D. (1995). Developmental constraints and patterns: Some pertinent distinctions. *Journal of Theoretical Biology* 173:231–240. (Ch. 9)
Risau, W. (1991a). Embryonic angiogenesis factors. *Pharmacology and Therapeutics* 51:371–376. (Ch. 6)
Risau, W. (1991b). Growth factors and matrix influences. In *The Development of the Vascular System,* ed. R. N. Feinber, G. K. Sherer, & R. Auerback, pp. 58–68. Issues Biomed, vol. 14. Basel: Karger. (Ch. 13)
Risau, W., & Lemmon, V. (1988). Changes in the vascular extracellular matrix during embryonic vasculogenesis and angiogenesis. *Developmental Biology* 125:441–450. (Chs. 4, 6)

Risau, W., Sariola, H., Zerwes, H. G., et al. (1988). Vasculogenesis and angiogenesis in embryonic stem cells derived from embryoid bodies. *Development* 102:471–478. (Ch. 6)

Roark, E. F., Keene, D. R., Haudenschild, C. C., Godyna, S., Little, C. D., & Argraves, W. S. (1995). The association of human fibulin-1 with elastic fibers: An immunohistological, ultrastructural, and RNA study. *Journal of Histochemistry and Ctyochemistry* 43:401–411. (Ch. 5)

Robb, J. S. (1965). *Comparative Basic Cardiology.* New York: Grune & Stratton. (Ch. 9)

Roberts, J. T., & Wearn, J. T. (1941). Quantitative changes in the capillary muscle relationship in human hearts during normal growth and hypetrophy. *American Heart Journal* 21:617–633. (Ch. 4)

Roberts, R., Marian, A. J., & Bachinski, L. (1995). Overview: Application of molecular biology to medical genetics. In *Developmental Mechanisms of Heart Disease,* ed. E. B. Clark, R. R. Markwald, & A. Takao, pp. 87–111. Mount Kisco, N.Y.: Futura. (Ch. 19)

Roberts, R. J., Bell, M., & Young, H. (1973). Studies on the skin of plaice (*Pleuronectes platessa* L.). II, The development of larval plaice skin. *Journal of Fish Biology* 5:103–108. (Ch. 12)

Rogers, T. B., and Lokuta, A. J. (1994). Angiotensin II signal transduction pathways in the cardiovascular system. *Trends in Cardiovascular Medicine* 4:110–116. (Ch. 5)

Rolph, T. P., Jones, C. T., & Parry, D. (1982). Ultrastructural and enzymatic development of fetal guinea pig heart. *American Journal of Physiology* 243:H87–H93. (Ch. 2)

Romanoff, A. L. (1960a). *The Avian Embryo: Structural and Functional Development.* New York: Macmillan. (Ch. 15)

Romanoff, A. L. (1960b). The Heart. In *The Avian Embryo: Structural and Functional Development,* ed. A. L. Romanoff, pp. 679–780. New York: Macmillan. (Ch. 17)

Romanoff, A. L. (1967). *Biochemistry of the Avian Embryo.* New York: John Wiley & Sons. (Chs. 7, 15)

Rombough, P. J. (1986). Mathematical model for predicting the dissolved oxygen requirements of steelhead (*Salmo gairdneri*) embryos and alevins in hatchery incubators. *Aquaculture* 59:119–137. (Ch. 17)

Rombough, P. J. (1988a). Growth, aerobic metabolism, and dissolved oxygen requirements of embryos and alevins of steelhead, *Salmo gairdneri. Canadian Journal of Zoology* 66:651–660. (Ch. 17)

Rombough, P. J. (1988b). Respiratory gas exchange, aerobic metabolism, and effects of hypoxia during early life. In *Fish Physiology,* ed. W. S. Hoar & D. J. Randall, vol. IIA, pp. 59–161. New York: Academic Press. (Chs. 12, 17, 18)

Rombough, P. J. (1992). Intravascular oxygen tensions in cutaneously respiring rainbow trout (*Oncorhynchus mykiss*) larvae. *Comparative Biochemistry and Physiology* 101A:23–27. (Chs. 12, 17)

Rombough, P. J. (1994). Cardiovascular development in fishes. *Physiologist,* abstract A-44. (Ch. 1)

Rombough, P. J., & Moroz, B. M. (1990). The scaling and potential importance of cutaneous and branchial surfaces in respiratory gas exchange in young chinook salmon (*Oncorhynchus tshawytscha*). *Journal of Experimental Biology* 154:1–12. (Ch. 12)

Rombough, P. J., & Ure, D. (1991). Partitioning of oxygen uptake between cutaneous and branchial surfaces in larval and young juvenile chinook salmon *Oncorhynchus tshawytscha. Physiological Zoology* 64:717–727. (Chs. 12, 17, 18)

Romer, A. S. (1962). *The Vertebrate Body.* Philadelphia: Saunders. (Ch. 9)

Rongish, B. J., Hinchman, G., Doty, M. K., Baldwin, H. S., & Tomanek, R. J. (1996). Relationship of the extracellular matrix to coronary neovascularization during development. *Journal of Molecular and Cellular Cardiology* 28:2203–2215. (Ch. 4)

Rongish, B. J., Torry, R. J., Tucker, D. C., & Tomanek, R. J. (1994). Neovascularization of embryonic rat hearts cultured in oculo closely mimics in utero coronary vessel development. *Journal of Vascular Research* 31:205–215. (Ch. 4)

Rose, V., & Clark, L. E. (1992). Etiology of congenital heart diseases. In *Neonatal Heart Disease,* ed. R. M. Freedom, L. N. Benson, & J. F. Smallhorn, pp. 3–13. New York: Springer Verlag. (Ch. 20)

Rosen, M. R., & Danilo, P., Jr. (1992). Developmental electrophysiology of the heart. In *Fetal and Neonatal Physiology,* ed. R. A. Polin & W. W. Fox, pp. 656–665. Philadelphia: Saunders. (Ch. 2)

Rosenbloom, J., Abrams, W. R., & Mecham, R. (1993). Extracellular matrix 4: The elastic fiber. *FASEB Journal* 7:1208–1218. (Ch. 5)

Rosenquist, G. C., & DeHaan, R. L. (1966). Migration of precardiac cells in the chick embryo: A radioautographic study. *Carnegie Institute of Washington Contributions to Embryology* 38:111–121. (Ch. 10)

Rosenthal, H., & Alderdice, D. F. (1976). Sublethal effects of environmental stressors, natural and pollutional, on marine fish eggs and larvae. *Journal of the Fisheries Research Board of Canada* 33:2047–2065. (Ch. 17)

Ross, S. A., Ahrens, R. A., & De Luca, L. M. (1994). Retinoic acid enhances adhesiveness, laminin, and integrin beta 1 synthesis, and retinoic acid receptor expression in F9 teratocarcinoma cells. *Journal of Cellular Physiology* 159:263–273. (Ch. 5)

Roughton, F. J. W., & Forster, R. E. (1957). Relative importance of diffusion and chemical reaction rates in determining rate of exchange of gases in the human lung, with special reference to time diffusing capacity of pulmonary membrane and volume of blood in the lung capillaries. *Journal of Applied Physiology* 11:290–302. (Ch. 18)

Rovner, A. S., McNally, E. M., & Leinwand, L. A. (1990). Complete cDNA sequence of rat atrial myosin light chain 1: Patterns of expression during development and with hypertension. *Nucleic Acids Research* 18:1581–1586. (Ch. 3)

Rubanyi, G. M., Freay, A. D., Kauser, K., et al. (1990). Mechanoreception by the endothelium: Mediators and mechanism of pressure- and flow-induced vascular responses. *Blood Vessels* 27:246–257. (Ch. 6)

Rubanyi, G. M., Romero, J. C., & Vanhoutte, P. M. (1986). Flow-induced release of endothelium-derived relaxing factor. *American Journal of Physiology* 250:H1145–H1149. (Ch. 6)

Rudolph, A. M. (1970). Changes in circulation after birth: Their importance in congenital heart disease. *Circulation* 41:343–348. (Ch. 20)

Rudolph, A. M., & Heymann, M. A. (1967). Circulation of the fetus in utero. *Circulation Research* 21:163–184. (Ch. 20)

Rudolph, A. M., & Heymann, M. A. (1970). Circulatory changes during growth in the fetal lamb. *Circulation Research* 26:289–299. (Ch. 16)

Rugendorff, A., Younossi-Hartenstein, A., & Hartenstein, V. (1994). Embryonic origin and differentiation of the *Drosophila* heart. *Roux's Archives of Developmental Biology* 203:266–280. (Ch. 1)

Rugh, R. (1956). Early development of the fish. In *Laboratory Manual of Vertebrate Embryology.* Minneapolis: Burgess. (Ch. 12)

Ruiter, D. J., Schlingemann, R. O., Westphal, J. R., et al. (1993). Angiogenesis in wound healing and tumor metastasis. *Behring Institut Mitteilungen* (West Germany) 92:258–272. (Ch. 6)

Runyan, R. B., & Markwald, R. R. (1983). Invasion of mesenchyme into three dimensional gels: A regional and temporal analysis of interaction in embryonic heart tissue. *Developmental Biology* 95:108–114. (Ch. 5)

Ruppert, E. E., & Balser, E. J. (1986). Nephridia in the larvae of hemichordates and echinoderms. *Biological Bulletin* 171:188–196. (Ch. 11)

Ruppert, E.E., & Barnes, R. D. (1994). *Invertebrate Zoology.* 6th ed. Philadelphia: Saunders. (Ch.11)

Ruppert, E. E., & Carle, K. J. (1983). Morphology of metazoan circulatory systems. *Zoomorphology* 103:193–208. (Ch. 11)

Ruppert, E. E., & Smith, P. R. (1988). The functional organization of filtration nephridia. *Biological Reviews* 63:231–258. (Ch. 11)

Ruzicka, D. L., & Schwartz, R. J. (1988). Sequential activation of a-actin genes during avian cardiogenesis: Vascular smooth muscle a-actin gene transcripts mark the onset of cardiomyocyte differentiation. *Journal of Cell Biology* 107:2575–2586. (Chs. 3, 10)

Rychter, Z. (1962). Experimental morphology of the aortic arches and the heart loop in chick embryo. *Advances in Morphogenesis* 2:333–371. (Chs. 6, 7)

Rychter, Z., & Lemez, L. (1965). Changes in localization in aortic arches of laminar blood streams of main venous trunks to heart after exclusion of vitelline vessels on second day of incubation. *Federaction Proceedings, Translation Supplement (United States),* 24:815–820. Originally published in *Ceskoslovenska Morfologie* (1964) 12:268–275. (Ch. 19)

Rychter, Z., & Oštádal, B. (1971). Mechanisms of the development of coronary arteries in chick embryo. *Folia Morphologica* (Prague) 19:113–124. (Ch. 4)

Sabik, J. F., Heinemann, M. K., Assad, R. S., et al. (1994). High dose steroids prevent placental dysfunction after fetal cardiac bypass. *Journal of Thoracic and Cariovascular Surgery* 107:116–125. (Ch. 20)

Sabry, M. A., & Dhoot, G. K. (1989). Identification and pattern of expression of a developmental isoform of troponin I in chicken and rat cardiac muscle. *Journal of Muscle Research and Cell Motility* 10:85–91. (Ch. 3)

Sagawa, K., Maughan, L., Suga, H., & Sunagawa, K. (1988a). Physiologic determinants of the ventricular pressure–volume relationship. In *Cardiac Contraction and the Pressure–Volume Relationship,* pp. 8–28. New York: Oxford University Press. (Ch. 7)

Sagawa, K., Maughan, L., Suga, H., & Sunagawa, K. (1988b). Energetics of the heart. In *Cardiac Contraction and the Pressure–Volume Relationship,* pp. 171–231. New York: Oxford University Press. (Ch. 7)

Saggin, L., Gorza, L., Ausoni, S., & Schiaffino, S. (1989). Troponin I switching in the developing heart. *Journal of Biological Chemistry* 264:16299–16302. (Ch. 3)

Sakamoto, Y., Ono, H., Haque, M. A., Pearson, J., & Tazawa, H. (1996). Two-dimensional cardiogenic ballistic movements of avian eggs. *Medical and Biological Engineering and Computing* 33:611–614. (Ch. 15)

Salazar del Rio, J. (1974). Influence of extrinsic factors on the development of the bulboventricular loop of the chick embryo. *Journal of Embryology and Experimental Morphology* 31: 199–206. (Ch. 10)

Salmivirta, M., Mali, M., Heino, J., Hermonen, J., & Jalkanen, M. (1994). A novel laminin-binding form of syndecan-1 (cell surface proteoglycan) produced by syndecan-1 cDNA-transfected NIH-3T3 cells. *Experimental Cell Research* 215:180–188. (Ch. 5)

Sanchez-Chapula, J., Elizalde, A., Navarro-Polanco, R., & Barajas, H. (1994). Differences in outward currents between neonatal and adult rabbit ventricular cells. *American Journal of Physiology* 266:H1184–H1194. (Ch. 2)

Sartorelli, V., Kurabayashi, M., & Kedes, L. (1993). Muscle-specific gene expression: A comparison of cardiac and skeletal muscle transcription strategies. *Circulation Research* 72:925–931. (Ch. 3)

Sasaki, T., Inui, M., Kimura, Y., Kuzuya, T., & Tada, M. (1992). Molecular mechanism of regulation of Ca^{2+} pump ATPase by phospholamban in cardiac sarcoplasmic reticulum. *Journal of Biological Chemistry* 267:1674–1679. (Ch. 2)

Satchell, G. H. (1991). *Physiology and Form of Fish Circulation.* Cambridge: Cambridge University Press. (Ch. 9)

Sater, A. K., & Jacobson, A. G. (1990a). The restriction of the heart morphogenetic field in *Xenopus laevis. Developmental Biology* 140:328–336. (Ch. 10)

Sater, A. K., & Jacobson, A. G. (1990b). The role of the dorsal lip in the induction of heart mesoderm in *Xenopus laevis. Development* 108:461–470. (Ch. 10)

Satin, J., Fujii, S., & DeHaan, R. L. (1988). Development of cardiac beat rate in early chick embryos is regulated by regional cues. *Developmental Biology* 129:103–113. (Ch. 10)

Sato, T. N., Tozawa, Y., Deutsch, U., Wolburg-Buchholz, K., Fujiwara, Y., Gendron-Maguire, M., Gridley, T., Wolburg, G., Risau, W., & Qin, Y. (1995). Distinct roles of the receptor tyrosine kinases *Tie-1* and *Tie-2* in blood vessel formation. *Nature* 376:70–74. (Ch. 1)

Sawai, S., Shimono, A., Wakamatsu, Y., Palmes, C., Hanaoka, K., & Kondoh, H. (1993). Defects of embryonic organogenesis resulting from targeted disruption of the *N-myc* gene in the mouse. *Development* 117:1445–1455. (Ch. 1)

Sawaya, P. (1947). Metabolismo respiratório de anfíbio Gymnophiona, *Typhlonectes compressicauda. Boletins da Faculdade de Filosofia, Ciéncias, e Letras, Universidade de Sãao Paulo, Serie Zoologia,* 12:51–56. (Ch. 13)

Scammon, R. E. (1911). Normal plates of the development of *Squalus acanthias.* In *Normentafeln zur Entwicklungs-geschichte der Wiebeltieve,* ed. F. Keibel, vol. 12, pp. 1–140. Jena: Gustav Fischer. (Ch. 12)

Scheid, P. (1987). Role of modelling in physiology. In *New Directions in Ecological Physiology,* ed. M. E. Feder, A. F. Bennett, W. W. Burggren, & R. B. Huey. New York: Cambridge University Press. (Ch. 18)

Scheid, P., Hook, C., & Piiper, J. (1986). Model for analysis of counter-current gas transfer in fish gills. *Respiration Physiology* 64:365–374. (Ch. 18)

Schier, J. J., & Adelstein, R. S. (1982). Structural and enzymatic comparison of human cardiac muscle myosins isolated from infants, adults, and patients with hypertrophic cardiomyopathy. *Journal of Clinical Investigation* 69:816–825. (Ch. 3)

Schindler, J. F., & Hamlett, W. C. (1993). Maternal–embryonic relations in viviparous teleosts. *Journal of Experimental Zoology* 266:378–393. (Ch. 18)

Schiro, J. A., Chan, B. M. C., & Roswit, W. T. (1991). Integrin a2β1 (VLA-2) mediates reorganization and contraction of collagen matrices by human cells. *Cell* 67:403–410. (Ch. 5)

Schlosshauer, B. (1991). Neurothelin: Molecular characteristics and development regulation in the chick CNS. *Development* 113:129–140. (Ch. 6)

Schlosshauer, B. (1993). The blood–brain barrier: Morphology, molecules, and neurothelin. *Bioassay* 5:341–346. (Ch. 6)

Schlosshauer, B., & Herzog, K. H. (1990). Neurothelin: An inducible cell surface glycoprotein of blood–brain barrier specific endothelial cells and distinct neurons. *Journal of Cell Biology* 110:1261–1274. (Ch. 6)

Schlumpf, M., & Lichtensteiger, W. (1979). Catecholamines in the yolk sac epithelium of the rat. *Anatomy and Embryology* (West Germany) 156:177–187. (Ch. 8)

Schmidt-Nielsen, K. (1984). *Scaling: Why Is Animal Size So Important?* Cambridge: Cambridge University Press. (Ch. 13)

Scholzen, T., Solursh, M., Suzuki, S., Reiter, R., Morgan, J. L., Buchberg, A. M., Siracusa, L. D. & Iozzo, R. V. (1994). The murine decorin: Complete cDNA cloning, genomic organization, chromosomal assignment, and expression during organogenesis and tissue differentiation. *Journal of Biological Chemistry* 269:28270–28281. (Ch. 5)

Schott, R. J., & Morrow, L. A. (1993). The role of growth factors in angiogenesis. In *Growth Factors and the Cardiovascular System,* ed. P. Cummins, pp. 148–168. Boston: Kluwer Academic. (Ch. 4)

Schumacker, H. B., Jr. (1992). *The Evolution of Cardiac Surgery.* Bloomington: Indiana University Press. (Ch. 20)

Schwartz, K., Lecarpentier, Y., Martin, J. L., Lompre, A. M., Mercadier, J. J., & Swynghedauw, B. (1981). Myosin isoenzymic distribution correlates with speed of myocardial contraction. *Journal of Molecular and Cellular Cardiology* 13:1071–1075. (Ch. 3)

Schwartz, M. S., & Liaw, L. (1993). Growth control and morphogenesis in the development and pathology of arteries. *Journal of Cardiovascular Pharmacology* 213:S31–S49. (Ch. 6)

Schweiki, D., Itin, A., Soffer, D., & Keshet, E. (1992). Vascular endothelial growth factor induced by hypoxia may mediate hypoxia-initiated angiogenesis. *Nature* 359:843–845. (Ch. 4)

Scott, J. N., & Jennes, L. (1988). Development of immunoreactive atrial natriuretic peptide in fetal hearts of spontaneously hypertensive and Wistar-Kyoto rats. *Anatomy and Embryology* (West Germany) 178:359–363. (Ch. 8)

Sedgewick, A. (1894). On the law of development commonly known as von Baer's law, and on the significance of ancestral rudiments in embryonic development. *Quarterly Journal of Microscopical Science* 36:35. (Ch. 9)

Segrove, F. (1941). The development of the serpulid *Pomatoceros triqueter* L. *Quarterly Journal of Microscopical Science* 82:467–540. (Ch. 11)

Seguchi, M., Harding, J. A., & Jarmakani, J. M. (1986). Developmental change in the function of sarcoplasmic reticulum. *Journal of Molecular and Cellular Cardiology* 18:189–195. (Ch. 2)

Seguchi, M., Jarmakani, J. M., George, B. L., & Harding, J. A. (1986). Effect of Ca^{2+} antagonists on mechanical function in the neonatal heart. *Pediatric Research* 20:838–842. (Ch. 2)

Senior, H. D. (1909). The development of the heart in shad (*Alosa sapadissima,* Wilson) with a note on the classification of teleostean embryos from a morphological standpoint. *American Journal of Anatomy* 9:211–262. (Ch. 12)

Seo, J. W., Shin, S. S., & Chi, J. G. (1985). Cardiovascular system in conjoined twins: An analysis of 14 Korean cases. *Teratology* 32:151–161. (Ch. 10)

Seymour, R. S. (1994). Oxygen diffusion through the jelly capsules of amphibian eggs. *Israel Journal of Zoology* 40:493–506. (Ch. 18)

Seymour, R. S., & Bradford, D. (1995). Respiration of amphibian eggs. *Physiological Zoology* 68:1–25. (Ch. 18)

Seymour, R. S., & Piiper, J. (1988). Aeration of the shell membranes of avian eggs. *Respiration Physiology* 71:101–116. (Ch. 18)

Seymour, R. S., & Roberts, J. D. (1991). Embryonic respiration and oxygen distribution in foamy and nonfoamy egg masses of the frog *Limnodynastes tasmaniensis. Physiological Zoology* 64:1322–1340. (Ch. 18)

Shafer, N. L., Smith, D. S., Babineau, H. P., Keller, B. B., & Clark, E. B. (1992). Vitelline arterial dimensions during primary cardiovascular development. *Pediatric Research* 31:23A. (Ch. 7)

Shalaby, F., Rossant, J., Yamaguchi, T. P., Gertsenstein, M., Wu, X.-F., Breltman, M. L., & Schuh, A. C. (1995). Failure of blood-island formation and vasculogenesis in *Flk-1*-deficient mice. *Nature* 376:62–66. (Ch. 1)

Sham, J. S. K., Cleeman, L., & Morad, M. (1992). Gating of the cardiac Ca^{2+} release channel: The role of Na^+ current and Na^+–Ca^{2+} exchange. *Science* 255:850–853. (Ch. 2)

Sham, J. S. K., Cleeman, L., & Morad, M. (1995). Functional coupling of Ca^{2+} channels and ryanodine receptors in cardiac myocytes. *Proceedings of the National Academy of Sciences USA* 92:121–125. (Ch. 2)

Sham, J. S. K., Hatem, S. N., & Morad, M. (1995). Species differences in the activity of the Na^+–Ca^{2+} exchanger in mammalian cardiac myocytes. *Journal of Physiology* 488(3):623–631. (Ch. 2)

Sharifi, B. G., LaFleur, D. W., Pirola, C. J., Forrester, J. S., & Fagin, J. A. (1992). Angiotensin II regulates tenascin gene expression in vascular smooth muscle cells. *Journal of Biological Chemistry* 267:23910–23915. (Ch. 5)

Shaul, P., Farrar, M. A., & Magness, R. R. (1993). Pulmonary endothelial nitric oxide production is developmentally regulated in the fetus and newborn. *American Journal of Physiology* 265:H1056–H1063. (Ch. 6)

Shaw, G. M., O'Malley, C. D., Wasserman, C. R., Tolarova, M. M., & Lammer, E. J. (1995). Maternal periconceptional use of multi-vitamins and reduced risk of conotruncal heart and limb reduction defects among offspring. *American Journal of Medical Genetics* 59:536–545. (Ch. 21)

Sheldon, C. A., Friedman, W. F., & Sybers, H. D. (1976). Scanning electron microscopy of fetal and neonatal lamb cardiac cells. *Journal of Molecular and Cellular Cardiology* 8:853–862. (Ch. 2)

Shelton, G. (1976). Gas exchange, pulmonary blood supply, and the partially divided amphibian heart. In *Perspectives in Experimental Biology,* ed. P. Spencer Davies, pp. 247–259. Oxford: Pergamon. (Ch. 13)

Shelton, G., & Jones, D. R. (1991). The physiology of the alligator heart: The cardiac cycle. *Journal of Experimental Biology* 158:539–564. (Ch. 14)

Shepard, M. P. (1955). Resistance and tolerance of young speckled trout (*Salvelinus fontinalis*) to oxygen lack, with special reference to low oxygen acclimation. *Journal of the Fisheries Research Board of Canada* 12:387–433. (Ch. 17)

Sheridan, D. J., Cullen, M. J., & Tynan, M. J. (1979). Qualitative and quantitative observations on ultrastructural changes during postnatal development in the cat myocardium. *Journal of Molecular and Cellular Cardiology* 11:1173–1181. (Ch. 2)

Shiino, S. M. (1968). Crustacea. In *Invertebrate Embryology,* ed. M. Kume & K. Dan, Trans. E. Dan, pp. 129–161. Belgrade: Prosveta. (Ch. 11)

Shull, M. M., and Lingrel, J. B. (1987). Multiple genes encode the human Na^+, K^+-ATPase catalytic subunit. *Proceedings of the National Academy of Sciences USA* 84:4039–4043. (Ch. 2)

Siefert, R. E., Carlson, A. R., & Herman, L. J. (1974). Effects of reduced oxygen concentrations on the early life stages of mountain whitefish, smallmouth bass, and white bass. *Progressive Fish-Culturist* 36:186–190. (Ch. 17)

Siefert, R. E., Spoor, W. A., & Syrett, R. F. (1973). Effects of reduced oxygen concentrations on northern pike (*Esox lucius*) embryos and larvae. *Journal of the Fisheries Research Board of Canada* 30:849–852. (Ch. 17)

Sill, A. M., Prenger V. L., Scheel, J. N., Boughman, J. A., Martin, G. R., Clark, E. B., & Brenner, J. I. (1993). A family study of congenital cardiovascular malformations: Preliminary report. *Mid Atlantic Pediatric Cardiology.* Charlottesville, VA. (abstract). (Ch. 21)

Simmerman, H. K. B., Collins, J. H., Theibert, J. L., Wegener, A. D., & Jones, L. R. (1986). Sequence analysis of phospholamban identification of phosphorylation sites and two major structural domains. *Journal of Biological Chemistry* 261:13333–13341. (Ch. 2)

Simpson, D. G., Terracio, L., Terracio, M., Price, R. L., Turner, D. C., & Borg, T. K. (1994). Modulation of cardiac myocyte phenotype *in vitro* by the composition and orientation of the extracellular matrix. *Journal of Cellular Physiology* 161:89–105. (Chs. 5, 10)

Singh, G., Supp, D. M., Schreiner, C., McNeish, J., Merker, H.-J., Copeland, N. G., Jenkins, N. A., Potter, S. S., & Scott, W. (1991). Legless insertional mutation: Morphological, molecular, and genetic characterization. *Genes and Development* 5:2245–2255. (Ch. 10)

Sinning, A. R., Krug, E. L., & Markwald, R. R. (1992). Multiple glycoproteins localize to a particulate form of extracellular matrix in regions of the embryonic heart where endothelial cells transform into mesenchyme. *Anatomical Record* 232:285–292. (Ch. 5)

Sissman, N. J. (1970). Developmental landmarks in cardiac morphogenesis: Comparative chronology. *American Journal of Cardiology* 25:141–148. (Ch. 7)

Slack, J. M. W. (1991). *From Egg to Embryo.* Cambridge: Cambridge University Press. (Ch. 13)

Slotkin, T. A., Lau, C., & Seidler, F. J. (1994). Beta-adrenergic receptor overexpression in the fetal rat: Distribution, receptor subtypes, and coupling to adrenergic cyclase activity via G-proteins. *Toxicology and Applied Pharmacology* 129:223–234. (Ch. 8)

Smith, C. W. (1993). Endothelial adhesion molecules and their role in inflammation. *Canadian Journal of Physiology and Pharmacology* 71:76–87. (Ch. 6)

Smith, F. M., & Jones, D. R. (1978). Localization of receptors causing hypoxic bradycardia in trout (*Salmo gairdneri*). *Canadian Journal of Zoology* 56:1260–1265. (Ch. 12)

Smith, H. E., & Page, E. (1977). Ultrastructural changes in rabbit heart mitochondria during the perinatal period: Neonatal transition to aerobic metabolism. *Developmental Biology* 57:109–117. (Ch. 2)

Smith, P. R. (1986). Development of the blood vascular system in *Sabellaria cementarium* (Annelida, Polychaeta). *Zoomorphology* 106:67–74. (Ch. 11)

Smith, P. R., & Chia, F.-S. (1985). Metamorphosis of the sabellariid polychaete *Sabellaria cementarium* Moore: A histological analysis. *Canadian Journal of Zoology* 63:2852–2866. (Ch. 11)

Smith, S. C., & Armstrong, J. B. (1991). Heart development in normal and cardiac-lethal mutant axolotls: A model for the control of vertebrate cardiogenesis. *Differentiation* 47:129–134. (Ch. 17)

Smith, S. C., & Armstrong, J. B. (1993). Reaction-diffusion control of heart development: Evidence for activation and inhibition in precardiac mesoderm. *Developmental Biology* 160:535–542. (Ch. 10)

Smolich, J. J., Walker, A. M., Campbell, G. R., & Adamson, T. M. (1989). Left and right ventricular myocardial morphometry in fetal neonatal and adult sheep. *American Journal of Physiology* 257:H1–H9. (Chs. 2, 4, 16)

Smyth, D. H. (1939). The central and reflex control of respiration in the frog. *Journal of Physiology* (London) 95:305–327. (Ch. 13)

Sol, P. (1981). *Development of Blood Pressure in the Atria of the Chick Embryo and Its Relationship with the Formation of the Interatrial Septum.* Lausanne, Switzerland: University of Lausanne School of Medicine, Institute of Histology and Embryology. (Ch. 7)

Solaro, R. J. (1991). Regulation of Ca^{2+}-signaling in cardiac myofilaments. *Medicine and Science in Sports and Exercise* 23:1145–1148. (Ch. 3)

Solaro, R. J., Kumar, P., Blanchard, E. M., & Martin, A. F. (1986). Differential effects of pH on calcium activation of myofilaments of adult and perinatal dog hearts. *Circulation Research* 58:721–729. (Ch. 3)

Solewski, W. (1949). The development of the blood vessels of the gills of the sea-trout, *Salmo trutta* L. *Bulletin International de l'Academie Polonaise des Sciences et de lettre, Classe des Sciences Naturele,* ser. B, 2:121–144. (Ch. 12)

Solnica-Krezel, L., Schier, A. F., & Driever, W. (1994). Efficient recovery of ENV-induced mutations from the zebrafish germline. *Genetics* 136:1401–1420. (Chs. 1, 12)

Specht, L. A., Pickel, V. M., Joh, T. H., & Reis, D. J. (1981). Light-microscopic immunocytochemical localization of tyrosine hydroxylase in prenatal rat brain. I, Early ontogeny. *Journal of Comparative Neurology* 199:233–253. (Ch. 8)

Spence, S. G., Argraves, W. S., Walters, L., Hungerford, J. E., & Little, C. D. (1992). Fibulin is localized at sites of epithelial-mesenchymal transitions in the early avian embryo. *Developmental Biology* 151:473–484. (Ch. 5)

Spicer, J. I. (1994). Ontogeny of cardiac function in the brine shrimp *Artemia franciscana* Kellog 1906 (Branchiopoda: Anostraca). *Journal of Experimental Zoology* 270:508–516. (Chs. 7, 11)

Spies, R. B. (1973). The blood system of the flagellirigid polychaete *Flabelliderma commersalis* (Moore). *Journal of Morphology* 139:465–490. (Ch. 11)

Spirito, P., Fu, Y.-M., Yu, Z.-X., Epstein, S. E., & Casscells, W. (1989). Immunohistochemical localization of basic and acidic fibroblast growth factors in the developing rat heart. *Circulation* 84:322–332. (Ch. 4)

Spiro, M. J., & Crowley, T. J. (1993). Increased rat myocardial type VI collagen in diabetes mellitus and hypertension. *Diabtologia* 36:93–98. (Ch. 5)

Spoor, W. A. (1977). Oxygen requirements of embryos and larvae of the largemouth bass, *Micropterus salmoides* (Lacépède). *Journal of Fish Biology* 11:77–86. (Ch. 17)

Spoor, W. A. (1984). Oxygen requirements of larvae of the smallmouth bass, *Micropterus dolomieui* Lacépède. *Journal of Fish Biology* 25:587–592. (Ch. 17)

Spranger, J., Benirschke, K., Hall, J. G., Lenz, W., Lowry, R. B., Opitz, J. M., Pinsky, L., Schwarzacher, H. G., & Smith, D. W. (1982). Errors of morphogenesis: Concepts and terms. *Journal of Pediatrics* 100:160–165. (Ch. 7)

Staehling-Hampton, K., Hoffmann, F. M., Baylies, M. K., Rushton, E., & Bate, M. (1994). *dpp* induces mesodermal gene expression in *Drosophila. Nature* 372:783–786. (Ch. 1)

Stainier, D. Y. R., & Fishman, M. C. (1992). Patterning the zebrafish heart tube: Acquisition of anteroposterion polarity. *Developmental Biology* 153:91–101. (Chs. 1, 10, 12)

Stainier, D. Y. R., & Fishman, M. C. (1994). The zebrafish as a model system to study cardiovascular development. *Trends in Cardiovascular Medicine* 4:207–212. (Ch. 1)

Stainier, D. Y. R., Lee, R. K., & Fishman, M. C. (1993). Cardiovascular development in the zebrafish. I, Myocardial fate map and heart tube formation. *Development* 119:31–40. (Chs. 1, 10, 12)

Stalsberg, H. (1970). Mechanism of dextral looping of the embryonic heart. *American Journal of Cardiology* 25:265–271. (Ch. 10)

Stanton, B. R., Perkins, A. S., Tessarollo, L., Sassoon, D. A., & Parada, L. F. (1992). Loss of *N-myc* function results in embryonic lethality and failure of the epithelial component of the embryo to develop. *Genes & Development* 6:2235–2247. (Ch. 1)

Stent, G. S., Thompson, W. J., & Calabrese, R. L. (1979). Neural control of heartbeat in the leech and in some other invertebrates. *Physiological Reviews* 59:101–136. (Ch. 11)

Stern, M. (1992). Theory of excitation–contraction coupling in cardiac muscle. *Biophysical Journal* 63:497–517. (Ch. 2)

Stewart, D. E., Kirby, M. L., Aronstam, R. S. (1986). Regulation of beta-adrenergic receptor density in the non-innervated and denervated embryonic chick heart. *Journal of Molecular and Cellular Cardiology* 18:469–475. (Ch. 8)

Stewart, J. R. (1993). Yolk sac placentation in reptiles: Structural innovation in a fundamental vertebrate fetal nutritional system. *Journal of Experimental Zoology* 266:431–449. (Ch. 18)

Stewart, J. R., & Blackburn, D. G. (1988). Reptilian placentation: Structural diversity and terminology. *Copeia* 1984:839–852. (Ch. 14)

St. Johnston, R. D., & Gelbart, W. M. (1987). *Decapentaplegic* transcripts are localized along the dorsal–ventral axis of the *Drosophila* embryo. *EMBO Journal* 6:2785–2791. (Ch. 1)

St. Johnston, R. D., & Nüsslein-Volhard, C. (1992). The origin of pattern and polarity in the *Drosophila* embryo. *Cell* 68:201–219. (Ch. 1)

Stockard, C. R. (1915). The origin of blood and vascular endothelium in embryos without a circulation of blood and in the normal embryo. *American Journal of Anatomy* 18:227–327. (Ch. 12)

Strathmann, R. R., & Chaffee, C. (1984). Constraints on egg masses. II, Effect of spacing, size, and number of eggs on ventilation of masses of embryos in jelly, adherent groups, or thin-walled capsules. *Journal of Experimental Marine Biology and Ecology* 84:85–93. (Ch. 18)

Studer, R., Reinecke, H., Bilger, J., Eschenhagen, T., Böhm, M., Hasenfus, G., Just, H., Holtz, J., & Drexler, H. (1994). Gene expression of the cardiac Na^{+}–Ca^{2+} exchanger in end-stage human heart failure. *Circulation Research* 75:443–453. (Ch. 2)

Sucov, H. M., Dyson, E., Gumeringer, C. L., Price, J., Chien, K. R., & Evans, R. M. (1994). RXRa mutant mice establish a genetic basis for vitamin A signaling in heart morphogenesis. *Genes & Development* 8:1007–1018. (Ch. 1)

Suga, H. (1974). Importance of atrial compliance in cardiac performance. *Circulation Research* 35:39–43. (Ch. 7)

Suga, H., Yamada, O., & Goto, Y. (1984). Energetics of ventricular contraction as traced in the pressure–volume diagram. *Federation Proceedings* 43:2411–2413. (Ch. 7)

Suga, H., Yamada, O., Goto, Y., & Igarashi, Y. (1986). Peak isovolumic pressure–volume relation of puppy left ventricle. *American Journal of Physiology* 250:H167–H172. (Ch. 7)

Sugi, Y., & Lough, J. (1992). Onset of expression and regional deposition of alpha-smooth and sarcomeric actin during avian heart development. *Developmental Dynamics* 193:116–124. (Ch. 3)

Sugi, Y., Sasse, J., Barron, M., & Lough, J. (1995). Developmental expression of fibroblast growth factor receptor-1 (cek-1; flg) during heart development. *Developmental Dynamics* 202:115–125. (Ch. 10)

Sumida, H., Nakamura, H., & Satow, Y. (1991). Relation of the distribution of fibronectin to the obliteration of the arch artery in the chick. *Archives of Histology and Cytology* 54:89–94. (Ch. 6)

Sunagawa, K., Maughan, W. L., & Sagawa, K. (1985). Optimal arterial resistance for the maximal stroke work studied in isolated canine left ventricle. *Circulation Research* 56:586–595. (Ch. 7)

Sunagawa, K., Sagawa, K., & Maughan, W. L. (1984). Ventricular interaction with the loading system. *Annals of Biomedical Engineering* 12:163–189. (Ch. 7)

Sunagawa, K., Sagawa, K., & Maughan, W. L. (1987). Ventricular interaction with the vascular system in terms of pressure–volume relationships. In *Ventricular Vascular Coupling: Clinical, Physiological, and Engineering Aspects,* ed. F. C. P. Yin, pp. 210–239. New York: Springer-Verlag. (Ch. 7)

Sushko, B. S. (1982). Microelectrode studies of oxygen tension and transport in loach *Misgurnus fossilis* eggs in helium–oxygen and nitrogen–oxygen media. *Fiziolohichny Zhurnal* (Kiev) 28:593–597. *Canadian Translations of Fisheries and Aquatic Sciences* 5211 (1986). (Ch. 18)

Suzuki, H., & Kashiwagi, H. (1989). Molecular biology of cytokine effects on vascular endothelial cells. *International Reviews of Experimental Pathology* 32:95–148. (Ch. 6)

Swain, D. P., & Pittman, R. N. (1989). Oxygen exchange in the microcirculation of hamster retractor muscle. *American Journal of Physiology* 256:H247–H255. (Ch. 18)

Sweadner, K. J. (1979). Two molecular forms of (Na^+ + K^+)-stimulated ATPase in brain: Separation and difference in affinity for strophanthidin. *Journal of Biological Chemistry* 254:6060–6067. (Ch. 2)

Sweeney, H. L., Bowman, B. F., & Stull, J. T. (1993). Myosin light chain phoshorylation in vertebrate striated muscle: Regulation and function. *American Journal of Physiology* 264:C1085–C1095. (Ch. 3)

Sweeney, H. L., Kushmerick, M. J., Mabuchi, K., Sreter, F. A., & Gergely, J. (1988). Myosin alkali light chain and heavy chain variations correlate with altered shortening velocity of isolated skeletal muscle fibers. *Journal of Biological Chemistry* 263:9034–9039. (Ch. 3)

Sweeny, L. J. (1981). Morphometric analysis of an experimental model of left heart hypoplasia in the chick. Ph.D. diss., University of Nebraska, Omaha. (Chs. 7, 20)

Swiderski, R. E., Daniels, K. J., Jensen, K. L., & Solursh, M. (1994). Type II collagen is transiently expressed during avian cardiac valve morphogenesis. *Developmental Dynamics* 200:294–304. (Ch. 5)

Sylvester, J. R., Nash, C. E., & Emberson, C. R. (1975). Salinity and oxygen tolerances of eggs and larvae of Hawaiian striped mullet, *Mugil cephalus* L. *Journal of Fish Biology* 7:621–629. (Ch. 17)

Taber, L. A., Keller, B. B., & Clark, E. B. (1992). Cardiac mechanics in the stage 16 chick embryo. *Journal of Biomechanical Engineering* 114:427–434. (Ch. 7)

Tada, M., Inui, M., Yamada, M., Kadoma, M., Kuzuka, T., Abe, H., & Kakiuchi, S. (1983). Effects of phospholamban phosphorylation catalyzed by adenosine 3′ 5′-monophosphate and calmodulin-dependent protein kinases on calcium transport ATPase of cardiac sarcoplasmic reticulum. *Journal of Molecular and Cellular Cardiology* 15:335–346. (Ch. 2)

Takeichi, M. (1990). Cadherins: A molecular family important in selective cell–cell adhesion. *Annual Reviews of Biochemistry* 59:237–252. (Ch. 6)

Tang, D. G., Chen, Y. Q., Neuman, P. J., et al. (1993). Identification of PECAM-1 in solid tumor cells and its potential involvement in tumor cell adhesion to endothelium. *Journal of Biological Chemistry* 30:22883–22894. (Ch. 6)

Tavolga, W. (1949). Embryonic development of the platyfish (*Platypoecilus*), the swordtail (*Xiphophorus*), and their hybrids. *Bulletin of the American Museum of Natural History* 94:161–230. (Ch. 12)

Taylor, E. W. (1992). Nervous control of the heart and cardiorespiratory interactions. In *Fish Physiology,* vol. 12B, *The Cardiovascular System,* ed. W. S. Hoar, D. J. Randall, & A. P. Farrell, pp. 343–389. San Diego: Academic Press. (Ch. 12)

Taylor, P. M., Rose, M. I., Yacoub, M. H., et al. (1992). Induction of vascular adhesion molecules during rejection of human cardiac allografts. *Transplantation* 54:451–457. (Ch. 6)

Tazawa, H. (1981a). Effects of O_2 and CO_2 in N_2, He and SF_6 on chick embryo blood pressure and heart rate. *Journal of Applied Physiology* 51:1017–1022. (Ch. 17)

Tazawa, H. (1981b). Measurement of blood pressure of chick embryo with an implanted needle catheter. *Journal of Applied Physiology* 51:1023–1026. (Chs. 7, 15)

Tazawa, H., Ar, A., Rahn, H., & Piiper, J. (1980). Repetitive and simultaneous sampling from the air cell and blood vessels in the chick embryo. *Respiration Physiology* 39:265–272. (Ch. 15)

Tazawa, H., Hashimoto, Y., & Doi, K. (1992). Blood pressure and heart rate of chick embryo (*Gallus domesticus*) within the egg: Responses to autonomic drugs. In *Phylogenic Models in Functional Coupling of the CNS and the Cardiovascular System,* ed. P. B. Hill, K. Kuwasawa, B. R. McMahon, & T. Kuramoto, pp. 86–96. Basel: Karger. (Ch. 15)

Tazawa, H., Hashimoto, Y., Takami, M., Yufu, Y., & Whittow, G. C. (1993). Simple, noninvasive system for measuring the heart rate of avian embryos and hatchlings by means of a piezoelectric film. *Medical and Biological Engineering and Computing* 31:129–134. (Ch. 15)

Tazawa, H., Hiraguchi, T., Asakura, T., Fujii, H., & Whittow, G. C. (1989a). Noncontact measurements of avian embryo heart rate by means of the laser speckle: Comparison with contact measurements. *Medical and Biological Engineering and Computing* 27:580–586. (Ch. 15)

Tazawa, H., Hiraguchi, T., Kuroda, O., Tullett, S. G., & Deeming, D. C. (1991). Embryonic heart rate during development of domesticated birds. *Physiological Zoology* 64:1002–1022. (Chs. 7, 15, 17)

Tazawa, H., & Johansen, K. (1987). Comparative model analysis of central shunts in vertebrate cardiovascular systems. *Comparative Biochemistry and Physiology* 86A:595–607. (Ch. 14)

Tazawa, H., Kuroda, O., & Whittow, G. C. (1991). Noninvasive determination of the embryonic heart rate during hatching in the brown noddy (*Anous stolidus*). *Auk* 108:594–601. (Ch. 15)

Tazawa, H., Lomholt, J. P., & Johansen, K. (1985). Direct measurement of allantoic blood flow in the chicken, *Gallus domesticus:* Responses to alteration in ambient temperature and P_{O_2} *Comparative Biochemistry and Physiology* 81A:641–642. (Ch. 15)

Tazawa, H., Mikami, T., & Yoshimoto, C. (1971). Respiratory properties of chicken embryonic blood during development. *Respiration Physiology* 13:160–170. (Ch. 15)

Tazawa, H., & Mochizuki, M. (1976). Estimation of contact time and diffusing capacity for oxygen in the chorioallantoic vascular plexus. *Respiration Physiology* 28:119–128. (Chs. 7, 15)

Tazawa, H., & Mochizuki, M. (1977). Oxygen analyses of chicken embryo blood. *Respiration Physiology* 31:203–215. (Ch. 15)

Tazawa, H., & Nakagawa, S. (1985). Response of egg temperature, heart rate, and blood pressure in the chick embryo to hypothermal stress. *Journal of Comparative Physiology* B155:195–200. (Chs. 15, 17)

Tazawa, H., Nakazawa, S., Okuda, A., & Whittow, G. C. (1988a). Short-term effects of altered shell conductance on oxygen uptake and hematological variables of late chicken embryos. *Respiration Physiology* 74:199–210. (Ch. 15)

Tazawa, H., Okuda, A., Nakagawa, S., & Whittow, G. C. (1989b). Metabolic responses of chicken embryos to graded, prolonged alterations in ambient temperature. *Comparative Biochemistry and Physiology* 92A:613–617. (Ch. 17)

Tazawa, H., & Rahn, H. (1986). Tolerance of chick embryos to low temperatures in reference to the heart rate. *Comparative Biochemistry and Physiology* 85A:531–534. (Chs. 15, 17)

Tazawa, H., & Rahn, H. (1987). Temperature and metabolism of chick embryos and hatchlings after prolonged cooling. *Journal of Experimental Zoology* 1:105–109. (Ch. 17)

Tazawa, H., Suzuki, Y., & Musashi, H. (1989). Simultaneous acquisition of ECG, BCG, and blood pressure from chick embryos in the egg. *Journal of Applied Physiology* 67:478–483. (Ch. 15)

Tazawa, H., & Takenaka, H. (1985). Cardiovascular shunt and model analysis in the chick embryo. In *Cardiovascular Shunts,* ed. K. Johansen & W. Burggren, pp. 178–198. Copenhagen: Munksgaard. (Ch. 15)

Tazawa, H., Wakayama, H., Turner, J. S., & Paganelli, C. V. (1988b). Metabolic compensation for gradual cooling in developing chick embryos. *Comparative Biochemistry and Physiology* 89A:125–129. (Ch. 17)

Tazawa, H., Watanabe, W., & Burggren, W. W. (1994). Embryonic heart rate in altricial birds, the pigeon (*Columba domestica*) and the bank swallow (*Riparia riparia*). *Physiological Zoology* 40:1448–1460. (Ch. 15)

Tazawa, H., & Whittow, G. C. (1994). Embryonic heart rate and oxygen pulse in two procellarii-form seabirds, *Diomedea immutabilis* and *Puffinus pacificus. Journal of Comparative Physiology* B163:642–648. (Ch. 15)

Tazawa, H., Yamaguchi, S., Yamada, M., & Doi, K. (1992). Embryonic heart rate of the domestic fowl (*Gallus domesticus*) in a quasiequilibrium state of altered ambient temperatures. *Comparative Biochemistry and Physiology* 101A:103–108. (Chs. 15, 17)

Teitel, D. F. (1988). Circulatory adjustments of postnatal life. *Seminars in Perinatology* 12:96–103. (Ch. 16)

Teitel, D. F., Klautz, R. J. M., Steendijk, P., Van Der Velde, E. T., Van Bel, F., & Baan, J. (1991). The end-systolic pressure–volume relationship in the newborn lamb: Effects of loading and inotropic interventions. *Pediatric Research* 29:473–482. (Ch. 7)

Terracio, L., & Borg, T. K. (1988). Factors affecting cardiac cell shape. *Heart Failure* 4:114–124. (Ch. 7)

Terracio, L., Rubin, K., Balog, E., Jyring, R., Gullberg, D., Carver, W., & Borg, T. K. (1991). Expression of collagen binding integrins during cardiac development and hypertrophy. *Circulation Research* 68:734–743. (Ch. 5)

Terracio, L., Simpson, D. G., Hilenski, L., Carver, W., Decker, R. S., Vinson, N., & Borg, T. K. (1990). Distribution of vinculin in the z-disk of striated muscle: Analysis by laser scanning confocal microscopy. *Journal of Cellular Physiology* 145:78–87. (Ch. 5)

Territo, P., & Burggren, W. W. (1995). The ontogeny of respiratory physiology with perfusion limitations: The effects of 2% carbon monoxide on oxygen consumption and utilization. *Physiological Zoology* 68(4):98. (Ch. 13)

Territo, P., & Burggren, W. W. (in press). Cardio-respiratory ontogeny during chronic carbon monoxide induced hypoxemia in the clawed frog *Xenopus laevis. Journal of Experimental Biology.* (Ch. 13)

Theiler, K. (1989). *The House Mouse: Atlas of Embryonic Development.* New York: Springer-Verlag. (Ch. 7)

Thierfelder, L., Watkins, H., MacRae, C., Lamas, R., McKenna, W., Vosberg, H. P., Seidman, J. G., & Seidman, C. E. (1994). Alpha-tropomyosin and cardiac troponin T mutations cause familial hypertrophic cardiomyopathy: A disease of the sarcomere. *Cell* 7:701–712. (Ch. 3)

Thompson, D'A. (1917). *On Growth and Form.* Cambridge: Cambridge University Press. (Ch. 9)

Thompson, W. R., Hu, N., & Clark, E. B. (1986). Acute effects of calcium channel blockade on cardiovascular function in the stage 24 chick embryo. *Teratology* 33:36C. (Ch. 7)

Thornburg, K. L., & Morton, M. J. (1983). Filling and arterial pressures as determinants of RV stroke volume in the sheep fetus. *American Journal of Physiology* 44:H656–H663. (Ch. 16)

Thornburg, K. L., & Morton, M. J. (1986). Filling and arterial pressures as determinants of left ventricular stroke volume in fetal lambs. *American Journal of Physiology* 51:H961–H968. (Ch. 16)

Thornburg, K. L., & Morton, M. J. (1994). Development of the cardiovascular system. In *Textbook of Fetal Physiology,* ed. G. D. Thorburn & R. Harding, pp. 95–130. Oxford: Oxford University Press. (Ch. 16)

Tibbits, G. F., Moyes, C. D., & Hove-Madsen, L. (1992). Excitation–contraction coupling in the teleost heart. In *Fish Physiology,* ed. W. S. Hoar, D. J. Randall, & A. P. Farrell, vol. 12A, pp. 267–304. San Diego: Academic Press. (Ch. 9)

Toews, D., & MacIntyre, D. (1977). Blood respiratory properties of a viviparous amphibian. *Nature* 266:464–465. (Ch. 18)

Toews, D., & MacIntyre, D. (1978). Respiration and circulation in an apodan amphibian. *Canadian Journal of Zoology* 56:998–1004. (Ch. 13)

Tohse, T., Meszaros, J., & Sperelakis, N. (1992). Developmental changes in long-opening behavior of L-type Ca^{2+} channels in embryonic chick heart cells. *Circulation Research* 71:376–384. (Ch. 2)

Tokimitsu, I., Kato, H., Wachi, H., & Tajima, S. (1994). Elastin synthesis is inhibited by angiotensin II but not by platelet-derived growth factor in arterial smooth muscle cells. *Biochimica et Biophysica Acta* 1207:68–73. (Ch. 5)

Tokuyasu, T. K., & Maher, P. A. (1987a). Immunocytochemical studies of cardiac myofibrillogenesis in early chick embryos. I, Presence of immunofluorescence titin spots in the premyofibril stages. *Journal of Cell Biology* 105:2781–2794. (Ch. 10)

Tokuyasu, T. K., & Maher, P. A. (1987b). Immunocytochemical studies of cardiac myofibrillogenesis in early chick embryos. II, Generation of a-actinin dots within titin spots at the time of the first myofibril formation. *Journal of Cell Biology* 105:2795–2802. (Ch. 10)

Tomanek, R. J., Haung, L., Suvarna, P., O'Brien, L. C., Ratajska, A., & Sandra, A. (1996). Coronary vascularization during development in the rat and its relationship to basic fibroblast growth factor. *Cardiovascular Research* 31:E116–E126. (Ch. 4)

Tomita, H., Nakazawa, M., Nishibatake, M., Ikeda, K., & Takao, A. (1989). Effect of isoproterenol on heart volume in chick embryos (in Japanese). *Acta Cardiol Paed Jpn* 5:286–290. (Ch. 8)

Tonissen, K. F., Drysdale, T. A., Lints, T. J., Harvey, R. P., & Krieg, P. A. (1994). *XNkx-2.5,* a *Xenopus* gene related to *Nkx-2.5* and *tinman:* Evidence for a conserved role in cardiac development. *Developmental Biology* 162:325–328. (Ch. 1)

Torrey, T. W., & Feduccia, A. (1979). *Morphogenesis of the Vertebrates.* 4th ed. New York: John Wiley & Sons. (Ch. 12)

Torry, R. J., & Rongish, R. J. (1992). Angiogenesis in the uterus: Potential regulation and relation to tumor angiogenesis. *American Journal of Reproductive Immunology* 27:171–179. (Ch. 6)

Toshimori, H., Toshimori, K., Oura, C., & Matsuo, H. (1987). Immunohistochemical study of atrial natriuretic polypeptides in the embryonic, fetal, and neonatal rat heart. *Cell and Tissue Research* 248:627–633. (Ch. 8)

Tota, B. (1983). Vascular and metabolic zonation of the ventricular myocardium of mammals and fishes. *Comparative Biochemistry and Physiology* 76A:423–437. (Chs. 9, 12)

Tota, B. (1989). Myoarchitecture and vascularization of the elasmobranch heart ventricle. *Journal of Experimental Zoology Supplement* 2:122–135. (Ch. 4)

Tota, B., Acierno, R., & Agnisola, C. (1991). Mechanical performance of the isolated and perfused heart of the haemoglobinless Antarctic icefish *Chionodraco hamatus* (Lonneberg): Effects of loading conditions and temperature. *Philosophical Transactions of the Royal Society of London,* ser. B, 332:191–198. (Ch. 9)

Tota, B., Cimini, V., Salvatore, G., & Zummo, G. (1983). Comparative study of the arterial and lacunary systems of the ventricular myocardium of elasmobranch and teleost fishes. *American Journal of Anatomy* 167:15–32. (Chs. 4, 9)

Tournier, J. M., Goldstein, G. A., Hall, D. E., Damsky, C. H., & Basbaum, C. B. (1992). Extracellular matrix proteins regulate morphologic and biochemical properties of tracheal gland serous cells through integrins. *American Journal of Respiratory Cell and Molecular Biology* 6:461–471. (Ch. 5)

Toyota, N. (1993). Expression of troponin C genes during development in the chicken. *International Journal of Developmental Biology* 37:531–537. (Ch. 3)

Toyota, N., & Shimada, Y. (1981). Differentiation of troponin in cardiac and skeletal muscles in chicken embryos as studied by immunofluorescence microscopy. *Journal of Cell Biology* 91:497–504. (Chs. 3, 10)

Toyota, N., Shimada, Y., & Bader D. (1989). Molecular cloning and expression of chicken cardiac troponin C. *Circulation Research* 65:1241–1246. (Ch. 3)

Trachtman, H., Futterweit, S., & Singhal, P. (1995). Nitric oxide modulates the synthesis of extracellular matrix proteins in cultured rat mesangial cells. *Biochemical and Biophysical Research Communications* 207:120–125. (Ch. 5)

Trend, S. G., & Bruce, N. W. (1989). Resistance of the rat embryo to elevated maternal epinephrine concentrations. *American Journal of Obstetrics and Gynecology* 160:498–501. (Ch. 8)

Trybus, K. M. (1994). Role of myosin light chains. *Journal of Muscle Research and Cell Motility* 15:587–594. (Ch. 3)

Tucker, D. C., Snider, C., & Woods, W. T., Jr. (1988). Pacemaker development in embryonic rat heart cultured in oculo. *Pediatric Research* 23:637–642. (Ch. 10)

Tucker, R. P., Hammarback, J. A., Jenrath, D. A., Mackie, E. J., & Xu, Y. (1993). Tenascin expression in the mouse: *In situ* localization and induction *in vitro* by bFGF. *Journal of Cell Science* 104:69–76. (Ch. 5)

Tullett, S. G., & Burton, F. G. (1982). Factors affecting the weight and water status of the chick at hatch. *British Poultry Science* 23:361–369. (Ch. 15)

Turner, C. L. (1940). Pericardial sac, trophotaeniae, and alimentary tract in the embryos of goodeid fishes. *Journal of Morphology* 67:271–289. (Ch. 12)

Turner, R. R., Beckstead, J. H., & Warnke, R. A. (1987). Endothelial cell phenotypic diversity. *American Journal of Clinical Pathology* 87:569–575. (Ch. 6)

Unger, E. F., Banai, S., Shou, M., Lazarous, D. F., Jaklitsch, M. T., Scheinowitz, M., Correa, R., Klingbeil, C., & Epstein, S. (1994). Basic fibroblast growth factor enhances myocardial collateral flow in a canine model. *American Journal of Physiology* 266:H1588–H1595. (Ch. 4)

Uynh, T. V., Chen, F., Wetzel, G. T., Friedman, W. F., & Klitzner, T. S. (1992). Developmental changes in membrane Ca^{2+} and K^+ currents in fetal, neonatal, and adult rabbit ventricular myocytes. *Circulation Research* 70:508–515. (Ch. 2)

VanBuren, P., Waller, G. S., Harris, D. E., Trybus, K. M., Warshaw, D. M., & Lowey, S. (1994). The essential light chain is required for full force production by skeletal muscle myosin. *Proceedings of the National Academy of Sciences* 91:12403–12407. (Ch. 3)

van Gronigen, J. P., Wenink, A. C. G., & Testers, L. H. M. (1991). Myocardial capillaries: Increase by splitting of existing vessels. *Anatomy and Embryology* (West Germany) 184:65–70. (Ch. 4)

Van Mierop, L. H. S. (1979). Morphological development of the heart. In *Handbook of Physiology,* sec. 2, *The Cardiovascular System,* vol. 1, *The Heart,* ed. R. M. Berne, N. Sperelakis, & S. R. Geiger, pp. 1–28. Bethesda, Md.: American Physiological Society. (Ch. 7)

Van Mierop, L. H. S., & Bertuch, C. J., Jr. (1967). Development of arterial blood pressure in the chick embryo. *American Journal of Physiology* 212:43–48. (Chs. 7, 15)

Van Vliet, B. N., & West, N. H. (1994). Phylogenetic trends in the baroreceptor control of arterial blood pressure. *Physiological Zoology* 67:1284–1304. (Ch. 9)

Vaxelaire, J. F., Laurent, S., Lacolley, P., Briand, V., Schmitt, H., & Michel, J. B. (1989). Atrial natriuretic peptide decreases contractility of cultured chick ventricular cell. *Life Science* 45:41–48. (Ch. 8)

Venema, R. C., & Kuo, J. F. (1993). Protein kinase C-mediated phosphorylation of troponin I and C-protein in isolated myocardial cells is associated with inhibition of myofibrillar actomyosin MgATPase. *Journal of Biological Chemistry* 268:2705–2711. (Ch. 3)

Venstrom, K. A., & Reichardt, L. F. (1993). Extracellular matrix 2: Role of extracellular matrix molecules and their receptors in the nervous system. *FASEB Journal* 7:996–100. (Ch. 6)

Vernidub, M. F. (1966). Composition of the red and white blood cells of embryos of Atlantic salmon and Baltic salmon (*Salmo salar* L.) and changes in this composition during the growth of the organism. *Trudy Murmanskogo Morskogo Biologicheskogo Instituta (Proceedings of the Murmansk Institute of Marine Biology),* pp. 139–162 (English translation). (Ch. 12)

Vernier, J. M. (1969). Table chronologique du développement embryonnaire de la truite arc-en-ciel, *Salmo gairdneri* Rich. 1836. *Annales d'Embryologie et de Morphogenèse* 2:495–520. (Ch. 12)

Vetter, R., & Will, H. (1986). Sarcolemmal Na–Ca exchange and sarcoplasmic reticulum uptake in developing chick heart. *Journal of Molecular and Cellular Cardiology* 18:1267–1275. (Ch. 7)

Vince, M. A., Clarke, J. V., & Reader, M. R. (1979). Heart rate response to egg rotation in the domestic fowl embryo. *British Poultry Science* 20:247–254. (Ch. 15)

Virágh, S., & Challice, C. E. (1973). Origin and differentiation of cardiac muscle cells in the mouse. *Journal of Ultrastructure Research* 42:1–24. (Ch. 10)

Virágh, S., & Challice, C. E. (1981). The origin of the epicardium and the embryonic myocardial circulation in the mouse. *Anatomical Record* 201:157–168. (Ch. 4)

Virágh, S., Gittenberger–de Groot, A. C., Poelmann, R. E., & Kalman, F. (1993). Early development of quail heart epicardium and associated vascular and glandular structures. *Anatomy and Embryology* (West Germany) 188:381–393. (Ch. 4)

Virágh, S., Kálmán, F., Gittenberger–de Groot, A. C., Poelmann, R. E., & Moorman, A. F. M. (1990). Angiogenesis and hematopoiesis in the epicardium of the vertebrate embryo heart. *Annals of the New York Academy of Sciences* 588:455–458. (Ch. 4)

Visschedijk, A. H. J. (1968). The air space and embryonic respiration. I, The pattern of gaseous exchange in the fertile egg during the closing stages of incubation. *British Poultry Science* 9:173–184. (Ch. 15)

Visschedijk, A. H. J., Girard, H., & Ar, A. (1988). Gas diffusion in the shell membranes of the hen's egg: Lateral diffusion in situ. *Journal of Comparative Physiology* B158:567–574. (Ch. 18)

Voboril, A., & Schiebler, T. H. (1969). Uber die Entwicklung des Gefässversorgung des Rattenherzen. *Zeitschrift für Anatomie und Entwicklungsgeschichte* 129:24–40. (Ch. 4)

Vogt, C. (1842). Embryologie des salmones. In *Histoire naturelle des poissons d'eau douce de l'Europe centrale,* ed. L. Agassiz. Neuchâtel: O. Petitpierre. (Ch. 12)

Voogt, P. A., Broertjes, J. J. S., & Oudejans, R. C. H. M. (1985). Vitellogenesis in sea star: Physiological and metabolic implications. *Comparative Biochemistry and Physiology* 80A:141–147. (Ch. 11)

Vornanen, M. (1992). Force–frequency relationship, contraction duration, and recirculating fraction of calcium in post-natally developing rat heart ventricles: Correlation with heart rate. *Acta Physiologica Scandinavica* 145:311–321. (Ch. 9)

Vuillemin, M., & Pexieder, T (1989a). Normal stages of cardiac organogenesis in the mouse. I, Development of the external shape of the heart. *American Journal of Anatomy* 184:101–113. (Ch. 7)

Vuillemin, M., & Pexieder, T. (1989b). Normal stages of cardiac organogenesis in the mouse. II, Development of the internal relief of the heart. *American Journal of Anatomy* 184:114–128. (Ch. 7)

Wagman, A. J., Hu, N., & Clark, E. B. (1990). Effect of changes in circulating blood volume on cardiac output and arterial and ventricular blood pressure in the stage 18, 24, and 29 chick embryo. *Circulation Research* 67:187–192. (Chs. 7, 8)

Wagner, P. D. (1977). Diffusion and chemical reaction in pulmonary gas exchange. *Physiological Reviews* 57:257–312. (Ch. 18)

Wagner, P. D., Saltzman, H. A., & West, J. B. (1974). Measurement of continuous distributions of ventilation–perfusion ratios: Theory. *Journal of Applied Physiology* 36:588–599. (Ch. 18)

Wahler, G. M. (1992). Developmental increases in the inwardly rectifying potassium current of rat ventricular myocytes. *American Journal of Physiology* 262:C1266–C1272. (Ch. 2)

Wahler, G. M., Dollinger, S. J., Smith, J. M., & Flemal, K. L. (1994). Time course of postnatal changes in rat heart action potential and in transient outward current is different. *American Journal of Physiology* 267:H1157–H1166. (Ch. 2)

Waitzman, N. J., Romano, P. S., & Scheffler, R. M. (1994). Estimates of the economic costs of birth defects. *Inquiry* 33:118–205. (Ch. 21)

Waldo, K. L., Kumski, D. H., & Kirby, M. L. (1994). Association of the cardiac neural crest with development of the coronary arteries in the chick embryo. *Anatomical Record* 239:315–331. (Ch. 4)

Waldo, K. L., Willner, W., & Kirby, M. L. (1990). Origin of the proximal coronary artery stems and a review of ventricular vascularization in the chick embryo. *American Journal of Anatomy* 188:109–120. (Ch. 4)

Wang, D. H., & Prewitt, R. L. (1993). Alterations of mature arterioles associated with chronically-reduced blood flow. *American Journal of Physiology* 264:H40–H44. (Ch. 6)

Wang, N., Butler, J. P., & Banzett, R. B. (1990). Gas exchange across avian eggshells oscillates in phase with heartbeat. *Journal of Applied Physiology* 69:1549–1552. (Ch. 15)

Wang, Y., & Coceani, F. (1992). Isolated pulmonary resistance vessels from fetal lambs: Contractile behavior and responses to indomethacin and endothelin-1. *Circulation Research* 72:320–330. (Ch. 8)

Wangensteen, O. D., & Rahn, H. (1970/71). Respiratory gas exchange by the avian embryo. *Respiration Physiology* 11:31–45. (Ch. 18)

Wangensteen, O. D., & Weibel, E. R. (1982). Morphometric evaluation of chorioallantoic oxygen transport in the chick embryo. *Respiration Physiology* 47:1–20. (Ch. 18)

Wangensteen, O. D., Wilson, D., & Rahn, H. (1970/71). Diffusion of gases across the shell of the hen's egg. *Respiration Physiology* 11:16–30. (Ch. 18)

Warburton, S. J., Hastings, D., & Wang, T. (1995). Responses to chronic hypoxia in embryonic alligators. *Journal of Experimental Zoology* 273:44–50. (Chs. 7, 14)

Wasserloos, E. (1911). Die Entwicklung der Kiemen bei *Cyclas cornea* und andern Acephalen des süssen Wassers. *Zoologische Jahrbucher, Abteilung für Anatomie* 31:171–288. (Ch. 11)

Wassersug, R. J., Paul, R. D., & Feder, M. E. (1981). Cardio-respiratory synchrony in anuran larvae (*Xenopus laevis, Pachymedusa dacnicolor,* and *Rana berlandieri*). *Comparative Biochemistry and Physiology* 70A:329–334. (Ch. 17)

Wassersug, R. J., & Seibert, E. A. (1975). Behavioral responses of amphibian larvae to variation in dissolved oxygen. *Copeia* 1975:86–103. (Ch. 17)

Watson, W. H., III, & Groome, J. R. (1989). Modulation of the *Limulus* heart. *American Zoologist* 29:1287–1303. (Ch. 11)

Wattanapermpool, J., Guo, X., & Solaro, R. J. (1995). The unique amino-terminal peptide of cardiac troponin I regulates myofibrillar activity only when it is phosphorylated. *Journal of Molecular and Cellular Cardiology* 27:1383–1391. (Ch. 3)

Webb, P. W., & Brett, J. R. (1972a). Respiratory adaptations of prenatal young in the ovary of two species of viviparous seaperch, *Rhacochilus vacca* and *Embiotoca lateralis. Journal of the Fisheries Research Board of Canada* 29:1525–1542. (Ch. 18)

Webb, P. W., & Brett, J. R. (1972b). Oxygen consumption of embryos and parents, and oxygen transfer characteristics within the ovary of two species of viviparous seaperch, *Rhacochilus vacca* and *Embiotoca lateralis. Journal of the Fisheries Research Board of Canada* 29:1543–1553. (Ch. 18)

Weber, A. (1908). Récherches sur quelques stades du développement de coeur de la raie. *Comptes Rendu Association d'Anatomie, Paris* 10:10–14. (Ch. 12)

Weber, R. E. (1994). Hemoglobin-based O_2 transfer in viviparous animals. *Israel Journal of Zoology* 40:541–550. (Ch. 18)

Weber, R. E., & Hartvig, M. (1984). Specific fetal hemoglobin underlies the fetal–maternal shift in blood oxygen affinity in a viviparous teleost *Zoarces viviparus. Molecular Physiology* 6:27–32. (Chs. 12, 17)

Wegener, A. D., Simmerman, H. K. B., Lindemann, J. P., & Jones, L. R. (1989). Phospholamban phosphorylation in intact ventricles: Phosphorylation of serine 16 and threonine 17 in response to beta-adrenergic stimulation. *Journal of Biological Chemistry* 264:11468–11474. (Ch. 2)

Weihs, D. (1980). Respiration and depth control as possible reasons for swimming of northern anchovy *Engraulis mordax* yolk sac larvae. *Fisheries Bulletin* 78:109–117. (Ch. 17)

Weiner, H. L., Weiner, L. H., & Swain J. (1987). The tissue distribution and developmental expression of the messenger RNA encoding angiogenic. *Journal of Cell Physiology* 102:267–277. (Ch. 6)

Weiss, J., Goldman, Y., & Morad, M. (1976). Electromechanical properties of the single cell–layered heart of the tunicate *Boltenia ovifera* (sea potato). *Journal General Physiology* 68:503–518. (Ch. 11)

Weitzberg, E. (1993). Circulatory responses of endotoxin shock and nitric oxide inhalation. *Acta Physiologica Scandinavica* 148(suppl.):611. (Ch. 6)

Wells, P. R. (1993). Morphometry of respiratory surfaces and partitioning oxygen uptake in Atlantic salmon, *Salmo salar* (L.) during development from larva to juvenile. M. S. thesis, Dalhousie University, Halifax, N. S., Canada. (Ch. 12)

Wells, P. R., & Pinder, A. W. (1996a). The respiratory development of Atlantic salmon. I, Morphometry of gills, yolk sac, and body surface. *Journal of Experimental Biology* 199:2725–2736. (Ch. 18)

Wells, P. R., & Pinder, A. W. (1996b). The respiratory development of Atlantic salmon. II, Partitioning of oxygen uptake among gills, yolk sac, and body surfaces. *Journal of Experimental Biology.* 199:2737–2744. (Ch. 18)

Welsh, U., & Rehkämper, G. (1987). Podocytes in the axial organ of echinoderms. *Journal of Zoology, London,* 213:45–50. (Ch. 11)

Werb, Z., Tremble, P. M., Behrendtsen, O., Crowley, E., & Damsky, C. H. (1989). Signal transduction through the fibronectin receptor induces collagenase and stromelysin gene expression. *Journal of Cell Biology* 109:877–889. (Ch. 5)

Werner, B. (1955). Ueber die Anatomie, die Entwicklung, und Biologie des Veligers und der Veliconcha von *Crepidula fornicata* L. (Gastropoda, Prosobranchia). *Helgoländer Wissenschaftliche Meeresuntersuchungen* 5:169–217. (Ch. 11)

Wernovsky, G., Hougen, T. J., Walsh, E. P., et al. (1988). Midterm results after arterial switch operation for transposition of great arteries with intact ventricular septum: Clinical, hemodynamic, echocardiographic, and electrophysiologic data. *Circulation* 77:1333–1344. (Ch. 20)

Wessels, A., Vermeulen, J. L., Viragh, S., Kalman, F., Lamers, W. H., & Moorman, A. F. (1991). Spatial distribution of "tissue-specific" antigens in the developing human heart and skeletal muscle. II, An immunohistochemical analysis of myosin heavy chain isoform expression patterns in the embryonic heart. *Anatomical Record* 229:355–368. (Ch. 3)

West, N. H., & Burggren, W. W. (1982). Gill and lung ventilatory responses to steady-state aquatic hypoxia and hyperoxia in the bullfrog tadpole. *Respiration Physiology* 47:165–176. (Ch. 17)

West, N. H., & Burggren, W. W. (1983). Reflex interactions between aerial and aquatic gas exchange organs in larval bullfrogs. *American Journal of Physiology* 244:R770–R777. (Ch. 17)

West, N. H., & Van Vliet, B. N. (1992). Sensory mechanisms regulating the respiratory and cardiovascular systems. In *Environmental Physiology of the Amphibians,* ed. M. E. Feder & W. W. Burggren, pp. 151–182. Chicago: University of Chicago Press. (Ch. 13)

Wetzel, G. T., Chen, F., & Klitzner, T. S. (1993). Ca^{2+} channel kinetics in acute isolated fetal, neonatal, and adult rabbit cardiac myocytes. *Circulation Research* 72:1065–1074. (Ch. 2)

Wetzel, G. T., Chen, F., & Klitzner, T. S. (1995). Na^{+}/Ca^{+} exchange and cell contraction in isolated neonatal and adult rabbit cardiac myocytes. *American Journal of Physiology* 268:H1723–H1733. (Ch. 2)

Whalen, R. G., & Sell, S. M. (1980). Myosin from fetal hearts contains the skeletal muscle embryonic light chain, *Nature* 286:731–733. (Ch. 3)

White, P. T. (1974). Experimental studies on the circulatory system of the late chick embryo. *Journal of Experimental Biology* 61:571–592. (Ch. 15)

Wiens, D., Jensen, L., Jasper, J., & Becker, J. (1995). Developmental expression of connexins in the chick embryo myocardium and other tissues. *Anatomical Record* 241:541–553. (Ch. 10)

Wiens, D., & Spooner, B. S. (1983). Actin isotype biosynthetic transitions in early cardiogenesis. *European Journal of Cell Biology* 30:60–66. (Ch. 10)

Wiens, D., Sullins, M., & Spooner, B. S. (1984). Precardiac mesoderm differentiation in vitro: Actin isotype synthetic transitions, myofibrillogenesis, initiation of heart beat, and the possible involvement of collagen. *Differentiation* 28:62–72. (Ch. 10)

Wier, W. G. (1992). $[Ca^{2+}]$ transients during excitation–contraction coupling of mammalian heart. In *The Heart and Cardiovascular System: Scientific Foundations,* ed. H. A. Fozzard, E. Haber, R. B. Jennings, A. M. Katz, & H. E. Morgan, pp. 1223–1248. New York: Raven Press. (Ch. 2)

Wilke, U. (1971). Die Feinstruktur des Glomerulus von *Glossobalanus minutus* Kowalevsky (Enteropneusta). *Cytobiologie* 5:439–447. (Ch. 11)

Wilkening, R. B., & Meschia, G. (1992). Comparative physiology of placental oxygen transport. *Placenta* 13:1–15. (Ch. 18)

Wilson, D. P. (1932). On the mitraria larvae of *Owenia fusiformis* delle Chiaje. *Philosophical Transactions of the Royal Society of London,* ser. B. 221:231–344. (Ch. 12)

Wilson, E., Mai, Q., Sudhir, K., Weiss, R. H., & Ives, H. E. (1993). Mechanical strain induces growth of vascular smooth muscle cells via autocrine action of PDGF. *Journal of Cell Biology* 123:741–747. (Ch. 5)

Wilting, J., Christ, B., Bkeloh, M., et al. (1993). In vivo effects of vascular endothelial growth factor on the chicken chorioallantoic membranes. *Cell and Tissue Research* 274:163–172. (Ch. 6)

Wladimiroff, J. W., Huisman, T. W. A., & Stewart, P. A. (1991). Fetal and umbilical flow velocity waveforms between 10 and 16 weeks of gestation; a preliminary study. *Obstetrics and Gynecology* 78:812–814. (Ch. 7)

Wladimiroff, J. W., Huisman, T. W. A., Stewart, P. A., & Stijnen, T. (1992). Normal fetal Doppler inferior vena cava, trans tricuspid and umbilical artery flow velocity waveforms between 11 and 16 weeks of gestation. *American Journal of Obstetrics and Gynecology* 3:921–924. (Chs. 7, 21)

Wolfe, B. L., Rich, C. B., Goud, H. D., Terpstra, A. J., Bashir, M., Rosenbloom, J., Sonennshein, G. E., & Foster, J. A. (1993). Insulin-like growth factor-I regulates transcription of the elastin gene. *Journal of Biological Chemistry* 268:12418–12426. (Ch. 5)

Wong, J., Fineman, J. R., & Heymann, M. A. (1994). The role of ET-1 and ET-1 receptor subtype in the regulation of fetal pulmonary vascular tone. *Pediatric Research* 35:664–670. (Ch. 6)

Wong, J., Fineman, J. R., Vanderford, P. A., et al. (1994). Endothelin-1 produces pulmonary vasoconstriction during pulmonary hypertension secondary to acute lung injury. *Critical Care Medicine* 22:A195. (Ch. 6)

Wong, J., Reddy, V. M., Hendricks-Munoz, K., et al. (in press). Altered endothelin-1 vasoactive responses in lambs with pulmonary hypertension and increased pulmonary blood flow. *Circulation.* (Ch. 6)

Wong, J., Vanderford, P. A., Fineman, J. R., et al. (1993a). Developmental effects of endothelin-1 on the pulmonary circulation in the intact sheep. *Pediatric Research* 35:664–671. (Ch. 6)

Wong, J., Vanderford, P. A., Fineman, J. R., et al. (1993b). Endothelin-1 produces pulmonary vasodilation in the intact newborn lambs. *American Journal of Physiology* 265:H1318–H1325. (Ch. 6)
Wong, J., Winters, J. W., Vanderford, P. A., et al. (1993c). ET β receptors agonists produce potent pulmonary vasodilation in the intact newborn lamb. *Journal of Cardiovascular Pharmacology* 25:207–215. (Ch. 6)
Wood, C. M., McDonald, D. G., Ingersoll, C. G., Mount, C. R., Johannsson, O. E., Landsberger, S., & Bergman, H. L. (1990). Effects of water acidity, calcium, and aluminum on whole body ions of brook trout (*Salvelinus fontinalis*) continuously exposed from fertilization to swim-up: A study by instrumental neutron activation analysis. *Canadian Journal of Fisheries and Aquatic Science* 47:1593–1603. (Ch. 17)
Woodbury, R. A., & Robertson, G. G. (1942). The one ventricle pump and the pulmonary arterial pressure of the turtle: The influence of artificial acceleration of the heart, changes in temperature, hemorrhage, and epinephrine. *American Journal of Physiology* 137:628–636. (Ch. 14)
Woodring, J. P. (1985). Circulatory systems. In *Fundamentals of Insect Physiology,* ed. M. S. Blum, pp. 5–57. New York: Wiley. (Ch. 11)
Wourms, J. P. (1977). Reproduction and development in chondrichthyan fishes. *American Zoologist* 17:379–410. (Ch. 17)
Wourms, J. P. (1994). The challenges of piscine viviparity. *Israel Journal of Zoology* 40:551–568. (Ch. 18)
Wunsch, A. M., Little, C. D., & Markwald, R. R. (1994). Cardiac endothelial heterogeneity defines valvular development as demonstrated by the diverse expression of JB3, an antigen of the endocardial cushion tissue. *Developmental Biology* 165:585–601. (Ch. 5)
Xu, Y., Pickoff, A. S., & Clarkson, C. W. (1991). Evidence for developmental changes in sodium channel inactivation gating and sodium channel block by phenytoin in rat cardiac myocytes. *Circulation Research* 69:644–656. (Ch. 2)
Xu, Y., Pickoff, A. S., & Clarkson, C. W. (1992). Developmental changes in the effects of lidocaine on sodium channels in rat cardiac myocytes. *Journal of Pharmacology and Experimental Therapeutics* 262:670–676. (Ch. 2)
Yamada, S., Yamamura, H. I., & Roeske, W. R. (1980). Ontogeny of mammalian cardiac alpha-1-adrenergic receptors. *European Journal of Pharmacology* 68:217–221. (Ch. 8)
Yamagishe, H., & Hirose, E. (1992). Nervous regulation of the myogenic heart in early juveniles of the isopod crustacean *Ligia exotica. Comparative Physiology* (Basel) 11:141–148. (Ch. 11)
Yamaguchi, K., Kawai, A., Mori, M., Asano, K., Takasugi, T., Umeda, A., Kawashiro, T., & Yokoyama, T. (1993). Uneven distribution of diffusing capacity: Limiting role for gas exchange. In *The Vertebrate Gas Transport Cascade: Adaptations to Environment and Mode of Life,* ed. J. E. P. W. Bicudo, pp. 132–141. Boca Raton, Fla.: CRC Press. (Ch. 18)
Yamamoto, K., & Moos, C. (1983). The C-proteins of rabbit red, white, and cardiac muscles. *Journal of Biological Chemistry* 258:8395–8401. (Ch. 3)
Yamazaki, Y., & Hirakow, R. (1994). Effect of growth factors on the differentiation of chick precardiac mesoderm in vitro. *Roux's Archives of Developmental Biology* 203:290–294. (Ch. 10)
Yanagisawa-Miwa, A., Uchida, Y., Nakamura, F., Tomaru, T., Kido, H., Kamijo, T., Sugimoto, T., Kaji, K., Utsuyama, M., Kurashima, C., & Ito, H. (1992). Salvage of infarcted myocardium by angiogenic action of basic fibroblast growth factor. Science 257:1401–1403. (Ch. 4)
Yang, J. T., Rayburn, H., & Hynes, R. O. (1993). Embryonic mesodermal defects in a_5 integrin-deficient mice. *Development* 119:1093–1105. (Chs. 1, 5)
Yang, J. T., Rayburn, H., & Hynes, R. O. (1995). Cell adhesion events mediated by a_4 integrins are essential in placental and cardiac development. *Development* 121:549–560. (Ch. 1)
Yasui, H., Nakazawa, M., Morishima, M., Miyagawa-Tomita, S., & Momma, K. (1995). Morphological observations on the pathogenetic process of transposition of the great arteries induced by retinoic acid in mice. *Circulation* 91:2478–2486. (Ch. 19)
Yellin, E. L., Peskin, C. S., & Frater, R. W. M. (1972). Pulsatile flow across the mitral valve: Hydraulic, electronic, and digital computer simulation. *American Society of Mechanical Engineers. Publication WA/BHF-10,* pp. 1–11. (Ch. 7)
Yin, F. C. P. (1987). *Ventricular/Vascular Coupling: Clinical, Physiological, and Engineering Aspects.* New York: Springer-Verlag. (Ch. 7)

Yokoyama, T., Copeland, N. G., Jenkins, N. A., Montgomery, C. A., Elder, F. F. B., & Overbeek, P. A. (1993). Reversal of left–right asymmetry: A situs inversus mutation. *Science* 260:679–682. (Ch. 10)

Yoshida, H., Manasek, F., & Arcilla, R. A. (1983). Intracardiac flow patterns in early embryonic life. *Circulation Research* 53:363–371. (Ch. 7)

Yoshigi, M., Hu, N., & Keller, B. B. (1996). Dorsal aortic impedance in the stage 24 chick embryo following acute changes in circulating blood volume. *American Journal of Physiology* 220:H1597–1606. (Ch. 7)

Yoshigi, M., & Keller, B. B. (1996). Linearity of pulsatile pressure-flow relations in the embryonic chick vascular system. *Circulation Research* 79:864–870. (Ch. 7)

Yosphe-Purer, Y., Endrich, J., & Davies, A. M. (1953). Estimation of the blood volumes of embryonated hen eggs at different ages. *American Journal of Physiology* 175:178–180. (Ch. 15)

Yost, H. J. (1990). Inhibition of proteoglycan synthesis eliminates left–right asymmetry in *Xenopus laevis* cardiac looping. *Development* 110:865–874. (Ch. 5)

Yost, H. J. (1992). Regulation of vertebrate left–right asymmetries by extracellular matrix. *Nature* 357:158–161. (Ch. 10)

Yost, H. J. (1994). Breaking symmetry: Left–right cardiac development in *Xenopus laevis.* In *Fourth International Symposium on Etiology and Morphogenesis of Congenital Heart Disease: Developmental Mechanisms,* ed. M. M. Markwald, E. B. Clark, & A. Takao, pp. 505–511. New York: Futura. (Ch. 13)

Yurchenco, P. D., & Schittny, J. C. (1990). Molecular architecture of basement membranes. *FASEB Journal* 4:1577–1590. (Ch. 5)

Yutzey, K. E., Rhee, J. T., & Bader, D. M. (1994). Expression of the atrial-specific myosin heavy chain AMHC1 and the establishment of anteroposterior polarity in the developing chicken heart. *Development* 120:871–883. (Ch. 10)

Zagris, N., Stavridis, V., & Chung, A. E. (1993). Appearance and distribution of entactin in the early chick heart. *Differentiation* 54:67–71. (Ch. 5)

Zahka, K. G., Hu, N., Brin, K. P., Yin, F. C. P., & Clark, E. B. (1989). Aortic impedance and hydraulic power in the chick embryo from stages 18 to 29. *Circulation Research* 64:1091–1095. (Chs. 7, 15)

Zahka, K. G., Hu, N., & Clark, E. B. (1987). Pulse wave velocity in the stage 18 to 29 chick embryo. *Circulation* 76:IV–454 (abstract). (Ch. 7)

Zak, R. (1974). Development and proliferative capacity of cardiac muscle cells. *Circulation Research* 35(2):II:17–26. (Ch. 5)

Zeller, R., Bloch, K. D., Williams, B. S., Arceci, R. J., & Seidman, C. E. (1987). Localized expression of the atrial natriuretic factor gene during cardiac embryogenesis. *Genes & Development* 1:693–698. (Ch. 8)

Zellers, T. M., McCormick, J., & Wu, Y. (1994). Interaction among ET-1, endothelium-derived nitric oxide, and prostacyclin in pulmonary arteries and veins. *American Journal of Physiology* 267:H139–H147. (Ch. 8)

Zhang, H. Y., Chu, M. L., Pan, T. C., Sasaki, T., Timpl, R., & Ekblom, P. (1995). Extracellular matrix protein fibulin-2 is expressed in the embryonic endocardial cushion tissue and is a prominent component of valves in adult heart. *Developmental Biology* 167:18–26. (Ch. 5)

Zhang, Q., & Whittow, G. C. (1992). The effect of incubation temperature on oxygen consumption and organ growth in domestic-fowl embryos. *Journal of Thermal Biology* 17:339–345. (Ch. 15)

Zhou, Q. Y., Quaife, C. J., & Palmiter, R. D. (1995). Targeted disruption of the tyrosine hydroxylase gene reveals that catecholamines are required for mouse fetal development. *Nature* 374:640–643. (Ch. 8)

Zimmerman, F. J., Hughes, S. F., Cuneo, B., & Benson, D. W., Jr. (1991). The effect of cardiac cycle length on ventricular end-diastolic pressure and maximum time derivative of pressure in the stage 24 chick embryo. *Pediatric Research* 29:338–341. (Ch. 7)

Zoran, M. J., & Ward, J. A. (1983). Parental egg care behavior and fanning activity for the organe chromide, *Etroplus maculatus. Environmental Biology of Fishes* 8:301–310. (Ch. 17)

Zot, A. S., & Potter, J. D. (1987). Structural aspects of troponin–tropomyosin regulation of skeletal muscle contraction. *Annual Reviews of Biophysics and Biochemistry* 16:535–559. (Ch. 3)

Zummo, G. (1983). Ultrastructural features of the embryonic heart of the elasmobranch *Scyllium stellare. Comparative Biochemistry and Physiology* 76A:459–464. (Ch. 12)

Systematic index

Note: For practical purposes, this systematic index is divided into invertebrates and vertebrates (which improperly classifies some of the more interesting animals that sit on the somewhat arbitrary intervening boundary). Major groups (e.g. classes) are alphabetical, with taxonomic hierarchy of minor groups indicated by level of indentation under major entry.

INVERTEBRATES

VERTEBRATES

Subject index

Note: For specific animals or groups of animals see the Systematic Index.